Springer Theses

Recognizing Outstanding Ph.D. Research

Aims and Scope

The series "Springer Theses" brings together a selection of the very best Ph.D. theses from around the world and across the physical sciences. Nominated and endorsed by two recognized specialists, each published volume has been selected for its scientific excellence and the high impact of its contents for the pertinent field of research. For greater accessibility to non-specialists, the published versions include an extended introduction, as well as a foreword by the student's supervisor explaining the special relevance of the work for the field. As a whole, the series will provide a valuable resource both for newcomers to the research fields described, and for other scientists seeking detailed background information on special questions. Finally, it provides an accredited documentation of the valuable contributions made by today's younger generation of scientists.

Theses are accepted into the series by invited nomination only and must fulfill all of the following criteria

- They must be written in good English.
- The topic should fall within the confines of Chemistry, Physics, Earth Sciences, Engineering and related interdisciplinary fields such as Materials, Nanoscience, Chemical Engineering, Complex Systems and Biophysics.
- The work reported in the thesis must represent a significant scientific advance.
- If the thesis includes previously published material, permission to reproduce this must be gained from the respective copyright holder.
- They must have been examined and passed during the 12 months prior to nomination.
- Each thesis should include a foreword by the supervisor outlining the significance of its content.
- The theses should have a clearly defined structure including an introduction accessible to scientists not expert in that particular field.

More information about this series at http://www.springer.com/series/8790

Jean-Baptiste Fouvry

Secular Evolution of Self-Gravitating Systems Over Cosmic Age

Doctoral Thesis accepted by
the Pierre-and-Marie-Curie University, France

Springer

Author
Dr. Jean-Baptiste Fouvry
Institute for Advanced Study
Princeton, NJ
USA

Supervisor
Prof. Christophe Pichon
Institut d'Astrophysique de Paris
Paris
France

ISSN 2190-5053 ISSN 2190-5061 (electronic)
Springer Theses
ISBN 978-3-319-66759-1 ISBN 978-3-319-66760-7 (eBook)
DOI 10.1007/978-3-319-66760-7

Library of Congress Control Number: 2017950273

Printed on acid-free paper

This Springer imprint is published by Springer Nature
The registered company is Springer International Publishing AG
The registered company address is: Gewerbestrasse 11, 6330 Cham, Switzerland

Supervisor's Foreword

In the expanding universe, an overdense perturbation passing a critical threshold will collapse onto itself and, through violent relaxation and mergers, rapidly converge towards a stationary, phase-mixed and highly symmetric state. This perturbation will convert its cold gas into stars and, depending on how much entropy has been radiated away, produce either a spiral or an elliptical galaxy, with a partially frozen orbital structure. The galaxy is then locked in a quasi-stationary state imposed by its mean gravitational field. Yet, such a self-gravitating relaxed system remains a highly complex object. It undergoes a wide variety of dynamical processes, depending on its environment and on its internal structure and "temperature", i.e. whether it is more pressure or centrifugally supported. Of particular interest are strongly responsive colder systems which, given time and kicks, have the opportunity to significantly reshuffle their orbital structure towards more likely configurations. Understanding this long-term reshuffling, called gravity-driven secular evolution, involves the use of an extended kinetic theory.

Since the seminal works of Einstein (1905) and Perrin, physicists have understood in the context of kinetic theory how ink slowly diffuses in a glass of water. The fluctuations of the stochastic forces acting on water molecules drive the diffusion of the ink in the fluid. This is the archetype of a process described by the fluctuation–dissipation theorem, which universally relates the rate of diffusion to the power spectrum of the fluctuating forces. For stars in galaxies, a similar process occurs but with two significant differences due to the long-range nature of the gravitational interaction. For the diffusion to be effective, stars need to resonate, i.e. present commensurable frequencies, otherwise they only follow the orbit imposed by their mean field. In addition, the amplitudes of the induced fluctuating forces are significantly boosted by collective effects, i.e. by the fact that, because of self-gravity, each star generates a wake in its neighbours. Capturing statistically both effects in self-gravitating systems is what Weinberg (2001) and Heyvaerts (2010) had achieved, opening a new window of opportunities for our understanding of the long-term behaviour of galaxies, and extending the work of Balescu (1960) and Lenard (1960) from homogeneous plasma physics. Jean-Baptiste Fouvry took these opportunities to investigate the cosmic implications of this universal fluctuation–dissipation theorem by considering simultaneously externally-driven and self-induced fluctuations on nested galactic scales, from stellar clusters orbiting massive black holes, galactic stellar discs to halos or even debris discs. This is a particularly timely topic: new instruments like GAIA and Gravity provide a wealth of information on the dynamical state of our Galaxy, past and present. We now also have a robust cosmic paradigm for the formation of structures: on galactic and sub-galactic scales, their long-term interaction with the environment can now be quantified in detail.

Jean-Baptiste's work stands out because it tackles galactic evolution in a truly novel way: it treats processes statistically while capturing the astrophysical differences at each scale. The key novel and unifying ingredient is a proper accounting of the self-gravity of these systems. In his research, he is able to gauge the role of nature versus nurture in establishing the galaxies' observed properties, using kinetic equations to capture both sources of fluctuations. He has for instance analytically shown how secular diffusion shapes the phase space structure of stellar discs. He quantified exactly how the efficiency of orbital diffusion is boosted by the square of the potential fluctuations (self-induced or external), hence the square of the gravitational susceptibility. This has far-reaching implications in various contexts, ranging from galaxies to black holes and protoplanetary discs. As the archetype of self-amplified diffusion, his work has implications well beyond astrophysics.

In this Ph.D., Jean-Baptiste derives the equation describing the secular diffusion of a self-gravitating collisionless system induced by an exterior perturbation, while assuming that the timescale corresponding to secular evolution is much larger than that corresponding to the natural frequencies of the system. He focuses on a two-dimensional formulation for a tepid galactic disc, while relying on the epicyclic approximation and assuming that only tightly wound transient spirals are sustained by the disc. It leads him to a quadrature for the diffusion coefficients which provides a straightforward understanding of the loci of maximal diffusion within the disc. His critical contribution is to pin down the optimal sampling condition one has to impose on the wavelet-like WKB basis describing the disc's self-gravitating response. Jean-Baptiste then implements his dressed Fokker–Planck formalism in angle–action variables to predict analytically the numerically observed signatures of the secular evolution of an isolated self-gravitating stellar Mestel disc: the shot-noise-driven formation of narrow ridges of resonant orbits. He studies the relative effects of the bulge, the halo, the disc temperature and the spectral properties of the shot noise. For instance, when the halo mass is decreased, he quantifies a transition between a regime of heating in the inner regions of the disc through the inner Lindblad resonance to a regime of radial migration of quasi-circular orbits via the corotation resonance in the outer part of the disc.

Jean-Baptiste investigates the secular evolution of an infinitely thin tepid isolated galactic disc made of a finite number of particles using the up-to-now-never-solved inhomogeneous Balescu–Lenard equation. Assuming once again that only tightly wound transient spirals are present in the disc, the WKB formalism leads him to a simple and tractable quadrature for the corresponding drift and diffusion coefficients. His investigation recovers nicely the loci of the ridge-like feature in action space, in agreement with simulations, but overestimates the timescale involved in its appearance. He shows how swing amplification resolves this discrepancy. While implementing numerically the matrix method, in order to model the induced gravitational polarisation, Jean-Baptiste takes special care to account for the amplification of potential fluctuations of mutually resonant orbits and the unwinding of the induced swing amplified transients. This is a numerical ordeal, as one must carry out singular and resonant integrations over phase space twice! He demonstrates that secular evolution of isolated discs is driven by resonances, and most importantly does not depend on the initial phases of the disc, as had been claimed in the literature. For a Mestel disc with $Q \sim 1.5$, he shows that the polarisation cloud around each star boosts up its secular effect by a factor of the order of a thousand or more, promoting accordingly the dynamical relevance of this self-induced collisional secular evolution.

He also carries out an original derivation of the kinetic equation describing the secular evolution of spatially inhomogeneous systems by relying on a functional integral formalism. His derivation provides an alternative to previous methods, either based on the resolution of the truncated BBGKY hierarchy or the Klimontovich equation. It should turn out to be fruitful to derive more accurate (possibly higher order) kinetic equations, e.g., accounting for collective effects as in the Balescu-Lenard case. He transposes his derivation to the kinetic theory of point vortices. He also computes the dressed Fokker–Planck diffusion and friction coefficients in inhomogeneous multicomponent self-gravitating systems (completing some work initiated by Jean Heyvaerts) which demonstrates how the Balescu-Lenard formalism captures the process of dynamical friction and induced heating self-consistently. This proof should convince the galactic community of the importance of this novel kinetic equation.

He investigates the secular thickening of a self-gravitating stellar galactic disc embedded in a fluctuating potential. Jean-Baptiste's theory predicts the formation of ridges of resonant orbits towards larger vertical actions, as found in simulations. Potential fluctuations within the disc statistically induce a vertical bending of a subset of resonant orbits, triggering the corresponding increase in vertical velocity dispersion. A simple ad hoc model of fading sequences of slowing bars explains how such a diffusion process also provides a possible mechanism allowing for the resonant thickening of galactic discs over secular timescales: the faster the breaking, the broader the effect. He also addresses the problem of GMC-induced thickening via the Balescu-Lenard formalism, which he also adapts to the quasi-Keplerian problem in the vicinity of a supermassive black hole. He demonstrates how this equation naturally explains the origin of the so-called Schwarzschild barrier and anomalous mass segregation. This is critical work in the flourishing context of the recently detected coalescence of massive black holes via gravitational waves, because it defines the framework in which to dynamically tackle the last parsec problem, i.e. to understand how massive black holes can be fed by surrounding clusters.

Paris, France
August 2017

Prof. Christophe Pichon

References

Balescu, R. (1960). *Phys. Fluids*, *3*, 52.
Einstein, A. (1905). *Annalen der Physik*, *322*, 549.
Heyvaerts, J. (2010). *MNRAS*, *407*, 355.
Lenard, A. (1960). *Annals of Physics*, *10*, 390.
Weinberg, M. D. (2001). *MNRAS*, *328*, 311.

Preface

Understanding the long-term evolution of self-gravitating astrophysical systems, such as for example stellar discs, is now a subject of renewed interest, motivated by the combination of two factors. On the one hand, we now have at our disposal the well established ΛCDM model for the formation of structures. When considered on galactic scales, depending on the nature of the accretion processes, interactions with the circumgalactic environment, may either be constructive (e.g., adiabatic gas accretion) or destructive (e.g., satellite infall). The statistical impacts of these cosmic perturbations on self-gravitating systems are now being quantified in detail. On the other hand, recent theoretical works now provide a precise description of the amplification of external disturbances and discreteness noise as well as their effects on a system's orbital structure over cosmic time, while properly accounting for the effect of self-gravity. These theories offer new physical insights on the dynamical processes at play in these self-gravitating systems on secular timescales.

These two complementary developments now allow us to address the pressing question of the respective roles of nature versus nurture in the establishment of the observed properties of self-gravitating systems. Numerous dynamical challenges are therefore ready to be re-examined in much greater detail than before. Examples include: the secular evolution of the metallicity dispersion relationship in galactic discs, the mechanisms of disc thickening via giant molecular clouds or spiral waves, the stellar dynamical evolution of galactic centres, etc. Characterising the secular evolution of such self-gravitating systems is a stimulating task, as it requires intricate theoretical models, complex numerical experiments and an accurate understanding of the involved physical processes.

The purpose of the present thesis is to describe such secular dynamics in contexts where self-gravity is deemed important. Two frameworks of diffusion, either external or internal, will be presented in detail. These approaches will be applied to various astrophysical systems to illustrate the particular relevance and ability of these approaches to describe the long-term evolution of self-gravitating systems. This thesis will first investigate the secular evolution of discrete razor-thin stellar discs and recover the formation of narrow ridges of resonant orbits in agreement with observations and numerical simulations, thanks to the first implementation of the Balescu–Lenard equation. The spontaneous thickening of stellar discs as a result of Poisson shot noise will also be investigated. These various approaches allow in particular for a self-consistent description of stellar migration and disc thickening. Finally, we will illustrate how the same formalisms allow us to describe the dynamics of stars orbiting a central supermassive black hole in galactic centres. Other processes of secular orbital restructuration will be discussed in less details.

Keywords Evolution of galaxies • Secular dynamics • Gravitation • Diffusion • Kinetic theory

Princeton, USA Jean-Baptiste Fouvry

Acknowledgements

First, I would like to express my most sincere gratitude to my supervisor, Christophe Pichon. His everyday involvement and constant support have been essential to carry out this work. With a lot of benevolence, he made me discover what research is all about, and I am deeply indebted to him.

I also extend my appreciation to James Binney for agreeing to co-supervise my work. His thoughts and remarks on the problems considered have always proved very insightful.

Special thanks also go to Pierre-Henri Chavanis. I greatly appreciated the time and effort he dedicated to our common projects, and I learned a lot from him on the fascinating properties of long-range interacting systems.

I must also express my gratitude to John Magorrian and Simon Prunet. I enjoyed our collaborations, and their different research backgrounds were perfect breeding grounds for our works.

I also thank Francis Bernardeau for presiding my jury, my two reporters Mir Abbas Jalali and Simon White for examining my manuscript, together with the members of my jury, Tal Alexander, Thierry Dauxois, Benoit Famaey and Clément Sire.

My Ph.D. took place at the Institut d'Astrophysique de Paris, and I want to thank its director and all the administrative and research staff for providing me with excellent working conditions. I also thank all my Ph.D. colleagues at IAP for their communicative dynamism, in particular, Rebekka Bieri for her genuine support and our stimulating discussions. Finally, I acknowledge the Corps des Mines for funding my Ph.D. for the past three years. I am very grateful for this opportunity.

Publications related to this thesis

This thesis has used material from the following first-authored publications.

- Fouvry, J.-B., Pichon, C., & Prunet, S. (2015). *MNRAS*, *449*, 1967.
  Secular resonant dressed orbital diffusion - I. Method and WKB limit for tepid discs
- Fouvry, J.-B., & Pichon, C. (2015). *MNRAS*, *449*, 1982.
  Secular resonant dressed orbital diffusion - II. Application to an isolated self-similar tepid galactic disc
- Fouvry, J.-B., Binney, J., & Pichon, C. (2015). *ApJ*, *806*, 117.
  Self-gravity, Resonances, and Orbital Diffusion in Stellar Disks
- Fouvry, J.-B., Pichon, C., & Chavanis, P.-H. (2015). *A&A*, *581*, A139.
  Secular diffusion in discrete self-gravitating tepid discs. I. Analytic solution in the tightly wound limit
- Fouvry, J.-B., Pichon, C., Magorrian, J., & Chavanis, P.-H. (2015). *A&A*, *584*, A129.
  Secular diffusion in discrete self-gravitating tepid discs. II. Accounting for swing amplification via the matrix method
- Fouvry, J.-B., Chavanis, P.-H., & Pichon, C. (2016). *Physica A*, *459*, 117.
  Functional integral approach to the kinetic theory of inhomogeneous systems
- Fouvry, J.-B., Chavanis, P.-H., & Pichon, C. (2016). *Phys. Lett. A*, *380*, 2589.
  Functional integral derivation of the kinetic equation of two-dimensional point vortices
- Fouvry, J.-B., Pichon, C., & Magorrian, J. (2016). *A&A*, *598*, A17.
  The secular evolution of discrete quasi-Keplerian systems. I. Kinetic theory of stellar clusters near black holes
- Fouvry, J.-B., Pichon, C., Chavanis, P.-H., & Monk, L. (2017). *MNRAS*, *471*, 2642.
  Resonant thickening of self-gravitating discs: imposed or self-induced orbital diffusion in the tightly wound limit

Contents

Chapter 1
Introduction

1.1 Context

The current paradigm for the formation of astrophysical structures is the Lambda Cold Dark Matter (ΛCDM) model (Springel et al. 2006). Initial quantum density fluctuations (Bardeen et al. 1986) in the non-baryonic dark matter appear right after the Big Bang and get stretched by the expansion of the universe. On the other hand, gravity drives a hierarchical clustering which leads to a strong increase of initial overdensities. These densities grow, separate, collapse and give birth to galaxy haloes. These structures form on the large scale a "cosmic web" of nodes, filaments, walls and voids (see, e.g., Frenk and White 2012 for a review), as can be seen in the top row of Fig. 1.1. At a later stage in the evolution of the universe, "dark energy", which now accounts for approximately 70% of the total energy content of the universe Planck Collaboration et al. (2014), comes into play to cause the originally slowing universal expansion to reaccelerate (Riess et al. 1998; Perlmutter et al. 1999). This late reacceleration tends to isolate even more the different regions of the clustering hierarchy, reducing the later merging rate of haloes, leading to a more dynamically quiescent period.

Contrary to dark matter, baryonic matter, which only makes up approximately 15% of the matter content of the universe, is not a collisionless fluid and therefore undergoes shocks as large scale structures develop. Indeed, baryons are accreted along the structures formed by the dark matter, cool within the haloes, and form stars (White and Rees 1978). This essentially leads to the formation of gravitationally bound objects, the galaxies. From such mechanisms, one expects principally two types of galaxies. Spirals form when haloes accrete gas that dissipates and leads to the formation of discs. On the other hand, ellipticals are mainly expected to form as results of collisions and mergers triggering active galactic nuclei (AGN) feedback that prevents further gas accretion (Toomre and Toomre 1972; Toomre 1977; Barnes and Hernquist 1992; Dubois et al. 2016). Figure 1.2 illustrates two examples of galactic morphologies observed in the local universe. This general paradigm of formation is widely accepted as broadly correct, but some of its predictions still appear as incon-

J.-B. Fouvry, *Secular Evolution of Self-Gravitating Systems Over Cosmic Age*,
Springer Theses, DOI 10.1007/978-3-319-66760-7_1

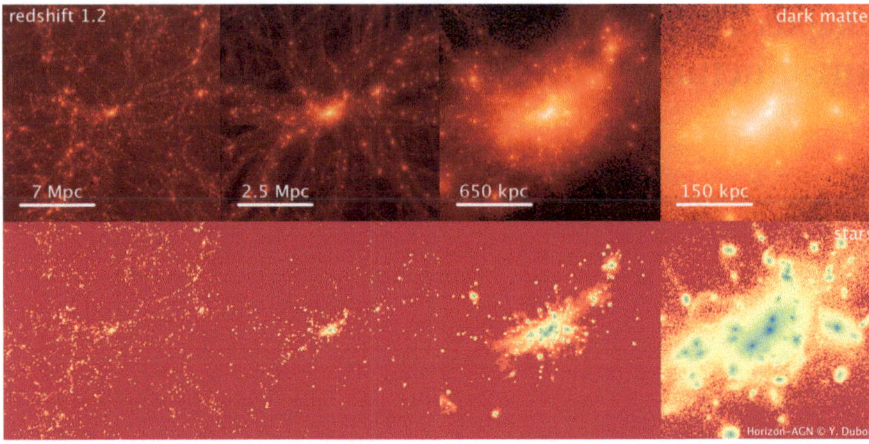

Fig. 1.1 Snapshot extracted from the Horizon-AGN cosmological hydrodynamical simulation (Dubois et al. 2014). It illustrates the dark matter density (*top row*) and stellar density (*bottom row*) centred on a massive halo at redshift $z = 1.2$ for various scales. One notes the formation of large scale structures (the "cosmic web") via the hierarchical clustering of primordial quantum fluctuations, stretched by the expansion of the universe and increased in contrast by self-gravity

Fig. 1.2 Two examples of galactic morphologies located in the nearby Virgo Cluster. *Left panel* Spiral galaxy NGC 4321 (=M100) (*credit* ESO). Spiral galaxies possess a disc shape and display spiral patterns. *Right panel* Giant elliptical galaxy NGC4486 (=M87) (*credit* Australian Astronomical Observatory). Elliptical galaxies possess very little substructures and have a roughly ellipsoidal shape

sistent with the observations (Silk and Mamon 2012; Kroupa 2012). See for example Appendix 4.D for a description of one of these tensions through the so-called cusp-core problem. One important outcome of these developments is that galaxies are not isolated islands distributed randomly in the universe, but rather follow and interact with the intricate cosmic web network (Bond et al. 1996; Pichon et al. 2011).

Galaxies are therefore complex structures at the interface of two scales: the large scale structure (the so-called intergalactic medium) and the small scale of their internal constituents, the interstellar medium and the stars. Galaxies are also at the interface of both destructive and constructive processes. Indeed, gas accretion onto a galaxy can either be smooth and constructive as with cold gas flows, or abrupt, violent, and destructive as with mergers or satellite infalls with inconsistent impact parameters. In the former case, the formation process sets the forming stars in a very specific configuration, with a large reservoir of free energy. Later, galaxies are subject to both stabilising and destabilising influences. The constant resupply of new stars from the quasi-circular gas orbits makes disc galaxies dynamically colder and more responsive, while a wide variety of heating mechanisms tend to increase the velocity dispersion of the stars, making galaxies dynamically hotter and less responsive. Finally, galaxies, because they were formed from cold gas, are originally created in a highly improbable state, i.e. a low entropy state of low velocity dispersion. These (thermodynamically improbable) states are maintained by symmetry, given that their initially axially symmetric distributions do not allow for efficient angular momentum exchange. The dynamics of these systems will aim at leaving these metastable states (quasi-steady on short timescales) towards more probable states of higher entropy. Understanding the long-term evolution of such self-gravitating systems requires to consider the joint contributions associated with on the one hand external effects, the cosmic environment, and on the other hand associated with internal effects, such as the system's own graininess and internal structure (e.g., bars or giant molecular clouds). Getting a better grasp of the secular dynamics of these systems involves therefore ranking the strengths of each of these sources of evolution, i.e. quantifying the effects of nurture and nature on their evolution, and weighing their respective efficiency.

The general purpose of the present thesis is to describe and understand the secular evolution of self-gravitating systems. See Kormendy (2013); Binney (2013); Sellwood (2014) for detailed reviews on recents developments in this respect. This thesis also aims at incrementing our theoretical knowledge of the self-interaction of self-gravitating systems, while providing explicit solutions to the recently published corresponding kinetic equations. Before entering the core of the thesis, let us first briefly describe the structure of disc galaxies and their associated dynamical components (Sect. 1.2), as well as the tools and mechanisms from Hamiltonian dynamics essential to characterise the evolution of such systems (Sect. 1.3).

1.2 Stellar Discs

Following Binney and Tremaine (2008), let us now review the important orders of magnitude for a spiral galaxy such as the Milky Way. As illustrated in Fig. 1.3, the Milky Way comprises various components. First, the stellar component of the Milky Way is made of about 10^{11} stars for a total mass of order $5 \times 10^{10} M_{\odot}$. Most of the stars belong to a disc of approximate radius 10 kpc. The Sun is located near its midplane at a radius of 8 kpc. Observations have indicated that the stellar disc may be constituted of two components, a thin and thick discs, of respective typical thickness 300 pc and 1 kpc. The thick disc is made of older stars with different chemical compositions and its luminosity is about 7% that of the thin disc. Stellar discs are said to be dynamically cold, as the random velocities of their constituents are much smaller than the mean ordered velocity, i.e. the mean quasi-circular motion. Chapters 3 and 4 will especially consider the secular dynamics of razor-thin stellar discs, while Chap. 5 will investigate possible mechanisms of secular thickening of stellar discs. While this is not illustrated in Fig. 1.3, let us also note that stellar discs also contain gas, atomic and molecular hydrogen and helium, forming the interstellar medium (ISM). The ISM

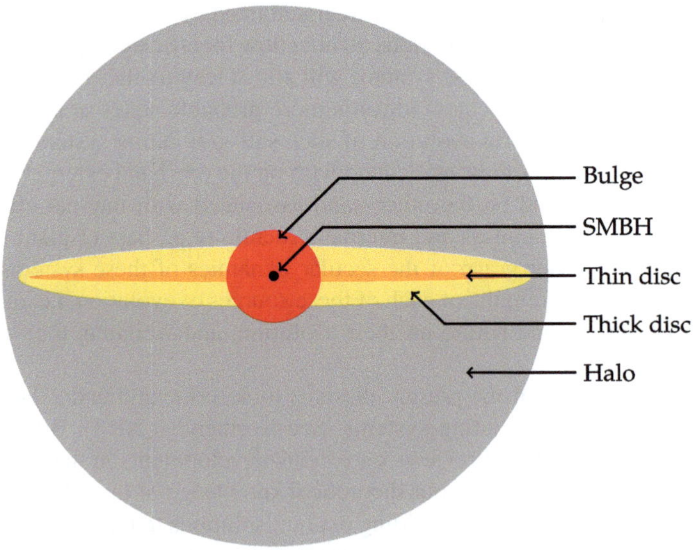

Fig. 1.3 Qualitative illustration of the main dynamical components of a spiral galaxy. Most of the galaxies are expected to contain a central supermassive black hole (see Chap. 6 for a detailed discussion of the secular dynamics of stars in the vicinity of such objects). The central region of the galaxy takes the form of a spherical component, the bulge. At larger radii, the stellar disc is roughly made of two distinct components, namely a thin disc of stars (see Chaps. 3 and 4 for a discussion of the secular dynamics of razor-thin stellar discs), and a thick disc of stars (see Chap. 5 for a discussion of the secular dynamics of thickened stellar discs). Finally, the disc is embedded in a spheroidal dark matter halo (see Appendix 4.D for a brief illustration of how the dynamics of such spherical systems may be described)

only makes up about 10% of the total stellar mass, and is therefore of little importance for the dynamics of the Milky Way. However, the transient giant molecular clouds, dense gas regions, remain important for the dynamics of a galaxy as they are the birth place of new stars. They impose as well the chemistry of the newly formed stars. In the centre of the disc, one finds an amorphous component, the bulge, of approximate mass $0.5 \times 10^{10} M_{\odot}$. Contrary to the disc, the bulge is a dynamically hot region, where random velocities are larger than the mean velocity. Let us finally note that the Milky Way's bulge, being triaxial, is sometimes called a bar. The secular effects associated with bars will only be briefly discussed in Chap. 5. At the centre of these regions is located a supermassive black hole of approximate mass $4 \times 10^{6} M_{\odot}$, called Sgr A*. See Chap. 6 for a description of the secular dynamics of stars near supermassive black holes. Finally, the most massive component of the Milky Way is its surrounding dark matter halo, with an approximate radius of 200 kpc and approximate mass of $10^{12} M_{\odot}$. In the context of galactic dynamics, the halo mainly only interacts with the stellar component through the joint gravitational potential they define. See Appendix 4.D for a brief illustration of how to describe the secular dynamics of dark matter haloes.

1.3 Hamiltonian Dynamics

Let us now present a short introduction to Hamiltonian dynamics, with a particular emphasis on the tools and processes essential for the secular evolution of self-gravitating systems. We refer the reader to Goldstein (1950); Arnold (1978); Binney and Tremaine (2008) for thorough presentations of Hamiltonian dynamics.

A $n-$dimensional dynamical system can be described by its Hamiltonian H expressed as a function of the canonical coordinates $(\boldsymbol{q}, \boldsymbol{p})$. These coordinates follow Hamilton's equations reading

$$\frac{\mathrm{d}\boldsymbol{q}}{\mathrm{d}t} = \frac{\partial H}{\partial \boldsymbol{p}} \;\; ; \;\; \frac{\mathrm{d}\boldsymbol{p}}{\mathrm{d}t} = -\frac{\partial H}{\partial \boldsymbol{q}} . \tag{1.1}$$

We define the configuration space of a system as the $n-$dimensional space with coordinates $(q_1, ..., q_n)$, and the associated momentum space as $(p_1, ..., p_n)$. More importantly, we define as phase space the $2n-$dimensional space with coordinates $(q_1, ..., q_n, p_1, ..., p_n) = (\boldsymbol{q}, \boldsymbol{p}) = \boldsymbol{w}$.

Let us consider two scalar functions $F_1(\boldsymbol{w})$ and $F_2(\boldsymbol{w})$ depending on the phase space coordinates. We define their Poisson bracket as

$$\big[F_1, F_2\big] = \frac{\partial F_1}{\partial \boldsymbol{q}} \cdot \frac{\partial F_2}{\partial \boldsymbol{p}} - \frac{\partial F_1}{\partial \boldsymbol{p}} \cdot \frac{\partial F_2}{\partial \boldsymbol{q}} . \tag{1.2}$$

Thanks to this notation, Hamilton's equation can be written as

$$\frac{\mathrm{d}\boldsymbol{w}}{\mathrm{d}t} = \big[\boldsymbol{w}, H\big] . \tag{1.3}$$

In addition, the phase space coordinates satisfy the canonical commutation relations

$$\big[w_\alpha, w_\beta\big] = \mathbf{J}_{\alpha\beta} \;\; \text{with} \;\; \mathbf{J} = \begin{pmatrix} \mathbf{0} & \mathbf{I} \\ -\mathbf{I} & \mathbf{0} \end{pmatrix} , \tag{1.4}$$

where we introduced the $2n \times 2n$ symplectic matrix $\mathbf{J}$, and where $\mathbf{0}$ and $\mathbf{I}$ are respectively the $n \times n$ zero and identity matrix.

Because Hamiltonian dynamics describes the system's dynamics in phase space, it allows for generalised change of coordinates. Phase space coordinates $\boldsymbol{W} = (\boldsymbol{Q}, \boldsymbol{P})$ are said to be canonical if they satisfy

$$\big[\boldsymbol{W}_\alpha, \boldsymbol{W}_\beta\big] = \mathbf{J}_{\alpha\beta} . \tag{1.5}$$

The essential property of canonical coordinates is that in any of these coordinates, Hamilton's equations conserve the same form. One has $\dot{\boldsymbol{W}} = [\boldsymbol{W}, H]$, where the Hamiltonian is expressed as a function of the new coordinates and the Poisson bracket involves as well derivatives w.r.t. the new coordinates. Let us also note that infinitesimal phase space volumes are conserved by canonical transformations so that $\mathrm{d}\boldsymbol{W} = \mathrm{d}\boldsymbol{w}$. Poisson brackets are also conserved through canonical transformations.

We define an integral of motion $I(\boldsymbol{w})$ to be any function of the phase space coordinates constant along the orbits. It is said to be isolating if for any value in the image of I, the region of phase space which reaches this value is a smooth manifold of dimension $2n-1$. For example, for Hamiltonians independent of time, the energy constitutes an isolating integral of motion. A system is said to be integrable (in the Liouville sense) if it possesses n independent integrals of motions, i.e. whose differentials are linearly independent in all points. For such integrable systems, one may then devise a set of canonical coordinates, the angle-action coordinates $(\boldsymbol{\theta}, \boldsymbol{J})$, such that the actions $\boldsymbol{J}$ are independent isolating integrals of motion. Within these coordinates, the Hamiltonian H becomes independent of the angles $\boldsymbol{\theta}$, so that $H = H(\boldsymbol{J})$. Hamilton's equations then read

$$\frac{\mathrm{d}\boldsymbol{\theta}}{\mathrm{d}t} = \frac{\partial H}{\partial \boldsymbol{J}} = \boldsymbol{\Omega}(\boldsymbol{J}) \;\; ; \;\; \frac{\mathrm{d}\boldsymbol{J}}{\mathrm{d}t} = 0 , \tag{1.6}$$

where we introduced as $\boldsymbol{\Omega}(\boldsymbol{J}) = \partial H / \partial \boldsymbol{J}$ the intrinsic frequencies of motion. In these coordinates, the motions are straight lines given by

$$\boldsymbol{\theta}(t) = \boldsymbol{\theta}_0 + \boldsymbol{\Omega}(\boldsymbol{J})\, t \;\; ; \;\; \boldsymbol{J}(t) = \text{cst.} \tag{1.7}$$

An additional property here is that the angles $\boldsymbol{\theta}$ are assumed to be $2\pi-$periodic, so that the actions $\boldsymbol{J}$ describe a $n-$dimensional torus in phase space on which the orbit lies. This is the crucial strength of the angle-action coordinates, which formally allows for a simple description of the complex trajectories in the physical phase space $(\boldsymbol{q}, \boldsymbol{p})$ as straight lines motions in the angle-action space $(\boldsymbol{\theta}, \boldsymbol{J})$. Unfortunately, angle-action coordinates are not always guaranteed to exist. In addition, even for integrable systems, simple analytical expressions for these coordinates are rarely available. In the upcoming chapters, we will illustrate examples of angle-action coordinates for razor-thin and thickened axisymmetric discs, $3D$ spherical systems, and Keplerian systems. Figure 1.4 offers a visualisation of angle-action coordinates for $1D$ harmonic oscillators. This is an important example when applying the epicyclic approximation in Chaps. 3 and 5. As emphasised in Eq. (1.7), once the angle-action coordinates have been constructed, individual motions then take the form of quasi-periodic motions along the tori defined by the actions $\boldsymbol{J}$. In Fig. 1.5, we illustrate two possible behaviours for the motion along this torus. These are resonant periodic motions or non-resonant quasiperiodic motions, depending on the properties of the intrinsic frequencies $\boldsymbol{\Omega}$.

Let us now assume that our system can be described statistically by a distribution function (DF) $F(\boldsymbol{w})$. Let us then present the differential equation satisfied by F as a consequence of the individual evolutions imposed by Hamilton's equation (1.3). As

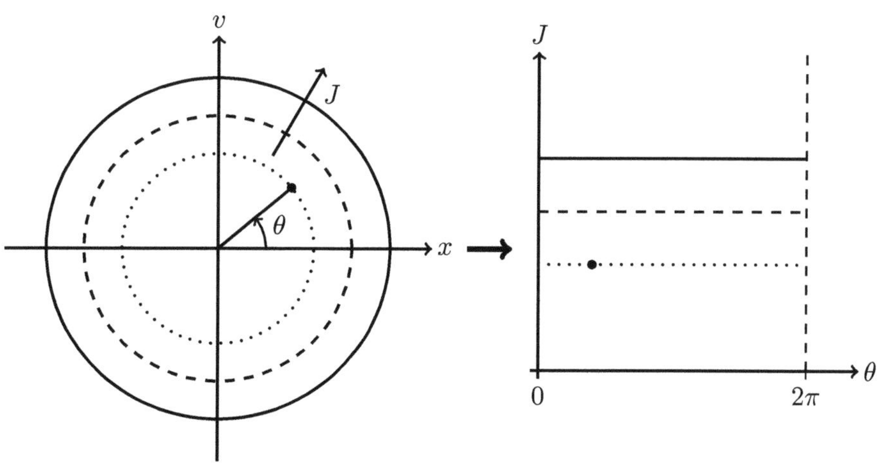

Fig. 1.4 Illustration of the phase space diagram of a harmonic oscillator. *Left panel* Illustration of the particles' trajectories in the physical phase space (x, v). The trajectories take the form of concentric circles along which particles move. Here, the action J should be seen as a label for the circle, while the angle θ should be seen as the position along the circle. *Right panel* Illustration of the trajectories in the angle-action space (θ, J). In these coordinates, the motions are straight lines. The action J is conserved, while the angle θ evolves linearly with time with the frequency $\Omega = \partial H / \partial J$

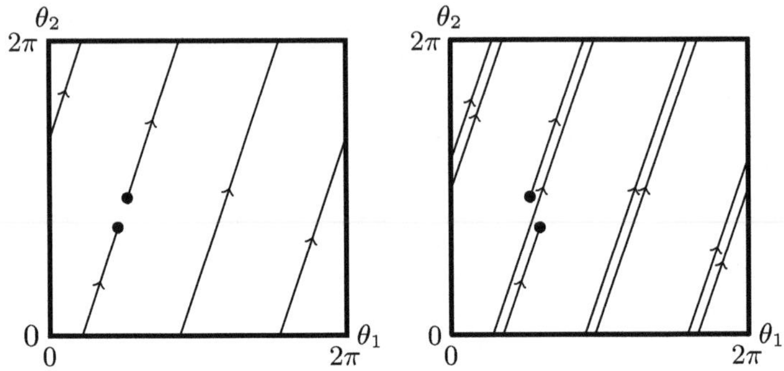

Fig. 1.5 Illustration of two integrable trajectories in angle space $\boldsymbol{\theta}=(\theta_1,\theta_2)$. An integrable trajectory is fully characterised by its actions $\boldsymbol{J}$, while the position of the particle along its orbit is described by the angles $\boldsymbol{\theta}$. Along the unperturbed motion, the actions are conserved, while the angles evolve linearly with time with the frequency $\boldsymbol{\Omega}$. *Left panel* Illustration of a degenerate trajectory for which there exists $\boldsymbol{n}\in\mathbb{Z}^2$ such that $\boldsymbol{n}\cdot\boldsymbol{\Omega}=0$, i.e. the frequencies are in a rational ratio. The trajectory is closed, periodic, and does not fill the angle space (see Chap. 6 for an illustration of how to study the secular evolution of degenerate systems). Right-panel Illustration of a non-degenerate trajectory, for which the trajectory is quasiperiodic and densely covers the angle domain

the DF evolves, probability must be conserved, so that the DF satisfies a continuity equation in phase space given by

$$\frac{\partial F}{\partial t}+\frac{\partial}{\partial \boldsymbol{w}}\cdot\big[F\,\dot{\boldsymbol{w}}\big]=0\,. \tag{1.8}$$

Using Hamilton's equation (1.3), this can equivalently be rewritten as

$$\begin{aligned}0&=\frac{\partial F}{\partial t}+\dot{\boldsymbol{w}}\cdot\frac{\partial F}{\partial \boldsymbol{w}}\\&=\frac{\partial F}{\partial t}+\big[F,H\big]\\&=\frac{\partial F}{\partial t}+\frac{\partial F}{\partial \boldsymbol{q}}\cdot\frac{\partial H}{\partial \boldsymbol{p}}-\frac{\partial F}{\partial \boldsymbol{p}}\cdot\frac{\partial H}{\partial \boldsymbol{q}}\\&=\frac{\mathrm{d}F}{\mathrm{d}t}\,,\end{aligned} \tag{1.9}$$

where we introduced as $\mathrm{d}F/\mathrm{d}t$ the rate of change of the local probability density along the motion. Equation (1.9) has numerous names and depending on the context can be referred to as Liouville's equation (when considering the full $N-$body DF of a system of N particles), collisionless Boltzmann equation (when restricted to a DF depending on only one particle coordinates), or Vlasov equation (when accounting for the self-consistency of the system's potential). See Hénon (1982) for a historical account of these various names. Equation (1.9) essentially captures the conservation of the system's probability during its diffusion.

Equation (1.9) becomes particularly simple when the system admits angle-action coordinates. It then reads

$$\frac{\partial F}{\partial t} + \boldsymbol{\Omega} \cdot \frac{\partial F}{\partial \boldsymbol{\theta}} = 0\,. \tag{1.10}$$

With such a rewriting, one can note that steady states of the collisionless Boltzmann equation are reached by DFs such that $F = F(\boldsymbol{J})$. This is Jeans theorem (Jeans 1915).

These steady states are of particular importance for self-gravitating systems. Indeed, they are very efficiently reached thanks to two complementary dynamical mechanisms. The first mechanism is phase mixing and is illustrated in Figs. 1.6 and 1.7. This mixing mechanism relies on the fact that any dependence of the intrinsic frequencies $\boldsymbol{\Omega}$ with the actions $\boldsymbol{J}$ introduces a shearing and dephasing in the angle coordinates. This leads to the appearance of ever finer structures in the system's DF, which, when coarse grained, converges to a steady state $F = F(\boldsymbol{J})$ independent of the angles. The second mechanism is the one of violent relaxation (Lynden-Bell 1967) illustrated in Fig. 1.8. This occurs for self-gravitating systems initially far from equilibrium. Such systems undergo a phase of violent and abrupt potential oscillations, during which the energy of individual particles is redistributed. This allows the system to reach very efficiently a steady state on a few dynamical times. These two processes motivate the use of the orbit-averaged approximation in Chap. 2. Secular dynamics can then mostly be seen as a slow evolution along quasi-stationary collisionless equilibria given by Jeans theorem.

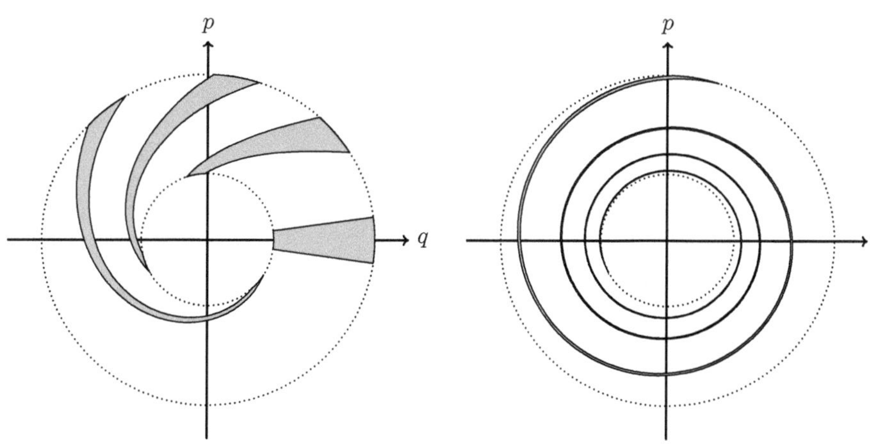

Fig. 1.6 Inspired from Fig. 4.27 of Binney and Tremaine (2008). Illustration of phase mixing undergone by a population of anharmonic oscillators (see left panel of Fig. 1.4). Each particle follows a circular trajectory in phase space, but the intrinsic frequency of motion decreases with the size of the circles. Because of this shearing in frequency, the particles dephase (*left panel*), leading to the appearance of ever finer structures in phase space (*right panel*). This is phase mixing. When coarse grained, these fine structures are washed out and the systems reaches a quasi-stationary mixed state

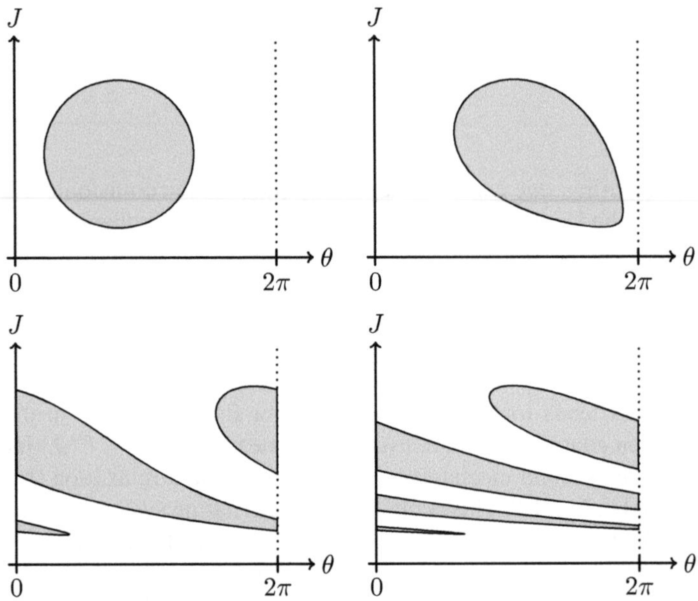

Fig. 1.7 Inspired from Fig. 2 of Lynden-Bell (1967). Illustration of phase mixing, similarly to Fig. 1.6, in angle-action space, for various times. Here, within the angle-action coordinates, as a result of the conservation of actions, trajectories are simple straight lines. Provided that the intrinsic frequencies $\boldsymbol{\Omega}=\partial H/\partial \boldsymbol{J}$ change with the actions, particles of different actions dephase. This phase mixing in the angles $\boldsymbol{\theta}$ is one of the main justifications for the consideration of orbit-averaged diffusion, i.e. the assumption that the system's mean DF depends only on the actions. This is at the heart of both diffusion equations presented in Chap. 2

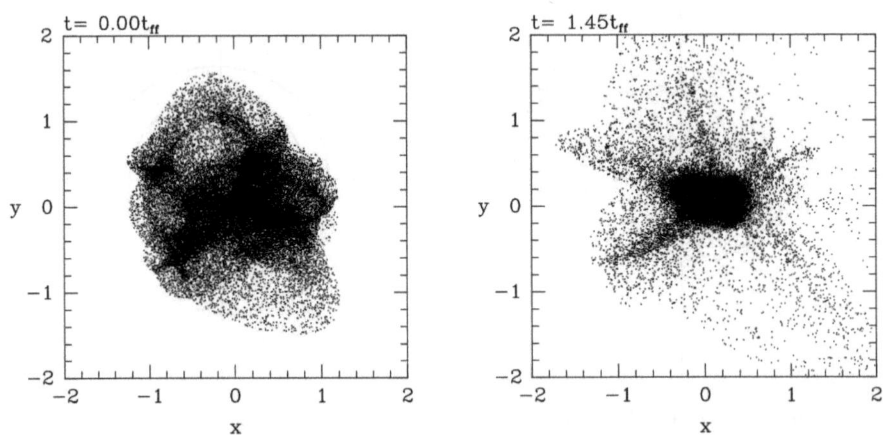

Fig. 1.8 Extracted from Fig. 4.28 of Binney and Tremaine (2008). Illustration of the mechanism of violent relaxation, during which an initially out-of-equilibrium self-gravitating system undergoes a phase of strong potential fluctuations allowing the system to rapidly reach a collisionless quasi-stationary state

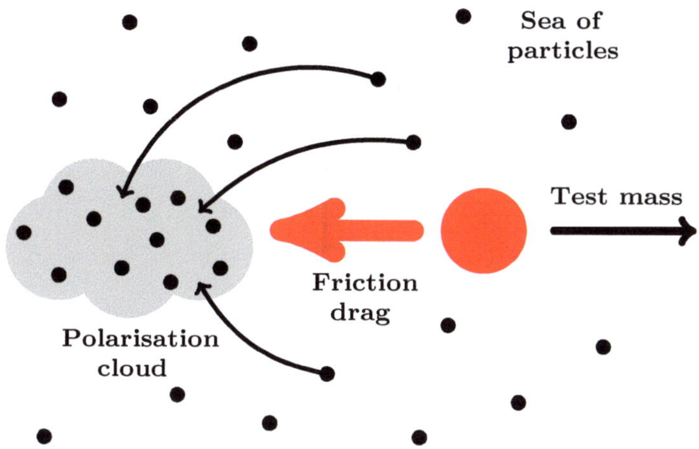

Fig. 1.9 Illustration of the homogeneous dynamical friction, as first introduced in Chandrasekhar (1943). We consider a test mass (illustrated with the red particle) moving to the right along a straight line, while embedded in a homogeneous "sea" of background particles (illustrated with *black* dots). Along its motion, the test particle is followed by a gravitational wake, also coined polarisation cloud, constituted of background stars. This polarisation cloud has two main effects. First, being located behind the test mass, it exerts a drag force on the test mass, hence the name dynamical friction. It also illustrates the importance of collective effects in self-gravitating systems. Because of these polarised background stars, the test mass is dressed. Its effective mass is increased, which hastens the secular diffusion. (See Chap. 4 for a detailed discussion on the importance of collective effects in cold dynamical systems.) Let us finally note that in real self-gravitating systems, the situation is more intricate if one accounts for the complexity of the trajectories, i.e. the fact that the system is inhomogeneous (Heyvaerts et al. 2017). In particular, there are situations where the polarisation can accelerate rather than drag

Let us now discuss one final important physical process occurring in self-gravitating systems as a result of their ability to amplify and respond to perturbations. This is the mechanism of dynamical friction, first introduced in the seminal work from Chandrasekhar (1943). This is illustrated in Fig. 1.9. See Nelson and Tremaine (1999) for a review. Let us consider a test mass travelling through of "sea" of background particles assumed to be infinite and homogeneous. Because of their interaction with the test mass, the background particles tend to accumulate behind the test mass, forming a gravitational wake, the polarisation cloud. Because it is located behind the test mass, this wake induces a drag force on the test mass. This is the dynamical friction. In addition, one can also note that along its motion, the test mass appears as dressed by its polarisation cloud. Collective effects, i.e. the fact that the system is self-gravitating, therefore lead to an increase of the effective mass of the test particle. A self-gravitating system can therefore (strongly) amplify perturbations. Collisions between dressed particles can have a qualitatively different outcome than collisions between bare ones (Weinberg 1998). This dressing is important in particular in cold dynamical systems such as stellar discs, see Chap. 4.

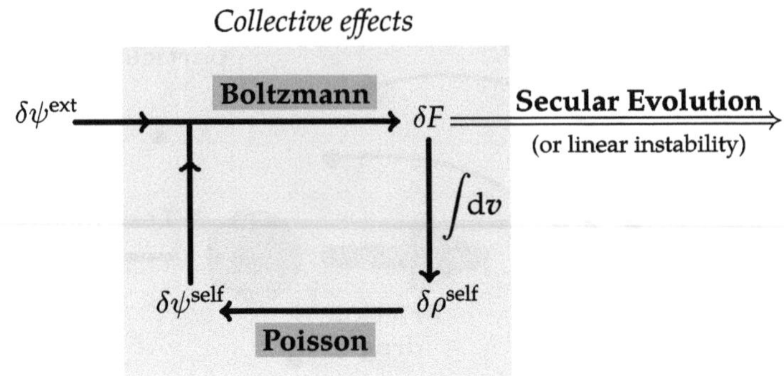

Fig. 1.10 Illustration of the role of collective effects in a self-gravitating system. Any external potential perturbation, $\delta\psi^{\mathrm{ext}}$, induces a response, δF, in the DF of a self-gravitating system, as captured by Boltzmann equation (1.9). The system being self-gravitating, a density perturbation, $\delta\rho^{\mathrm{self}} = \int \mathrm{d}\boldsymbol{v} F$, is directly associated with this response. Poisson's equation then relates this density perturbation to a potential perturbation $\delta\psi^{\mathrm{self}}$. Following this self-consistent amplification, the system must now respond to the total perturbation $\delta\psi^{\mathrm{tot}} = \delta\psi^{\mathrm{ext}} + \delta\psi^{\mathrm{self}}$. Accounting for the additional potential perturbation $\delta\psi^{\mathrm{self}}$ is accounting for collective effects

The gist of collective effects in self-gravitating systems is summarised in Fig. 1.10. Any external potential perturbation, $\delta\psi^{\mathrm{ext}}$, imposed to a self-gravitating system induces a response in the distribution function of this system. The properties of this response are generically given by Boltzmann equation (1.9). However, because the system is self-gravitating (e.g., stars in a stellar disc have a mass), one can associate a perturbation in density, $\delta\rho^{\mathrm{self}} = \int \mathrm{d}\boldsymbol{v} F$, to this response. From Poisson's equation, one may then determine the potential perturbation, $\delta\psi^{\mathrm{self}}$, self-generated by the system. As a result of this amplification loop, the system must now respond to the conjunction of two potential perturbations, $\delta\psi^{\mathrm{tot}} = \delta\psi^{\mathrm{ext}} + \delta\psi^{\mathrm{self}}$. This additional perturbation is associated with the system's self-gravity, and describes how the system may perturb itself. External perturbations are said to be dressed by collective effects. The upcoming chapters will emphasise the essential role played by this self-induced amplification in dynamically cold self-gravitating systems.

After having briefly laid the required elements of Hamiltonian dynamics needed to address the secular dynamics of self-gravitating systems, Chap. 2 will rely on these remarks to present the formalisms appropriate to describe their long-term evolution.

1.4 Overview

This thesis discusses approaches to the long-term evolution of self-gravitating systems. It also illustrates applications to various classes of astrophysical systems to recover some of the features they develop on secular timescales. Two main types of secular evolution are considered depending on the sources of perturbations and fluctuations in the system. This dichotomy, around which this thesis is organised, allows for the detailed description of the secular dynamics of large classes of astrophysical systems. This thesis is composed of five main chapters. First, Chap. 2 presents the main theoretical tools required for the description of such secular dynamics, and derives the associated diffusion equations. Chapter 3 focuses on the secular dynamics of razor-thin stellar discs, and emphasises how their secular dynamics may be significantly simplified by relying on a tailored WKB approximation. Chapter 4 considers the same razor-thin discs and emphasises how a proper accounting of the disc's self-gravitating amplification allows for a precise description of their diffusion features. Chapter 5 focuses on the dynamics of thickened stellar discs, simplifies their dynamics via a new thickened WKB approximation, and investigates various possible sources of secular thickening. Chapter 6 focuses on quasi-Keplerian systems (such as galactic centres) and details how their intrinsic dynamical degeneracies can be dealt with. Finally, in Chap. 7, we present the conclusions of the thesis and outline possible follow-up works. Let us briefly sum up below the content of each chapter.

In Chap. 2, we present the main formalisms capturing the secular dynamics of self-gravitating systems. We successively consider two types of diffusion: collisionless and collisional. The first type of collisionless diffusion corresponds to cases where the source of fluctuations is induced by an external perturber. We investigate the interplay between the spectral properties of the external perturbations and the internal orbital structure of the system. The second type of collisional diffusion is associated with cases where the source of evolution is due to the system's own intrinsic graininess. Self-gravitating systems being inhomogeneous, we especially emphasise how this approach allows for the description of distant resonant encounters. They are shown to be the drivers of the evolution in systems made of a finite number of particles. Throughout these derivations, we also underline how these diffusion equations account for the system's self-gravity, i.e. its ability to amplify perturbations. This proves essential for cold dynamical systems such as stellar discs.

In Chap. 3, we consider a first class of astrophysical systems, razor-thin stellar discs. The main aim of this chapter is to illustrate the use of a tailored WKB approximation (i.e. limited to radially tightly wound perturbations) to explicitly and straightforwardly compute the properties of the diffusion occurring in such systems. When applied to isolated discrete stellar discs, we illustrate how the two diffusion formalisms (collisionless and collisional) allow for the recovery of the shot-noise-driven formation of narrow ridges observed in numerical simulations. We also discuss one discrepancy obtained in the applications of these formalisms, namely the mismatch of the diffusion timescales. This is interpreted as being due to the neglect of some contributions to the disc's self-gravity (namely the loosely wound contributions), which are accounted for in Chap. 4.

The heart of Chap. 4 is to illustrate, in the context of razor-thin discs, how one can fully account for self-gravity through a proper numerical calculation. Relying on the collisional Balescu-Lenard equation, we show how this formalism recovers in detail the diffusion features observed in secular simulations of stable self-gravitating razor-thin discs. We emphasise that collective effects cause cool discs to have 2−body relaxation time much shorter than naively expected. We also argue that this anomalous relaxation introduces small scale structures in the disc, which destabilise it at the collisionless level. Resorting to our own simulations, we also investigate in detail some generic properties of such systems, such as the scaling of the system's diffusion with the number of particles, as well as the presence of unstable secular dynamical phase transitions.

In Chap. 5, we extend the results of Chap. 3 to thickened stellar discs. We illustrate how one may devise a thickened WKB approximation offering straightforward estimations of the collisionless and collisional diffusion fluxes. We show how these two formalisms allow for the qualitative recovery of the diffusion features observed in numerical simulations of stable thickened stellar discs, with the caveat of a diffusion timescale discrepancy due to the neglect of the contributions of loosely wound perturbations to the disc's self-gravity. We also investigate some other possible mechanisms of secular thickening such as series of central decaying bars, or the joint evolution of giant molecular clouds. This illustrates how different perturbation mechanisms can lead to different signatures in the disc's diffusion.

Chapter 6 develops this diffusion formalism for quasi-Keplerian systems, such as galactic centres. Because these systems are dominated by one central object, their constituents approximately follow closed Keplerian orbits. These systems are dynamically degenerate. We detail how such degeneracies can be dealt with to derive the associated kinetic equation. We show how this new diffusion equation captures the mechanism of resonant relaxation between Keplerian wires. We also emphasise how this approach sheds new light on some important diffusion properties of these systems. We focus in particular on understanding the Schwarzschild barrier, which strongly damps the rate with which stars can diffuse resonantly towards the central black hole.

References

Arnold, V. (1978). *Mathematical methods of classical mechanics*. New York: Springer.
Bardeen, J. M., Bond, J. R., Kaiser, N., & Szalay, A. S. (1986). *ApJ*, *304*, 15.
Barnes, J. E., & Hernquist, L. (1992). *ARA&A*, *30*, 705.
Binney, J., & Tremaine, S. (2008). *Galactic dynamics* (2nd ed.). Princeton University Press.
Binney, J. (2013). *Dynamics of secular evolution*, (p. 259). Cambridge University Press.
Bond, J. R., Kofman, L., & Pogosyan, D. (1996). *Nature*, *380*, 603.
Chandrasekhar, S. (1943). *ApJ*, *97*, 255.
Dubois, Y., Pichon, C., Welker, C., et al. (2014). *MNRAS*, *444*, 1453.
Dubois, Y., Peirani, S., Pichon, C., et al. (2016). *MNRAS*, *463*, 3948.
Frenk, C. S., & White, S. D. M. (2012). *Annalen der Physik*, *524*, 507.
Goldstein, H. (1950). *Classical mechanics*. Addison-Wesley.
Hénon, M. (1982). *A&A*, *114*, 211.
Heyvaerts, J., Fouvry, J.-B., Chavanis, P.-H., & Pichon, C. (2017). *MNRAS*, *469*, 4193.
Jeans, J. (1915). *MNRAS*, *76*, 70.
Kormendy, J. (2013). *Secular evolution in disk galaxies*, (p.1). Cambridge University Press.
Kroupa, P. (2012). *PASA*, *29*, 395.
Lynden-Bell, D. (1967). *MNRAS*, *136*, 101.
Nelson, R. W., & Tremaine, S. (1999). *MNRAS*, *306*, 1.
Perlmutter, S., Aldering, G., Goldhaber, G., et al. (1999). *ApJ*, *517*, 565.

Pichon, C., Pogosyan, D., Kimm, T., et al. (2011). *MNRAS*, *418*, 2493.
Planck Collaboration, Ade, P. A. R., Aghanim, N., et al. (2014). *A&A*, *571*, A16.
Riess, A. G., Filippenko, A. V., Challis, P., et al. (1998). *AJ*, *116*, 1009.
Sellwood, J. A. (2014). *Rev. Mod. Phys.*, *86*, 1.
Silk, J., & Mamon, G. A. (2012). *Res. Astron. Astrophys.*, *12*, 917.
Springel, V., Frenk, C. S., & White, S. D. M. (2006). *Nature*, *440*, 1137.
Toomre, A. (1977). *Evolution of galaxies and stellar populations*, (p. 401). New Haven: Yale University Observatory.
Toomre, A., & Toomre, J. (1972). *ApJ*, *178*, 623.
Weinberg, M. D. (1998). *MNRAS*, *297*, 101.
White, S. D. M., & Rees, M. J. (1978). *MNRAS*, *183*, 341.

Chapter 2
Secular Diffusion

2.1 Introduction

The previous chapter described the typical fate of self-gravitating systems, which is summed up in Fig. 2.1. As a result of both phase mixing (see Fig. 1.4) and violent relaxation (see Fig. 1.5), self-gravitating systems very efficiently reach quasi-stationary states for the collisionless mean field dynamics. The systems are virialised and the mean potentials do not strongly fluctuate anymore. Stars follow their orbit set up by the mean field potential and are typically uniformly distributed in phase along each of them. Yet, as gravity is a long-range interaction, self-gravitating systems have the ability to amplify and dress perturbations (see, e.g., Fig. 1.7). These collective effects have two main consequences. They may first lead to the spontaneous growth of dynamical instabilities if ever the system was dynamically unstable. Moreover, even for genuinely stable systems, these effects can also lead to polarisation, i.e. a dressing of perturbations and therefore a boost in the amplitude of the fluctuations in the system. This self-gravitating amplification is especially important for cold dynamical systems, i.e. within which most of the gravitational support comes from centrifugal forces and for which the velocity dispersion is low. This makes the system strongly responsive. This is for example important for stellar discs, where new stars, born on the cold orbits of the gas, are constantly being supplied to the system.

Once the system has reached a quasi-stationary state through these various mixing processes, the mean collisionless dynamics maintains stationarity and such a quiescent system can now only slowly evolve on long timescales.[1] This is the timescale for secular evolution, which will be our main interest here. At this stage, only additional fluctuations can drive the system's evolution. Such considerations fall within

The work presented in this chapter is based on Fouvry et al. (2015a, b, 2016a, b).

[1] Another possibility allowing self-gravitating systems to reach more probable and hotter configurations is for them to spontaneously develop an instability, such as a bar (Hohl 1971), leading as well to an efficient rearrangement of the orbital structure. Such outcomes are not investigated in the present thesis.

J.-B. Fouvry, *Secular Evolution of Self-Gravitating Systems Over Cosmic Age*,
Springer Theses, DOI 10.1007/978-3-319-66760-7_2

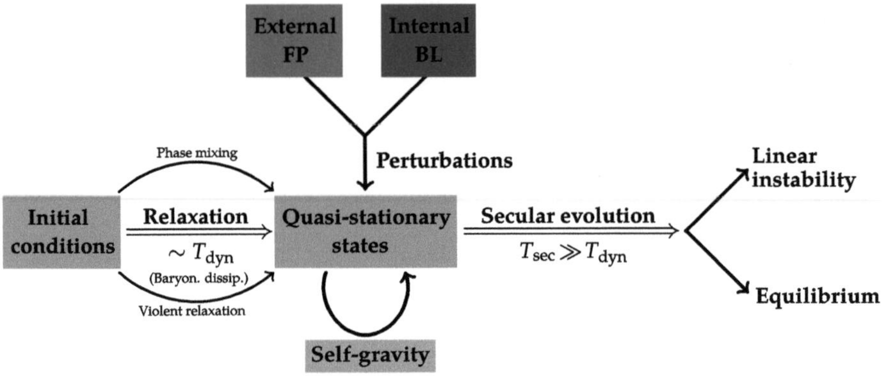

Fig. 2.1 Illustration of the typical fate of a self-gravitating system. On the first few dynamical times, the system's evolution is governed by the Vlasov-Poisson system (see Eq. (1.9)), and as a result of violent relaxation and phase mixing, the system reaches a quasi-stationary state, generically difficult to predict. Following this relaxation, the mean system does not evolve anymore on short timescales. On longer timescales, the fluctuations present in the system, amplified by self-gravity, induce a long-term distorsion of the system's orbital structure: it is the regime of secular evolution. These perturbations may either be external (via the so-called dressed Fokker-Planck equation, see Sect. 2.2) or internal (via the inhomogeneous Balescu-Lenard equation, see Sect. 2.3). Finally, as a result of the secular evolution, the system may either become dynamically unstable or may reach an equilibrium state, if it exists

the general framework of the fluctuation-dissipation theorem, for which fluctuations dissipate and lead to a diffusion in the system. Let us now introduce an important dichotomy on which the two upcoming sections rely. There are two main channels to induce fluctuations in a system. Fluctuations of the first type are induced by external stochastic perturbations, whose contributions will be felt by the system and will lead therein to slow orbital distortions. As will be discussed in detail in the next section, the efficiency of such secular dynamics is dictated in particular by the match between the temporal frequencies of these perturbations and the system's natural intrinsic frequencies. We call this framework the collisionless framework. Another source of fluctuations is also present in any system made of a finite number N of particles: these are finite$-N$ effects, also called Poisson shot noise. This graininess can not only be triggered by the finite number of constituents in the system, but can also originate from the variety of its components, e.g., the existence of a mass spectrum of components. As a direct consequence of the finite number of particles, the system's self-induced potential is not perfectly smooth, and therefore fluctuates around its mean quasi-stationary value. These unavoidable and non-vanishing fluctuations may then act as the source of a secular irreversible evolution. We call this framework the collisional framework, in the sense that it relies on encounters between the finite number of particles. Let us finally note that whatever the source of the perturbations, these fluctuations are dressed by collective effects. A proper accounting of the importance of the gravitational polarisation is at the heart of the upcoming derivations. On the long-term, secular dynamics may lead the system towards a dynamical instability (see, e.g., Fig. 4.6) or towards an equilibrium state, if it exists.

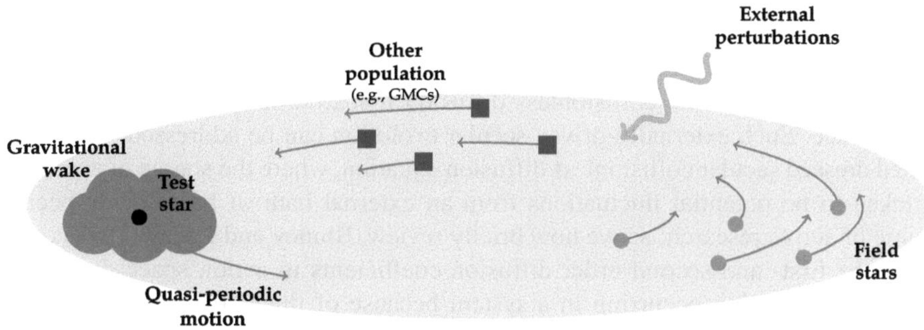

Fig. 2.2 Illustration of the different components involved in the secular dynamics of a test star in a stellar disc. To describe the complex quasiperiodic motion of the test star, one has to rely on angle-action coordinates. The disc being self-gravitating, any perturbation is dressed by collective effects and surrounded by a gravitational wake. The long-term orbital diffusion of the test star may then be induced either by external perturbations (e.g., cosmic environment) or internal perturbations (e.g., finite$-N$ effects)

The main elements involved in the secular evolution of a self-gravitating system can also be considered directly in physical space. This is illustrated in Fig. 2.2, where we consider the example of the diffusion of a test star in a stellar disc. Because a stellar disc is inhomogeneous, the trajectories of individual stars are intricate. Assuming their motions to be quasiperiodic, one can describe the associated orbits thanks to angle-action coordinates (see Eq. (1.6)). The disc being self-gravitating, any test star is surrounded by a "polarisation cloud" associated with collective effects (see Fig. 1.7). The test star may then diffuse as a result of two types of fluctuations. These perturbations may first be external, imposed for example by the cosmological environment. The long-term effects of these externally-induced perturbations can be described by the secular collisionless diffusion equation and will be considered in Sect. 2.2. There exist also intrinsic perturbations due to the finite number of field stars or due to the discreteness of another component of the disc (e.g., giant molecular clouds). Such an internally-induced evolution can be described by the inhomogeneous Balescu-Lenard equation and will be considered in Sect. 2.3.

This dichotomy is essential for all the upcoming sections. It allows us to distinguish secular evolution induced by the system's environment from secular evolution induced by the system's internal properties. It is therefore an useful tool to disentangle the respective contributions from nurture and nature in driving the evolution of a self-gravitating system. The aim of the present chapter is to detail the relevant formalisms allowing for the description of long-term evolutions induced by (internal or external) potential fluctuations. The following chapters will illustrate applications of these formalisms to various astrophysical systems. Let us first focus in Sect. 2.2 on the collisionless framework, where the dynamics is driven by external perturbations. Then, in Sect. 2.3, we will consider the collisional framework of diffusion, sourced by the discreteness of these self-gravitating systems.

2.2 Collisionless Dynamics

Let us first describe the collisionless diffusion that external potential fluctuations may induce. Such externally-driven secular evolution can be addressed via the so-called dressed secular collisionless diffusion equation, where the source of evolution is taken to be potential fluctuations from an external bath. It has already been a theme of active research, as we now briefly review. Binney and Lacey (1988) computed the first- and second-order diffusion coefficients in action space describing the orbital diffusion occurring in a system because of fluctuations in the gravitational potential. This first approach however did not account for collective effects, i.e. the ability of the system to dress and amplify perturbations. Weinberg (1993) emphasised the importance of self-gravity for the non-local and collective relaxation of stellar systems. Weinberg (2001a, b) considered similar secular evolutions while accounting for the self-gravitating amplification of perturbations, and studied the impacts of the properties of the noise processes. Ma and Bertschinger (2004) relied on a quasilinear approach to investigate the diffusion of dark matter induced by cosmological fluctuations. Pichon and Aubert (2006) sketched a time-decoupling approach to solve the collisionless Boltzmann equation in the presence of external perturbations and applied it to a statistical study of the effect of dynamical flows through dark matter haloes on secular timescales. The approach developed therein is close to the one presented in Fouvry et al. (2015b). Chavanis (2012a) considered the evolution of homogeneous collisionless systems when forced by an external perturbation, while Nardini et al. (2012) investigated similarly the effects of stochastic forces on the long-term evolution of long-range interacting systems.

In the upcoming section, let us follow Fouvry et al. (2015b) and present a derivation of the appropriate secular resonant collisionless dressed diffusion equation. This derivation is based on a quasilinear timescale decoupling of the collisionless Boltzmann equation. This yields two evolution equations, one for the fast dynamical evolution and amplification of perturbations within the system, and one for the secular evolution of the system's mean DF.

2.2.1 Evolution Equations

Let us consider a collisionless self-gravitating quasi-stationary system undergoing external stochastic perturbations. The mean system being quasi-stationary, we introduce its quasi-stationary Hamiltonian H_0, associated with the mean potential ψ_0. We assume that throughout its evolution, the system remains integrable, so that one can always define an angle-action mapping $(\boldsymbol{x}, \boldsymbol{v}) \mapsto (\boldsymbol{\theta}, \boldsymbol{J})$ appropriate for the Hamiltonian H_0. Thanks to Jeans theorem (Jeans 1915), the mean DF of the system, F, depends only on the actions, so that $F = F(\boldsymbol{J}, t)$. We suppose that an external source is perturbing the system, and we expand the system's total DF and Hamiltonian as

$$\begin{cases} F^{\mathrm{tot}}(\boldsymbol{J}, \boldsymbol{\theta}, t) = F(\boldsymbol{J}, t) + \delta F(\boldsymbol{J}, \boldsymbol{\theta}, t)\,, \\ H^{\mathrm{tot}}(\boldsymbol{J}, \boldsymbol{\theta}, t) = H_0(\boldsymbol{J}, t) + \delta\psi^{\mathrm{e}}(\boldsymbol{J}, \boldsymbol{\theta}, t) + \delta\psi^{\mathrm{s}}(\boldsymbol{J}, \boldsymbol{\theta}, t)\,. \end{cases} \tag{2.1}$$

In the decompositions from Eq. (2.1), one should pay attention to the presence of two types of potential perturbations. Here, $\delta\psi^{\mathrm{e}}$ corresponds to an external stochastic perturbation, while $\delta\psi^{\mathrm{s}}$ corresponds to the self-response of the system induced by its self-gravity (Weinberg 2001a). This additional perturbation is crucial to capture the system's gravitational susceptibility, i.e. its ability to amplify perturbations. We place ourselves in the limit of small perturbations, so that $\delta F \ll F$, and $\delta\psi^{\mathrm{e}}, \delta\psi^{\mathrm{s}} \ll \psi_0$. Assuming that the system evolves in a collisionless fashion, its dynamics is fully described by the collisionless Boltzmann equation (1.9) reading

$$\frac{\partial F^{\mathrm{tot}}}{\partial t} + \left[F^{\mathrm{tot}}, H^{\mathrm{tot}}\right] = 0\,, \tag{2.2}$$

where $[\,.\,,\,.\,]$ stands for the Poisson bracket as defined in Eq. (1.2). Let us then inject the decomposition from Eq. (2.1) into Eq. (2.2) to get

$$\frac{\partial F}{\partial t} + \frac{\partial \delta F}{\partial t} + \left[F, H_0\right] + \left[F, \delta\psi^{\mathrm{e}} + \delta\psi^{\mathrm{s}}\right] + \left[\delta F, H_0\right] + \left[\delta F, \delta\psi^{\mathrm{e}} + \delta\psi^{\mathrm{s}}\right] = 0\,. \tag{2.3}$$

Because we assumed the mean DF to be quasi-stationary, i.e. $F = F(\boldsymbol{J}, t)$, one has $\left[F, H_0\right] = 0$, since $H_0 = H_0(\boldsymbol{J}, t)$. Let us now take an average of Eq. (2.3) w.r.t. the angles $\boldsymbol{\theta}$. In Eq. (2.3), all the terms linear in the perturbations vanish, and we get a secular evolution equation for the mean DF F as

$$\frac{\partial F}{\partial t} = \frac{\partial}{\partial \boldsymbol{J}} \cdot \left[\int \frac{\mathrm{d}\boldsymbol{\theta}}{(2\pi)^d}\, \delta F\, \frac{\partial [\delta\psi^{\mathrm{e}} + \delta\psi^{\mathrm{s}}]}{\partial \boldsymbol{\theta}} \right], \tag{2.4}$$

where d is the dimension of the physical space, e.g., $d = 2$ for a razor-thin disc. At this stage, let us note that $\partial F / \partial t$ can be considered as a second-order term as it is the product of two fluctuations. Keeping only first-order terms in Eq. (2.3) (quasilinear approximation), one finally gets a second evolution equation of the form

$$\frac{\partial \delta F}{\partial t} + \boldsymbol{\Omega} \cdot \frac{\partial \delta F}{\partial \boldsymbol{\theta}} - \frac{\partial F}{\partial \boldsymbol{J}} \cdot \frac{\partial [\delta\psi^{\mathrm{e}} + \delta\psi^{\mathrm{s}}]}{\partial \boldsymbol{\theta}} = 0\,, \tag{2.5}$$

where we used the assumptions from Eq. (2.1) to rewrite the Poisson brackets. We also introduced the mean orbital frequencies $\boldsymbol{\Omega} = \partial H_0 / \partial \boldsymbol{J}$. The two evolution equations (2.4) and (2.5) are the two coupled evolution equations from which one can obtain the secular collisionless diffusion equation. Equation (2.5) describes the evolution of the perturbation δF on dynamical timescales, while Eq. (2.4) describes the long-term evolution of the quasi-stationary DF F. Let us now solve Eq. (2.5) to describe the dynamical amplification of perturbations. Its solution, when injected in Eq. (2.4), will then allow for the description of the secular evolution of the system's mean quasi-stationary DF.

As the angles $\boldsymbol{\theta}$ are $2\pi-$periodic, let us define the discrete Fourier transform w.r.t. these variables as

$$X(\boldsymbol{\theta}, \boldsymbol{J}) = \sum_{\boldsymbol{m}\in\mathbb{Z}^d} X_{\boldsymbol{m}}(\boldsymbol{J})\,\mathrm{e}^{\mathrm{i}\boldsymbol{m}\cdot\boldsymbol{\theta}} \quad ; \quad X_{\boldsymbol{m}}(\boldsymbol{J}) = \int \frac{\mathrm{d}\boldsymbol{\theta}}{(2\pi)^d}\, X(\boldsymbol{\theta}, \boldsymbol{J})\,\mathrm{e}^{-\mathrm{i}\boldsymbol{m}\cdot\boldsymbol{\theta}}\,, \tag{2.6}$$

so that Eq. (2.5) immediately becomes

$$\frac{\partial \delta F_{\boldsymbol{m}}}{\partial t} + \mathrm{i}\boldsymbol{m}\cdot\boldsymbol{\Omega}\,\delta F_{\boldsymbol{m}} - \mathrm{i}\boldsymbol{m}\cdot\frac{\partial F}{\partial \boldsymbol{J}}\left[\delta\psi^{\mathrm{e}}_{\boldsymbol{m}} + \delta\psi^{\mathrm{s}}_{\boldsymbol{m}}\right] = 0\,. \tag{2.7}$$

We now introduce the assumption of timescale decoupling, also coined Bogoliubov's ansatz. Indeed, let us assume that the fluctuations (i.e. δF, $\delta\psi^{\mathrm{e}}$, and $\delta\psi^{\mathrm{s}}$) evolve rapidly on dynamical timescales, while the mean orbit-averaged quantities (such as F) only evolve on secular timescales, i.e. over many dynamical times. As a consequence, in Eq. (2.7), we may push the secular time to infinity, while assuming in the meantime that $\partial F/\partial \boldsymbol{J} = \mathrm{cst}$. Forgetting transient terms and bringing the initial time to $-\infty$ to consider only the forced regime of evolution, Eq. (2.7) can then be solved explicitly as

$$\delta F_{\boldsymbol{m}}(\boldsymbol{J}, t) = \int_{-\infty}^{t} \mathrm{d}\tau\, \mathrm{e}^{-\mathrm{i}\boldsymbol{m}\cdot\boldsymbol{\Omega}(t-\tau)}\, \mathrm{i}\boldsymbol{m}\cdot\frac{\partial F}{\partial \boldsymbol{J}}\left[\delta\psi^{\mathrm{e}}_{\boldsymbol{m}} + \delta\psi^{\mathrm{s}}_{\boldsymbol{m}}\right](\boldsymbol{J}, \tau)\,. \tag{2.8}$$

We define the temporal Fourier transform with the convention

$$\widehat{f}(\omega) = \int_{-\infty}^{+\infty} \mathrm{d}t\, f(t)\,\mathrm{e}^{\mathrm{i}\omega t} \quad ; \quad f(t) = \frac{1}{2\pi}\int_{-\infty}^{+\infty} \mathrm{d}\omega\, \widehat{f}(\omega)\,\mathrm{e}^{-\mathrm{i}\omega t}\,. \tag{2.9}$$

Taking the temporal Fourier transform of Eq. (2.7), we immediately get

$$\delta\widehat{F}_{\boldsymbol{m}}(\boldsymbol{J}, \omega) = -\frac{\boldsymbol{m}\cdot\partial F/\partial \boldsymbol{J}}{\omega - \boldsymbol{m}\cdot\boldsymbol{\Omega}}\left[\delta\widehat{\psi}^{\mathrm{e}}{}_{\boldsymbol{m}}(\boldsymbol{J}, \omega) + \delta\widehat{\psi}^{\mathrm{s}}{}_{\boldsymbol{m}}(\boldsymbol{J}, \omega)\right], \tag{2.10}$$

so that we expressed the DF's perturbations in terms of the potential fluctuations.

2.2.2 *Matrix Method*

The next step of the calculation is to account for the system's self-gravity, i.e. the fact that the perturbing DF δF should be consistent with the self-induced potential perturbation $\delta\psi^{\mathrm{s}}$ and its associated density $\delta\rho^{\mathrm{s}}$. One has

$$\delta\rho^{\mathrm{s}}(\boldsymbol{x}) = \int \mathrm{d}\boldsymbol{v}\, \delta F(\boldsymbol{x}, \boldsymbol{v})\,. \tag{2.11}$$

In Eq. (2.11), the potential and density perturbations are connected through Poisson's equation $\Delta\delta\psi^{\mathrm{s}} = 4\pi G\delta\rho^{\mathrm{s}}$. The method to deal with this self-consistency constraint is to follow Kalnajs matrix method (Kalnajs 1976). Let us introduce a representative biorthogonal basis of potential and densities $\psi^{(p)}$ and $\rho^{(p)}$ satisfying

$$\Delta\psi^{(p)} = 4\pi G\rho^{(p)} \quad ; \quad \int \mathrm{d}\boldsymbol{x}\, \psi^{(p)*}(\boldsymbol{x})\, \rho^{(q)}(\boldsymbol{x}) = -\delta_p^q \,. \tag{2.12}$$

We will then use these basis elements to represent any potential and density disturbances in the system. The potential perturbations $\delta\psi^{\mathrm{s}}$ and $\delta\psi^{\mathrm{e}}$ may therefore be written as

$$\delta\psi^{\mathrm{s}}(\boldsymbol{x}, t) = \sum_p a_p(t)\, \psi^{(p)}(\boldsymbol{x}) \quad ; \quad \delta\psi^{\mathrm{e}}(\boldsymbol{x}, t) = \sum_p b_p(t)\, \psi^{(p)}(\boldsymbol{x}) \,, \tag{2.13}$$

and we introduce as $c_p = a_p + b_p$ the total potential perturbation. The linearity of Poisson's equation immediately ensures that one also has the decomposition $\delta\rho^{\mathrm{s}}(\boldsymbol{x}, t) = \sum_p a_p(t)\, \rho^{(p)}(\boldsymbol{x})$. Multiplying Eq. (2.11) by $\psi^{(p)*}(\boldsymbol{x})$ and integrating over $\mathrm{d}\boldsymbol{x}$, we get

$$a_p(t) = -\sum_{\boldsymbol{m}} \int \mathrm{d}\boldsymbol{x}\mathrm{d}\boldsymbol{v}\, \delta F_{\boldsymbol{m}}(\boldsymbol{J}, t)\, \mathrm{e}^{\mathrm{i}\boldsymbol{m}\cdot\boldsymbol{\theta}}\, \psi^{(p)*}(\boldsymbol{J}) \,. \tag{2.14}$$

The transformation to angle-action coordinates $(\boldsymbol{x}, \boldsymbol{v}) \mapsto (\boldsymbol{\theta}, \boldsymbol{J})$ is canonical so that it conserves infinitesimal volumes, i.e. one has $\mathrm{d}\boldsymbol{x}\mathrm{d}\boldsymbol{v} = \mathrm{d}\boldsymbol{\theta}\mathrm{d}\boldsymbol{J}$. Equation (2.14) can then be rewritten as

$$a_p(t) = -(2\pi)^d \sum_{\boldsymbol{m}} \int \mathrm{d}\boldsymbol{J}\, \delta F_{\boldsymbol{m}}(\boldsymbol{J}, t)\, \psi_{\boldsymbol{m}}^{(p)*}(\boldsymbol{J}) \,, \tag{2.15}$$

where $\psi_{\boldsymbol{m}}^{(p)}(\boldsymbol{J})$ stands for the Fourier transformed basis elements in angles following Eq. (2.6). Thanks to Eq. (2.10) and taking a temporal Fourier transform, we finally obtain

$$\widehat{a}_p(\omega) = (2\pi)^d \sum_q \widehat{c}_q(\omega) \sum_{\boldsymbol{m}} \int \mathrm{d}\boldsymbol{J}\, \frac{\boldsymbol{m}\cdot\partial F/\partial\boldsymbol{J}}{\omega - \boldsymbol{m}\cdot\boldsymbol{\Omega}}\, \psi_{\boldsymbol{m}}^{(p)*}(\boldsymbol{J})\, \psi_{\boldsymbol{m}}^{(q)}(\boldsymbol{J}) \,. \tag{2.16}$$

Let us finally introduce the system's response matrix $\widehat{\mathbf{M}}$ as

$$\widehat{\mathbf{M}}_{pq}(\omega) = (2\pi)^d \sum_{\boldsymbol{m}} \int \mathrm{d}\boldsymbol{J}\, \frac{\boldsymbol{m}\cdot\partial F/\partial\boldsymbol{J}}{\omega - \boldsymbol{m}\cdot\boldsymbol{\Omega}}\, \psi_{\boldsymbol{m}}^{(p)*}(\boldsymbol{J})\, \psi_{\boldsymbol{m}}^{(q)}(\boldsymbol{J}) \,, \tag{2.17}$$

so that Eq. (2.16) becomes

$$\widehat{\boldsymbol{a}}(\omega) = \widehat{\mathbf{M}}(\omega)\cdot\widehat{\mathbf{c}}(\omega) \,. \tag{2.18}$$

One should note that the response matrix depends only on the mean state of the system, since $\partial F/\partial \boldsymbol{J}$ only evolves on secular timescales, the perturbing and self-gravitating potentials are absent, and the basis elements $\psi^{(p)}$ from Eq. (2.12) are chosen once for all. Assuming that the mean system is linearly stable, so that $[\mathbf{I}-\widehat{\mathbf{M}}(\omega)]^{-1}$ admits no poles in the upper complex plane, one can invert Eq. (2.18) to get

$$\widehat{\boldsymbol{c}}(\omega) = \big[\mathbf{I}-\widehat{\mathbf{M}}(\omega)\big]^{-1}\!\cdot\widehat{\boldsymbol{b}}(\omega)\,, \tag{2.19}$$

where $\mathbf{I}$ stands for the identity matrix. Equation (2.19) is a crucial relation, which allows us to express the total perturbations as a function of the external perturbation only, whose statistical properties may be characterised. Equation (2.19) describes the short timescale (dynamical) response of the system and the associated self-gravitating amplification.

2.2.3 Diffusion Coefficients and Statistical Average

Let us now describe how these solutions may be used in Eq. (2.4) to describe the secular evolution of the system. The l.h.s. of Eq. (2.4) requires us to evaluate an expression of the form

$$\frac{1}{(2\pi)^d}\int \mathrm{d}\boldsymbol{\theta}\,\delta F(\boldsymbol{J},\boldsymbol{\theta},t)\,\frac{\partial[\delta\psi^{\mathrm{e}}+\delta\psi^{\mathrm{s}}]}{\partial\boldsymbol{\theta}} = -\sum_{\boldsymbol{m}}\delta F_{\boldsymbol{m}}\,\mathrm{i}\boldsymbol{m}\big[\delta\psi^{\mathrm{e}\,*}_{\boldsymbol{m}}+\delta\psi^{\mathrm{s}\,*}_{\boldsymbol{m}}\big]\,, \tag{2.20}$$

where we used the fact that $\delta\psi_{-\boldsymbol{m}}=\delta\psi^{*}_{\boldsymbol{m}}$. Thanks to the resolution from Eq. (2.8), we may now rewrite Eq. (2.4) as

$$\frac{\partial F}{\partial t} = \frac{\partial}{\partial \boldsymbol{J}}\cdot\bigg[\sum_{\boldsymbol{m}}\boldsymbol{m}\,D_{\boldsymbol{m}}(\boldsymbol{J},t)\,\boldsymbol{m}\cdot\frac{\partial F}{\partial \boldsymbol{J}}\bigg]\,, \tag{2.21}$$

where the diffusion coefficients $D_{\boldsymbol{m}}(\boldsymbol{J},t)$ are given by

$$D_{\boldsymbol{m}}(\boldsymbol{J},t) = \sum_{p,q}\psi^{(p)}_{\boldsymbol{m}}(\boldsymbol{J})\,\psi^{(q)*}_{\boldsymbol{m}}(\boldsymbol{J})\,c^{*}_{q}(t)\int_{-\infty}^{t}\mathrm{d}\tau\,\mathrm{e}^{-\mathrm{i}\boldsymbol{m}\cdot\boldsymbol{\Omega}(t-\tau)}c_{p}(\tau)\,. \tag{2.22}$$

The amplification relation from Eq. (2.19) allows us to rewrite Eq. (2.22) as a function of the external perturbation $\widehat{\boldsymbol{b}}$ only, to get

$$\begin{aligned}D_{\boldsymbol{m}}(\boldsymbol{J},t) = \frac{1}{(2\pi)^2}\sum_{p,q}\sum_{p_1,q_1}\psi^{(p)}_{\boldsymbol{m}}(\boldsymbol{J})\psi^{(q)*}_{\boldsymbol{m}}(\boldsymbol{J})\int\mathrm{d}\omega\,\mathrm{e}^{\mathrm{i}\omega t}\big[\mathbf{I}-\widehat{\mathbf{M}}(\omega)\big]^{-1*}_{qq_1}\widehat{b}^{*}_{q_1}(\omega)\\ \times\int_{-\infty}^{t}\mathrm{d}\tau\,\mathrm{e}^{-\mathrm{i}\boldsymbol{m}\cdot\boldsymbol{\Omega}(t-\tau)}\int\mathrm{d}\omega'\,\mathrm{e}^{-\mathrm{i}\omega'\tau}\big[\mathbf{I}-\widehat{\mathbf{M}}(\omega')\big]^{-1}_{pp_1}\widehat{b}_{p_1}(\omega')\,.\end{aligned} \tag{2.23}$$

The final step of the derivation is to consider statistical averages over various realisations of the perturbations, i.e. to consider only the mean response of the system. Let us denote as $\big\langle\,.\,\big\rangle$ the ensemble average operation on such different realisations. When applying this average, we assume that the response matrix $\widehat{\mathbf{M}}$, as well as the DF F and its gradients $\partial F/\partial \boldsymbol{J}$, do not change significantly from one realisation to another. Thanks to these assumptions, Eq. (2.21) becomes

$$\frac{\partial F}{\partial t}=\frac{\partial}{\partial \boldsymbol{J}}\cdot\bigg[\sum_{\boldsymbol{m}}\boldsymbol{m}\,\big\langle D_{\boldsymbol{m}}(\boldsymbol{J},t)\big\rangle\,\boldsymbol{m}\cdot\frac{\partial F}{\partial \boldsymbol{J}}\bigg]. \tag{2.24}$$

Let us now suppose that the external perturbations are stationary in time, so that one can introduce the corresponding temporal autocorrelation function $\mathbf{C}$ as

$$\mathbf{C}_{k\ell}(t_1-t_2)=\big\langle b_k(t_1)\,b_\ell^*(t_2)\big\rangle\,, \tag{2.25}$$

where it is assumed that the exterior perturbation is of zero mean. When Fourier transformed, Eq. (2.25) becomes

$$\big\langle\widehat{b}_k(\omega)\,\widehat{b}_\ell^*(\omega')\big\rangle=2\pi\delta_{\mathrm{D}}(\omega-\omega')\,\widehat{\mathbf{C}}_{k\ell}(\omega)\,. \tag{2.26}$$

One can now immediately rewrite the averaged diffusion coefficients from Eq. (2.24) as

$$\big\langle D_{\boldsymbol{m}}(\boldsymbol{J},t)\big\rangle=\frac{1}{2\pi}\sum_{p,q}\psi_{\boldsymbol{m}}^{(p)}(\boldsymbol{J})\,\psi_{\boldsymbol{m}}^{(q)*}(\boldsymbol{J})\!\int\!\mathrm{d}\omega\!\int_{-\infty}^{0}\!\mathrm{d}\tau'\,\mathrm{e}^{-\mathrm{i}(\omega-\boldsymbol{m}\cdot\boldsymbol{\Omega})\tau'}\Big[\big[\mathbf{I}-\widehat{\mathbf{M}}\big]^{-1}\!\cdot\widehat{\mathbf{C}}\cdot\big[\mathbf{I}-\widehat{\mathbf{M}}\big]^{-1}\Big]_{pq}\!(\omega)\,, \tag{2.27}$$

where we relied on the hermiticity of the response matrix $\widehat{\mathbf{M}}^*=\widehat{\mathbf{M}}^{\mathrm{t}}$. One should note that after the ensemble average, the diffusion coefficients become (explicitly) independent of t (while they still depend on the secular timescale via the slow variations of F). To shorten temporarily the notations, let us introduce the notation $\widehat{\mathbf{L}}=\big[\mathbf{I}-\widehat{\mathbf{M}}\big]^{-1}\!\cdot\widehat{\mathbf{C}}\cdot\big[\mathbf{I}-\widehat{\mathbf{M}}\big]^{-1}$. In Eq. (2.27), one must then evaluate a double integral of the form

$$\begin{aligned}\frac{1}{2\pi}\int_{-\infty}^{+\infty}\!\mathrm{d}\omega\,\widehat{\mathbf{L}}(\omega)\!\int_{-\infty}^{0}\!\mathrm{d}\tau'\,\mathrm{e}^{-\mathrm{i}(\omega-\boldsymbol{m}\cdot\boldsymbol{\Omega})\tau'}&=\frac{\mathrm{i}}{2\pi}\int_{-\infty}^{+\infty}\!\mathrm{d}\omega\,\frac{\widehat{\mathbf{L}}(\omega)}{\omega-\boldsymbol{m}\cdot\boldsymbol{\Omega}}\\&=\frac{\mathrm{i}}{2\pi}\mathcal{P}\!\int_{-\infty}^{+\infty}\!\mathrm{d}\omega\,\frac{\widehat{\mathbf{L}}(\omega)}{\omega-\boldsymbol{m}\cdot\boldsymbol{\Omega}}+\frac{1}{2}\widehat{\mathbf{L}}(\boldsymbol{m}\cdot\boldsymbol{\Omega})\,,\end{aligned} \tag{2.28}$$

where to perform the integration over τ', we kept only the boundary term for $\tau'=0$, by adding a small imaginary part to the frequency ω, so that $\omega=\omega+\mathrm{i}0^+$, which ensures the convergence for $\tau\to-\infty$. To evaluate the last integral over ω, we also relied on Plemelj formula

$$\frac{1}{x \pm \mathrm{i}0^+} = \mathcal{P}\left(\frac{1}{x}\right) \mp \mathrm{i}\pi\delta_{\mathrm{D}}(x)\,, \tag{2.29}$$

where $\mathcal{P}$ stands for Cauchy principal value. The last step of the derivation is to note that the contributions associated with the principal value in Eq. (2.28) have no impact on the secular diffusion equation. Indeed, Eq. (2.22) gives us that the diffusion coefficients are such that $D_{-\boldsymbol{m}}(\boldsymbol{J}) = D^*_{\boldsymbol{m}}(\boldsymbol{J})$. Since we are summing on all vectors $\boldsymbol{m} \in \mathbb{Z}^d$, we may then rewrite Eq. (2.24) as

$$\frac{\partial F}{\partial t} = \frac{\partial}{\partial \boldsymbol{J}} \cdot \left[\sum_{\boldsymbol{m}} \boldsymbol{m} \left\langle \mathrm{Re}[D_{\boldsymbol{m}}(\boldsymbol{J}, t)] \right\rangle \boldsymbol{m} \cdot \frac{\partial F}{\partial \boldsymbol{J}} \right]. \tag{2.30}$$

Equations (2.17) and (2.25) impose $\widehat{\mathbf{M}}^* = \widehat{\mathbf{M}}^{\mathrm{t}}$, and $\widehat{\mathbf{C}}^* = \widehat{\mathbf{C}}^{\mathrm{t}}$, so that the matrix $\widehat{\mathbf{L}}$ defined above Eq. (2.28) is also hermitian. We finally recover the collisionless secular dressed secular diffusion equation (sometimes also called "dressed Fokker-Planck equation") as

$$\frac{\partial F}{\partial t} = \frac{\partial}{\partial \boldsymbol{J}} \cdot \left[\sum_{\boldsymbol{m}} \boldsymbol{m}\, D_{\boldsymbol{m}}(\boldsymbol{J})\, \boldsymbol{m} \cdot \frac{\partial F}{\partial \boldsymbol{J}} \right], \tag{2.31}$$

where the anisotropic diffusion coefficients are given by

$$D_{\boldsymbol{m}}(\boldsymbol{J}) = \frac{1}{2} \sum_{p,q} \psi^{(p)}_{\boldsymbol{m}}(\boldsymbol{J})\, \psi^{(q)*}_{\boldsymbol{m}}(\boldsymbol{J}) \left[\left[\mathbf{I} - \widehat{\mathbf{M}}\right]^{-1} \cdot \widehat{\mathbf{C}} \cdot \left[\mathbf{I} - \widehat{\mathbf{M}}\right]^{-1} \right]_{pq} (\omega = \boldsymbol{m} \cdot \boldsymbol{\Omega})\,. \tag{2.32}$$

Let us finally introduce the total diffusion flux $\mathcal{F}_{\mathrm{tot}}$ as

$$\mathcal{F}_{\mathrm{tot}} = \sum_{\boldsymbol{m}} \boldsymbol{m}\, D_{\boldsymbol{m}}(\boldsymbol{J})\, \boldsymbol{m} \cdot \frac{\partial F}{\partial \boldsymbol{J}}\,, \tag{2.33}$$

so that Eq. (2.31) becomes

$$\frac{\partial F}{\partial t} = \mathrm{div}(\mathcal{F}_{\mathrm{tot}})\,. \tag{2.34}$$

With this convention, $-\mathcal{F}_{\mathrm{tot}}$ corresponds to the direction along which individual particles diffuse. Equation (2.31) is the main result of this section.

Let us now briefly discuss the physical content of Eq. (2.31). First, because it is written as the divergence of a flux, the total number of stars is conserved during the diffusion. One can also note that the diffusion coefficients $D_{\boldsymbol{m}}(\boldsymbol{J})$ from Eq. (2.32) capture the joint and coupled contributions from the external perturbations (via the autocorrelation matrix $\widehat{\mathbf{C}}$) and from the self-gravitating susceptibility of the system (via the response matrix $\widehat{\mathbf{M}}$). The total diffusion coefficients appear therefore as a collaboration between the strength of the external perturbations and the local strength of the system's amplification. As Eq. (2.31) describes a resonant diffusion, the external perturbing power spectrum and the system's susceptibility have to be evaluated

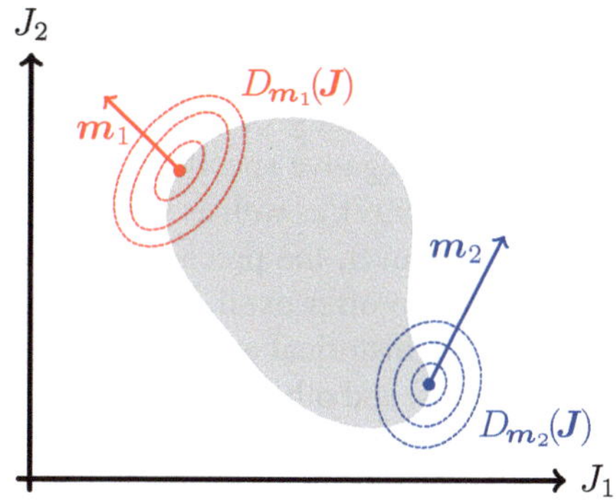

Fig. 2.3 Illustration of the strong anisotropy of the diffusion in action space captured by Eq. (2.31). The background *grey* domain illustrates the region where the system's DF, F, is present. For a given resonance vector $\boldsymbol{m}$, one can compute the associated diffusion coefficients $D_{\boldsymbol{m}}(\boldsymbol{J})$, whose level contours are represented with *dotted* colored lines. In the region where $D_{\boldsymbol{m}}(\boldsymbol{J})$ is maximum, following Eq. (2.31), one expects the associated flux to be aligned with the direction of $\boldsymbol{m}$. As a consequence, depending on which resonance vector locally dominates the diffusion, the DF's diffusion can occur along significantly different directions

at the local intrinsic frequency $\omega = \boldsymbol{m} \cdot \boldsymbol{\Omega}$. In this sense, this diffusion equation is appropriate to capture the nature of a collisionless system, via its natural frequencies and susceptibility, as well as its nurture, via the structure of the power spectrum of the external perturbations.

In addition, one can also note that the diffusion Eq. (2.31) takes the form of a strongly anisotropic diffusion equation in action space. It is anisotropic not only because the diffusion coefficients $D_{\boldsymbol{m}}(\boldsymbol{J})$ depend on the position in action space, but also because the diffusion associated with one resonance vector $\boldsymbol{m}$ corresponds to a diffusion in the preferential direction of the vector $\boldsymbol{m}$. For a given resonance $\boldsymbol{m}$, the diffusion is maximum along $\boldsymbol{m}$ and vanishes in the orthogonal directions. A qualitative illustration of the properties of Eq. (2.31) is given in Fig. 2.3. Finally, note that Eq. (2.31) is indeed an illustration of the fluctuation-dissipation theorem. The autocorrelation of the fluctuating potential drives the diffusion of the system's orbital structure.

2.3 Self-induced Collisional Dynamics

In the previous section, we considered the collisionless case where a secular diffusion is induced by external perturbations. However, a given self-gravitating system, even when isolated, may also undergo a secular evolution as a result of its own intrinsic graininess. This is a collisional evolution sourced by finite$-N$ effects.

The dynamics and thermodynamics of systems with long-range interactions has recently been a subject of active research (Campa et al. 2009; Campa et al. 2014), which led to a much better understanding of the equilibrium properties of these systems, their specificities such as negative specific heats (Antonov 1962; Lynden-Bell and Wood 1968; Lynden-Bell 1999), as well as various kinds of phase transitions and ensemble inequivalences. However, the precise description of their dynamical evolution remains to be improved to offer explicit predictions. We refer the reader to Chavanis (2010, 2013a, b) for a historical account of the development of kinetic theories of plasmas, stellar systems, and other systems with long-range interactions, but let us briefly recall here the main milestones.

The first kinetic theory focusing on the statistical description of the evolution of a large number of particles was considered by Boltzmann in the case of dilute neutral gases (Boltzmann 1872). For such systems, particles do not interact except during strong local collisions. The gas is assumed to be spatially homogeneous and Boltzmann equation describes the evolution of the system's velocity distribution $f(\boldsymbol{v}, t)$ as a result of strong collisions. This kinetic equation satisfies a H-theorem, associated with an increase of Boltzmann's entropy.

Boltzmann's approach was extended to charged gases (plasmas) by Landau (Landau 1936). For plasmas, particles interact via long-range Coulombian forces, but because of electroneutrality and Debye shielding (Debye and Hückel 1923a, b), these interactions are screened on a length scale of the order of the Debye length, and collisions become essentially local. Neutral plasmas are spatially homogeneous, so that the kinetic equation describes again the evolution of the velocity DF $f(\boldsymbol{v}, t)$, driven by close electrostatic encounters. Because the encounters are weak, one can expand the Boltzmann equation in the limit of small deflections and perform a linear trajectory approximation. In the weak coupling approximation, this leads to the so-called Landau equation. The Landau equation exhibits two formal divergences: one at small scales due to the neglect of strong collisions and one logarithmic divergence at large scales due to the neglect of collective effects, i.e. the dressing of particles by their polarisation cloud (a particle of a given charge has the tendency to be surrounded by a cloud of particles of opposite charges). Landau regularised these divergences by introducing a lower cut-off at the impact parameter producing a deflection of 90° (this is the Landau length) as well as an upper cut-off at the Debye length.

Collective effects were later rigorously taken into account in Balescu (1960) and Lenard (1960), leading to the Balescu-Lenard equation for plasmas. The Balescu-Lenard equation is similar to the Landau equation, except that it includes the square of the dielectric function in the denominator of the potential of interaction in Fourier space. This dieletric function first appeared as a probe of the dynamical stability of plasmas based on the linearised Vlasov equation (Vlasov 1938, 1945). In the Balescu-Lenard equation, the dielectric function accounts for Debye shielding and removes the large scale logarithmic divergence present in the Landau equation. The Landau equation is recovered from the Balescu-Lenard equation by replacing the dressed potential of interaction by its bare expression, i.e. by replacing the dielectric function by unity. In addition, the Balescu-Lenard equation, as given originally by Balescu and Lenard, exhibits a local resonance condition, encapsulated in a Dirac

δ_D−function. For such systems, resonant contributions are the drivers of the secular evolution. Integrating over this resonance condition leads to the original form of the kinetic equation given by Landau.

In parallel to the developments of kinetic equations for plasmas, the secular evolution of self-gravitating systems was also investigated. Self-gravitating systems are spatially inhomogeneous, but the first kinetic theories (Jeans 1929; Chandrasekhar 1942; Chandrasekhar 1943a, b) were all based on the assumption that collisions (i.e. close encounters) between stars can be treated with a local approximation, as if the system were infinite and homogeneous. Relying on the idea that a given star undergoes a large number of weak deflections, Chandrasekhar (1949) developed an analogy with Brownian motion. He started from a Fokker-Planck writing of the diffusion equation and computed the diffusion and friction coefficients relying on a binary collision theory. This led to a kinetic equation, often called Fokker-Planck equation in astrophysics, which is the gravitational equivalent of the Landau equation from plasmas. This equation exhibits similarly two divergences: one at small scales due to the mishandling of strong collisions, and one at large scales due to the local approximation, i.e. the assumption that the system is infinite and homogeneous. In the treatment of Chandrasekhar, strong collisions are taken into account without having to introduce a cut-off, so that the small scale divergence is regularised at the gravitational Landau length. The large scale divergence is usually regularised by introducing a cut-off at the Jeans length, which is the gravitational equivalent of the Debye length. This gravitational Landau equation is often considered to be relevant to describe the collisional dynamics of spherical systems such as globular clusters. Let us however note that the associated treatment based on the local approximation remains unsatisfactory, in particular because of the unavoidable appearance of a logarithmic divergence at large scales. In addition, within this framework, one cannot account for collective effects, i.e. the dressing of stars by their polarisation cloud, i.e. the fact that the gravitational force being attractive, a given star has the tendency to be surrounded by a cloud of stars. This increases its effective gravitational mass and reduces the collisional relaxation time.

In order to fully account for these properties, the kinetic theory of self-gravitating systems was recently generalised to fully inhomogeneous systems, either when collective effects are neglected (Polyachenko and Shukhman 1982; Chavanis 2010, 2013b) leading to the inhomogeneous Landau equation, or when they are accounted for leading to the inhomogeneous Balescu-Lenard equation (Heyvaerts 2010; Chavanis 2012b).[2] These kinetic equations, presented and discussed in detail in the upcoming section, are valid at order $1/N$, where N is the number of stars in the system. Having accounted for the finite extension of the system, these equations no longer present a divergence at large scales. In order to deal with the system's inhomogeneity, they are written in angle-action coordinates (see Sect. 1.3), which allow for the description of stars' intricate dynamics in spatially inhomogeneous

---

[2]The inhomogeneous Balescu-Lenard equation with angle-action variables was previously derived in Luciani and Pellat (1987); Mynick (1988), where the self-gravitating dressing of perturbations is introduced at a formal level without an explicit representation.

and multi-periodic systems. These equations involve similarly a resonance condition encapsulated in a Dirac $\delta_{\mathrm{D}}-$function (see Fig. 2.5), which generalises the one present in the homogeneous Balescu-Lenard equation. Finally, in order to capture collective effects, the inhomogeneous Balescu-Lenard equation also involves the system's response matrix (see Eq. (2.17)) expressed in angle-action variables. This generalises the dielectric function appearing in the homogeneous Balescu-Lenard equation for plasmas. This dressing accounts for anti-shielding, i.e. the fact that the gravitational mass of a star is enhanced by its polarisation, leading to a reduction of the relaxation time. The upcoming chapters will emphasise how these powerful and predictive kinetic equations may be used in the astrophysical context to probe complex secular regimes.

There are two standard methods to derive kinetic equations for a $N-$body system with long-range pairwise interactions. The first approach is based on Liouville's equation for the $N-$body distribution function of the system. One has to write the first two equations of the Bogoliubov-Born-Green-Kirkwood-Yvon (BBGKY) hierarchy. The hierarchy is then closed by considering only contributions of order $1/N$. One may then solve the second equation of the BBGKY hierarchy to express the $2-$body correlation function in terms of the system's $1-$body DF. One finally substitutes this expression in the first equation of the BBGKY hierarchy to obtain the closed self-consistent kinetic equation satisfied by the $1-$body DF. The same results can also be obtained thanks to projection operator techniques. The second method relies on the Klimontovich equation (Klimontovich 1967), which describes the dynamics of the system's DF written as a sum of δ_{D} functions. This exact DF is then decomposed in two parts, a smooth component and fluctuations. One can then write two evolution equations, one for the smooth mean component, and one for the fluctuations. This coupled system is then closed by neglecting non-linear terms in the evolution of the fluctuations (quasilinear approximation). The final step in this approach is to solve the equation for the fluctuations to express their properties as a function of the underlying smooth component. Injecting this result in the first evolution equation for the smooth part, one obtains a self-consistent kinetic equation. These two methods are physically equivalent, while technically different. Finally, we recently presented in Fouvry et al. (2016a, b) a third approach based on a functional rewriting of the evolution equations. This approach starts from the first two equations of the BBGKY hierarchy truncated at order $1/N$. Introducing auxiliary fields, the evolution of the two coupled dynamical quantities, $1-$body DF and $2-$body autocorrelation, can then be rewritten as a traditional functional integral. By functionally integrating over the $2-$body autocorrelation, one obtains a new contraint connecting the $1-$body DF and the auxiliary fields. When inverted, this constraint finally allows for the derivation of the closed non-linear kinetic equation satisfied by the $1-$body DF. Finally, in Heyvaerts et al. (2017), we recently showed how the Balescu-Lenard equation could be recovered via the direct computation of the first- and second-order diffusion coefficients, following a Fokker-Planck approach (Binney and Tremaine 2008).

In the upcoming sections, we will follow Chavanis (2012b) and present a derivation of the inhomogeneous Balescu-Lenard equation based on the resolution of the Klimontovich equation. We decided to present this derivation in the main text, in order to emphasise the various similarities it shares with the previous collisionless diffusion equation. In Appendix 2.A, we present the derivation of the BBGKY hierarchy. This allows us to revisit in Appendix 2.B the derivation of the inhomogeneous Balescu-Lenard equation first presented by Heyvaerts (2010) and based on the direct resolution of the BBGKY hierarchy. Finally, in Appendix 2.C, we consider the third approach to the derivation of kinetic equations based on a functional integral rewriting.

2.3.1 Evolution Equations

Let us consider an isolated system made of N particles of individual mass $\mu = M_{\mathrm{tot}}/N$, where M_{tot} is the total active mass of the system, embedded in a physical space of dimension d. We note as $(\boldsymbol{x}_i, \boldsymbol{v}_i)$ the position and velocity of particle i in an inertial frame. The individual dynamics of these particles is entirely described by Hamilton's equations which read

$$\mu \frac{\mathrm{d}\boldsymbol{x}_i}{\mathrm{d}t} = \frac{\partial H}{\partial \boldsymbol{v}_i} \quad ; \quad \mu \frac{\mathrm{d}\boldsymbol{v}_i}{\mathrm{d}t} = -\frac{\partial H}{\partial \boldsymbol{x}_i} , \tag{2.35}$$

where the Hamiltonian of the system contains all the binary interactions between particles as

$$H = \frac{\mu}{2} \sum_{i=1}^{N} \boldsymbol{v}_i^2 + \mu^2 \sum_{i<j}^{N} U(|\boldsymbol{x}_i - \boldsymbol{x}_j|) . \tag{2.36}$$

In Eq. (2.36), we introduced the binary potential of interaction $U(|\boldsymbol{x}_i - \boldsymbol{x}_j|)$, given by $U(|\boldsymbol{x}|) = -G/|\boldsymbol{x}|$ in the gravitational context. While capturing the exact dynamics of the system, one major drawback of Eq. (2.35) is that one has to deal with a set of N coupled differential equations. In Appendix 2.A, we show how these equations may be rewritten as an ordered hierarchy of evolution equations, the BBGKY hierarchy. Such a rewriting is at the heart of the derivation of the inhomogeneous Balescu-Lenard equation proposed in Heyvaerts (2010) and revisited in Appendix 2.B. Here, we intend to follow a different route, and rewrite Hamilton's equations (2.35) as a single evolution equation in phase space. To do so, let us introduce the discrete distribution function $F_{\mathrm{d}}(\boldsymbol{x}, \boldsymbol{v}, t)$ as

$$F_{\mathrm{d}}(\boldsymbol{x}, \boldsymbol{v}, t) = \mu \sum_{i=1}^{N} \delta_{\mathrm{D}}(\boldsymbol{x} - \boldsymbol{x}_i(t)) \, \delta_{\mathrm{D}}(\boldsymbol{v} - \boldsymbol{v}_i(t)) . \tag{2.37}$$

Let us also introduce the associated self-consistent potential ψ_{d} as

$$\psi_{\mathrm{d}}(\boldsymbol{x}, \boldsymbol{v}, t) = \int \mathrm{d}\boldsymbol{x}' \mathrm{d}\boldsymbol{v}' \, U(|\boldsymbol{x} - \boldsymbol{x}'|) \, F_{\mathrm{d}}(\boldsymbol{x}', \boldsymbol{v}', t) \, . \tag{2.38}$$

One can show that F_{d} satisfies the Klimontovich equation (Klimontovich 1967), given by

$$\frac{\partial F_{\mathrm{d}}}{\partial t} + \big[F_{\mathrm{d}}, H_{\mathrm{d}} \big] = 0 \, , \tag{2.39}$$

where we introduced the Hamiltonian H_{d} as

$$H_{\mathrm{d}}(\boldsymbol{x}, \boldsymbol{v}, t) = \frac{1}{2} \boldsymbol{v}^2 + \psi_{\mathrm{d}}(\boldsymbol{x}, t) \, . \tag{2.40}$$

At this stage, note that the Klimontovich equation (2.39) captures the exact same dynamics as Hamilton's equations (2.35), while being defined on a phase space of dimension $2d$. Let us assume that the system's DF and potential may be decomposed as the sum of a smooth component and a fluctuating one, so that

$$\begin{cases} F_{\mathrm{d}} = F + \delta F \, , \\ \psi_{\mathrm{d}} = \psi_0 + \delta\psi \, . \end{cases} \tag{2.41}$$

Let us emphasise how similar the decompositions from Eqs. (2.1) and (2.41) are. In addition to this decomposition, we assume that the smooth component F only evolves on secular timescales, while the fluctuating component δF evolves much faster on dynamical timescales. We also assume that the mean potential is integrable, so that there exists angle-action coordinates $(\boldsymbol{\theta}, \boldsymbol{J})$ appropriate for the smooth quasi-stationary potential ψ_0. Thanks to Jeans theorem, the system's mean DF being quasi-stationary, it can be written as $F(\boldsymbol{x}, \boldsymbol{v}, t) = F(\boldsymbol{J}, t)$. Performing the same timescale decoupling and quasilinear approximation as in Eq. (2.3), one gets two evolution equations from Eq. (2.39). First a secular evolution equation for F as

$$\frac{\partial F}{\partial t} = \frac{\partial}{\partial \boldsymbol{J}} \cdot \left[\int \frac{\mathrm{d}\boldsymbol{\theta}}{(2\pi)^d} \, \delta F \frac{\partial \delta\psi}{\partial \boldsymbol{\theta}} \right] , \tag{2.42}$$

and an evolution equation for the perturbation δF as

$$\frac{\partial \delta F}{\partial t} + \boldsymbol{\Omega} \cdot \frac{\partial \delta F}{\partial \boldsymbol{\theta}} - \frac{\partial F}{\partial \boldsymbol{J}} \cdot \frac{\partial \delta\psi}{\partial \boldsymbol{\theta}} = 0 \, . \tag{2.43}$$

These two evolution equations govern the evolution of the smooth DF F and the fluctuations δF at order $1/N$. They are the direct counterparts of Eqs. (2.4) and (2.5). Here, the potential fluctuations are not due to an external forcing, but to the intrinsic finite$-N$ Poisson shot noise. As was assumed in Eq. (2.7), we place ourselves within the adiabatic approximation so that the time variations of F may be neglected on the timescales for which the fluctuations δF and $\delta\psi$ evolve. In order to be valid, such an approximation requires to have $N \gg 1$. Finally, as in Eq. (2.19), we assume that the DF F remains Vlasov-stable throughout its evolution, so that its evolution is only governed by correlations and not by dynamical instabilities.

2.3.2 *Fast Timescale Amplification*

The first step of our calculation is to study the short timescale evolution equation (2.43), during which perturbations build up. As in Eq. (2.6), let us perform a Fourier transform w.r.t. to the angles $\boldsymbol{\theta}$. Let us also define the Laplace transform of the fluctuations with the convention

$$\widetilde{f}(\omega) = \int_0^{+\infty} \mathrm{d}t \, f(t) \, \mathrm{e}^{\mathrm{i}\omega t} \quad ; \quad f(t) = \frac{1}{2\pi} \int_{\mathcal{B}} \mathrm{d}\omega \, \widetilde{f}(\omega) \, \mathrm{e}^{-\mathrm{i}\omega t} , \tag{2.44}$$

where the Bromwich contour $\mathcal{B}$ in the complex ω−plane should pass above all the poles of the integrand, i.e. $\mathrm{Im}[\omega]$ should be large enough. The Fourier-Laplace transform of the DF's fluctuations δF is therefore given by

$$\delta\widetilde{F}_{\boldsymbol{m}_1}(\boldsymbol{J}_1, \omega_1) = \int \frac{\mathrm{d}\boldsymbol{\theta}_1}{(2\pi)^d} \int_0^{+\infty} \mathrm{d}t \, \mathrm{e}^{-\mathrm{i}(\boldsymbol{m}_1 \cdot \boldsymbol{\theta}_1 - \omega_1 t)} \, \delta F(\boldsymbol{\theta}_1, \boldsymbol{J}_1, t) \, . \tag{2.45}$$

One can perform a similar transformation for the potential fluctuations $\delta\psi$. Let us define the Fourier transform of the initial value of the DF as

$$\delta\widehat{F}_{\boldsymbol{m}_1}(\boldsymbol{J}_1, 0) = \int \frac{\mathrm{d}\boldsymbol{\theta}_1}{(2\pi)^d} \, \mathrm{e}^{-\mathrm{i}(\boldsymbol{m}_1 \cdot \boldsymbol{\theta}_1)} \, \delta F(\boldsymbol{\theta}_1, \boldsymbol{J}_1, 0) \, . \tag{2.46}$$

Relying on Bogoliubov's ansatz, $F = \mathrm{cst.}$, we multiply Eq. (2.43) by $\int \mathrm{d}\boldsymbol{\theta}_1/(2\pi)^d \int \mathrm{d}t \, \mathrm{e}^{-\mathrm{i}\boldsymbol{m}_1 \cdot \boldsymbol{\theta}_1 - \omega_1 t}$ to get

$$\delta\widetilde{F}_{\boldsymbol{m}_1}(\boldsymbol{J}_1, \omega_1) = \frac{\boldsymbol{m}_1 \cdot \partial F / \partial \boldsymbol{J}_1}{\boldsymbol{m}_1 \cdot \boldsymbol{\Omega}_1 - \omega_1} \, \delta\widetilde{\psi}_{\boldsymbol{m}_1}(\boldsymbol{J}_1, \omega_1) + \frac{\delta\widehat{F}_{\boldsymbol{m}_1}(\boldsymbol{J}_1, 0)}{\mathrm{i}(\boldsymbol{m}_1 \cdot \boldsymbol{\Omega}_1 - \omega_1)} \, . \tag{2.47}$$

Equation (2.47) relates the fluctuations in the potential $\delta\psi$ to the induced response δF in the system's DF. One now has to account for the fact that these perturbations are self-consistently generated by the system itself, i.e. $\delta\psi$ corresponds to potential fluctuations generated by the perturbing density $\delta\rho$ associated with the DF δF. To do so, we follow the matrix method introduced in Sect. 2.2.2. Relying on basis elements $(\psi^{(p)}, \rho^{(p)})$ as introduced in Eq. (2.12), we follow Eq. (2.13) and decompose the self-induced potential perturbations $\delta\psi$ as

$$\delta\psi(\boldsymbol{\theta}_1, \boldsymbol{J}_1, t) = \sum_p a_p(t)\, \psi^{(p)}(\boldsymbol{\theta}_1, \boldsymbol{J}_1) \quad ; \quad \delta\widetilde{\psi}_{\boldsymbol{m}_1}(\boldsymbol{J}_1, \omega_1) = \sum_p \widetilde{a}_p(\omega_1)\, \psi^{(p)}_{\boldsymbol{m}_1}(\boldsymbol{J}_1)\,, \tag{2.48}$$

where $\widehat{a}_p(\omega)$ stands for the Laplace transform of the basis coefficients and $\psi^{(p)}_{\boldsymbol{m}_1}(\boldsymbol{J}_1)$ for the Fourier transformed basis elements as introduced in Eq. (2.15). In order to capture this self-consistency, we follow the same method as presented in Eq. (2.16). We start from $\delta\widetilde{\rho} = \int \mathrm{d}\boldsymbol{v}\, \delta\widetilde{F}(\boldsymbol{x}, \boldsymbol{v})$, multiply this relation by $\psi^{(p)*}(\boldsymbol{x})$, integrate it w.r.t. $\boldsymbol{x}$, and rely on the fact that $\mathrm{d}\boldsymbol{x}\mathrm{d}\boldsymbol{v} = \mathrm{d}\boldsymbol{\theta}\mathrm{d}\boldsymbol{J}$ as the transformation $(\boldsymbol{x}, \boldsymbol{v}) \mapsto (\boldsymbol{\theta}, \boldsymbol{J})$ is canonical. Equation (2.47) finally gives

$$\widetilde{a}_p(\omega_1) = -(2\pi)^d \sum_q \big[\mathbf{I} - \widehat{\mathbf{M}}(\omega_1)\big]^{-1}_{pq} \sum_{\boldsymbol{m}_2} \int \mathrm{d}\boldsymbol{J}_2\, \frac{\delta\widehat{F}_{\boldsymbol{m}_2}(\boldsymbol{J}_2, 0)}{\mathrm{i}(\boldsymbol{m}_2 \cdot \boldsymbol{\Omega}_2 - \omega_1)}\, \psi^{(q)*}_{\boldsymbol{m}_2}(\boldsymbol{J}_2)\,. \tag{2.49}$$

In Eq. (2.49), we recover the role played by the system's susceptibility through the system's response matrix $\widehat{\mathbf{M}}$ introduced in Eq. (2.17). Note as well that we assumed the system to be stable, so that one could indeed compute the matrix $\big[\mathbf{I} - \widehat{\mathbf{M}}(\omega_1)\big]^{-1}$. Let us now introduce the system's dressed susceptibility coefficients $1/\mathcal{D}_{\boldsymbol{m}_1, \boldsymbol{m}_2}$ as

$$\frac{1}{\mathcal{D}_{\boldsymbol{m}_1, \boldsymbol{m}_2}(\boldsymbol{J}_1, \boldsymbol{J}_2, \omega)} = \sum_{p,q} \psi^{(p)}_{\boldsymbol{m}_1}(\boldsymbol{J}_1) \big[\mathbf{I} - \widehat{\mathbf{M}}(\omega)\big]^{-1}_{pq} \psi^{(q)*}_{\boldsymbol{m}_2}(\boldsymbol{J}_2)\,, \tag{2.50}$$

so that Eq. (2.49) when multiplied by $\psi^{(p)}_{\boldsymbol{m}_1}(\boldsymbol{J}_1)$ and summed over "p" gives

$$\delta\widetilde{\psi}_{\boldsymbol{m}_1}(\boldsymbol{J}_1, \omega_1) = -(2\pi)^d \sum_{\boldsymbol{m}_2} \int \mathrm{d}\boldsymbol{J}_2\, \frac{1}{\mathcal{D}_{\boldsymbol{m}_1, \boldsymbol{m}_2}(\boldsymbol{J}_1, \boldsymbol{J}_2, \omega_1)}\, \frac{\delta\widehat{F}_{\boldsymbol{m}_2}(\boldsymbol{J}_2, 0)}{\mathrm{i}(\boldsymbol{m}_2 \cdot \boldsymbol{\Omega}_2 - \omega_1)}\,. \tag{2.51}$$

Equation (2.51) gives the Laplace transform of the response potential as a function of the initial conditions in the DF's fluctuations. It describes the dynamical amplification of the perturbations occurring in the system.

2.3.3 *Estimating the Collision Operator*

Thanks to Eq. (2.51), one may now proceed to the evaluation of the collision operator in the r.h.s. of Eq. (2.42). As was argued in Eq. (2.24), let us emphasise that here we are interested in the system's mean evolution averaged over various realisations. We may then take the ensemble average of the evolution equation (2.42). When taking this average, we assume that the response matrix $\widehat{\mathbf{M}}$ as well as the DF F and its gradients $\partial F/\partial \boldsymbol{J}$ do not change significantly from one realisation to another. Equation (2.42) becomes

$$\frac{\partial F}{\partial t} = \frac{\partial}{\partial \boldsymbol{J}_1} \cdot \big[\mathcal{F}_{\mathrm{tot}}(\boldsymbol{J}_1)\big]\,, \tag{2.52}$$

where we introduced the total diffusion flux $\mathcal{F}_{\mathrm{tot}}$ as

$$\mathcal{F}_{\mathrm{tot}}(\boldsymbol{J}) = \int \frac{\mathrm{d}\boldsymbol{\theta}}{(2\pi)^d} \left\langle \delta F \, \frac{\partial \delta\psi}{\partial \boldsymbol{\theta}} \right\rangle , \tag{2.53}$$

where $\langle \, \cdot \, \rangle$ stands for the ensemble average operation. Taking a Fourier transform w.r.t. the angles as well as an inverse Laplace transform, Eq. (2.53) gives

$$\mathcal{F}_{\mathrm{tot}}(\boldsymbol{J}_1) = - \sum_{\boldsymbol{m}_1} \int_{\mathcal{B}_1} \frac{\mathrm{d}\omega_1}{2\pi} \int_{\mathcal{B}_2} \frac{\mathrm{d}\omega_2}{2\pi} \, \mathrm{i}\boldsymbol{m}_1 \, \mathrm{e}^{-\mathrm{i}\omega_1 t} \, \mathrm{e}^{-\mathrm{i}\omega_2 t} \left\langle \delta \widetilde{F}_{\boldsymbol{m}_1}(\boldsymbol{J}_1, \omega_1) \, \delta\widetilde{\psi}_{-\boldsymbol{m}_1}(\boldsymbol{J}_1, \omega_2) \right\rangle , \tag{2.54}$$

where $\mathcal{B}_1$ (resp. $\mathcal{B}_2$) stands for the Bromwich contour associated with the inverse Laplace transform w.r.t. ω_1 (resp. ω_2). Let us now rely on Eqs. (2.47) and (2.51) to express Eq. (2.54) only as a function of the fluctuations in the initial conditions captured by $\delta\widehat{F}$. Equation (2.54) can be written as

$$\mathcal{F}_{\mathrm{tot}}(\boldsymbol{J}_1) = \mathcal{F}_{\mathrm{tot}}^{(I)}(\boldsymbol{J}_1) + \mathcal{F}_{\mathrm{tot}}^{(II)}(\boldsymbol{J}_1) , \tag{2.55}$$

where the two components $\mathcal{F}_{\mathrm{tot}}^{(I)}(\boldsymbol{J}_1)$ and $\mathcal{F}_{\mathrm{tot}}^{(II)}(\boldsymbol{J}_1)$ are respectively given by

$$\begin{aligned}
\mathcal{F}_{\mathrm{tot}}^{(I)}(\boldsymbol{J}_1) = {} & \mathrm{i} \sum_{\boldsymbol{m}_1} \boldsymbol{m}_1 \int_{\mathcal{B}_1} \frac{\mathrm{d}\omega_1}{2\pi} \mathrm{e}^{-\mathrm{i}\omega_1 t} \int_{\mathcal{B}_2} \frac{\mathrm{d}\omega_2}{2\pi} \mathrm{e}^{-\mathrm{i}\omega_2 t} \, (2\pi)^{2d} \frac{\boldsymbol{m}_1 \cdot \partial F / \partial \boldsymbol{J}_1}{\boldsymbol{m}_1 \cdot \boldsymbol{\Omega}_1 - \omega_1} \sum_{\boldsymbol{m}_2, \boldsymbol{m}_3} \int \mathrm{d}\boldsymbol{J}_2 \mathrm{d}\boldsymbol{J}_3 \bigg\{ \frac{1}{\boldsymbol{m}_2 \cdot \boldsymbol{\Omega}_2 - \omega_1} \\
& \times \frac{1}{\mathcal{D}_{\boldsymbol{m}_1, \boldsymbol{m}_2}(\boldsymbol{J}_1, \boldsymbol{J}_2, \omega_1)} \frac{1}{\mathcal{D}_{-\boldsymbol{m}_1, \boldsymbol{m}_3}(\boldsymbol{J}_1, \boldsymbol{J}_3, \omega_2)} \frac{1}{\boldsymbol{m}_3 \cdot \boldsymbol{\Omega}_3 - \omega_2} \left\langle \delta\widehat{F}_{\boldsymbol{m}_2}(\boldsymbol{J}_2, 0) \, \delta\widehat{F}_{\boldsymbol{m}_3}(\boldsymbol{J}_3, 0) \right\rangle \bigg\} , \\
\mathcal{F}_{\mathrm{tot}}^{(II)}(\boldsymbol{J}_1) = {} & - \mathrm{i} \sum_{\boldsymbol{m}_1} \boldsymbol{m}_1 \int_{\mathcal{B}_1} \frac{\mathrm{d}\omega_1}{2\pi} \mathrm{e}^{-\mathrm{i}\omega_1 t} \int_{\mathcal{B}_2} \frac{\mathrm{d}\omega_2}{2\pi} \mathrm{e}^{-\mathrm{i}\omega_2 t} \, (2\pi)^d \frac{1}{\boldsymbol{m}_1 \cdot \boldsymbol{\Omega}_1 - \omega_1} \\
& \times \sum_{\boldsymbol{m}_2} \int \mathrm{d}\boldsymbol{J}_2 \frac{1}{\mathcal{D}_{-\boldsymbol{m}_1, \boldsymbol{m}_2}(\boldsymbol{J}_1, \boldsymbol{J}_2, \omega_2)} \frac{1}{\boldsymbol{m}_2 \cdot \boldsymbol{\Omega}_2 - \omega_2} \left\langle \delta\widehat{F}_{\boldsymbol{m}_1}(\boldsymbol{J}_1, 0) \, \delta\widehat{F}_{\boldsymbol{m}_2}(\boldsymbol{J}_2, 0) \right\rangle .
\end{aligned} \tag{2.56}$$

In order to evaluate the two expressions from Eq. (2.56), one needs to compute the statistical expectation of the product $\langle \delta\widehat{F}_{\boldsymbol{m}_1}(\boldsymbol{J}_1, 0) \, \delta\widehat{F}_{\boldsymbol{m}_2}(\boldsymbol{J}_2, 0) \rangle$ that we will now evaluate.

Let us recall here that the fluctuations δF introduced in Eq. (2.41) are given by $\delta F = F_{\mathrm{d}} - F$, i.e. stand for the difference between the actual discrete DF F_{d} and the smooth mean field one F. Starting from the expression (2.37) of the discrete distribution function F_{d} and temporarily dropping the time dependence, $t = 0$, to shorten the notations, one can write

$$\left\langle \delta F(\boldsymbol{\theta}_1, \boldsymbol{J}_1) \, \delta F(\boldsymbol{\theta}_2, \boldsymbol{J}_2) \right\rangle = \mu^2 \sum_{i,j}^{N} \left\langle \delta_{\mathrm{D}}(\boldsymbol{\theta}_1 - \boldsymbol{\theta}_i) \, \delta_{\mathrm{D}}(\boldsymbol{J}_1 - \boldsymbol{J}_i) \, \delta_{\mathrm{D}}(\boldsymbol{\theta}_2 - \boldsymbol{\theta}_j) \, \delta_{\mathrm{D}}(\boldsymbol{J}_2 - \boldsymbol{J}_j) \right\rangle - F(\boldsymbol{J}_1) \, F(\boldsymbol{J}_2) . \tag{2.57}$$

Here, we relied on the fact the fluctuations are of zero mean so that $\langle \delta F \rangle = 0$. Let us now evaluate the first term from Eq. (2.57) which reads

$$\begin{aligned}
\mu^2 \sum_{i,j}^{N} \big\langle \delta_{\mathrm{D}}(\boldsymbol{\theta}_1-\boldsymbol{\theta}_i)\, \delta_{\mathrm{D}}(\boldsymbol{J}_1-\boldsymbol{J}_i)\, \delta_{\mathrm{D}}(\boldsymbol{\theta}_2-\boldsymbol{\theta}_j)\, \delta_{\mathrm{D}}(\boldsymbol{J}_2-\boldsymbol{J}_j) \big\rangle& \\
&= \mu^2 \sum_{i}^{N} \big\langle \delta_{\mathrm{D}}(\boldsymbol{\theta}_1-\boldsymbol{\theta}_i)\, \delta_{\mathrm{D}}(\boldsymbol{J}_1-\boldsymbol{J}_i)\, \delta_{\mathrm{D}}(\boldsymbol{\theta}_1-\boldsymbol{\theta}_2)\, \delta_{\mathrm{D}}(\boldsymbol{J}_1-\boldsymbol{J}_2) \big\rangle \\
&+ \mu^2 \sum_{i\neq j}^{N} \big\langle \delta_{\mathrm{D}}(\boldsymbol{\theta}_1-\boldsymbol{\theta}_i)\, \delta_{\mathrm{D}}(\boldsymbol{J}_1-\boldsymbol{J}_i)\, \delta_{\mathrm{D}}(\boldsymbol{\theta}_2-\boldsymbol{\theta}_j)\, \delta_{\mathrm{D}}(\boldsymbol{J}_2-\boldsymbol{J}_j) \big\rangle \\
&= \mu\, F(\boldsymbol{J}_1)\, \delta_{\mathrm{D}}(\boldsymbol{\theta}_1-\boldsymbol{\theta}_2)\, \delta_{\mathrm{D}}(\boldsymbol{J}_1-\boldsymbol{J}_2) + F(\boldsymbol{J}_1)\, F(\boldsymbol{J}_2)\,, \qquad (2.58)
\end{aligned}$$

where, to get the last line, we assumed that the particles are initially uncorrelated and used the fact that $\big\langle F_{\mathrm{d}} \big\rangle = F$. Injecting Eq. (2.58) into Eq. (2.57), we get the relation

$$\big\langle \delta F(\boldsymbol{\theta}_1, \boldsymbol{J}_1)\, F(\boldsymbol{\theta}_2, \boldsymbol{J}_2) \big\rangle = \mu\, F(\boldsymbol{J}_1)\, \delta_{\mathrm{D}}(\boldsymbol{\theta}_1-\boldsymbol{\theta}_2)\, \delta_{\mathrm{D}}(\boldsymbol{J}_1-\boldsymbol{J}_2)\,. \qquad (2.59)$$

Finally, taking the Fourier transform of Eq. (2.59), one gets the needed correlations in the initial conditions as

$$\big\langle \delta \widehat{F}_{\boldsymbol{m}_1}(\boldsymbol{J}_1, 0)\, \delta \widehat{F}_{\boldsymbol{m}_2}(\boldsymbol{J}_2, 0) \big\rangle = \frac{\mu}{(2\pi)^d}\, \delta_{\boldsymbol{m}_1}^{-\boldsymbol{m}_2}\, \delta_{\mathrm{D}}(\boldsymbol{J}_1-\boldsymbol{J}_2)\, F(\boldsymbol{J}_1)\,. \qquad (2.60)$$

The two components of the diffusion flux from Eq. (2.56) then become

$$\begin{aligned}
\boldsymbol{\mathcal{F}}_{\mathrm{tot}}^{(I)}(\boldsymbol{J}_1) = & -\mathrm{i}\mu (2\pi)^d \sum_{\boldsymbol{m}_1} \boldsymbol{m}_1 \int_{\mathcal{B}_1} \frac{\mathrm{d}\omega_1}{2\pi} \mathrm{e}^{-\mathrm{i}\omega_1 t} \int_{\mathcal{B}_2} \frac{\mathrm{d}\omega_2}{2\pi} \mathrm{e}^{-\mathrm{i}\omega_2 t} \frac{\boldsymbol{m}_1\cdot\partial F/\partial \boldsymbol{J}_1}{\boldsymbol{m}_1\cdot\boldsymbol{\Omega}_1-\omega_1} \\
& \times \sum_{\boldsymbol{m}_2} \int \mathrm{d}\boldsymbol{J}_2 \frac{1}{\boldsymbol{m}_2\cdot\boldsymbol{\Omega}_2-\omega_1} \frac{1}{\mathcal{D}_{\boldsymbol{m}_1,\boldsymbol{m}_2}(\boldsymbol{J}_1, \boldsymbol{J}_2, \omega_1)} \frac{1}{\mathcal{D}_{-\boldsymbol{m}_1,-\boldsymbol{m}_2}(\boldsymbol{J}_1, \boldsymbol{J}_2, \omega_2)} \frac{F(\boldsymbol{J}_2)}{\boldsymbol{m}_2\cdot\boldsymbol{\Omega}_2+\omega_2}\,, \\
\boldsymbol{\mathcal{F}}_{\mathrm{tot}}^{(II)}(\boldsymbol{J}_1) = & \, \mathrm{i}\mu \sum_{\boldsymbol{m}_1} \boldsymbol{m}_1 \int_{\mathcal{B}_1} \frac{\mathrm{d}\omega_1}{2\pi} \mathrm{e}^{-\mathrm{i}\omega_1 t} \int_{\mathcal{B}_2} \frac{\mathrm{d}\omega_2}{2\pi} \mathrm{e}^{-\mathrm{i}\omega_2 t} \frac{1}{\boldsymbol{m}_1\cdot\boldsymbol{\Omega}_1-\omega_1} \\
& \times \frac{1}{\mathcal{D}_{-\boldsymbol{m}_1,-\boldsymbol{m}_1}(\boldsymbol{J}_1, \boldsymbol{J}_1, \omega_2)} \frac{F(\boldsymbol{J}_1)}{\boldsymbol{m}_1\cdot\boldsymbol{\Omega}_1+\omega_2}\,. \qquad (2.61)
\end{aligned}$$

Let us now proceed to the successive evaluations of both terms in Eq. (2.61). Let us first evaluate the term $\boldsymbol{\mathcal{F}}_{\mathrm{tot}}^{(I)}(\boldsymbol{J}_1)$, which corresponds to the diffusion component of the kinetic equation. Here, the difficulty is to deal with the resonant poles appearing in Eq. (2.61). We follow the argument presented in Eq. (51.17) of Pitaevskii and Lifshitz (2012) and note that considering only contributions that do not decay in time, one can perform the substitution

$$\frac{1}{\boldsymbol{m}_2\cdot\boldsymbol{\Omega}_2-\omega_1} \frac{1}{\boldsymbol{m}_2\cdot\boldsymbol{\Omega}_2+\omega_2} \longrightarrow (2\pi)^2\, \delta_{\mathrm{D}}(\omega_1+\omega_2)\, \delta_{\mathrm{D}}(\boldsymbol{m}_2\cdot\boldsymbol{\Omega}_2-\omega_1)\,. \qquad (2.62)$$

This substitution allows us to perform the integrations w.r.t. ω_1 and ω_2 in Eq. (2.61), so that $\boldsymbol{\mathcal{F}}_{\mathrm{tot}}^{(I)}$ becomes

$$\mathcal{F}_{\mathrm{tot}}^{(I)}(\boldsymbol{J}_1) = \mathrm{i}\mu(2\pi)^d \sum_{\boldsymbol{m}_1} \boldsymbol{m}_1 \int_{\mathcal{B}_1} \mathrm{d}\omega_1 \sum_{\boldsymbol{m}_2} \int \mathrm{d}\boldsymbol{J}_2 \, \frac{\delta_{\mathrm{D}}(\boldsymbol{m}_2 \cdot \boldsymbol{\Omega}_2 - \omega_1)}{\omega_1 - \boldsymbol{m}_1 \cdot \boldsymbol{\Omega}_1} \, \frac{\boldsymbol{m}_1 \cdot \partial F/\partial \boldsymbol{J}_1 \, F(\boldsymbol{J}_2)}{|\mathcal{D}_{\boldsymbol{m}_1,\boldsymbol{m}_2}(\boldsymbol{J}_1, \boldsymbol{J}_2, \boldsymbol{m}_1 \cdot \boldsymbol{\Omega}_1 - \omega_1)|^2} \, , \tag{2.63}$$

where we relied on the relation $1/\mathcal{D}_{-\boldsymbol{m}_1,-\boldsymbol{m}_2}(\boldsymbol{J}_1, \boldsymbol{J}_2, -\omega) = 1/\mathcal{D}^*_{\boldsymbol{m}_1,\boldsymbol{m}_2}(\boldsymbol{J}_1, \boldsymbol{J}_2, \omega)$ (see note [83] in Chavanis (2012b)). In Eq. (2.63), we may finally perform the integration w.r.t. ω_1 by lowering the contour $\mathcal{B}_1$ to the lower axis and using the Landau prescription $\boldsymbol{m}_1 \cdot \boldsymbol{\Omega}_1 \to \boldsymbol{m}_1 \cdot \boldsymbol{\Omega}_1 - \mathrm{i}0^+$ associated with the fact that the contour $\mathcal{B}_1$ has to pass above the pole. We finally rely on Plemelj formula from Eq. (2.29). Because $\mathcal{F}_{\mathrm{tot}}^{(I)}$ is a real quantity, only the Dirac delta remains, and Eq. (2.63) finally becomes

$$\mathcal{F}_{\mathrm{tot}}^{(I)}(\boldsymbol{J}_1) = \mu\pi(2\pi)^d \sum_{\boldsymbol{m}_1,\boldsymbol{m}_2} \boldsymbol{m}_1 \int \mathrm{d}\boldsymbol{J}_2 \, \frac{\delta_{\mathrm{D}}(\boldsymbol{m}_1 \cdot \boldsymbol{\Omega}_1 - \boldsymbol{m}_2 \cdot \boldsymbol{\Omega}_2)}{|\mathcal{D}_{\boldsymbol{m}_1,\boldsymbol{m}_2}(\boldsymbol{J}_1, \boldsymbol{J}_2, \boldsymbol{m}_1 \cdot \boldsymbol{\Omega}_1)|^2} \, \boldsymbol{m}_1 \cdot \frac{\partial F}{\partial \boldsymbol{J}_1} \, F(\boldsymbol{J}_2) \, . \tag{2.64}$$

Coming back to Eq. (2.61), let us now evaluate the second flux component $\mathcal{F}_{\mathrm{tot}}^{(II)}(\boldsymbol{J}_1)$. This term is associated with the drift component of the kinetic equation. To perform the integrations over ω_1 and ω_2, we distort once again the Bromwich contours $\mathcal{B}_1$ and $\mathcal{B}_2$ towards negative imaginary parts, while still remaining above all the singularities of the integrand. When deformed to large negative imaginary parts, the exponential terms $\mathrm{e}^{-\mathrm{i}\omega_1 t}$ and $\mathrm{e}^{-\mathrm{i}\omega_2 t}$ tend to 0, so that one should only account for the contributions from the poles. For $\mathcal{B}_1$, we note that there is only one pole in $\omega_1 = \boldsymbol{m}_1 \cdot \boldsymbol{\Omega}_1$ and that this pole is located along the real axis. One should also pay a careful attention to the direction of integration, so that here one has $\int \mathrm{d}\omega_1 f(\omega_1)/(\omega_1 - \omega_0) = -2\mathrm{i}\pi f(\omega_0)$. For the integration w.r.t. ω_2, one first notes an obvious pole along the real axis in $\omega_2 = -\boldsymbol{m}_1 \cdot \boldsymbol{\Omega}_1$. In addition, because the system is assumed to be stable, all the singularities associated with the susceptibility coefficients $\omega_2 \mapsto 1/\mathcal{D}(\omega_2)$ are located below the real axis. Such poles will then be multiplied by a decaying in time exponential. Considering only contributions which do not decay in time, we restrict ourselves only to the real pole in $\omega_2 = -\boldsymbol{m}_1 \cdot \boldsymbol{\Omega}_1$, and pay as well a careful attention to the sign of the residues. Equation (2.61) gives

$$\begin{aligned} \mathcal{F}_{\mathrm{tot}}^{(II)}(\boldsymbol{J}_1) &= \mathrm{i}\mu \sum_{\boldsymbol{m}_1} \boldsymbol{m}_1 \frac{1}{\mathcal{D}_{-\boldsymbol{m}_1,-\boldsymbol{m}_1}(\boldsymbol{J}_1, \boldsymbol{J}_1, -\boldsymbol{m}_1 \cdot \boldsymbol{\Omega}_1 + \mathrm{i}0^+)} \, F(\boldsymbol{J}_1) \\ &= \mu \sum_{\boldsymbol{m}_1} \boldsymbol{m}_1 \, \mathrm{Im} \left[\frac{1}{\mathcal{D}_{\boldsymbol{m}_1,\boldsymbol{m}_1}(\boldsymbol{J}_1, \boldsymbol{J}_1, \boldsymbol{m}_1 \cdot \boldsymbol{\Omega}_1 + \mathrm{i}0^+)} \right] F(\boldsymbol{J}_1) \, , \end{aligned} \tag{2.65}$$

where one should pay attention to the small positive imaginary part $\mathrm{i}0^+$ which was added following Landau prescription $\omega_2 \to \omega_2 + \mathrm{i}0^+$. This emphasises the fact that the contour $\mathcal{B}_2$ has to pass above the pole. In Eq. (2.65), we also note that the two time dependences introduced by the two inverse Laplace transforms cancelled out, so that $\mathcal{F}_{\mathrm{tot}}^{(II)}$ does not explicitly depend on time. To get the second relation in Eq. (2.65), we performed the change $\boldsymbol{m}_1 \to -\boldsymbol{m}_1$, and relied on the fact that $\mathcal{F}_{\mathrm{tot}}^{(II)}$ is a real quantity, hence the imaginary part. The calculation of the imaginary part in Eq. (2.65) is presented in Eq. (2.155). We refer to this calculation, so that we can finally rewrite Eq. (2.65) as

$$\mathcal{F}_{\mathrm{tot}}^{(II)}(\boldsymbol{J}_1) = -\mu\pi(2\pi)^d \sum_{\boldsymbol{m}_1,\boldsymbol{m}_2} \boldsymbol{m}_1 \int \mathrm{d}\boldsymbol{J}_2 \, \frac{\delta_{\mathrm{D}}(\boldsymbol{m}_1\cdot\boldsymbol{\Omega}_1 - \boldsymbol{m}_2\cdot\boldsymbol{\Omega}_2)}{|\mathcal{D}_{\boldsymbol{m}_1,\boldsymbol{m}_2}(\boldsymbol{J}_1, \boldsymbol{J}_2, \boldsymbol{m}_1\cdot\boldsymbol{\Omega}_1)|^2} \, F(\boldsymbol{J}_1) \, \boldsymbol{m}_2\cdot\frac{\partial F}{\partial \boldsymbol{J}_2} \, . \tag{2.66}$$

We have now evaluated the two components of the diffusion flux $\mathcal{F}_{\mathrm{tot}}$ from Eq. (2.52). We have therefore derived a closed kinetic equation, the inhomogeneous Balescu-Lenard equation, that will be presented in detail in the upcoming section.

2.3.4 The Balescu-Lenard Equation

Combining Eqs. (2.64) and (2.66), one can estimate the total diffusion flux $\mathcal{F}_{\mathrm{tot}}$. Equation (2.52) immediately gives the associated closed diffusion equation. This is the inhomogeneous Balescu-Lenard equation which reads

$$\begin{aligned} \frac{\partial F}{\partial t} = \pi(2\pi)^d \mu \frac{\partial}{\partial \boldsymbol{J}_1}\cdot \Bigg[\sum_{\boldsymbol{m}_1,\boldsymbol{m}_2} \boldsymbol{m}_1 \int \mathrm{d}\boldsymbol{J}_2 \, & \frac{\delta_{\mathrm{D}}(\boldsymbol{m}_1\cdot\boldsymbol{\Omega}_1 - \boldsymbol{m}_2\cdot\boldsymbol{\Omega}_2)}{\big|\mathcal{D}_{\boldsymbol{m}_1,\boldsymbol{m}_2}(\boldsymbol{J}_1, \boldsymbol{J}_2, \boldsymbol{m}_1\cdot\boldsymbol{\Omega}_1)\big|^2} \\ & \times \left(\boldsymbol{m}_1\cdot\frac{\partial}{\partial \boldsymbol{J}_1} - \boldsymbol{m}_2\cdot\frac{\partial}{\partial \boldsymbol{J}_2}\right) F(\boldsymbol{J}_1, t)\, F(\boldsymbol{J}_2, t) \Bigg] . \end{aligned} \tag{2.67}$$

Let us now detail the physical content of this diffusion equation. Let us first note that Eq. (2.67) is written as the divergence of a flux, so that it conserves the total number of particles. The presence of the prefactor $\mu = M_{\mathrm{tot}}/N$ illustrates the fact that the Balescu-Lenard equation was obtained thanks to a kinetic development at order $1/N$. It captures first-order contributions associated with finite$-N$ effects. In Eq. (2.67), one should note in particular the presence of a resonance condition encapsulated by the Dirac delta $\delta_{\mathrm{D}}(\boldsymbol{m}_1\cdot\boldsymbol{\Omega}_1 - \boldsymbol{m}_2\cdot\boldsymbol{\Omega}_2)$, where $\boldsymbol{m}_1, \boldsymbol{m}_2 \in \mathbb{Z}^d$ are resonance vectors. This is associated with an integration over the dummy variable $\boldsymbol{J}_2$ scanning action space looking for locations where the resonance condition is satisfied. Figure 2.4 illustrates the gist of these resonant encounters, when the motion of the particles is directly considered in angle-action space. Similarly, Fig. 2.5 illustrates qualitatively such a non-local resonance condition in the case of a razor-thin disc. One should note that such resonant encounters are non-local in the sense that they do not require the resonating orbits to be close in position nor in action space. Equation (2.67) finally involves the dressed susceptibility coefficients $1/\mathcal{D}_{\boldsymbol{m}_1,\boldsymbol{m}_2}(\boldsymbol{J}_1, \boldsymbol{J}_2, \omega)$ introduced in Eq. (2.50). They encode the strength of the self-gravitating amplification within the system. Let us finally note that Eq. (2.67) scales like $1/(N\mathcal{D}^2)$, so that increasing N or increasing the heat content of the system have the same effect by slowing down the diffusion.

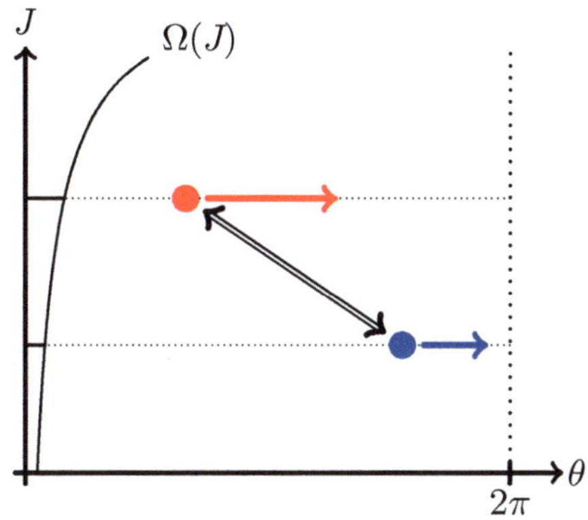

Fig. 2.4 Illustration in angle-action space of the resonance condition $\delta_{\mathrm{D}}(\boldsymbol{m}_1\cdot\boldsymbol{\Omega}_1-\boldsymbol{m}_2\cdot\boldsymbol{\Omega}_2)$ occurring in the Balescu-Lenard equation (2.67). In angle-action space, the trajectories of the particles are straight lines, with an intrinsic frequency $\Omega(J)$. This frequency depends only on the actions and is illustrated by the left curve. Here, the frequency associated with the *red* particle is twice the one of the *blue* particle: the particles are in resonance. These resonant encounters in angle-action space are the ones captured by the Balescu-Lenard equation (2.67)

Equation (2.67) can be rewritten as an anisotropic Fokker-Planck diffusion equation by introducing the relevant drift and diffusion coefficients. It becomes

$$\frac{\partial F}{\partial t}=\frac{\partial}{\partial \boldsymbol{J}_1}\cdot\bigg[\sum_{\boldsymbol{m}_1}\boldsymbol{m}_1\bigg(A_{\boldsymbol{m}_1}(\boldsymbol{J}_1)\,F(\boldsymbol{J}_1)+D_{\boldsymbol{m}_1}(\boldsymbol{J}_1)\,\boldsymbol{m}_1\cdot\frac{\partial F}{\partial \boldsymbol{J}_1}\bigg)\bigg], \tag{2.68}$$

where $A_{\boldsymbol{m}_1}(\boldsymbol{J}_1)$ and $D_{\boldsymbol{m}_1}(\boldsymbol{J}_1)$ are respectively the drift and diffusion coefficients associated with a given resonance vector $\boldsymbol{m}_1$. As the Balescu-Lenard equation describes the self-consistent evolution of the DF F, the drift and diffusion coefficients depend secularly on F. This dependence was not written out explicitly to simplify the notations. Following Eq. (2.67), the drift coefficients are given by

$$A_{\boldsymbol{m}_1}(\boldsymbol{J}_1)=-\pi(2\pi)^d\mu\sum_{\boldsymbol{m}_2}\int\mathrm{d}\boldsymbol{J}_2\,\frac{\delta_{\mathrm{D}}(\boldsymbol{m}_1\cdot\boldsymbol{\Omega}_1-\boldsymbol{m}_2\cdot\boldsymbol{\Omega}_2)}{\big|\mathcal{D}_{\boldsymbol{m}_1,\boldsymbol{m}_2}(\boldsymbol{J}_1,\boldsymbol{J}_2,\boldsymbol{m}_1\cdot\boldsymbol{\Omega}_1)\big|^2}\,\boldsymbol{m}_2\cdot\frac{\partial F}{\partial \boldsymbol{J}_2}, \tag{2.69}$$

while the diffusion coefficients are given by

$$D_{\boldsymbol{m}_1}(\boldsymbol{J}_1)=\pi(2\pi)^d\mu\sum_{\boldsymbol{m}_2}\int\mathrm{d}\boldsymbol{J}_2\,\frac{\delta_{\mathrm{D}}(\boldsymbol{m}_1\cdot\boldsymbol{\Omega}_1-\boldsymbol{m}_2\cdot\boldsymbol{\Omega}_2)}{\big|\mathcal{D}_{\boldsymbol{m}_1,\boldsymbol{m}_2}(\boldsymbol{J}_1,\boldsymbol{J}_2,\boldsymbol{m}_1\cdot\boldsymbol{\Omega}_1)\big|^2}\,F(\boldsymbol{J}_2)\,. \tag{2.70}$$

Finally, let us introduce the total diffusion flux $\boldsymbol{\mathcal{F}}_{\mathrm{tot}}(\boldsymbol{J})$ as

$$\boldsymbol{\mathcal{F}}_{\mathrm{tot}}(\boldsymbol{J})=\sum_{\boldsymbol{m}}\boldsymbol{m}\bigg(A_{\boldsymbol{m}}(\boldsymbol{J})\,F(\boldsymbol{J})+D_{\boldsymbol{m}}(\boldsymbol{J})\,\boldsymbol{m}\cdot\frac{\partial F}{\partial \boldsymbol{J}}\bigg), \tag{2.71}$$

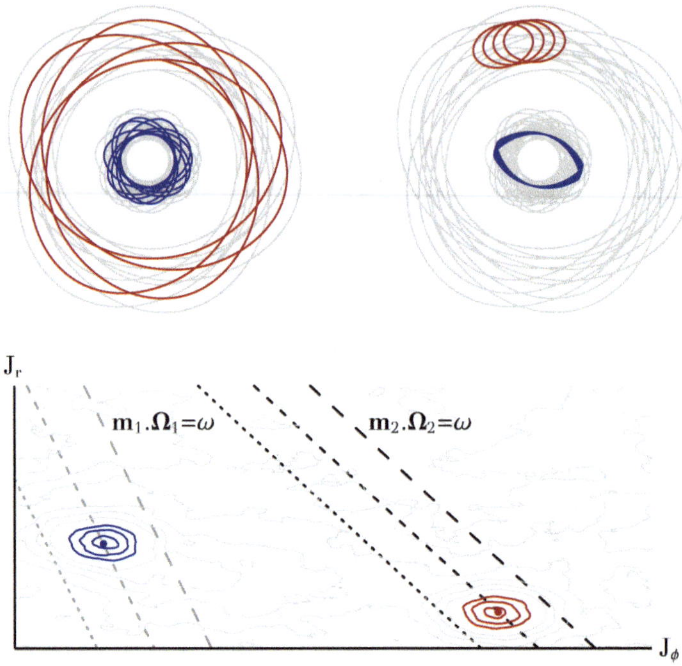

Fig. 2.5 Illustration of the resonance condition $\delta_{\rm D}(\boldsymbol{m}_1 \cdot \boldsymbol{\Omega}_1 - \boldsymbol{m}_2 \cdot \boldsymbol{\Omega}_2)$ of the Balescu-Lenard equation (2.67) in the case of a razor-thin stellar disc. *Top-left panel* A set of two resonant orbits in the inertial frame. *Top-right panel* The same two orbits in the rotating frame in which they are in resonance-here through an ILR-COR coupling (see Fig. 3.4). *Bottom panel* Fluctuations in action space of the system's DF sourced by finite$-N$ effects, exhibiting overdensities for the *blue* and *red* orbits. The *dashed* lines correspond to three contour levels of the intrinsic frequency respectively associated with the resonance vector $\boldsymbol{m}_1$ (*grey* lines) and $\boldsymbol{m}_2$ (*black* lines). The two sets of orbits satisfy the resonance condition $\boldsymbol{m}_1 \cdot \boldsymbol{\Omega}_1 - \boldsymbol{m}_2 \cdot \boldsymbol{\Omega}_2 = 0$, and therefore lead to a secular diffusion of the system's orbital structure according to the Balescu-Lenard equation (2.67). Let us emphasise that resonant orbits need not be caught in the same resonance ($\boldsymbol{m}_1 \neq \boldsymbol{m}_2$), be close in position space nor in action space

so that the Balescu-Lenard equation becomes

$$\frac{\partial F}{\partial t} = \mathrm{div}(\mathcal{F}_{\rm tot}) \,. \tag{2.72}$$

Here, with this convention, $-\mathcal{F}_{\rm tot}$ corresponds to the direction along which individual particles diffuse.

2.3.5 *The Bare Case: The Landau Equation*

When collective effects are neglected, the Balescu-Lenard equation (2.67) becomes the Landau equation (Polyachenko and Shukhman 1982; Chavanis 2007, 2010, 2013b) reading

$$\frac{\partial F}{\partial t} = \pi(2\pi)^d \mu \frac{\partial}{\partial \boldsymbol{J}_1} \cdot \bigg[\sum_{\boldsymbol{m}_1,\boldsymbol{m}_2} \boldsymbol{m}_1 \int \mathrm{d}\boldsymbol{J}_2 \, \delta_{\mathrm{D}}(\boldsymbol{m}_1 \cdot \boldsymbol{\Omega}_1 - \boldsymbol{m}_2 \cdot \boldsymbol{\Omega}_2) \, \big| A_{\boldsymbol{m}_1,\boldsymbol{m}_2}(\boldsymbol{J}_1, \boldsymbol{J}_2) \big|^2 \times \bigg(\boldsymbol{m}_1 \cdot \frac{\partial}{\partial \boldsymbol{J}_1} - \boldsymbol{m}_2 \cdot \frac{\partial}{\partial \boldsymbol{J}_2} \bigg) F(\boldsymbol{J}_1, t) \, F(\boldsymbol{J}_2, t) \bigg] . \tag{2.73}$$

In Eq. (2.73), the dressed susceptibility coefficients $1/|\mathcal{D}_{\boldsymbol{m}_1,\boldsymbol{m}_2}(\boldsymbol{J}_1, \boldsymbol{J}_2, \omega)|^2$ from Eq. (2.50) are replaced by the bare ones $|A_{\boldsymbol{m}_1,\boldsymbol{m}_2}(\boldsymbol{J}_1, \boldsymbol{J}_2)|^2$. These are defined as the Fourier transform of the interaction potential U (Lynden-Bell 1994; Pichon 1994; Chavanis 2013b), so that

$$A_{\boldsymbol{m}_1,\boldsymbol{m}_2}(\boldsymbol{J}_1, \boldsymbol{J}_2) = \frac{1}{(2\pi)^{2d}} \int \mathrm{d}\boldsymbol{\theta}_1 \mathrm{d}\boldsymbol{\theta}_2 \, U(\boldsymbol{x}(|\boldsymbol{\theta}_1, \boldsymbol{J}_1) - \boldsymbol{x}(\boldsymbol{\theta}_2, \boldsymbol{J}_2)|) \, \mathrm{e}^{-\mathrm{i}(\boldsymbol{m}_1 \cdot \boldsymbol{\theta}_1 - \boldsymbol{m}_2 \cdot \boldsymbol{\theta}_2)} . \tag{2.74}$$

In addition, these coefficients satisfy the symmetry relations

$$A_{\boldsymbol{m}_2,\boldsymbol{m}_1}(\boldsymbol{J}_2, \boldsymbol{J}_1) = A_{-\boldsymbol{m}_1,-\boldsymbol{m}_2}(\boldsymbol{J}_1, \boldsymbol{J}_2) = A^*_{\boldsymbol{m}_1,\boldsymbol{m}_2}(\boldsymbol{J}_1, \boldsymbol{J}_2) . \tag{2.75}$$

Note that the kinetic equations (2.67) and (2.73) share the same overall structure.

2.3.6 *The Multi-Component Case*

A crucial strength of the Balescu-Lenard formalism, already emphasised in Heyvaerts (2010) and Chavanis (2013b), is that this formalism also allows for a self-consistent description of the simultaneous evolution of multiple populations of various masses. Let us now detail the structure of such a multi-component diffusion equation. (See Appendix 6.B for an illustration of how the multi-component Balescu-Lenard equation may be derived in the specific context of quasi-Keplerian systems.) Here, we consider a system made of multiple components, indexed by the letters "a" and "b". The particles of the component "a" have an individual mass μ_{a} and follow the DF F^{a}. Each DF F^{a} is normalised such that $\int \mathrm{d}\boldsymbol{x} \mathrm{d}\boldsymbol{v} \, F^{\mathrm{a}} = M^{\mathrm{a}}_{\mathrm{tot}}$, where $M^{\mathrm{a}}_{\mathrm{tot}}$ is the total active mass of component "a". The evolution of each DF is then given by

$$\frac{\partial F^{\mathrm{a}}}{\partial t} = \pi(2\pi)^d \frac{\partial}{\partial \boldsymbol{J}_1} \cdot \bigg[\sum_{\boldsymbol{m}_1,\boldsymbol{m}_2} \boldsymbol{m}_1 \int \mathrm{d}\boldsymbol{J}_2 \, \frac{\delta_{\mathrm{D}}(\boldsymbol{m}_1 \cdot \boldsymbol{\Omega}_1 - \boldsymbol{m}_2 \cdot \boldsymbol{\Omega}_2)}{\big|\mathcal{D}_{\boldsymbol{m}_1,\boldsymbol{m}_2}(\boldsymbol{J}_1, \boldsymbol{J}_2, \boldsymbol{m}_1 \cdot \boldsymbol{\Omega}_1)\big|^2} \times \sum_{\mathrm{b}} \bigg\{ \mu_{\mathrm{b}} \, F^{\mathrm{b}}(\boldsymbol{J}_2) \, \boldsymbol{m}_1 \cdot \frac{\partial F^{\mathrm{a}}}{\partial \boldsymbol{J}_1} - \mu_{\mathrm{a}} \, F^{\mathrm{a}}(\boldsymbol{J}_1) \, \boldsymbol{m}_2 \cdot \frac{\partial F^{\mathrm{b}}}{\partial \boldsymbol{J}_2} \bigg\} \bigg] . \tag{2.76}$$

In the multi-component case, the susceptibility coefficients are still given by Eq. (2.50). However, the response matrix now encompasses all the active components of the system, so that

$$\widehat{\mathbf{M}}_{pq}(\omega) = (2\pi)^d \sum_{\boldsymbol{m}} \int \mathrm{d}\boldsymbol{J} \, \frac{\boldsymbol{m} \cdot \partial(\sum_{\mathrm{b}} F^{\mathrm{b}})/\partial \boldsymbol{J}}{\omega - \boldsymbol{m} \cdot \boldsymbol{\Omega}} \psi_{\boldsymbol{m}}^{(p)*}(\boldsymbol{J}) \, \psi_{\boldsymbol{m}}^{(q)}(\boldsymbol{J}) \, . \tag{2.77}$$

Similarly to Eq. (2.68), the multi-component Balescu-Lenard equation may also be written as an anisotropic diffusion equation, so that

$$\frac{\partial F^{\mathrm{a}}}{\partial t} = \frac{\partial}{\partial \boldsymbol{J}_1} \cdot \bigg[\sum_{\boldsymbol{m}_1} \boldsymbol{m}_1 \sum_{\mathrm{b}} \bigg\{ \mu_{\mathrm{a}} \, A_{\boldsymbol{m}_1}^{\mathrm{b}}(\boldsymbol{J}_1) \, F^{\mathrm{a}}(\boldsymbol{J}_1) + \mu_{\mathrm{b}} \, D_{\boldsymbol{m}_1}^{\mathrm{b}}(\boldsymbol{J}_1) \, \boldsymbol{m}_1 \cdot \frac{\partial F^{\mathrm{a}}}{\partial \boldsymbol{J}_1} \bigg\} \bigg] . \tag{2.78}$$

In Eq. (2.78), we introduced the multi-component drift and diffusion coefficients $A_{\boldsymbol{m}_1}^{\mathrm{b}}(\boldsymbol{J}_1)$ and $D_{\boldsymbol{m}_1}^{\mathrm{b}}(\boldsymbol{J}_1)$. They depend on the location $\boldsymbol{J}_1$ in action space, the considered resonance $\boldsymbol{m}_1$, and the component "b", whose DF is the underlying DF used to estimate them. In analogy with Eq. (2.69), the multi-component drift coefficients are given by

$$A_{\boldsymbol{m}_1}^{\mathrm{b}}(\boldsymbol{J}_1) = -\pi(2\pi)^d \sum_{\boldsymbol{m}_2} \int \mathrm{d}\boldsymbol{J}_2 \, \frac{\delta_{\mathrm{D}}(\boldsymbol{m}_1 \cdot \boldsymbol{\Omega}_1 - \boldsymbol{m}_2 \cdot \boldsymbol{\Omega}_2)}{\big|\mathcal{D}_{\boldsymbol{m}_1,\boldsymbol{m}_2}(\boldsymbol{J}_1, \boldsymbol{J}_2, \boldsymbol{m}_1 \cdot \boldsymbol{\Omega}_1)\big|^2} \, \boldsymbol{m}_2 \cdot \frac{\partial F^{\mathrm{b}}}{\partial \boldsymbol{J}_2} , \tag{2.79}$$

while the diffusion ones, similarly to Eq. (2.70), read

$$D_{\boldsymbol{m}_1}^{\mathrm{b}}(\boldsymbol{J}_1) = \pi(2\pi)^d \sum_{\boldsymbol{m}_2} \int \mathrm{d}\boldsymbol{J}_2 \, \frac{\delta_{\mathrm{D}}(\boldsymbol{m}_1 \cdot \boldsymbol{\Omega}_1 - \boldsymbol{m}_2 \cdot \boldsymbol{\Omega}_2)}{\big|\mathcal{D}_{\boldsymbol{m}_1,\boldsymbol{m}_2}(\boldsymbol{J}_1, \boldsymbol{J}_2, \boldsymbol{m}_1 \cdot \boldsymbol{\Omega}_1)\big|^2} \, F^{\mathrm{b}}(\boldsymbol{J}_2) \, . \tag{2.80}$$

One should pay attention to the fact that the multi-component drift and diffusion coefficients from Eqs. (2.79) and (2.80) do not have the same dimension as the single component ones. In order to emphasise the process of mass segregation, let us finally rewrite Eq. (2.78) as

$$\frac{\partial F^{\mathrm{a}}}{\partial t} = \frac{\partial}{\partial \boldsymbol{J}_1} \cdot \bigg[\sum_{\boldsymbol{m}_1} \boldsymbol{m}_1 \bigg\{ \mu_{\mathrm{a}} \, A_{\boldsymbol{m}_1}^{\mathrm{tot}}(\boldsymbol{J}_1) \, F^{\mathrm{a}}(\boldsymbol{J}_1) + D_{\boldsymbol{m}_1}^{\mathrm{tot}}(\boldsymbol{J}_1) \, \boldsymbol{m}_1 \cdot \frac{\partial F^{\mathrm{a}}}{\partial \boldsymbol{J}_1} \bigg\} \bigg] , \tag{2.81}$$

where we introduced the total drift and diffusion coefficients $A_{\boldsymbol{m}_1}^{\mathrm{tot}}(\boldsymbol{J}_1)$ and $D_{\boldsymbol{m}_1}^{\mathrm{tot}}(\boldsymbol{J}_1)$ as

$$A_{\boldsymbol{m}_1}^{\mathrm{tot}}(\boldsymbol{J}_1)=\sum_{\mathrm{b}} A_{\boldsymbol{m}_1}^{\mathrm{b}}(\boldsymbol{J}_1) \quad ; \quad D_{\boldsymbol{m}_1}^{\mathrm{tot}}(\boldsymbol{J}_1)=\sum_{\mathrm{b}} \mu_{\mathrm{b}}\, D_{\boldsymbol{m}_1}^{\mathrm{b}}(\boldsymbol{J}_1)\,. \tag{2.82}$$

In Eq. (2.81), let us note that the only differences between the different components is the presence of the mass prefactor μ_{a} in front of the total drift coefficient. This leads to the process of mass segregation, when a spectrum of mass is involved.

2.3.7 H-Theorem

Following closely Heyvaerts (2010), let us define the system's entropy $S(t)$ as

$$S(t)=-\int \mathrm{d}\boldsymbol{J}_1\, s(F(\boldsymbol{J}_1,t))\,, \quad \text{with} \quad s(x)=x\log(x)\,, \tag{2.83}$$

where $s(x)$ corresponds to Boltzmann's entropy function. Differentiating Eq. (2.83) once w.r.t. to time yields

$$\frac{\mathrm{d}S}{\mathrm{d}t}=-\int \mathrm{d}\boldsymbol{J}_1\, s'(F(\boldsymbol{J}_1))\,\frac{\partial F}{\partial t}\,. \tag{2.84}$$

Let us follow the definition of the total diffusion flux $\mathcal{F}_{\mathrm{tot}}(\boldsymbol{J}_1)$ from Eq. (2.71) to rewrite $\mathcal{F}_{\mathrm{tot}}(\boldsymbol{J}_1)$ as

$$\mathcal{F}_{\mathrm{tot}}(\boldsymbol{J}_1)=\sum_{\boldsymbol{m}_1,\boldsymbol{m}_2} \boldsymbol{m}_1 \int \mathrm{d}\boldsymbol{J}_2\, \alpha_{\boldsymbol{m}_1,\boldsymbol{m}_2}(\boldsymbol{J}_1,\boldsymbol{J}_2)\left[\boldsymbol{m}_1\cdot\frac{\partial}{\partial \boldsymbol{J}_1}-\boldsymbol{m}_2\cdot\frac{\partial}{\partial \boldsymbol{J}_2}\right] F(\boldsymbol{J}_1)\,F(\boldsymbol{J}_2)\,, \tag{2.85}$$

with $\alpha_{\boldsymbol{m}_1,\boldsymbol{m}_2}(\boldsymbol{J}_1,\boldsymbol{J}_2)$ given by

$$\alpha_{\boldsymbol{m}_1,\boldsymbol{m}_2}(\boldsymbol{J}_1,\boldsymbol{J}_2)=\pi(2\pi)^d \mu\,\frac{\delta_{\mathrm{D}}(\boldsymbol{m}_1\cdot\boldsymbol{\Omega}_1-\boldsymbol{m}_2\cdot\boldsymbol{\Omega}_2)}{|\mathcal{D}_{\boldsymbol{m}_1,\boldsymbol{m}_2}(\boldsymbol{J}_1,\boldsymbol{J}_2,\boldsymbol{m}_1\cdot\boldsymbol{\Omega}_1)|^2}\geq 0\,. \tag{2.86}$$

Integrating Eq. (2.84) by parts and ignoring boundary terms, one gets

$$\frac{\mathrm{d}S}{\mathrm{d}t}=\int \mathrm{d}\boldsymbol{J}_1\, s''(F(\boldsymbol{J}_1))\,\frac{\partial F}{\partial \boldsymbol{J}_1}\cdot\mathcal{F}_{\mathrm{tot}}(\boldsymbol{J}_1)\,. \tag{2.87}$$

Thanks to the rewriting from Eq. (2.85), one can express Eq. (2.87) as

$$\frac{\mathrm{d}S}{\mathrm{d}t}=\sum_{\boldsymbol{m}_1,\boldsymbol{m}_2}\int \mathrm{d}\boldsymbol{J}_1 \mathrm{d}\boldsymbol{J}_2\, \alpha_{\boldsymbol{m}_1,\boldsymbol{m}_2}(\boldsymbol{J}_1,\boldsymbol{J}_2)\, s_1''\,(\boldsymbol{m}_1\cdot\boldsymbol{F}_1')\left[F_2\,(\boldsymbol{m}_1\cdot\boldsymbol{F}_1')-F_1\,(\boldsymbol{m}_2\cdot\boldsymbol{F}_2')\right], \tag{2.88}$$

where we used the shortened notations $s_i''=s''(F(\boldsymbol{J}_i))$, $F_i=F(\boldsymbol{J}_i)$, and $\boldsymbol{F}_i'=\partial F/\partial \boldsymbol{J}_i$. Equation (2.88) can be symmetrised via the substitutions $\boldsymbol{m}_1 \leftrightarrow \boldsymbol{m}_2$ and $\boldsymbol{J}_1 \leftrightarrow \boldsymbol{J}_2$. As $\alpha_{\boldsymbol{m}_2,\boldsymbol{m}_1}(\boldsymbol{J}_2,\boldsymbol{J}_1)=\alpha_{\boldsymbol{m}_1,\boldsymbol{m}_2}(\boldsymbol{J}_1,\boldsymbol{J}_2)$, Eq. (2.88) finally becomes

$$\frac{\mathrm{d}S}{\mathrm{d}t}=\frac{1}{2}\sum_{\boldsymbol{m}_1,\boldsymbol{m}_2}\int \mathrm{d}\boldsymbol{J}_1\mathrm{d}\boldsymbol{J}_2\,\alpha_{\boldsymbol{m}_1,\boldsymbol{m}_2}(\boldsymbol{J}_1,\boldsymbol{J}_2)\bigg[F_2 s_1''(\boldsymbol{m}_1\cdot\boldsymbol{F}_1')^2-(\boldsymbol{m}_1\cdot\boldsymbol{F}_1')(\boldsymbol{m}_2\cdot\boldsymbol{F}_2')(F_1 s_1''+F_2 s_2'')+F_1 s_2''(\boldsymbol{m}_2\cdot\boldsymbol{F}_2')^2\bigg]. \tag{2.89}$$

As Boltzmann's entropy function[3] satisfies $s''(x)=1/x$, the square bracket in Eq. (2.89) may immediately be factored as

$$\frac{1}{F_1F_2}\bigg[F_2(\boldsymbol{m}_1\cdot\boldsymbol{F}_1')-F_1(\boldsymbol{m}_2\cdot\boldsymbol{F}_2')\bigg]^2\geq 0\,, \tag{2.90}$$

so that one finally gets $\mathrm{d}S/\mathrm{d}\tau\geq 0$. The Balescu-Lenard equation (2.67) therefore satisfies Boltzmann's H-theorem. This entropy increase corresponds to heat generation as the orbital structure of the system secularly rearranges itself driven by self-induced collisional effects. The previous demonstration naturally extends both to the Landau equation (2.73), but also more interestingly to the multi-component Balescu-Lenard equation (2.76). Indeed, defining the system's total entropy S_{tot} as

$$S_{\mathrm{tot}}(t)=-\int\mathrm{d}\boldsymbol{J}_1\sum_{\mathrm{a}}\frac{1}{\mu_{\mathrm{a}}}s(F^{\mathrm{a}}(\boldsymbol{J}_1,t))\,, \tag{2.91}$$

and following the same approach, one can show that for $s''(x)=1/x$, one has $\mathrm{d}S_{\mathrm{tot}}/\mathrm{d}t\geq 0$. Let us finally note that this does not necessarily imply that the entropy of each individual component increases.

2.4 Conclusion

In this chapter, we presented two important sources of diffusion to induce secular evolution in self-gravitating systems. The first source, presented in Sect. 2.2, considers the case of a collisionless system undergoing external perturbations. The second source, presented in Sect. 2.3, is captured by the Balescu-Lenard equation, which describes the long-term effects of finite$-N$ fluctuations on isolated discrete self-gravitating systems. In our two derivations, we emphasised the strong similarities existing between the two approaches, as can be seen in particular in their similar decoupled evolution equations. Let us finally underline that both Eqs. (2.31) and (2.67) share the properties that they describe strongly anisotropic diffusion in action space (see Fig. 2.3), account for the system's internal susceptibility (via the response matrix from Eq. (2.17) and the associated gravitational polarisation, see

[3] Any double primitive of $1/x$ would work as well.

Fig. 1.6). Because they are sourced by different fluctuations, either external or internal, these two orbital diffusion processes provide the ideal frameworks in which to study the secular evolution of self-gravitating systems.

The rest of the thesis is focused on illustrating for various astrophysical systems how these formalisms allow for a detailed description of their secular dynamics. In Chap. 3, we will consider the case of razor-thin stellar discs. In order to obtain simple quadratures for the diffusion fluxes, we will develop a razor-thin WKB formalism (i.e. restriction to radially tightly wound perturbations) providing a straightforward understanding of the regions of maximum amplification within the disc. We will illustrate how the functional form of the diffusion coefficients explains the self-induced formation of resonant ridges in the disc's DF, as observed in numerical simulations. In Chap. 4, we will resort to the same razor-thin stellar discs, but will devote our efforts to correctly account for the disc's self-gravity and the associated strong amplification. This will be shown to significantly hasten the diffusion in the disc. In addition, in Appendix 4.D, we will illustrate how the same method may also be applied to study the long-term dynamics of $3D$ spherical systems such as dark matter haloes. This framework provides a promising way to investigate the secular transformation of dark matter haloes' cusps into cores. In Chap. 5, we will extend our WKB approximation to apply it to thickened stellar discs. We will investigate various possible mechanisms of thickening such as the disc's internal Poisson shot noise, a series of central decaying bars, or the joint evolution of giant molecular clouds within the disc. Finally, in Chap. 6, we will consider the case of quasi-Keplerian systems, such as galactic centres, for which the presence of a dominating central body imposes a degenerate Keplerian dynamics. Once tailored for such systems, we will detail in particular how the Balescu-Lenard formalism recovers the process of "resonant relaxation" specific to these systems.

2.4.1 Future Works

The previous formalisms could be generalised in various ways.

In Appendix 2.C, we presented a new method based on a functional approach to derive the inhomogeneous Landau equation. Because of the simplicity of the required calculations, this throws new light on the complex dynamical processes at play. One could hope to generalise this calculation to account for collective effects and recover the inhomogeneous Balescu-Lenard equation. Such a calculation is expected to be more demanding, as it will involve a Fredholm type equation, such as Eq. (2.125). Similarly, we showed in Fouvry et al. (2016b) how the same functional approach could also be transposed to the kinetic theory of two-dimensional point vortices (Chavanis 2012c, d). One should investigate other physical systems for which this approach could also be successful. Finally, it would be of particular interest to apply this method to derive a closed kinetic equation when higher order correlation terms are accounted for. This could for example allow us to describe the dynamics of $1D$ homogeneous systems, for which the $1/N$ Balescu-Lenard collision term vanishes by symmetry (Eldridge and Feix 1963; Kadomtsev and Pogutse 1970). This is also the case for the Hamiltonian Mean Field model (HMF) in the homogeneous limit (Chavanis et al. 2005; Bouchet and Dauxois 2005; Rocha Filho et al. 2014).

Inspired by Pichon and Aubert (2006), the previous approaches could also be extended and developed for open systems, by accounting for possible sources and sinks of particles. Similarly, it could also prove interesting to investigate the Balescu-Lenard equation in a context where the system's number of particles gets to evolve during the secular evolution, to describe for example the progressive dissolution of overdensities, etc. Similarly, as can be seen in the proposed derivations, all these formalisms rely on the fundamental assumption of integrability, i.e. on the existence of angle-action coordinates. It would be of interest to investigate as well how such approaches could be tailored to deal with chaotic behaviours and their associated diffusions. Finally, one could also investigate within these frameworks the role that gas may play on the dynamical properties of the system. Indeed, one crucial property of gas is that it cannot shell-cross so that it shocks. This typically means that the gas component is dynamically much colder than its stellar counterpart, which alters the system's dynamical susceptibility.

Appendix

A Derivation of the BBGKY Hierarchy

In this Appendix, let us briefly recover the fundamental equations of the BBGKY hierarchy. This decomposition is at the heart of the derivation of the Balescu-Lenard equation presented in Heyvaerts (2010). Such a derivation with similar notations is presented in Fouvry et al. (2016a). Let us consider an isolated system made of N identical particles of individual mass $\mu = M_{\rm tot}/N$, where $M_{\rm tot}$ is the total mass of the system. The individual dynamics of these particles is exactly described by Hamilton's equation which read

$$\mu \frac{\mathrm{d}\boldsymbol{x}_i}{\mathrm{d}t} = \frac{\partial H}{\partial \boldsymbol{v}_i} \quad ; \quad \mu \frac{\mathrm{d}\boldsymbol{v}_i}{\mathrm{d}t} = -\frac{\partial H}{\partial \boldsymbol{x}_i} , \tag{2.92}$$

where $(\boldsymbol{x}_i, \boldsymbol{v}_i)$ corresponds to the position and velocity of particle i. The total Hamiltonian H appearing in Eq. (2.92) encompasses all the binary interactions between particles, so that

$$H = \frac{\mu}{2} \sum_{i=1}^{N} \boldsymbol{v}_i^2 + \mu^2 \sum_{i<j}^{N} U(|\boldsymbol{x}_i - \boldsymbol{x}_j|) , \tag{2.93}$$

where $U(|\boldsymbol{x}|)$ stands for the interaction potential, e.g., $U(|\boldsymbol{x}|) = -G/|\boldsymbol{x}|$ in the gravitational context. As will be underlined in Chap. 6 when considering quasi-Keplerian systems, one can easily add an external potential to this Hamiltonian, and the associated hierarchy equations are straightforward to deduce. While Eq. (2.92) captures the individual dynamics of the system's components, we are interested in a statistical description of our system. As a consequence, let us introduce the system's $N-$body probability distribution function (PDF) $P_N(\boldsymbol{x}_1, \boldsymbol{v}_1, ..., \boldsymbol{x}_N, \boldsymbol{v}_N, t)$ which gives the probability of finding at time t particle 1 at position $\boldsymbol{x}_1$ with velocity $\boldsymbol{v}_1$, particle 2 at position $\boldsymbol{x}_2$ with velocity $\boldsymbol{v}_2$, etc. We choose the convention that

$$\int \mathrm{d}\Gamma_1 \mathrm{d}\Gamma_2 ... \mathrm{d}\Gamma_N \, P_N(\Gamma_1, ... \Gamma_N, t) = 1 \,, \tag{2.94}$$

where we introduced the phase space coordinates $\Gamma_i = (\boldsymbol{x}_i, \boldsymbol{v}_i)$, so that $\mathrm{d}\Gamma_i = \mathrm{d}\boldsymbol{x}_i \mathrm{d}\boldsymbol{v}_i$. The evolution of P_N is given by Liouville's equation (see Eq. (1.9)) which reads

$$\frac{\partial P_N}{\partial t} + \sum_{i=1}^{N} \left[\boldsymbol{v}_i \cdot \frac{\partial P_N}{\partial \boldsymbol{x}_i} + \mu \boldsymbol{\mathcal{F}}_i^{\mathrm{tot}} \cdot \frac{\partial P_N}{\partial \boldsymbol{v}_i} \right] = 0 \,, \tag{2.95}$$

where we introduced the total force, $\boldsymbol{\mathcal{F}}_i^{\mathrm{tot}}$, exerted on particle i as

$$\boldsymbol{\mathcal{F}}_i^{\mathrm{tot}} = \sum_{j \neq i}^{N} \boldsymbol{\mathcal{F}}_{ij} = - \sum_{j \neq i}^{N} \frac{\partial U_{ij}}{\partial \boldsymbol{x}_i} \,. \tag{2.96}$$

In Eq. (2.96), $\boldsymbol{\mathcal{F}}_{ij}$ stands for the force exerted by particle j on particle i. It satisfies $\boldsymbol{\mathcal{F}}_{ij} = -\partial U_{ij} / \partial \boldsymbol{x}_i$, where we wrote the interaction potential as $U_{ij} = U(|\boldsymbol{x}_i - \boldsymbol{x}_j|)$. At this stage, let us insist on the fact that Liouville's equation (2.95) is an exact equation, which encompasses the same information as Hamilton's equation (2.92). It is however defined on the (large) space of configurations $(\Gamma_1, ..., \Gamma_N)$. In order to reduce the dimension of the space where the evolution equations are defined, let us introduce the reduced PDFs P_n for $1 \leq n \leq N$ as

$$P_n(\Gamma_1, ..., \Gamma_n, t) = \int \mathrm{d}\Gamma_{n+1} ... \mathrm{d}\Gamma_N \, P_N(\Gamma_1, ..., \Gamma_N, t) \,. \tag{2.97}$$

Relying on the symmetry of P_N w.r.t. permutations of its arguments, we may integrate Eq. (2.95) w.r.t. $\mathrm{d}\Gamma_{n+1} ... \mathrm{d}\Gamma_N$ to obtain the evolution equation satisfied by P_n. This gives the general equation of the BBGKY hierarchy which reads

$$\frac{\partial P_n}{\partial t} + \sum_{i=1}^{n} \boldsymbol{v}_i \cdot \frac{\partial P_n}{\partial \boldsymbol{x}_i} + \sum_{i=1}^{n} \sum_{k=1, k \neq i}^{n} \mu \boldsymbol{\mathcal{F}}_{ik} \cdot \frac{\partial P_n}{\partial \boldsymbol{v}_i} + (N-n) \sum_{i=1}^{n} \int \mathrm{d}\Gamma_{n+1} ... \mathrm{d}\Gamma_N \, \mu \boldsymbol{\mathcal{F}}_{i,n+1} \cdot \frac{\partial P_{n+1}}{\partial \boldsymbol{v}_i} = 0 \,. \tag{2.98}$$

One can note that Eq. (2.98) is defined on the smaller space $(\Gamma_1, ..., \Gamma_n)$. The three first terms only involve the first n particles, while the last collision term involves the reduced P_{n+1} of higher order, i.e. the BBGKY hierarchy is not closed. In order to simplify the prefactors present in Eq. (2.98), let us introduce the reduced distribution functions f_n as

$$f_n(\Gamma_1, ..., \Gamma_n, t) = \mu^n \frac{N!}{(N-n)!} P_n(\Gamma_1, ..., \Gamma_n, t) \,. \tag{2.99}$$

The hierarchy from Eq. (2.98) immediately becomes

$$\frac{\partial f_n}{\partial t} + \sum_{i=1}^{n} \boldsymbol{v}_i \cdot \frac{\partial f_n}{\partial \boldsymbol{x}_i} + \sum_{i=1}^{n} \sum_{k=1,k\neq i}^{n} \mu \boldsymbol{\mathcal{F}}_{ik} \cdot \frac{\partial f_n}{\partial \boldsymbol{v}_i} + \sum_{i=1}^{n} \int \mathrm{d}\Gamma_{n+1} \, \boldsymbol{\mathcal{F}}_{i,n+1} \cdot \frac{\partial f_{n+1}}{\partial \boldsymbol{v}_i} = 0 \,. \tag{2.100}$$

Equation (2.100) corresponds to the traditional writing of the BBGKY hierarchy. In order to emphasise the importance of the contributions arising from correlations between particles, let us introduce the cluster representation of the reduced distribution functions. We therefore define the 2−body correlation g_2 as

$$f_2(\Gamma_1, \Gamma_2) = f_1(\Gamma_1) f_1(\Gamma_2) + g_2(\Gamma_1, \Gamma_2) \,, \tag{2.101}$$

where the dependences w.r.t. t were not written out explicitly to simplify the notations. Similarly, we introduce the 3−body autocorrelation g_3, so that f_3 reads

$$f_3(\Gamma_1, \Gamma_2, \Gamma_3) = f_1(\Gamma_1) f_1(\Gamma_2) f_1(\Gamma_3) + f_1(\Gamma_1) g_2(\Gamma_2, \Gamma_3) + f_1(\Gamma_2) g_2(\Gamma_1, \Gamma_3) + f_1(\Gamma_3) g_2(\Gamma_1, \Gamma_2) + g_3(\Gamma_1, \Gamma_2, \Gamma_3) \,. \tag{2.102}$$

Thanks to the convention from Eq. (2.94), it is straightforward to check that one has the normalisations

$$\int \mathrm{d}\Gamma_1 \, f_1(\Gamma_1) = \mu N \,; \int \mathrm{d}\Gamma_1 \mathrm{d}\Gamma_2 \, g_2(\Gamma_1, \Gamma_2) = -\mu^2 N \,; \int \mathrm{d}\Gamma_1 \mathrm{d}\Gamma_2 \mathrm{d}\Gamma_3 \, g_3(\Gamma_1, \Gamma_2, \Gamma_3) = 2\mu^3 N \,. \tag{2.103}$$

As the mass of the individual particles is given by $\mu = M_{\mathrm{tot}}/N$, one immediately gets the scalings w.r.t. the number of particles as $|f_1| \sim 1$, $|g_2| \sim 1/N$, and $|g_3| \sim 1/N^2$. Thanks to these decompositions, the two first equations of the BBGKY hierarchy from Eq. (2.100) respectively become

$$\frac{\partial f_1}{\partial t} + \boldsymbol{v}_1 \cdot \frac{\partial f_1}{\partial \boldsymbol{x}_1} + \bigg[\int \mathrm{d}\Gamma_2 \, \boldsymbol{\mathcal{F}}_{12} f_1(\Gamma_2) \bigg] \cdot \frac{\partial f_1}{\partial \boldsymbol{v}_1} + \int \mathrm{d}\Gamma_2 \, \boldsymbol{\mathcal{F}}_{12} \cdot \frac{\partial g_2(\Gamma_1, \Gamma_2)}{\partial \boldsymbol{v}_1} = 0 \,, \tag{2.104}$$

and

$$\begin{aligned} \frac{1}{2} \frac{\partial g_2}{\partial t} + \boldsymbol{v}_1 \cdot \frac{\partial g_1}{\partial \boldsymbol{x}_1} + \bigg[\int \mathrm{d}\Gamma_3 \boldsymbol{\mathcal{F}}_{13} f_1(\Gamma_3) \bigg] \cdot \frac{\partial g_2}{\partial \boldsymbol{v}_1} + \mu \boldsymbol{\mathcal{F}}_{12} \cdot \frac{\partial f_1}{\partial \boldsymbol{v}_1} f_1(\Gamma_2) + \bigg[\int \mathrm{d}\Gamma_3 \, \boldsymbol{\mathcal{F}}_{13} g_2(\Gamma_2, \Gamma_3) \bigg] \cdot \frac{\partial f_1}{\partial \boldsymbol{v}_1} \\ + \mu \boldsymbol{\mathcal{F}}_{12} \cdot \frac{\partial g_2}{\partial \boldsymbol{v}_1} + \bigg[\int \mathrm{d}\Gamma_3 \, \boldsymbol{\mathcal{F}}_{13} \, \frac{\partial g_3(\Gamma_1, \Gamma_2, \Gamma_3)}{\partial \boldsymbol{v}_1} \bigg] + (1 \leftrightarrow 2) = 0 \,, \end{aligned} \tag{2.105}$$

where $(1 \leftrightarrow 2)$ stands for the permutation of indices 1 and 2, and applies to all preceding terms. When considering the long-term evolution induced by discreteness effects, one may perform a truncation at order $1/N$ of the two equations (2.104) and (2.105). This requires to rely on the scalings from Eq. (2.103), as well as on the fact that $\mu \sim 1/N$ and $\mathcal{F}_{ij} \sim 1$. In Eq. (2.104), all the terms are at least of order $1/N$, so that they should all be conserved. In Eq. (2.105), all the terms on the first line are of order $1/N$ and have to be conserved, while all the terms on the second line are of order $1/N^2$ and may therefore be neglected.[4] In addition to these truncations, and in order to consider quantities of order 1, let us introduce the system's $1-$body DF F and the $2-$body correlation function $\mathcal{C}$ as

$$F = f_1 \quad ; \quad \mathcal{C} = \frac{g_2}{\mu} \, . \tag{2.106}$$

It is straightforward to note that $|F| \sim 1$ and $|\mathcal{C}| \sim 1$. When truncated at order $1/N$, Eqs. (2.104) and (2.105) finally take the form

$$\frac{\partial F}{\partial t} + \boldsymbol{v}_1 \cdot \frac{\partial F}{\partial \boldsymbol{x}_1} + \bigg[\int \mathrm{d}\Gamma_2 \, \boldsymbol{\mathcal{F}}_{12} F(\Gamma_2) \bigg] \cdot \frac{\partial F}{\partial \boldsymbol{v}_1} + \mu \int \mathrm{d}\Gamma_2 \, \boldsymbol{\mathcal{F}}_{12} \cdot \frac{\partial \mathcal{C}(\Gamma_1, \Gamma_2)}{\partial \boldsymbol{v}_1} = 0 \, , \tag{2.107}$$

and

$$\begin{aligned} & \frac{1}{2} \frac{\partial \mathcal{C}}{\partial t} + \boldsymbol{v}_1 \cdot \frac{\partial \mathcal{C}}{\partial \boldsymbol{x}_1} + \bigg[\int \mathrm{d}\Gamma_3 \, \boldsymbol{\mathcal{F}}_{13} F(\Gamma_3) \bigg] \cdot \frac{\partial \mathcal{C}}{\partial \boldsymbol{v}_1} + \boldsymbol{\mathcal{F}}_{12} \cdot \frac{\partial F}{\partial \boldsymbol{v}_1} F(\Gamma_2) \\ & + \bigg[\int \mathrm{d}\Gamma_3 \, \boldsymbol{\mathcal{F}}_{13} \mathcal{C}(\Gamma_2, \Gamma_3) \bigg] \cdot \frac{\partial F}{\partial \boldsymbol{v}_1} + (1 \leftrightarrow 2) = 0 \, . \end{aligned} \tag{2.108}$$

These two coupled evolution equations (2.107) and (2.108) only involve the system's $1-$body DF F and its $2-$body autocorrelation function $\mathcal{C}$. They are the two central equations, from which one may derive the inhomogeneous Balescu-Lenard equation, as proposed in Heyvaerts (2010) and revisited in Appendix 2.B. The second and third terms in Eq. (2.107) correspond to the Vlasov advection term associated with the mean self-consistent potential generated by the stars. The fourth term is a collisional correction, as it scales like $\mu \propto 1/N$, and captures the effects of $2-$body correlations on the dynamics of the $1-$body DF. Similarly, in Eq. (2.108), the second and third terms are associated with the mean Vlasov advection due to the system's mean potential. The fourth term is a source term, which depends only on the $1-$body DF and sources the dynamics of the $2-$body correlation. Finally, the last term in Eq. (2.108) corresponds to the collective effects and is associated with the fact that the system can amplify and dress perturbations.

[4]There is a subtlety with the first term on the second line of Eq. (2.105). While being of order $1/N^2$, it can become arbitrarily large when particle 2 approaches particle 1, due to the divergence of the interaction force at small separation. This term describes strong collisions and is not accounted for in the present formalism of resonance-driven diffusion.

B Derivation of the Balescu-Lenard Equation via the BBGKY Hierarchy

In this Appendix, we revisit the derivation of the Balescu-Lenard equation (2.67) following the method presented in Heyvaerts (2010). This method, based on the direct resolution of the BBGKY hierarchy, is complementary to the second approach subsequently proposed by Chavanis (2012b), based on the Klimontovich equation and already presented in Sect. 2.3.

As already shown in Appendix 2.A, at order $1/N$, the dynamics of a self-gravitating system made of N identical particles is fully characterised by its 1−body DF F and 2−body autocorrelation $\mathcal{C}$. These two dynamical quantities are coupled by the two first truncated equations of the BBGKY hierarchy, namely Eqs. (2.107) and (2.108). They can be rewritten as

$$\frac{\partial F}{\partial t}+\bigg[\boldsymbol{v}_1\cdot\frac{\partial}{\partial \boldsymbol{x}_1}+\bigg[\!\int\! \mathrm{d}\Gamma_2\, \boldsymbol{\mathcal{F}}_{12}F(\Gamma_2)\bigg]\cdot\frac{\partial}{\partial \boldsymbol{v}_1}\bigg]F=-\mu\!\int\! \mathrm{d}\Gamma_2\, \boldsymbol{\mathcal{F}}_{12}\cdot\frac{\partial \mathcal{C}(\Gamma_1,\Gamma_2)}{\partial \boldsymbol{v}_1}\,, \tag{2.109}$$

and

$$\begin{aligned}&\frac{\partial \mathcal{C}}{\partial t}+\bigg[\boldsymbol{v}_1\cdot\frac{\partial}{\partial \boldsymbol{x}_1}+\bigg[\!\int\! \mathrm{d}\Gamma_3\, \boldsymbol{\mathcal{F}}_{13}F(\Gamma_3)\bigg]\cdot\frac{\partial}{\partial \boldsymbol{v}_1}\bigg]\mathcal{C}+\bigg[\boldsymbol{v}_2\cdot\frac{\partial}{\partial \boldsymbol{x}_2}+\bigg[\!\int\! \mathrm{d}\Gamma_3\, \boldsymbol{\mathcal{F}}_{23}F(\Gamma_3)\bigg]\cdot\frac{\partial}{\partial \boldsymbol{v}_2}\bigg]\mathcal{C}\\&+\bigg[\!\int\! \mathrm{d}\Gamma_3\, \boldsymbol{\mathcal{F}}_{13}\mathcal{C}(\Gamma_2,\Gamma_3)\bigg]\cdot\frac{\partial F}{\partial \boldsymbol{v}_1}+\bigg[\!\int\! \mathrm{d}\Gamma_3\, \boldsymbol{\mathcal{F}}_{23}\mathcal{C}(\Gamma_1,\Gamma_3)\bigg]\cdot\frac{\partial F}{\partial \boldsymbol{v}_2}=-\boldsymbol{\mathcal{F}}_{12}\cdot\bigg[\frac{\partial}{\partial \boldsymbol{v}_1}-\frac{\partial}{\partial \boldsymbol{v}_2}\bigg]F(\Gamma_1)F(\Gamma_2)\,.\end{aligned} \tag{2.110}$$

B.1 Solving for the Autocorrelation

The first step of the derivation is to solve Eq. (2.110) to obtain $\mathcal{C}=\mathcal{C}[F]$. Injecting this solution in the r.h.s. of Eq. (2.109), one finally obtains a closed kinetic equation involving F only.

First, one can note that Eq. (2.110) is linear in the correlation $\mathcal{C}$, symmetric in 1 and 2, and its r.h.s. is a source term $S_2(\Gamma_1,\Gamma_2,t)$ depending only on the 1− body DF F. It reads

$$S_2(\Gamma_1,\Gamma_2,t)=-\boldsymbol{\mathcal{F}}_{12}\cdot\bigg[\frac{\partial}{\partial \boldsymbol{v}_1}-\frac{\partial}{\partial \boldsymbol{v}_2}\bigg]F(\Gamma_1)F(\Gamma_2)\,. \tag{2.111}$$

Equation (2.110) can be solved for $\mathcal{C}(\Gamma_1,\Gamma_2,t)$ by working out the Green's function $\mathcal{G}^{(2)}(\Gamma_1,\Gamma_2,\Gamma_1',\Gamma_2',\tau)$ of the linear differential operator in its l.h.s. Indeed, the solution for $\mathcal{C}(\Gamma_1,\Gamma_2,t)$ can be written as

$$\mathcal{C}(\Gamma_1,\Gamma_2,t)=\int_0^{+\infty}\!\mathrm{d}\tau\!\int\!\mathrm{d}\Gamma_1'\mathrm{d}\Gamma_2'\,\mathcal{G}^{(2)}(\Gamma_1,\Gamma_2,\Gamma_1',\Gamma_2',\tau)\,S_2(\Gamma_1',\Gamma_2',t-\tau)\,. \tag{2.112}$$

Injecting Eq. (2.112) into Eq. (2.110), one gets the propagation equation satisfied by $\mathcal{G}^{(2)}$. It reads

$$\frac{\partial \mathcal{G}^{(2)}}{\partial \tau}+\bigg[\boldsymbol{v}_1\cdot\frac{\partial}{\partial \boldsymbol{x}_1}+\bigg[\int \mathrm{d}\Gamma_3\, \boldsymbol{\mathcal{F}}_{13}F(\Gamma_3)\bigg]\cdot\frac{\partial}{\partial \boldsymbol{v}_1}\bigg]\mathcal{G}^{(2)}+\bigg[\boldsymbol{v}_2\cdot\frac{\partial}{\partial \boldsymbol{x}_2}+\bigg[\int \mathrm{d}\Gamma_3\, \boldsymbol{\mathcal{F}}_{23}F(\Gamma_3)\bigg]\cdot\frac{\partial}{\partial \boldsymbol{v}_2}\bigg]\mathcal{G}^{(2)}$$
$$+\bigg[\int \mathrm{d}\Gamma_3\, \boldsymbol{\mathcal{F}}_{13}\mathcal{G}^{(2)}(\Gamma_3,\Gamma_2,\Gamma_1',\Gamma_2',\tau)\bigg]\cdot\frac{\partial F}{\partial \boldsymbol{v}_1}+\bigg[\int \mathrm{d}\Gamma_3\, \boldsymbol{\mathcal{F}}_{23}\mathcal{G}^{(2)}(\Gamma_1,\Gamma_3,\Gamma_1',\Gamma_2',\tau)\bigg]\cdot\frac{\partial F}{\partial \boldsymbol{v}_2}=0\,, \tag{2.113}$$

where we assumed that the source term $S_2(t)$ was effectively turned on only for $t \geq 0$, so that $S_2(t<0)=0$. Moreover, the Green's function initially satisfies $\mathcal{G}^{(2)}(\Gamma_1,\Gamma_2,\Gamma_1',\Gamma_2',0)=\delta_{\mathrm{D}}(\Gamma_1-\Gamma_1')\,\delta_{\mathrm{D}}(\Gamma_2-\Gamma_2')$. Once the autocorrelation has been expressed as a function of F, i.e. $\mathcal{C}=\mathcal{C}[F]$, one may finally proceed to the evaluation of the collision operator $\mathscr{C}[F]$ appearing in the r.h.s. of Eq. (2.109), which reads

$$\mathscr{C}[F]=-\mu\int \mathrm{d}\Gamma_2\, \boldsymbol{\mathcal{F}}_{12}\cdot\frac{\partial \mathcal{C}[F](\Gamma_1,\Gamma_2)}{\partial \boldsymbol{v}_1}\,. \tag{2.114}$$

When considering Eq. (2.113), it is crucial to note that this propagation equation acts separately on the variables (Γ_1,Γ_1') and (Γ_2,Γ_2'). The initial condition of $\mathcal{G}^{(2)}$ is also separable. Let us then solve Eq. (2.113) by factoring the 2−body Green's function $\mathcal{G}^{(2)}$ as the product of two 1−body Green's function $\mathcal{G}^{(1)}$, so that

$$\mathcal{G}^{(2)}(\Gamma_1,\Gamma_2,\Gamma_1',\Gamma_2',\tau)=\mathcal{G}^{(1)}(\Gamma_1,\Gamma_1',\tau)\,\mathcal{G}^{(1)}(\Gamma_2,\Gamma_2',\tau)\,, \tag{2.115}$$

where the 1−body Green's function $\mathcal{G}^{(1)}$ satisfies the linearised 1−body Vlasov equation, namely

$$\frac{\partial \mathcal{G}^{(1)}(\Gamma_1,\Gamma_1',\tau)}{\partial \tau}+\bigg[\boldsymbol{v}_1\cdot\frac{\partial}{\partial \boldsymbol{x}_1}+\bigg[\int \mathrm{d}\Gamma_2\, \boldsymbol{\mathcal{F}}_{12}\,F(\Gamma_2)\bigg]\cdot\frac{\partial}{\partial \boldsymbol{v}_1}\bigg]\mathcal{G}^{(1)}(\Gamma_1,\Gamma_1',\tau)$$
$$+\bigg[\int \mathrm{d}\Gamma_2\, \mathcal{G}^{(1)}(\Gamma_2,\Gamma_1',\tau)\boldsymbol{\mathcal{F}}_{12}\bigg]\cdot\frac{\partial F}{\partial \boldsymbol{v}_1}=0\,, \tag{2.116}$$

with the initial condition $\mathcal{G}^{(1)}(\Gamma_1,\Gamma_1',0)=\delta_{\mathrm{D}}(\Gamma_1-\Gamma_1')$.[5] Because of the causality requirement, one needs to solve Eq. (2.116) only for $\tau \geq 0$. To do so, we rely on Bogoliubov's ansatz, which assumes that the system's 1−body DF F only evolves on a slow secular timescale, while the fluctuations and correlations evolve on a fast dynamical timescale. As a consequence, in Eq. (2.116), which describes the evolution of fluctuations, one may assume F to be frozen. Because of this decoupling, the correlations at a given time t can be seen as functionals of F evaluated at the very same time. To solve Eq. (2.116), let us perform a Laplace transform following the conventions from Eq. (2.44). One gets

---

[5] Heyvaerts (2010) interestingly notes that if one was to account for contributions associated with strong collisions, such as in the first term of the second line of Eq. (2.105), the previous property of separability would not hold anymore.

$$- \mathrm{i}\omega \widetilde{\mathcal{G}}^{(1)}(\Gamma_1, \Gamma_1', \omega) + \left[\boldsymbol{v}_1 \cdot \frac{\partial}{\partial \boldsymbol{x}_1} + \left[\int \mathrm{d}\Gamma_2 \, \boldsymbol{\mathcal{F}}_{12} \, F(\Gamma_2) \right] \cdot \frac{\partial}{\partial \boldsymbol{v}_1} \right] \widetilde{\mathcal{G}}^{(1)}(\Gamma_1, \Gamma_1', \omega)$$
$$+ \left[\int \mathrm{d}\Gamma_2 \, \widetilde{\mathcal{G}}^{(1)}(\Gamma_2, \Gamma_1', \omega) \boldsymbol{\mathcal{F}}_{12} \right] \cdot \frac{\partial F}{\partial \boldsymbol{v}_1} = \delta_{\mathrm{D}}(\Gamma_1 - \Gamma_1') \, . \tag{2.117}$$

B.2 Application to Inhomogeneous Systems

Let us now assume that the system's mean potential is integrable, so that the physical phase space coordinates $(\boldsymbol{x}, \boldsymbol{v})$ may be remapped to angle-action ones $(\boldsymbol{\theta}, \boldsymbol{J})$. Such a mapping allows for a simple description of the intricate trajectories of individual particles. This change of coordinates is canonical and the infinitesimal volumes are conserved, i.e. $\mathrm{d}\Gamma = \mathrm{d}\boldsymbol{x}\mathrm{d}\boldsymbol{v} = \mathrm{d}\boldsymbol{\theta}\mathrm{d}\boldsymbol{J}$. Thanks to the adiabatic approximation (Heyvaerts 2010; Chavanis 2012b, 2013b), let us also assume that the system's 1−body DF is a quasi-stationary solution of the collisionless dynamics, so that $F(\boldsymbol{\theta}, \boldsymbol{J}, t) = F(\boldsymbol{J}, t)$. The angle-action coordinates satisfy two important additional properties. First, the derivatives along the mean motion take the simple form

$$\boldsymbol{v}_1 \cdot \frac{\partial}{\partial \boldsymbol{x}_1} + \left[\int \mathrm{d}\Gamma_2 \, \boldsymbol{\mathcal{F}}_{12} \, F(\Gamma_2) \right] \cdot \frac{\partial}{\partial \boldsymbol{v}_1} = \boldsymbol{\Omega}_1 \cdot \frac{\partial}{\partial \boldsymbol{\theta}_1} \, , \tag{2.118}$$

where $\boldsymbol{\Omega}_1$ are the intrinsic frequencies of motion associated with the mean potential. Secondly, the Poisson brackets are invariant under the change of coordinates $(\boldsymbol{x}, \boldsymbol{v}) \mapsto (\boldsymbol{\theta}, \boldsymbol{J})$, so that for any functions $L_1(\boldsymbol{x}, \boldsymbol{v})$ and $L_2(\boldsymbol{x}, \boldsymbol{v})$, one has

$$\frac{\partial L_1}{\partial \boldsymbol{x}} \cdot \frac{\partial L_2}{\partial \boldsymbol{v}} - \frac{\partial L_1}{\partial \boldsymbol{v}} \cdot \frac{\partial L_2}{\partial \boldsymbol{x}} = \frac{\partial L_1}{\partial \boldsymbol{\theta}} \cdot \frac{\partial L_2}{\partial \boldsymbol{J}} - \frac{\partial L_1}{\partial \boldsymbol{J}} \cdot \frac{\partial L_2}{\partial \boldsymbol{\theta}} \, . \tag{2.119}$$

With these transformations, Eq. (2.117) becomes

$$-\mathrm{i}\omega \widetilde{\mathcal{G}}^{(1)}(\Gamma_1, \Gamma_1', \omega) + \boldsymbol{\Omega}_1 \cdot \frac{\partial \widetilde{\mathcal{G}}^{(1)}(\Gamma_1, \Gamma_1', \omega)}{\partial \boldsymbol{\theta}_1} - \int \mathrm{d}\Gamma_2 \, \widetilde{\mathcal{G}}^{(1)}(\Gamma_2, \Gamma_1', \omega) \frac{\partial U_{12}}{\partial \boldsymbol{\theta}_1} \cdot \frac{\partial F}{\partial \boldsymbol{J}_1} = \delta_{\mathrm{D}}(\Gamma_1 - \Gamma_1') \, . \tag{2.120}$$

Following the convention from Eq. (2.6), let us multiply Eq. (2.117) by $1/(2\pi)^d$ $\mathrm{e}^{-\mathrm{i}\boldsymbol{m}_1 \cdot \boldsymbol{\theta}_1}$ and integrate it w.r.t. $\boldsymbol{\theta}_1$ to get

$$- \mathrm{i}\omega \, \widetilde{\mathcal{G}}^{(1)}_{\boldsymbol{m}_1}(\boldsymbol{J}_1, \Gamma_1', \omega) + \mathrm{i}\boldsymbol{m}_1 \cdot \boldsymbol{\Omega}_1 \widetilde{\mathcal{G}}^{(1)}_{\boldsymbol{m}_1}(\boldsymbol{J}_1, \Gamma_1', \omega)$$
$$- (2\pi)^d \, \mathrm{i}\boldsymbol{m}_1 \cdot \frac{\partial F}{\partial \boldsymbol{J}_1} \sum_{\boldsymbol{m}_2} \int \mathrm{d}\boldsymbol{J}_2 \, \widetilde{\mathcal{G}}^{(1)}_{\boldsymbol{m}_2}(\boldsymbol{J}_2, \Gamma_1', \omega) \, A_{\boldsymbol{m}_1, \boldsymbol{m}_2}(\boldsymbol{J}_1, \boldsymbol{J}_2) = \frac{\mathrm{e}^{-\mathrm{i}\boldsymbol{m}_1 \cdot \boldsymbol{\theta}_1'}}{(2\pi)^d} \delta_{\mathrm{D}}(\boldsymbol{J}_1 - \boldsymbol{J}_1') \, , \tag{2.121}$$

where the bare susceptibility coefficients $A_{\boldsymbol{m}_1, \boldsymbol{m}_2}(\boldsymbol{J}_1, \boldsymbol{J}_2)$ were introduced in Eq. (2.74). Equation (2.121) can easily be rewritten as

$$\widetilde{\mathcal{G}}^{(1)}_{\boldsymbol{m}_1}(\boldsymbol{J}_1,\Gamma'_1,\omega)+(2\pi)^d \frac{\boldsymbol{m}_1\cdot\partial F/\partial \boldsymbol{J}_1}{\omega-\boldsymbol{m}_1\cdot\boldsymbol{\Omega}_1}\sum_{\boldsymbol{m}_2}\!\int\!\mathrm{d}\boldsymbol{J}_2\,\widetilde{\mathcal{G}}^{(1)}_{\boldsymbol{m}_2}(\boldsymbol{J}_2,\Gamma'_1,\omega)\,A_{\boldsymbol{m}_1,\boldsymbol{m}_2}(\boldsymbol{J}_1,\boldsymbol{J}_2)$$
$$=\frac{\mathrm{i}}{(2\pi)^d}\frac{\mathrm{e}^{-\mathrm{i}\boldsymbol{m}_1\cdot\boldsymbol{\theta}'_1}}{\omega-\boldsymbol{m}_1\cdot\boldsymbol{\Omega}_1}\,\delta_{\mathrm{D}}(\boldsymbol{J}_1-\boldsymbol{J}'_1)\,. \tag{2.122}$$

At this stage, let us note that Eq. (2.122) takes the form of a Fredholm equation, as the Green's function appears twice on the l.h.s, in particular once as an integral term. The method to solve such an equation is to rely on Kalnajs matrix method (Kalnajs 1976). Let us therefore introduce a basis of potential and densities $(\psi^{(p)},\rho^{(p)})$ as in Eq. (2.12), thanks to which the potential perturbations may be decomposed. Let us first develop the interaction potential U on these elements. We consider the function $\boldsymbol{x}_1\mapsto U(|\boldsymbol{x}_1-\boldsymbol{x}_2|)$ and decompose it on the basis elements $\psi^{(p)}(\boldsymbol{x}_1)$. This takes the form $U(|\boldsymbol{x}_1-\boldsymbol{x}_2|)=\sum_p u_p(\boldsymbol{x}_2)\,\psi^{(p)}(\boldsymbol{x}_1)$, where the coefficients $u_p(\boldsymbol{x}_2)$ are given by

$$u_p(\boldsymbol{x}_2)=-\!\int\!\mathrm{d}\boldsymbol{x}_1\,U(|\boldsymbol{x}_1-\boldsymbol{x}_2|)\,\rho^{(p)*}(\boldsymbol{x}_1)=-\psi^{(p)*}(\boldsymbol{x}_2)\,. \tag{2.123}$$

Because they were defined as the Fourier transform in angles of the interaction potential, the bare susceptibility coefficients from Eq. (2.74) can immediately be rewritten as

$$A_{\boldsymbol{m}_1,\boldsymbol{m}_2}(\boldsymbol{J}_1,\boldsymbol{J}_2)=-\sum_p\psi^{(p)}_{\boldsymbol{m}_1}(\boldsymbol{J}_1)\,\psi^{(p)*}_{\boldsymbol{m}_2}(\boldsymbol{J}_2)\,. \tag{2.124}$$

In order to invert the l.h.s. of Eq. (2.122), let us perform on $\widetilde{\mathcal{G}}^{(1)}_{\boldsymbol{m}_1}$ the same operations than the ones operating on $\widetilde{\mathcal{G}}^{(1)}_{\boldsymbol{m}_2}$. This amounts to multiplying Eq. (2.122) by $(2\pi)^d\sum_{\boldsymbol{m}_1}\int\mathrm{d}\boldsymbol{J}_1\,\psi^{(q)*}_{\boldsymbol{m}_1}(\boldsymbol{J}_1)$, so that it becomes

$$\begin{aligned}&\bigg[(2\pi)^d\sum_{\boldsymbol{m}_1}\!\int\!\mathrm{d}\boldsymbol{J}_1\,\psi^{(q)*}_{\boldsymbol{m}_1}(\boldsymbol{J}_1)\,\widetilde{\mathcal{G}}^{(1)}_{\boldsymbol{m}_1}(\boldsymbol{J}_1,\Gamma'_1,\omega)\bigg]\\
&-\sum_p\bigg[(2\pi)^d\sum_{\boldsymbol{m}_1}\!\int\!\mathrm{d}\boldsymbol{J}_1\,\frac{\boldsymbol{m}_1\cdot\partial F/\partial\boldsymbol{J}_1}{\omega-\boldsymbol{m}_1\cdot\boldsymbol{\Omega}_1}\psi^{(p)}_{\boldsymbol{m}_1}(\boldsymbol{J}_1)\,\psi^{(q)*}_{\boldsymbol{m}_1}(\boldsymbol{J}_1)\bigg]\bigg[(2\pi)^d\sum_{\boldsymbol{m}_2}\!\int\!\mathrm{d}\boldsymbol{J}_2\,\widetilde{\mathcal{G}}^{(1)}_{\boldsymbol{m}_2}(\boldsymbol{J}_2,\Gamma'_1,\omega)\,\psi^{(p)*}_{\boldsymbol{m}_2}(\boldsymbol{J}_2)\bigg]\\
&=\sum_{\boldsymbol{m}_1}\frac{\mathrm{i}\,\mathrm{e}^{-\mathrm{i}\boldsymbol{m}_1\cdot\boldsymbol{\theta}'_1}}{\omega-\boldsymbol{m}_1\cdot\boldsymbol{\Omega}'_1}\,\psi^{(q)*}_{\boldsymbol{m}_1}(\boldsymbol{J}'_1)\,.\end{aligned} \tag{2.125}$$

In order to clarify Eq. (2.125), let us introduce the notations

$$K_p(\Gamma'_1,\omega)=(2\pi)^d\sum_{\boldsymbol{m}}\!\int\!\mathrm{d}\boldsymbol{J}\,\widetilde{\mathcal{G}}^{(1)}_{\boldsymbol{m}}(\boldsymbol{J},\Gamma'_1,\omega)\,\psi^{(p)*}_{\boldsymbol{m}}(\boldsymbol{J})\quad;\quad L_p(\Gamma'_1,\omega)=\sum_{\boldsymbol{m}}\frac{\mathrm{i}\,\mathrm{e}^{-\mathrm{i}\boldsymbol{m}\cdot\boldsymbol{\theta}'_1}}{\omega-\boldsymbol{m}\cdot\boldsymbol{\Omega}'_1}\,\psi^{(p)*}_{\boldsymbol{m}}(\boldsymbol{J}'_1)\,. \tag{2.126}$$

Recalling also the expression of the response matrix from Eq. (2.17), we may finally rewrite Eq. (2.125) under the shortened form

$$K_p(\Gamma_1', \omega) - \sum_q \widehat{\mathbf{M}}_{pq}(\omega)\, K_q(\Gamma_1', \omega) = L_p(\Gamma_1', \omega)\,. \tag{2.127}$$

Assuming that the system considered is dynamically stable, so that $[\mathbf{I}-\widehat{\mathbf{M}}(\omega)]$ can be inverted, Eq. (2.127) finally leads to

$$K_p(\Gamma_1', \omega) = \sum_q \big[\mathbf{I}-\widehat{\mathbf{M}}(\omega)\big]^{-1}_{pq}\, L_q(\Gamma_1', \omega)\,. \tag{2.128}$$

Thanks to Eq. (2.128), one can finally rewrite Eq. (2.122) as

$$\widetilde{\mathcal{G}}^{(1)}_{\boldsymbol{m}_1}(\boldsymbol{J}_1, \Gamma_1', \omega) = \frac{1}{(2\pi)^d} \frac{\mathrm{i}\, \mathrm{e}^{-\mathrm{i}\boldsymbol{m}_1\cdot\boldsymbol{\theta}_1'}}{\omega - \boldsymbol{m}_1\cdot\boldsymbol{\Omega}_1} \delta_{\mathrm{D}}(\boldsymbol{J}_1 - \boldsymbol{J}_1') + \frac{\boldsymbol{m}_1\cdot\partial F/\partial \boldsymbol{J}_1}{\omega - \boldsymbol{m}_1\cdot\boldsymbol{\Omega}_1} \sum_{\boldsymbol{m}_1'} \frac{1}{\mathcal{D}_{\boldsymbol{m}_1,\boldsymbol{m}_1'}(\boldsymbol{J}_1, \boldsymbol{J}_1', \omega)} \frac{\mathrm{i}\, \mathrm{e}^{-\mathrm{i}\boldsymbol{m}_1'\cdot\boldsymbol{\theta}_1'}}{\omega - \boldsymbol{m}_1'\cdot\boldsymbol{\Omega}_1'}\,, \tag{2.129}$$

where the dressed susceptibility coefficients, $1/\mathcal{D}_{\boldsymbol{m}_1,\boldsymbol{m}_1'}$, have been introduced in Eq. (2.50). Thanks to the inverse Fourier transform from Eq. (2.6), one can finally obtain the expression of $\widetilde{\mathcal{G}}^{(1)}(\Gamma_1, \Gamma_1', \omega)$ as

$$\begin{aligned}
\widetilde{\mathcal{G}}^{(1)}(\Gamma_1, \Gamma_1', \omega) &= \sum_{\boldsymbol{m}_1,\boldsymbol{m}_1'} \frac{\mathrm{i}\,\mathrm{e}^{\mathrm{i}(\boldsymbol{m}_1\cdot\boldsymbol{\theta}_1 - \boldsymbol{m}_1'\cdot\boldsymbol{\theta}_1')}}{\omega - \boldsymbol{m}_1\cdot\boldsymbol{\Omega}_1} \left[\frac{\delta_{\boldsymbol{m}_1}^{\boldsymbol{m}_1'}}{(2\pi)^d} \delta_{\mathrm{D}}(\boldsymbol{J}_1 - \boldsymbol{J}_1') + \frac{\boldsymbol{m}_1\cdot\partial F/\partial \boldsymbol{J}_1}{(\omega - \boldsymbol{m}_1'\cdot\boldsymbol{\Omega}_1')\, \mathcal{D}_{\boldsymbol{m}_1,\boldsymbol{m}_1'}(\boldsymbol{J}_1, \boldsymbol{J}_1', \omega)} \right] \\
&= \sum_{\boldsymbol{m}_1,\boldsymbol{m}_1'} \widetilde{\mathcal{G}}^{(1)}_{\boldsymbol{m}_1,\boldsymbol{m}_1'}(\boldsymbol{J}_1, \boldsymbol{J}_1', \omega)\, \mathrm{e}^{\mathrm{i}(\boldsymbol{m}_1\cdot\boldsymbol{\theta}_1 - \boldsymbol{m}_1'\cdot\boldsymbol{\theta}_1')}\,.
\end{aligned} \tag{2.130}$$

B.3 Rewriting the Collision Operator

Thanks to the explicit expression of the 1−body Green's function from Eq. (2.130), one may now proceed to the evaluation of the collision operator from Eq. (2.114). Let us first rely on Bogoliubov's ansatz, so that in Eq. (2.112) we may do the replacement $S_2(\Gamma_1', \Gamma_2', t-\tau) \to S_2(\Gamma_1', \Gamma_2', t)$. Relying on the factorisation of the Green's function from Eq. (2.115) and the inverse Laplace transform from Eq. (2.44), the collision operator takes the form

$$\begin{aligned}
\mathscr{C}[F] = &\int_0^{+\infty} \!\mathrm{d}\tau \int \mathrm{d}\Gamma_2 \mathrm{d}\Gamma_1' \mathrm{d}\Gamma_2' \int_{\mathcal{B}} \frac{\mathrm{d}\omega}{2\pi} \int_{\mathcal{B}'} \frac{\mathrm{d}\omega'}{2\pi}\, \mathrm{e}^{-\mathrm{i}(\omega+\omega')\tau} \\
&\times \mu\, \boldsymbol{\mathcal{F}}_{12}\cdot\frac{\partial}{\partial \boldsymbol{v}_1} \bigg\{ \widetilde{\mathcal{G}}^{(1)}(\Gamma_1, \Gamma_1', \omega)\, \widetilde{\mathcal{G}}^{(1)}(\Gamma_2, \Gamma_2', \omega') \bigg[\boldsymbol{\mathcal{F}}_{1'2'}\cdot \bigg[\frac{\partial}{\partial \boldsymbol{v}_1'} - \frac{\partial}{\partial \boldsymbol{v}_2'} \bigg] F(\Gamma_1')\, F(\Gamma_2') \bigg] \bigg\}\,,
\end{aligned} \tag{2.131}$$

where the Laplace transformed 1−body Green's function was obtained in Eq. (2.130). The rest of the calculations is now to rewrite Eq. (2.131) under a simpler form. Let us first rely on the properties from Eq. (2.119) to rewrite the various terms appearing in Eq. (2.131) in angle-action space. One has

$$
\begin{aligned}
\mathcal{F}_{12}\cdot\frac{\partial \widetilde{\mathcal{G}}^{(1)}(\Gamma_1)}{\partial \boldsymbol{v}_1} &= -\frac{\partial U_{12}}{\partial \boldsymbol{\theta}_1}\cdot\frac{\partial \widetilde{\mathcal{G}}^{(1)}(\Gamma_1)}{\partial \boldsymbol{J}_1} + \frac{\partial U_{12}}{\partial \boldsymbol{J}_1}\cdot\frac{\partial \widetilde{\mathcal{G}}^{(1)}(\Gamma_1)}{\partial \boldsymbol{\theta}_1} \\
&= -\frac{\partial}{\partial \boldsymbol{J}_1}\cdot\bigg[\int \frac{\mathrm{d}\boldsymbol{\theta}_1}{(2\pi)^d}\,\frac{\partial U_{12}}{\partial \boldsymbol{\theta}_1}\,\widetilde{\mathcal{G}}^{(1)}(\Gamma_1)\bigg],
\end{aligned}
\tag{2.132}
$$

where we used the shortened notation $\widetilde{\mathcal{G}}^{(1)}(\Gamma_1)=\widetilde{\mathcal{G}}^{(1)}(\Gamma_1,\Gamma_1',\omega)$. To obtain the second line of Eq. (2.132), we relied on Schwarz's theorem. We also relied on the fact that during the secular diffusion, the 1−body DF is of the form $F=F(\boldsymbol{J}_1,t)$, allowing us to perform an angle average w.r.t. $\boldsymbol{\theta}_1$. Similarly, one can write

$$
\mathcal{F}_{1'2'}\cdot\bigg[\frac{\partial}{\boldsymbol{v}_1'}-\frac{\partial}{\boldsymbol{v}_2'}\bigg]F(\Gamma_1')F(\Gamma_2') = -\bigg[\frac{\partial U_{1'2'}}{\partial \boldsymbol{\theta}_1'}\cdot\frac{\partial F}{\partial \boldsymbol{J}_1'}\,F(\boldsymbol{J}_2')+\frac{\partial U_{2'1'}}{\partial \boldsymbol{\theta}_2'}\cdot\frac{\partial F}{\partial \boldsymbol{J}_2'}\,F(\boldsymbol{J}_1')\bigg].
\tag{2.133}
$$

Let us now use the two previous rewritings, as well as Eq. (2.130), to rewrite the collision operator from Eq. (2.131) in angle-action space. After integrating w.r.t. $\boldsymbol{\theta}_1$, $\boldsymbol{\theta}_2$, $\boldsymbol{\theta}_1'$, and $\boldsymbol{\theta}_2'$, it reads

$$
\begin{aligned}
\mathscr{C}[F] = &\int_0^{+\infty}\!\mathrm{d}\tau\!\int\!\mathrm{d}\boldsymbol{J}_2\mathrm{d}\boldsymbol{J}_1'\mathrm{d}\boldsymbol{J}_2'\int_{\mathcal{B}}\frac{\mathrm{d}\omega}{2\pi}\int_{\mathcal{B}'}\frac{\mathrm{d}\omega'}{2\pi}\,\mathrm{e}^{-\mathrm{i}(\omega+\omega')\tau}\,\mu(2\pi)^{3d} \\
&\times\frac{\partial}{\partial \boldsymbol{J}_1}\cdot\bigg[\sum_{\boldsymbol{m}_1,\boldsymbol{m}_2}\sum_{\boldsymbol{m}_1',\boldsymbol{m}_2'}\widetilde{\mathcal{G}}^{(1)}_{\boldsymbol{m}_1,\boldsymbol{m}_1'}(\omega)\,\widetilde{\mathcal{G}}^{(1)}_{\boldsymbol{m}_2,\boldsymbol{m}_2'}(\omega')\,\boldsymbol{m}_1 A_{-\boldsymbol{m}_1,\boldsymbol{m}_2} \\
&\times\bigg[A_{\boldsymbol{m}_1',-\boldsymbol{m}_2'}\,\boldsymbol{m}_1'\cdot\frac{\partial F}{\partial \boldsymbol{J}_1'}\,F(\boldsymbol{J}_2') + A_{\boldsymbol{m}_2',-\boldsymbol{m}_1'}\,\boldsymbol{m}_2'\cdot\frac{\partial F}{\partial \boldsymbol{J}_2'}\,F(\boldsymbol{J}_1')\bigg]\bigg],
\end{aligned}
\tag{2.134}
$$

where we used the shortened notations $\widetilde{\mathcal{G}}^{(1)}_{\boldsymbol{m}_1,\boldsymbol{m}_1'}(\omega)=\widetilde{\mathcal{G}}^{(1)}_{\boldsymbol{m}_1,\boldsymbol{m}_1'}(\boldsymbol{J}_1,\boldsymbol{J}_1',\omega)$ and $A_{\boldsymbol{m}_1,\boldsymbol{m}_2}=A_{\boldsymbol{m}_1,\boldsymbol{m}_2}(\boldsymbol{J}_1,\boldsymbol{J}_2)$. Let us now use the explicit expression of the Fourier coefficients of the 1−body Green's function from Eq. (2.130). Equation (2.134) becomes

$$
\begin{aligned}
\mathscr{C}[F] = -&\int_0^{+\infty}\!\mathrm{d}\tau\!\int\!\mathrm{d}\boldsymbol{J}_2\mathrm{d}\boldsymbol{J}_1'\mathrm{d}\boldsymbol{J}_2'\int_{\mathcal{B}}\frac{\mathrm{d}\omega}{2\pi}\int_{\mathcal{B}'}\frac{\mathrm{d}\omega'}{2\pi}\,\mathrm{e}^{-\mathrm{i}(\omega+\omega')\tau}\,\mu(2\pi)^{d} \\
&\times\frac{\partial}{\partial \boldsymbol{J}_1}\cdot\bigg[\sum_{\boldsymbol{m}_1,\boldsymbol{m}_2}\sum_{\boldsymbol{m}_1',\boldsymbol{m}_2'}\frac{1}{\omega-\omega_1}\,\frac{1}{\omega'-\omega_2}\,\boldsymbol{m}_1 A_{-\boldsymbol{m}_1,\boldsymbol{m}_2} \\
&\times\bigg[\delta_{\boldsymbol{m}_1}^{\boldsymbol{m}_1'}\,\delta_{\mathrm{D}}(\boldsymbol{J}_1-\boldsymbol{J}_1')+(2\pi)^d\frac{\boldsymbol{m}_1\cdot\partial F/\partial \boldsymbol{J}_1}{(\omega-\omega_1')\mathcal{D}_{\boldsymbol{m}_1,\boldsymbol{m}_1'}(\omega)}\bigg]\bigg[\delta_{\boldsymbol{m}_2}^{\boldsymbol{m}_2'}\,\delta_{\mathrm{D}}(\boldsymbol{J}_2-\boldsymbol{J}_2')+(2\pi)^d\frac{\boldsymbol{m}_2\cdot\partial F/\partial \boldsymbol{J}_2}{(\omega'-\omega_2')\mathcal{D}_{\boldsymbol{m}_2,\boldsymbol{m}_2'}(\omega')}\bigg] \\
&\times\bigg[A_{\boldsymbol{m}_1',-\boldsymbol{m}_2'}\,\boldsymbol{m}_1'\cdot\frac{\partial F}{\partial \boldsymbol{J}_1'}\,F(\boldsymbol{J}_2') + A_{\boldsymbol{m}_2',-\boldsymbol{m}_1'}\,\boldsymbol{m}_2'\cdot\frac{\partial F}{\partial \boldsymbol{J}_2'}\,F(\boldsymbol{J}_1')\bigg]\bigg],
\end{aligned}
\tag{2.135}
$$

where we used the shortened notations $1/\mathcal{D}_{\boldsymbol{m}_1,\boldsymbol{m}_1'}(\omega)=1/\mathcal{D}_{\boldsymbol{m}_1,\boldsymbol{m}_1'}(\boldsymbol{J}_1,\boldsymbol{J}_1',\omega)$, as well as $\omega_{1/2}=\boldsymbol{m}_{1/2}\cdot\boldsymbol{\Omega}_{1/2}$ and $\omega_{1/2}'=\boldsymbol{m}_{1/2}'\cdot\boldsymbol{\Omega}_{1/2}'$. The next step of the calculation is to deal with the integration and sum w.r.t. $\boldsymbol{J}_2$ and $\boldsymbol{m}_2$. One can write

$$\sum_{m_2}\int \mathrm{d}J_2 \,\frac{A_{-m_1,m_2}}{\omega'-\omega_2}\left[\delta_{m_2}^{m_2'}\,\delta_{\mathrm{D}}(J_2-J_2')+(2\pi)^d\frac{m_2\cdot\partial F/\partial J_2}{(\omega'-\omega_2')\,\mathcal{D}_{m_2,m_2'}(\omega')}\right]=-\frac{1}{\omega'-\omega_2'}\frac{1}{\mathcal{D}_{-m_1,m_2'}(\omega')}\,,\tag{2.136}$$

where we relied on the intrinsic definition of the dressed susceptibility coefficients $1/\mathcal{D}_{m_1,m_2}$ given by

$$\frac{1}{\mathcal{D}_{m_1,m_2}(J_1,J_2,\omega)}=-A_{m_1,m_2}(J_1,J_2)-(2\pi)^d\sum_{m_3}\int\mathrm{d}J_3\,\frac{m_3\cdot\partial F/\partial J_3}{\omega-m_3\cdot\Omega_3}\frac{A_{m_1,m_3}(J_1,J_3)}{\mathcal{D}_{m_3,m_2}(J_3,J_1,\omega)}\,.\tag{2.137}$$

Equation (2.137) is straightforward to obtain thanks to the basis decompositions of the susceptibility coefficients from Eqs. (2.50) and (2.124), and the definition of the response matrix from Eq. (2.17). Equation (2.135) becomes

$$\begin{aligned}
\mathscr{C}[F]=&\int_0^{+\infty}\mathrm{d}\tau\int\mathrm{d}J_1'\mathrm{d}J_2'\int_{\mathcal{B}}\frac{\mathrm{d}\omega}{2\pi}\int_{\mathcal{B}'}\frac{\mathrm{d}\omega'}{2\pi}\,\mathrm{e}^{-\mathrm{i}(\omega+\omega')\tau}\mu(2\pi)^d\\
&\times\frac{\partial}{\partial J_1}\cdot\Bigg[\sum_{m_1}\sum_{m_1',m_2'}\frac{1}{\omega-\omega_1}\frac{1}{\omega'-\omega_2'}m_1\frac{1}{\mathcal{D}_{-m_1,m_2'}(\omega')}\\
&\times\left[\delta_{m_1}^{m_1'}\,\delta_{\mathrm{D}}(J_1-J_1')+(2\pi)^d\frac{m_1\cdot\partial F/\partial J_1}{(\omega-\omega_1')\mathcal{D}_{m_1,m_1'}(\omega)}\right]\\
&\times\left[A_{m_1',-m_2'}\,m_1'\cdot\frac{\partial F}{\partial J_1'}\,F(J_2')+A_{m_2',-m_1'}\,m_2'\cdot\frac{\partial F}{\partial J_2'}\,F(J_1')\right]\Bigg].
\end{aligned}\tag{2.138}$$

The next step of the calculation is to perform the integration and the sum w.r.t. J_1' and m_1'. These only act on the two last lines of Eq. (2.138). As previously, to perform this calculation, we rely on the intrinsic definition of the dressed susceptibility coefficients from Eq. (2.137). One has to deal with two distinct contributions: the first one $\mathscr{C}_1[F]$ associated with the term in $m_1'\cdot\partial F/\partial J_1'\,F(J_2)$ and the second one $\mathscr{C}_2[F]$ associated with the term in $m_2'\cdot\partial F/\partial J_2'F(J_1')$. The first contribution $\mathscr{C}_1[F]$ takes the form

$$\begin{aligned}
\mathscr{C}_1[F]&=\sum_{m_1'}\int\mathrm{d}J_1'\left[\delta_{m_1}^{m_1'}\,\delta_{\mathrm{D}}(J_1-J_1')+(2\pi)^d\frac{m_1\cdot\partial F/\partial J_1}{(\omega-\omega_1')\mathcal{D}_{m_1,m_1'}(\omega)}\right]A_{m_1',-m_2'}\,m_1'\cdot\frac{\partial F}{\partial J_1'}\,F(J_2')\\
&=-\frac{1}{\mathcal{D}_{m_1,-m_2'}(\omega)}\,m_1\cdot\frac{\partial F}{\partial J_1}\,F(J_2')\,.
\end{aligned}\tag{2.139}$$

The second contribution $\mathscr{C}_2[F]$ takes the form

$$\begin{aligned}
\mathscr{C}_2[F]&=\sum_{m_1'}\int\mathrm{d}J_1'\left[\delta_{m_1}^{m_1'}\,\delta_{\mathrm{D}}(J_1-J_1')+(2\pi)^d\frac{m_1\cdot\partial F/\partial J_1}{(\omega-\omega_1')\mathcal{D}_{m_1,m_1'}(\omega)}\right]A_{m_2',-m_1'}\,m_2'\cdot\frac{\partial F}{\partial J_2'}\,F(J_1')\\
&=A_{m_2',-m_1}m_2'\cdot\frac{\partial F}{\partial J_2'}F(J_1)+m_1\cdot\frac{\partial F}{\partial J_1}\,m_2'\cdot\frac{\partial F}{\partial J_2'}(2\pi)^d\sum_{m_1'}\int\mathrm{d}J_1'\,\frac{F(J_1')\,A_{m_2',-m_1'}}{(\omega-\omega_1')\mathcal{D}_{m_1,m_1'}(\omega)}\,.
\end{aligned}\tag{2.140}$$

Let us now rewrite Eq. (2.138) by relying on the matrix method, i.e. by using the basis elements $\psi^{(p)}$. The bare and dressed susceptibility coefficients take the form

$$A_{\boldsymbol{m}_1,\boldsymbol{m}_2}(\boldsymbol{J}_1,\boldsymbol{J}_2) = -\psi^{(\alpha)}_{\boldsymbol{m}_1}(\boldsymbol{J}_1)\,\psi^{(\alpha)*}_{\boldsymbol{m}_2}(\boldsymbol{J}_2) \;;\quad \frac{1}{\mathcal{D}_{\boldsymbol{m}_1,\boldsymbol{m}_2}(\boldsymbol{J}_1,\boldsymbol{J}_2,\omega)} = \psi^{(\alpha)}_{\boldsymbol{m}_1}(\boldsymbol{J}_1)\,\varepsilon^{-1}_{\alpha\beta}(\omega)\,\psi^{(\beta)*}_{\boldsymbol{m}_2}(\boldsymbol{J}_2)\,, \tag{2.141}$$

where we introduced the matrix $\varepsilon(\omega) = \mathbf{I} - \widehat{\mathbf{M}}(\omega)$, with $\widehat{\mathbf{M}}(\omega)$ the response matrix from Eq. (2.17) and $\mathbf{I}$ the identity matrix. In Eq. (2.141) and the following, all the sums over the greek indices are implied. Let us finally define the matrix $\mathbf{H}(\omega)$ as

$$H_{\alpha\beta}(\omega) = (2\pi)^d \sum_{\boldsymbol{m}} \int \mathrm{d}\boldsymbol{J}\, \frac{F(\boldsymbol{J})}{\omega - \boldsymbol{m}\cdot\boldsymbol{\Omega}}\, \psi^{(\alpha)*}_{\boldsymbol{m}}(\boldsymbol{J})\, \psi^{(\beta)*}_{-\boldsymbol{m}}(\boldsymbol{J})\,. \tag{2.142}$$

Gathering the two contributions from Eqs. (2.139) and (2.140), and after some straightforward calculations, one can rewrite Eq. (2.138) as

$$\begin{aligned}
\mathscr{C}[F] = -\int_0^{+\infty}\!\mathrm{d}\tau \int_{\mathcal{B}} \frac{\mathrm{d}\omega}{2\pi} \int_{\mathcal{B}'} \frac{\mathrm{d}\omega'}{2\pi}\, \mathrm{e}^{-\mathrm{i}(\omega+\omega')\tau}\, \mu \frac{\partial}{\partial \boldsymbol{J}_1}\cdot \Bigg[\sum_{\boldsymbol{m}_1} \frac{1}{\omega - \omega_1}\, \boldsymbol{m}_1 \\
\times \Bigg\{ \psi^{(\alpha)}_{-\boldsymbol{m}_1}(\boldsymbol{J}_1)\, \varepsilon^{-1}_{\alpha\beta}(\omega')\, H_{\beta\delta}(\omega')\, \varepsilon^{-1}_{\gamma\delta}(\omega)\, \psi^{(\gamma)}_{\boldsymbol{m}_1}(\boldsymbol{J}_1)\, \boldsymbol{m}_1 \cdot \frac{\partial F}{\partial \boldsymbol{J}_1} \\
+ \psi^{(\alpha)}_{-\boldsymbol{m}_1}(\boldsymbol{J}_1) \left[\varepsilon^{-1}_{\alpha\gamma}(\omega') - \delta_{\alpha\gamma} \right] \psi^{(\gamma)*}_{-\boldsymbol{m}_1}(\boldsymbol{J}_1)\, F(\boldsymbol{J}_1) \\
+ \psi^{(\alpha)}_{-\boldsymbol{m}_1}(\boldsymbol{J}_1)\, \varepsilon^{-1}_{\alpha\gamma}(\omega')\, \varepsilon^{-1}_{\delta\lambda}(\omega)\, H_{\lambda\gamma}(\omega)\, \psi^{(\delta)}_{\boldsymbol{m}_1}(\boldsymbol{J}_1)\, \boldsymbol{m}_1 \cdot \frac{\partial F}{\partial \boldsymbol{J}_1} \\
- \psi^{(\alpha)}_{-\boldsymbol{m}_1}(\boldsymbol{J}_1)\, \varepsilon^{-1}_{\delta\lambda}(\omega)\, H_{\lambda\alpha}(\omega)\, \psi^{(\delta)}_{\boldsymbol{m}_1}(\boldsymbol{J}_1)\, \boldsymbol{m}_1 \cdot \frac{\partial F}{\partial \boldsymbol{J}_1} \Bigg\} \Bigg].
\end{aligned} \tag{2.143}$$

At this stage, let us perform the integration w.r.t. τ and ω' in Eq. (2.143). It generically takes the form

$$\int_0^{+\infty}\!\mathrm{d}\tau \int_{\mathcal{B}'} \frac{\mathrm{d}\omega'}{2\pi}\, \mathrm{e}^{-\mathrm{i}(\omega+\omega')\tau}\, g(\omega,\omega')\,. \tag{2.144}$$

The integration over τ is straightforward provided that $\omega+\omega'$ has a negative imaginary part. Introducing $p>0$, we perform the substitution $\omega+\omega' \to \omega+\omega'-\mathrm{i}p$, and evaluate the integration over τ as

$$(2.144) = \lim_{p\to 0} \int_{\mathcal{B}'} \frac{\mathrm{d}\omega'}{2\pi}\, \frac{-\mathrm{i}}{\omega+\omega'-\mathrm{i}p}\, g(\omega,\omega')\,. \tag{2.145}$$

As the system is supposed to be stable, the poles of the function $\omega' \mapsto g(\omega,\omega')$ are all in the lower-half complex plane, and the Bromwich contour $\mathcal{B}'$ has to pass above all these singularities. The only pole in ω' which remains is then $\omega' = -\omega + \mathrm{i}p$, which is in the upper plane. The integration on ω' is then carried out thanks to the residue theorem by closing the contour $\mathcal{B}'$ in the upper half complex plane – this is possible because the integrand decreases sufficiently fast at infinity like $1/|\omega'|^2$. One gets

$$(2.144) = \lim_{p\to 0} g(\omega, -\omega+\mathrm{i}p) \,. \tag{2.146}$$

Let us now consider the integration w.r.t. ω in Eq. (2.143). First, one can note that the fourth term of Eq. (2.143) vanishes when integrated upon ω. Indeed, by construction the Bromwich contour $\mathcal{B}$ has to pass above all the singularities of the functions of $+\omega$. The contour $\mathcal{B}$ can then be closed in the upper half complex plane and, because it surrounds no singularities, gives a vanishing result for this term. Equation (2.143), when rearranged, becomes

$$\begin{aligned}\mathscr{C}[F] = \lim_{p\to 0} -\!\int_{\mathcal{B}} \frac{\mathrm{d}\omega}{2\pi}\, \mu \frac{\partial}{\partial \boldsymbol{J}_1}\cdot\bigg[\sum_{\boldsymbol{m}_1} \frac{1}{\omega-\omega_1}\boldsymbol{m}_1 \\ \times \bigg\{\psi^{(\alpha)}_{-\boldsymbol{m}_1}(\boldsymbol{J}_1)\left[\varepsilon^{-1}_{\alpha\gamma}(-\omega+\mathrm{i}p)-\delta_{\alpha\gamma}\right]\psi^{(\gamma)*}_{-\boldsymbol{m}_1}(\boldsymbol{J}_1)\, F(\boldsymbol{J}_1) \\ +\psi^{(\alpha)}_{-\boldsymbol{m}_1}(\boldsymbol{J}_1)\, \varepsilon^{-1}_{\alpha\beta}(-\omega+\mathrm{i}p)\, \varepsilon^{-1}_{\gamma\delta}(\omega)\, \psi^{(\gamma)}_{\boldsymbol{m}_1}(\boldsymbol{J}_1)\left[H_{\beta\delta}(-\omega+\mathrm{i}p)+H_{\delta\beta}(\omega)\right]\boldsymbol{m}_1\cdot\frac{\partial F}{\partial \boldsymbol{J}_1}\bigg\}\bigg] \,.\end{aligned} \tag{2.147}$$

Let us now evaluate the term within brackets in the second term of Eq. (2.147). It reads

$$\left[H_{\beta\delta}(-\omega+\mathrm{i}p)+H_{\delta\beta}(\omega)\right] = (2\pi)^d \sum_{\boldsymbol{m}_2}\int \mathrm{d}\boldsymbol{J}_2\, \psi^{(\delta)*}_{\boldsymbol{m}_2}(\boldsymbol{J}_2)\, \psi^{(\beta)*}_{-\boldsymbol{m}_2}(\boldsymbol{J}_2)\, F(\boldsymbol{J}_2)\left[\frac{1}{\omega-\omega_2}-\frac{1}{\omega-(\omega_2+\mathrm{i}p)}\right], \tag{2.148}$$

where we used the notation $\omega_2 = \boldsymbol{m}_2\cdot\boldsymbol{\Omega}(\boldsymbol{J}_2)$. As one takes the limit $p\to 0$, a naive reading of Eq. (2.148) would indicate that Eq. (2.148) vanishes. However, one should be careful with the two poles $\omega=\omega_2$ and $\omega=\omega_2+\mathrm{i}p$, as these two poles are on opposite sides of the prescribed integration contour $\mathcal{B}$. Indeed, when lowering the integration $\mathcal{B}$ to the real axis, the pole $\omega=\omega_2$ remains below the contour, while the one in $\omega=\omega_2+\mathrm{i}p$ is above it. Relying on Plemelj formula from Eq. (2.29), one can rewrite Eq. (2.148) as

$$\begin{aligned}\left[H_{\beta\delta}(-\omega+\mathrm{i}p)+H_{\delta\beta}(\omega)\right] &= (2\pi)^d \sum_{\boldsymbol{m}_2}\int \mathrm{d}\boldsymbol{J}_2\, \psi^{(\delta)*}_{\boldsymbol{m}_2}(\boldsymbol{J}_2)\, \psi^{(\beta)*}_{-\boldsymbol{m}_2}(\boldsymbol{J}_2)\, F(\boldsymbol{J}_2)\left[\frac{1}{\omega-\omega_2+\mathrm{i}0}-\frac{1}{\omega-\omega_2-\mathrm{i}0}\right] \\ &= -2\pi\mathrm{i}(2\pi)^d \sum_{\boldsymbol{m}_2}\int \mathrm{d}\boldsymbol{J}_2\, \psi^{(\delta)*}_{\boldsymbol{m}_2}(\boldsymbol{J}_2)\, \psi^{(\beta)*}_{-\boldsymbol{m}_2}(\boldsymbol{J}_2)\, F(\boldsymbol{J}_2)\, \delta_{\mathrm{D}}(\omega-\omega_2) \,.\end{aligned} \tag{2.149}$$

When lowering the contour $\mathcal{B}$ to the real axis, one can also compute the integration w.r.t. ω for the first term in Eq. (2.147). Because the system is stable, the poles of $\epsilon^{-1}_{\alpha\gamma}(-\omega+\mathrm{i}p)$ are all located in the upper half plane and there remains only one pole on the real axis in $\omega=\omega_1$. The Bromwich contour $\mathcal{B}$ is then closed in the lower half plane and only encloses this second pole. Paying attention to the direction of integration, the residue gives a factor $-2\mathrm{i}\pi$, and Eq. (2.147) becomes

$$\mathscr{C}[F] = \mathrm{i}\mu \frac{\partial}{\partial \boldsymbol{J}_1} \cdot \Bigg[\sum_{\boldsymbol{m}_1} \boldsymbol{m}_1 \, \psi^{(\alpha)}_{-\boldsymbol{m}_1}(\boldsymbol{J}_1) \left[\varepsilon^{-1}_{\alpha\gamma}(-\omega_1 + \mathrm{i}0) - \delta_{\alpha\gamma} \right] \psi^{(\gamma)*}_{-\boldsymbol{m}_1}(\boldsymbol{J}_1) \, F(\boldsymbol{J}_1)$$
$$+ (2\pi)^d \sum_{\boldsymbol{m}_1, \boldsymbol{m}_2} \boldsymbol{m}_1 \int \mathrm{d}\boldsymbol{J}_2 \left[\psi^{(\alpha)}_{-\boldsymbol{m}_1}(\boldsymbol{J}_1) \, \varepsilon^{-1}_{\alpha\beta}(-\omega_2) \, \psi^{(\beta)*}_{-\boldsymbol{m}_2}(\boldsymbol{J}_2) \right]$$
$$\times \left[\psi^{(\gamma)}_{\boldsymbol{m}_1}(\boldsymbol{J}_1) \, \varepsilon^{-1}_{\gamma\delta}(\omega_2) \, \psi^{(\delta)*}_{\boldsymbol{m}_2}(\boldsymbol{J}_2) \right] \frac{\boldsymbol{m}_1 \cdot \partial F / \partial \boldsymbol{J}_1 \, F(\boldsymbol{J}_2)}{\omega_2 - \omega_1 + \mathrm{i}0} \Bigg], \tag{2.150}$$

where one should pay attention to the small positive imaginary part in the pole $1/(\omega_2 - \omega_1 + \mathrm{i}0)$ associated with the fact that the contour $\mathcal{B}$ passes above the pole $\omega = \omega_1$. Relying on the expression of the susceptibility coefficients from Eq. (2.141), one can rewrite Eq. (2.150) as

$$\mathscr{C}[F] = \mathrm{i}\mu \frac{\partial}{\partial \boldsymbol{J}_1} \cdot \Bigg[- \sum_{\boldsymbol{m}_1} \boldsymbol{m}_1 \left(\frac{1}{\mathcal{D}_{\boldsymbol{m}_1, \boldsymbol{m}_1}(\boldsymbol{J}_1, \boldsymbol{J}_1, \omega_1 + \mathrm{i}0)} + A_{\boldsymbol{m}_1, \boldsymbol{m}_1}(\boldsymbol{J}_1, \boldsymbol{J}_1) \right) F(\boldsymbol{J}_1)$$
$$+ (2\pi)^d \sum_{\boldsymbol{m}_1, \boldsymbol{m}_2} \boldsymbol{m}_1 \int \mathrm{d}\boldsymbol{J}_2 \frac{1}{\mathcal{D}_{-\boldsymbol{m}_1, -\boldsymbol{m}_2}(\boldsymbol{J}_1, \boldsymbol{J}_2, -\omega_2)} \frac{1}{\mathcal{D}_{\boldsymbol{m}_1, \boldsymbol{m}_2}(\boldsymbol{J}_1, \boldsymbol{J}_2, \omega_2)} \frac{\boldsymbol{m}_1 \cdot \partial F / \partial \boldsymbol{J}_1 \, F(\boldsymbol{J}_2)}{\omega_2 - \omega_1 + \mathrm{i}0} \Bigg], \tag{2.151}$$

where we performed the change $\boldsymbol{m}_1 \to -\boldsymbol{m}_1$ in the first term. Note that $A_{\boldsymbol{m}_1, \boldsymbol{m}_1}(\boldsymbol{J}_1, \boldsymbol{J}_1)$ is real, thanks to Eq. (2.141). Let us now rely on the fact that the collision term $\mathscr{C}[F]$ is real. As a consequence, because of the prefactor "i" in Eq. (2.150), we may restrict ourselves only to the imaginary part of the terms within brackets. The first term requires us to study

$$\mathrm{Im}\left[\frac{1}{\mathcal{D}_{\boldsymbol{m}_1, \boldsymbol{m}_1}(\boldsymbol{J}_1, \boldsymbol{J}_1, \omega_1 + \mathrm{i}0)} \right] = \frac{1}{2\mathrm{i}} \psi^{(\alpha)}_{\boldsymbol{m}_1}(\boldsymbol{J}_1) \left[\varepsilon^{-1}_{\alpha\beta}(\omega_1 + \mathrm{i}0) - \varepsilon^{-1*}_{\beta\alpha}(\omega_1 + \mathrm{i}0) \right] \psi^{(\beta)*}_{\boldsymbol{m}_1}(\boldsymbol{J}_1) \, . \tag{2.152}$$

In order to compute the term within brackets, we rely on the identity (Heyvaerts 2010)

$$\varepsilon^{-1} - (\varepsilon^{-1})^{\dagger} = \varepsilon^{-1} (\varepsilon^{\dagger} - \varepsilon) \, (\varepsilon^{\dagger})^{-1} \, . \tag{2.153}$$

The term within parenthesis in Eq. (2.153) can be evaluated and reads

$$\left[\varepsilon^{\dagger} - \varepsilon \right]_{\gamma\delta}(\omega_1 + \mathrm{i}0) = -(2\pi)^d \sum_{\boldsymbol{m}_2} \int \mathrm{d}\boldsymbol{J}_2 \, \boldsymbol{m}_2 \cdot \frac{\partial F}{\partial \boldsymbol{J}_2} \, \psi^{(\gamma)*}_{\boldsymbol{m}_2}(\boldsymbol{J}_2) \, \psi^{(\delta)}_{\boldsymbol{m}_2}(\boldsymbol{J}_2) \left[\left(\frac{1}{\omega_1 - \omega_2 + \mathrm{i}0} \right)^{*} - \frac{1}{\omega_1 - \omega_2 + \mathrm{i}0} \right]$$
$$= -2\pi\mathrm{i}(2\pi)^d \sum_{\boldsymbol{m}_2} \int \mathrm{d}\boldsymbol{J}_2 \, \delta_{\mathrm{D}}(\omega_1 - \omega_2) \, \boldsymbol{m}_2 \cdot \frac{\partial F}{\partial \boldsymbol{J}_2} \, \psi^{(\gamma)*}_{\boldsymbol{m}_2}(\boldsymbol{J}_2) \, \psi^{(\delta)}_{\boldsymbol{m}_2}(\boldsymbol{J}_2) \, . \tag{2.154}$$

Combining Eqs. (2.152) and (2.154), one finally gets the relation

$$\mathrm{Im}\left[\frac{1}{\mathcal{D}_{\boldsymbol{m}_1, \boldsymbol{m}_1}(\boldsymbol{J}_1, \boldsymbol{J}_1, \omega_1 + \mathrm{i}0)} \right] = -\pi (2\pi)^d \sum_{\boldsymbol{m}_2} \int \mathrm{d}\boldsymbol{J}_2 \frac{\delta_{\mathrm{D}}(\omega_1 - \omega_2)}{|\mathcal{D}_{\boldsymbol{m}_1, \boldsymbol{m}_2}(\boldsymbol{J}_1, \boldsymbol{J}_2, \omega_1)|^2} \, \boldsymbol{m}_2 \cdot \frac{\partial F}{\partial \boldsymbol{J}_2} \, . \tag{2.155}$$

This contribution corresponds to the drift term in the Balescu-Lenard equation. To evaluate the second term in Eq. (2.151), we rely on the relation $1/\mathcal{D}_{-m_1,-m_2}(\boldsymbol{J}_1, \boldsymbol{J}_2, -\omega) = 1/\mathcal{D}^*_{m_1,m_2}(\boldsymbol{J}_1, \boldsymbol{J}_2, \omega)$ (see note [83] in Chavanis (2012b)). Thanks to Plemelj formula, it immediately gives the contribution

$$\mathrm{Im}\bigg[(2\pi)^d \sum_{\boldsymbol{m}_1,\boldsymbol{m}_2} \boldsymbol{m}_1 \int \mathrm{d}\boldsymbol{J}_2 \, \frac{1}{\mathcal{D}_{-\boldsymbol{m}_1,-\boldsymbol{m}_2}(\boldsymbol{J}_1, \boldsymbol{J}_2, -\omega_2)} \frac{1}{\mathcal{D}_{\boldsymbol{m}_1,\boldsymbol{m}_2}(\boldsymbol{J}_1, \boldsymbol{J}_2, \omega_2)} \frac{\boldsymbol{m}_1 \cdot \partial F/\partial \boldsymbol{J}_1 \, F(\boldsymbol{J}_2)}{\omega_2 - \omega_1 + \mathrm{i}0}\bigg]$$
$$= -\pi (2\pi)^d \sum_{\boldsymbol{m}_1,\boldsymbol{m}_2} \boldsymbol{m}_1 \int \mathrm{d}\boldsymbol{J}_2 \, \frac{\delta_{\mathrm{D}}(\omega_1 - \omega_2)}{|\mathcal{D}_{\boldsymbol{m}_1,\boldsymbol{m}_2}(\boldsymbol{J}_1, \boldsymbol{J}_2, \omega_1)|^2} \, \boldsymbol{m}_1 \cdot \frac{\partial F}{\partial \boldsymbol{J}_1} \, F(\boldsymbol{J}_2) \, . \tag{2.156}$$

This contribution corresponds to the diffusion term in the Balescu-Lenard equation. Gathering the two contributions from Eqs. (2.155) and (2.156), and paying a careful to the signs of the various terms, one gets the final expression of the collision term $\mathscr{C}[F]$ as

$$\mathscr{C}[F] = \pi (2\pi)^d \mu \frac{\partial}{\partial \boldsymbol{J}_1} \cdot \bigg[\sum_{\boldsymbol{m}_1,\boldsymbol{m}_2} \boldsymbol{m}_1 \, \frac{\delta_{\mathrm{D}}(\boldsymbol{m}_1 \cdot \boldsymbol{\Omega}_1 - \boldsymbol{m}_2 \cdot \boldsymbol{\Omega}_2)}{|\mathcal{D}_{\boldsymbol{m}_1,\boldsymbol{m}_2}(\boldsymbol{J}_1, \boldsymbol{J}_2, \boldsymbol{m}_1 \cdot \boldsymbol{\Omega}_1)|^2} \bigg(\boldsymbol{m}_1 \cdot \frac{\partial}{\partial \boldsymbol{J}_1} - \boldsymbol{m}_2 \cdot \frac{\partial}{\partial \boldsymbol{J}_2} \bigg) F(\boldsymbol{J}_1) \, F(\boldsymbol{J}_2) \bigg] . \tag{2.157}$$

This allows us to recover the inhomogeneous Balescu-Lenard equation (2.67).

C Functional Approach to the Landau Equation

The previous sections presented two complementary derivations of the Balescu-Lenard equation, respectively based on the Klimontovich equation and the BBGKY hierarchy. In this Appendix, let us present an alternative approach based on a functional integral rewriting of the dynamics. In a little-known seven-page paper, Jolicoeur and Le Guillou (1989) presented how the general functional integral framework (Faddeev and Slavnov 1993) was suited for the study of classical kinetic theory. Using this formalism and starting from Liouville's equation, they recovered the BBGKY hierarchy. More importantly, they illustrated how this approach allows for a simple derivation of the homogeneous Balescu-Lenard equation (Balescu 1960; Lenard 1960) of plasma physics. In the context of inhomogeneous systems, we presented in Fouvry et al. (2016a) how this same functional approach may be used to recover the inhomogeneous Landau equation (2.73). Relying on the analogy between self-gravitating systems and $2D$ systems of point vortices (Chavanis 2002), a similar derivation in the context of $2D$ hydrodynamics was also presented in Fouvry et al. (2016b). In order to offer some new insights on the content of the collisional kinetic equations (2.67) and (2.73), we will now present this alternative derivation in the context of inhomogeneous systems.

C.1 Functional Integral Formalism

As previously, let us consider a system made of N identical particles. At order $1/N$, its dynamics is fully described by the two first truncated equations of the BBGKY hierarchy (2.107) and (2.108), which involve the system's $1-$body DF F and the $2-$body autocorrelation $\mathcal{C}$. The first step of the present derivation is to rewrite these two coupled evolution equations under a functional form. As an illustration of this method, let us consider a dynamical quantity f depending on time t and defined on a phase space Γ. We assume that this quantity follows an evolution equation of the form $[\partial_t + L] f = 0$, where L is a differential operator. Let us now introduce an auxiliary field λ defined on the same space than f to rewrite the evolution constraint of f as a functional integral of the form (see Jolicoeur and Le Guillou (1989); Fouvry et al. (2016a) for more details)

$$1 = \int \mathcal{D}f \mathcal{D}\lambda \, \exp\bigg[\mathrm{i} \int \mathrm{d}t \mathrm{d}\Gamma \, \lambda [\partial_t + L] f \bigg] . \tag{2.158}$$

In Eq. (2.158), we define the action $S[F, \lambda] = \mathrm{i} \int \mathrm{d}t \mathrm{d}\Gamma \, \lambda [\partial_t + L] f$ as the argument of the exponential.[6] It is important to note that the evolution equation satisfied by f corresponds to the quantity by which the auxiliary field λ is multiplied in the action.

When considering the two coupled evolution equations (2.107) and (2.108), one may proceed to a similar transformation. Let us define the phase space coordinates as $\Gamma = (\boldsymbol{x}, \boldsymbol{v})$. By introducing two auxiliary fields $\lambda_1(t, \Gamma_1)$ and $\lambda_2(t, \Gamma_1, \Gamma_2)$, respectively associated with F and $\mathcal{C}$, Eqs. (2.107) and (2.108) can be rewritten under the compact functional form

$$1 = \int \mathcal{D}F \mathcal{D}\mathcal{C} \mathcal{D}\lambda_1 \mathcal{D}\lambda_2 \, \exp\bigg\{ \mathrm{i} \int \mathrm{d}t \mathrm{d}\Gamma_1 \, \lambda_1 (A_1 F + B_1 \mathcal{C}) + \frac{\mathrm{i}}{2} \int \mathrm{d}t \mathrm{d}\Gamma_1 \mathrm{d}\Gamma_2 \, \lambda_2 (A_2 \mathcal{C} + D_2 \mathcal{C} + S_2) \bigg\} . \tag{2.159}$$

In Eq. (2.159), we introduced the operators A_1, B_1, A_2, D_2, and S_2 as

$$\begin{aligned}
A_1 F &= \bigg[\frac{\partial}{\partial t} + \boldsymbol{v}_1 \cdot \frac{\partial}{\partial \boldsymbol{x}_1} + \bigg[\int \mathrm{d}\Gamma_2 \, \boldsymbol{\mathcal{F}}_{12} F(\Gamma_2) \bigg] \cdot \frac{\partial}{\partial \boldsymbol{v}_1} \bigg] , \\
B_1 \mathcal{C} &= \mu \int \mathrm{d}\Gamma_2 \, \boldsymbol{\mathcal{F}}_{12} \cdot \frac{\partial \mathcal{C}(\Gamma_1, \Gamma_2)}{\partial \boldsymbol{v}_1} , \\
A_2 \mathcal{C} &= \bigg[\frac{\partial}{\partial t} + \boldsymbol{v}_1 \cdot \frac{\partial}{\partial \boldsymbol{x}_1} + \boldsymbol{v}_2 \cdot \frac{\partial}{\partial \boldsymbol{x}_2} + \int \mathrm{d}\Gamma_3 \, F(\Gamma_3) \bigg[\boldsymbol{\mathcal{F}}_{13} \cdot \frac{\partial}{\partial \boldsymbol{v}_1} + \boldsymbol{\mathcal{F}}_{23} \cdot \frac{\partial}{\partial \boldsymbol{v}_2} \bigg] \bigg] \mathcal{C}(\Gamma_1, \Gamma_2) , \\
D_2 \mathcal{C} &= \bigg[\int \mathrm{d}\Gamma_3 \, \boldsymbol{\mathcal{F}}_{13} \mathcal{C}(\Gamma_2, \Gamma_3) \bigg] \cdot \frac{\partial F}{\partial \boldsymbol{v}_1} + (1 \leftrightarrow 2) , \\
S_2 &= F(\Gamma_2) \, \boldsymbol{\mathcal{F}}_{12} \cdot \frac{\partial F}{\partial \boldsymbol{v}_1} + (1 \leftrightarrow 2) .
\end{aligned} \tag{2.160}$$

[6] This should not be mixed up with the angle-action coordinates $(\boldsymbol{\theta}, \boldsymbol{J})$ from inhomogeneous dynamics.

In Eq. (2.159), we did not write explicitly the dependence w.r.t. t to simplify the notations. In the expression of $B_1\mathcal{C}$, let us emphasise the presence of the small factor $\mu = M_{\mathrm{tot}}/N$, which illustrates the fact that we relied on a kinetic development at order $1/N$. Finally, the prefactor $1/2$ in Eq. (2.159) was only added for later convenience and does not play any role for the final expression of the evolution equation, since it was added as a global prefactor. Let us recall here the physical content of the various terms appearing in Eq. (2.159). Here, $A_1 F$ corresponds to the 1−body Vlasov advection term, and $B_1\mathcal{C}$ to the $1/N$ sourcing of the evolution of the 1−body DF under the effect of the 2−body autocorrelation. Similarly, $A_2\mathcal{C}$ encompasses the usual 2−body Vlasov advection term, $D_2\mathcal{C}$ corresponds to the dressing of particles by collective effects, and S_2 is a source term depending only on F, which sources the dynamics of $\mathcal{C}$.

Relying on basic manipulation, one can rewrite Eq. (2.159) as

$$1 = \int \mathcal{D}F\mathcal{D}\mathcal{C}\mathcal{D}\lambda_1\mathcal{D}\lambda_2 \exp\bigg\{ \mathrm{i}\!\int\! \mathrm{d}t\mathrm{d}\Gamma_1\, \lambda_1(\Gamma_1)A_1F(\Gamma_1) + \frac{\mathrm{i}}{2}\!\int\! \mathrm{d}t\mathrm{d}\Gamma_1\mathrm{d}\Gamma_2\, \lambda_2(\Gamma_1,\Gamma_2)\, G(\Gamma_1,\Gamma_2)$$
$$- \frac{\mathrm{i}}{2}\!\int\! \mathrm{d}t\mathrm{d}\Gamma_1\mathrm{d}\Gamma_2\, \mathcal{C}(\Gamma_1,\Gamma_2)\, E(\Gamma_1,\Gamma_2) \bigg\}, \qquad (2.161)$$

where it is crucial to note that all the dependences w.r.t. $\mathcal{C}$ were gathered in the prefactor of the second line. In Eq. (2.161), we introduced the quantity $G(\Gamma_1, \Gamma_2)$ as

$$G(\Gamma_1,\Gamma_2) = \boldsymbol{\mathcal{F}}_{12}\cdot\bigg[F(\Gamma_2)\frac{\partial F}{\partial \boldsymbol{v}_1} - F(\Gamma_1)\frac{\partial F}{\partial \boldsymbol{v}_2} \bigg], \qquad (2.162)$$

for which we used the relation $\boldsymbol{\mathcal{F}}_{21} = -\boldsymbol{\mathcal{F}}_{12}$. In Eq. (2.161), we also introduced the quantity $E(\Gamma_1, \Gamma_2)$ as

$$E(\Gamma_1,\Gamma_2) = A_2\lambda_2(\Gamma_1,\Gamma_2) + \!\int\! \mathrm{d}\Gamma_3 \Big[\boldsymbol{\mathcal{F}}_{13}\lambda_2(\Gamma_2,\Gamma_3) + \boldsymbol{\mathcal{F}}_{23}\lambda_2(\Gamma_1,\Gamma_3) \Big]\cdot\frac{\partial F}{\partial \boldsymbol{v}_3} + \mu\boldsymbol{\mathcal{F}}_{12}\cdot\bigg[\frac{\partial \lambda_1}{\partial \boldsymbol{v}_1} - \frac{\partial \lambda_1}{\partial \boldsymbol{v}_2} \bigg]. \qquad (2.163)$$

Equation (2.163) was obtained thanks to an integration by parts. In order to invert the time derivative $\partial\mathcal{C}/\partial t$ present in the term $\lambda_2 A_2\mathcal{C}$ from Eq. (2.159), we assumed $t \in [0;\, T]$, where T is an arbitrary upper temporal bound, along with the boundary conditions $\mathcal{C}(t=0)=0$ (the system is supposed to be initially uncorrelated) and $\lambda_2(T)=0$ (we are free to impose a condition on λ_2). As presented in Fouvry et al. (2016a), let us now neglect collective effects. This amounts to neglecting the contributions associated with the term $D_2\mathcal{C}$ in Eq. (2.159), so that Eq. (2.163) becomes

$$E(\Gamma_1,\Gamma_2) = A_2\lambda_2(\Gamma_1,\Gamma_2) + \mu\boldsymbol{\mathcal{F}}_{12}\cdot\bigg[\frac{\partial \lambda_1}{\partial \boldsymbol{v}_1} - \frac{\partial \lambda_1}{\partial \boldsymbol{v}_2} \bigg]. \qquad (2.164)$$

In order to obtain a closed kinetic equation involving F only, the traditional approach would be to start from Eq. (2.159) and proceed in the following way. By functionally integrating Eq. (2.159) w.r.t. λ_2, one gets a constraint of the form $A_2\mathcal{C} + D_2\mathcal{C} + S_2 = 0$, which couples F and $\mathcal{C}$. This constraint must then be inverted to give $\mathcal{C}=\mathcal{C}[F]$. One then uses this substitution in Eq. (2.159), and functionally integrates this equation w.r.t. λ_1, to obtain a kinetic equation involving F only. This gives the Balescu-Lenard equation (or the Landau equation when collective effects are not accounted for). This approach is identical to the direct resolution of the BBGKY hierarchy presented in Appendix 2.B.

However, based on the rewriting from Eq. (2.161), Jolicoeur and Le Guillou (1989) suggested a different strategy. One may indeed first integrate functionally Eq. (2.161) w.r.t. $\mathcal{C}$, to obtain a constraint of the form $E[F, \lambda_1, \lambda_2]=0$. Once inverted, this offers a relation of the form $\lambda_2=\lambda_2[F, \lambda_1]$. One may then substitute this relation in Eq. (2.161), to obtain a functional equation which only involves F and λ_1. The final step is then to functionally integrate this equation w.r.t. λ_1, to obtain a closed kinetic equation involving F only. Let us now show how this alternative approach allows for the derivation of the inhomogeneous Landau equation.

C.2 Application to Inhomogeneous Systems

As in Sect. 2.3, let us assume that the system's mean potential is integrable, so that one may always remap the physical phase space coordinates $(\boldsymbol{x}, \boldsymbol{v})$ to the angle-action ones $(\boldsymbol{\theta}, \boldsymbol{J})$. Relying on the adiabatic approximation (Heyvaerts 2010; Chavanis 2012b, 2013b), we assume that the 1−body DF is a quasi-stationary solution of the Vlasov equation, so that $F(\boldsymbol{\theta}, \boldsymbol{J})=F(\boldsymbol{J})$, where the dependence w.r.t. t has not been written out to shorten the notations. Since λ_1 is the auxiliary field associated with F, one also has $\lambda_1(\boldsymbol{\theta}, \boldsymbol{J})=\lambda_1(\boldsymbol{J})$, while the second auxiliary field $\lambda_2(\boldsymbol{\theta}_1, \boldsymbol{J}_1, \boldsymbol{\theta}_2, \boldsymbol{J}_2)$ still fully depends on all angle-action coordinates. Let us note that for homogeneous systems, the system's invariance by translation would impose $\lambda_2(\boldsymbol{x}_1, \boldsymbol{v}_1, \boldsymbol{x}_2, \boldsymbol{v}_2)=\lambda_2(\boldsymbol{x}_1-\boldsymbol{x}_2, \boldsymbol{v}_1, \boldsymbol{v}_2)$. Relying on the angle-action properties from Eqs. (2.118) and (2.119), one may now rewrite the various operators appearing in Eq. (2.161). Equation (2.160) gives

$$A_1 F = \frac{\partial F}{\partial t} . \tag{2.165}$$

Similarly, Eq. (2.162) can be rewritten as

$$G(\Gamma_1, \Gamma_2) = -\bigg[F(\boldsymbol{J}_2) \frac{\partial U_{12}}{\partial \boldsymbol{\theta}_1} \cdot \frac{\partial F}{\partial \boldsymbol{J}_1} + F(\boldsymbol{J}_1) \frac{\partial U_{21}}{\partial \boldsymbol{\theta}_2} \cdot \frac{\partial F}{\partial \boldsymbol{J}_2} \bigg] . \tag{2.166}$$

Finally, the constraint $E(\Gamma_1, \Gamma_2)=0$ from Eq. (2.164) takes the form

$$\frac{\partial \lambda_2}{\partial t} + \boldsymbol{\Omega}_1 \cdot \frac{\partial \lambda_2}{\partial \boldsymbol{\theta}_1} + \boldsymbol{\Omega}_2 \cdot \frac{\partial \lambda_2}{\partial \boldsymbol{\theta}_2} - \mu \bigg[\frac{\partial U_{12}}{\partial \boldsymbol{\theta}_1} \cdot \frac{\partial \lambda_1}{\partial \boldsymbol{J}_1} + \frac{\partial U_{21}}{\partial \boldsymbol{\theta}_2} \cdot \frac{\partial \lambda_1}{\partial \boldsymbol{J}_2} \bigg] = 0 . \tag{2.167}$$

C.3 Inverting the Constraint

In order to invert Eq. (2.167), we once again rely on Bogoliubov's ansatz by assuming that the fluctuations (such as $\mathcal{C}$ and λ_2) evolve much faster than the mean dynamical orbit-averaged quantities (such as F and λ_1). As a consequence, on the timescale on which λ_2 evolves, one can assume F and λ_1 to be frozen, while on the timescale of secular evolution, one can assume λ_2 to be equal to the asymptotic value associated with the current value of F and λ_1. As defined in Eq. (2.6), let us perform a Fourier transform w.r.t. the angles $\boldsymbol{\theta}$. We decompose the interaction potential U_{12} as

$$U_{12} = U(\boldsymbol{x}(\boldsymbol{\theta}_1, \boldsymbol{J}_1) - \boldsymbol{x}(\boldsymbol{\theta}_2, \boldsymbol{J}_2)) = \sum_{\boldsymbol{m}_1, \boldsymbol{m}_2} A_{\boldsymbol{m}_1, \boldsymbol{m}_2}(\boldsymbol{J}_1, \boldsymbol{J}_2)\, \mathrm{e}^{\mathrm{i}(\boldsymbol{m}_1\cdot\boldsymbol{\theta}_1 - \boldsymbol{m}_2\cdot\boldsymbol{\theta}_2)}, \tag{2.168}$$

where the bare susceptibility coefficients $A_{\boldsymbol{m}_1, \boldsymbol{m}_2}(\boldsymbol{J}_1, \boldsymbol{J}_2)$ were already introduced in Eq. (2.74). Multiplying Eq. (2.167) by $1/(2\pi)^{2d} \mathrm{e}^{\mathrm{i}(\boldsymbol{m}_1\cdot\boldsymbol{\theta}_1 - \boldsymbol{m}_2\cdot\boldsymbol{\theta}_2)}$ and integrating it w.r.t. $\boldsymbol{\theta}_1$ and $\boldsymbol{\theta}_2$, we obtain

$$\frac{\partial \lambda_{-\boldsymbol{m}_1, \boldsymbol{m}_2}}{\partial t} - \mathrm{i}\Delta\omega \lambda_{-\boldsymbol{m}_1, \boldsymbol{m}_2} = -\mathrm{i}\mu A^*_{\boldsymbol{m}_1, \boldsymbol{m}_2} \left[\boldsymbol{m}_1 \cdot \frac{\partial \lambda_1}{\partial \boldsymbol{J}_1} - \boldsymbol{m}_2 \cdot \frac{\partial \lambda_1}{\partial \boldsymbol{J}_2} \right], \tag{2.169}$$

where we used the shortening notations $\lambda_{-\boldsymbol{m}_1, \boldsymbol{m}_2} = \lambda_{-\boldsymbol{m}_1, \boldsymbol{m}_2}(\boldsymbol{J}_1, \boldsymbol{J}_2)$, $A_{\boldsymbol{m}_1, \boldsymbol{m}_2} = A_{\boldsymbol{m}_1, \boldsymbol{m}_2}(\boldsymbol{J}_1, \boldsymbol{J}_2)$, and $\Delta\omega = \boldsymbol{m}_1 \cdot \boldsymbol{\Omega}_1 - \boldsymbol{m}_2 \cdot \boldsymbol{\Omega}_2$. Thanks to the boundary condition $\lambda_2(T) = 0$ introduced in Eq. (2.163), and relying on the adiabatic approximation that λ_1 is frozen, one can straightforwardly solve the differential equation (2.169) as

$$\lambda_{-\boldsymbol{m}_1, \boldsymbol{m}_2}(t) = \mu A^*_{\boldsymbol{m}_1, \boldsymbol{m}_2} \left[\boldsymbol{m}_1 \cdot \frac{\partial \lambda_1}{\partial \boldsymbol{J}_1} - \boldsymbol{m}_2 \cdot \frac{\partial \lambda_1}{\partial \boldsymbol{J}_2} \right] \frac{1 - \mathrm{e}^{\mathrm{i}\Delta\omega(t-T)}}{\Delta\omega}. \tag{2.170}$$

In order to consider only the forced regime of evolution, let us now assume that the arbitrary temporal bound T is large compared to the considered time t. Therefore, we place ourselves in the limit $T \to +\infty$. Let us finally recall the formula

$$\lim_{T \to +\infty} \frac{\mathrm{e}^{\mathrm{i}T\Delta\omega} - 1}{\Delta\omega} = \mathrm{i}\pi\delta_{\mathrm{D}}(\Delta\omega), \tag{2.171}$$

so that Eq. (2.170) immediately gives

$$\lim_{T \to +\infty} \lambda_{-\boldsymbol{m}_1, \boldsymbol{m}_2}(t) = \mathrm{i}\pi\mu A^*_{\boldsymbol{m}_1, \boldsymbol{m}_2} \left[\boldsymbol{m}_1 \cdot \frac{\partial \lambda_1}{\partial \boldsymbol{J}_1} - \boldsymbol{m}_2 \cdot \frac{\partial \lambda_1}{\partial \boldsymbol{J}_2} \right] \delta_{\mathrm{D}}(\boldsymbol{m}_1 \cdot \boldsymbol{\Omega}_1 - \boldsymbol{m}_2 \cdot \boldsymbol{\Omega}_2). \tag{2.172}$$

Thanks to Bogoliubov's ansatz, we therefore inverted the constraint $E[F, \lambda_1, \lambda_2] = 0$ from Eq. (2.167), to obtain $\lambda_2 = \lambda_2[F, \lambda_1]$.

C.4 Recovering the Landau Collision Operator

Let us now substitute the inverted expression from Eq. (2.172) into the functional integral from Eq. (2.161), which then only involves F and λ_1. The remaining action term $S[F, \lambda_1]$ reads

$$S[F,\lambda_1] = \mathrm{i}\int \mathrm{d}t\mathrm{d}\Gamma_1\, \lambda_1 A_1 F + \frac{\mathrm{i}}{2}\int \mathrm{d}t\mathrm{d}\Gamma_1\mathrm{d}\Gamma_2\, \lambda_2[F,\lambda_1]\, G(\Gamma_1,\Gamma_2)\,. \tag{2.173}$$

Thanks to the expressions of A_1 and G from Eqs. (2.165) and (2.166), and using a Fourier transform in angles as in Eq. (2.6), one can rewrite Eq. (2.173) as

$$\begin{aligned} S[F,\lambda_1] =& \mathrm{i}\int \mathrm{d}t\mathrm{d}\Gamma_1\, \lambda_1(\Gamma_1)\, \frac{\partial F}{\partial t} \\ &+ \frac{\mathrm{i}}{2}\int \mathrm{d}t\mathrm{d}\Gamma_1\mathrm{d}\Gamma_2 \sum_{\boldsymbol{m}_1,\boldsymbol{m}_2} \mathrm{Im}\bigg[A_{\boldsymbol{m}_1,\boldsymbol{m}_2}\lambda_{-\boldsymbol{m}_1,\boldsymbol{m}_2}\bigg]\bigg[\boldsymbol{m}_1\cdot\frac{\partial F}{\partial \boldsymbol{J}_1}F(\boldsymbol{J}_2) - \boldsymbol{m}_2\cdot\frac{\partial F}{\partial \boldsymbol{J}_2}F(\boldsymbol{J}_1)\bigg]. \end{aligned} \tag{2.174}$$

Thanks to the inversion from Eq. (2.172), one immediately has

$$\mathrm{Im}\bigg[A_{\boldsymbol{m}_1,\boldsymbol{m}_2}\lambda_{-\boldsymbol{m}_1,\boldsymbol{m}_2}\bigg] = \pi\mu\delta_{\mathrm{D}}(\boldsymbol{m}_1\cdot\boldsymbol{\Omega}_1 - \boldsymbol{m}_2\cdot\boldsymbol{\Omega}_2)\, \big|A_{\boldsymbol{m}_1,\boldsymbol{m}_2}\big|^2 \bigg[\boldsymbol{m}_1\cdot\frac{\partial \lambda_1}{\partial \boldsymbol{J}_1} - \boldsymbol{m}_2\cdot\frac{\partial \lambda_1}{\partial \boldsymbol{J}_2}\bigg]. \tag{2.175}$$

Injecting this result in Eq. (2.174), one gets

$$\begin{aligned} S[F,\lambda_1] =& \mathrm{i}\int \mathrm{d}t\mathrm{d}\Gamma_1\, \lambda_1 \frac{\partial F}{\partial t} + \frac{\mathrm{i}}{2}\int \mathrm{d}t\mathrm{d}\Gamma_1\mathrm{d}\Gamma_2 \sum_{\boldsymbol{m}_1,\boldsymbol{m}_2} \pi\mu\delta_{\mathrm{D}}(\boldsymbol{m}_1\cdot\boldsymbol{\Omega}_1 - \boldsymbol{m}_2\cdot\boldsymbol{\Omega}_2)\big|A_{\boldsymbol{m}_1,\boldsymbol{m}_2}\big|^2 \\ &\times \bigg[\boldsymbol{m}_1\cdot\frac{\partial \lambda_1}{\partial \boldsymbol{J}_1} - \boldsymbol{m}_2\cdot\frac{\partial \lambda_1}{\partial \boldsymbol{J}_2}\bigg]\bigg[\boldsymbol{m}_1\cdot\frac{\partial F}{\partial \boldsymbol{J}_1}F(\boldsymbol{J}_2) - \boldsymbol{m}_2\cdot\frac{\partial F}{\partial \boldsymbol{J}_2}F(\boldsymbol{J}_1)\bigg]. \end{aligned} \tag{2.176}$$

The final step of the calculation is to rewrite the second term of Eq. (2.176) under the form $\int \mathrm{d}t\mathrm{d}\Gamma_1\lambda_1(\Gamma_1)\ldots$ This is a straightforward calculation, which requires to use an integration by parts and to permute accordingly the indices $1 \leftrightarrow 2$. Equation (2.176) can finally be rewritten as

$$\begin{aligned} S[F,\lambda_1] = \mathrm{i}\int \mathrm{d}t\mathrm{d}\Gamma_1\, \lambda_1(\Gamma_1) \bigg\{ \frac{\partial F}{\partial t} - \pi(2\pi)^d \mu \frac{\partial}{\partial \boldsymbol{J}_1}\cdot\bigg[\sum_{\boldsymbol{m}_1,\boldsymbol{m}_2} \boldsymbol{m}_1 \int \mathrm{d}\boldsymbol{J}_2\, \delta_{\mathrm{D}}(\boldsymbol{m}_1\cdot\boldsymbol{\Omega}_1 - \boldsymbol{m}_2\cdot\boldsymbol{\Omega}_2)\big|A_{\boldsymbol{m}_1,\boldsymbol{m}_2}\big|^2 \\ \times \bigg[\boldsymbol{m}_1\cdot\frac{\partial F}{\partial \boldsymbol{J}_1}F(\boldsymbol{J}_2) - \boldsymbol{m}_2\cdot\frac{\partial F}{\partial \boldsymbol{J}_2}F(\boldsymbol{J}_1)\bigg]\bigg]\bigg\}, \end{aligned} \tag{2.177}$$

where the additional prefactor $(2\pi)^d$ comes from the transformation $\int \mathrm{d}\Gamma_2\, f(\boldsymbol{J}_2) = (2\pi)^d \int \mathrm{d}\boldsymbol{J}_2\, f(\boldsymbol{J}_2)$. Integrating functionally Eq. (2.177) w.r.t. λ_1, one finally obtains a closed form expression for the kinetic equation as

$$\frac{\partial F}{\partial t} = \pi(2\pi)^d \mu \frac{\partial}{\partial \boldsymbol{J}_1} \cdot \left[\sum_{\boldsymbol{m}_1, \boldsymbol{m}_2} \boldsymbol{m}_1 \int \mathrm{d}\boldsymbol{J}_2 \, \delta_{\mathrm{D}}(\boldsymbol{m}_1 \cdot \boldsymbol{\Omega}_1 - \boldsymbol{m}_2 \cdot \boldsymbol{\Omega}_2) \big| A_{\boldsymbol{m}_1, \boldsymbol{m}_2}(\boldsymbol{J}_1, \boldsymbol{J}_2) \big|^2 \right. \\ \left. \times \left(\boldsymbol{m}_1 \cdot \frac{\partial}{\partial \boldsymbol{J}_1} - \boldsymbol{m}_2 \cdot \frac{\partial}{\partial \boldsymbol{J}_2} \right) F(\boldsymbol{J}_1, t) F(\boldsymbol{J}_2, t) \right] . \tag{2.178}$$

As a conclusion, relying on a functional integral formalism, we were able to exactly recover the inhomogeneous Landau equation (2.73). Such a new calculation provides additional insights on the origin of these diffusion equations. A natural next step would be to show how it may be used to account for collective effects and recover the inhomogeneous Balescu-Lenard equation (2.67). Such a derivation is expected to be more involved, as one will have to deal with a self-consistent Fredholm equation associated with the polarisation dressing of the potential fluctuations (similar to the one obtained in Eq. (2.122)). As illustrated in the two derivations from Sect. 2.3 and Appendix 2.B, this requires to rely on Kalnajs matrix method (Kalnajs 1976) and to introduce potential-density basis elements. Jolicoeur and Le Guillou (1989) managed to develop such a self-consistent calculation in the homogeneous context of plasma physics, where both the resonance condition and the Fredholm equation are simpler. The generalisation of this method to inhomogeneous systems will be the subject of a future work. Finally, because of its alternative point of view, this approach may also turn out fruitful to tackle the question of obtaining closed kinetic equations when higher order correlation terms are taken into account.

References

Antonov, V. A. (1962). *Vest. Leningr. Gos. Univ*, *7*, 135.
Balescu, R. (1960). *Phys. Fluids*, *3*, 52.
Binney, J., & Tremaine, S. (2008). *Galactic dynamics* (2nd ed.). Princeton University Press.
Binney, J., & Lacey, C. (1988). *MNRAS*, *230*, 597.
Boltzmann, L. (1872). *Wien. Ber.*, *66*, 275.
Bouchet, F., & Dauxois, T. (2005). *Phys. Rev. E*, *72*, 045103.
Campa, A., Dauxois, T., Fanelli, D., & Ruffo, S. (2014). *Physics of long-range interacting systems*. Oxford University Press.
Campa, A., Dauxois, T., & Ruffo, S. (2009). *Phys. Rep.*, *480*, 57.
Chandrasekhar, S. (1942). *Principles of stellar dynamics*. University of Chicago Press.
Chandrasekhar, S. (1943a). *ApJ*, *97*, 255.
Chandrasekhar, S. (1943b). *ApJ*, *97*, 263.
Chandrasekhar, S. (1949). *Rev. Mod. Phys.*, *21*, 383.
Chavanis, P.-H. (2002). *Dynamics and thermodynamics of systems with long-range interactions*. Springer.
Chavanis, P.-H. (2007). *Physica A*, *377*, 469.
Chavanis, P.-H. (2010). *J. Stat. Mech. Theor. Exp.*, *5*, 19.
Chavanis, P.-H. (2012a). *Eur. Phys. J. Plus*, *127*, 19.
Chavanis, P.-H. (2012b). *Physica A*, *391*, 3680.
Chavanis, P.-H. (2012c). *Physica A*, *391*, 3657.

Chavanis, P.-H. (2012d). *J. Stat. Mech. Theor. Exp.*, *2*, 02019.
Chavanis, P.-H. (2013a). *Eur. Phys. J. Plus*, *128*, 126.
Chavanis, P.-H. (2013b). *A&A*, *556*, A93.
Chavanis, P.-H., Vatteville, J., & Bouchet, F. (2005). *Eur. Phys. J. B*, *46*, 61.
Debye, P., & Hückel, E. (1923a). *Physik Z.*, *24*, 185.
Debye, P., & Hückel, E. (1923b). *Physik Z.*, *24*, 305.
Eldridge, O. C., & Feix, M. (1963). *Phys. Fluids*, *6*, 398.
Faddeev, L., & Slavnov, A. (1993). *Gauge fields: an introduction to quantum theory*. Addison-Wesley.
Fouvry, J.-B., Pichon, C., & Chavanis, P.-H. (2015a). *A&A*, *581*, A139.
Fouvry, J.-B., Pichon, C., & Prunet, S. (2015b). *MNRAS*, *449*, 1967.
Fouvry, J.-B., Chavanis, P.-H., & Pichon, C. (2016a). *Physica A*, *459*, 117.
Fouvry, J.-B., Chavanis, P.-H., & Pichon, C. (2016b). *Phys. Lett. A*, *380*, 2589.
Heyvaerts, J. (2010). *MNRAS*, *407*, 355.
Heyvaerts, J., Fouvry, J.-B., Chavanis, P.-H., & Pichon, C. (2017). *MNRAS*, *469*, 4193.
Hohl, F. (1971). *ApJ*, *168*, 343.
Jeans, J. (1929). *Astronomy and cosmogony*. Cambridge University Press.
Jeans, J. (1915). *MNRAS*, *76*, 70.
Jolicoeur, T., & Le Guillou, J. C. (1989). *Phys. Rev. A*, *40*, 5815.
Kadomtsev, B. B., & Pogutse, O. P. (1970). *Phys. Rev. Lett.*, *25*, 1155.
Kalnajs, A. J. (1976). *ApJ*, *205*, 745.
Klimontovich, I. (1967). *The statistical theory of non-equilibrium processes in a plasma*. Cambridge: MIT Press.
Landau, L. (1936). *Phys. Z. Sowj. Union*, *10*, 154.
Lenard, A. (1960). *Annals of Physics*, *10*, 390.
Luciani, J. F., & Pellat, R. (1987). *Journal de Physique*, *48*, 591.
Lynden-Bell, D. (1994). *Lectures on stellar dynamics*. Berlin: Springer.
Lynden-Bell, D. (1999). *Physica A*, *263*, 293.
Lynden-Bell, D., & Wood, R. (1968). *MNRAS*, *138*, 495.
Ma, C.-P., & Bertschinger, E. (2004). *ApJ*, *612*, 28.
Mynick, H. E. (1988). *J. Plasma Phys.*, *39*, 303.
Nardini, C., Gupta, S., Ruffo, S., Dauxois, T., & Bouchet, F. (2012). *J. Stat. Mech. Theor. Exp.*, *12*, 12010.
Pichon, C. (1994). Ph.D. thesis, University of Cambridge.
Pichon, C., & Aubert, D. (2006). *MNRAS*, *368*, 1657.
Pitaevskii, L., & Lifshitz, E. (2012). *Physical kinetics*, Vol. 10. Elsevier Science.
Polyachenko, V. L., & Shukhman, I. G. (1982). *Soviet Ast.*, *26*, 140.
Rocha Filho, T. M., Santana, A. E., Amato, M. A., & Figueiredo, A. (2014). *Phys. Rev. E*, *90*, 032133.
Vlasov, A. (1945). J. Phys. (U.S.S.R.), *9*, 25.
Vlasov, A. (1938). *Zh. Eksp. Teor. Fiz.*, *8*, 291.
Weinberg, M. D. (1993). *ApJ*, *410*, 543.
Weinberg, M. D. (2001a). *MNRAS*, *328*, 311.
Weinberg, M. D. (2001b). *MNRAS*, *328*, 321.

Chapter 3
Razor-Thin Discs

3.1 Introduction

Most stars, perhaps all, are born in stellar discs. Major mergers destroyed some of these discs quite early in the history of the universe, but some have survived up to the present day, including the Milky Way. Understanding the secular dynamics of stellar discs appears therefore as an essential ingredient of cosmology, as the discs' cosmological environments are now firmly established in the ΛCDM paradigm (Planck Collaboration et al. 2014). Self-gravitating stellar discs are cold responsive dynamical systems in which rotation provides an important reservoir of free energy and where orbital resonances play a key role. The availability of free energy leads to some stimuli being strongly amplified, while resonances tend to localise their dissipation, with the net result that even a very small perturbation can lead to discs evolving to significantly distinct equilibria. Stellar discs are submitted to various sources of gravitational noise, such as Poisson shot noise arising from the finite number of stars in the disc, or from the finite number of giant molecular clouds in the interstellar medium or sub-haloes orbiting around the galaxy. Spiral arms in the gas distribution also provide another source of fluctuations, while the central bar of the disc offers another source of stimulus more systematic than noisy. The history of a stellar disc likely comprises the joint responses to all these various stimuli.

One can find in the solar neighbourhood at least three illustrations of such effects. First, the random velocity of each coveal cohort of stars increases with the cohort's age (Wielen 1977; Aumer and Binney 2009). In addition, the velocity distribution around the Sun exhibits several "streams" of stars (Dehnen 1998). Each of these streams contains stars of various ages and chemistries, which are all responding to some stimulus in a similar fashion (Famaey et al. 2005). Finally, in the two-dimensional action space (J_ϕ, J_r), where J_ϕ stands for the angular momentum and J_r for a measure of the star's radial excursion (see Sect. 3.2), the distribution of stars

---

The work presented in this chapter is based on Fouvry and Pichon (2015), Fouvry et al. (2015a, b, c).

J.-B. Fouvry, *Secular Evolution of Self-Gravitating Systems Over Cosmic Age*,
Springer Theses, DOI 10.1007/978-3-319-66760-7_3

shows elongated features. The density of stars is indeed depressed near $J_r=0$, i.e. near circular orbits, but enhanced at larger J_r, such that the whole disturbed region forms a curve that is consistent with being given by a resonant condition such as $2\Omega_\phi - \Omega_r = \text{cst.}$ (Sellwood 2010; McMillan 2013). Such features are called resonant ridges and will play an important role in our upcoming discussions of the secular dynamics of razor-thin stellar discs, as already argued for example in Sellwood and Carlberg (2014).

Direct numerical simulations of razor-thin stellar discs are very challenging because their two-dimensional geometry combined with their responsiveness causes discreteness noise to be important unless a large number of particles is employed. It is only recently that it became possible to simulate a disc with a sufficient number of particles for Poisson shot noise to be dynamically unimportant for many orbital times, such as in the simulations presented in Sellwood (2012). In addition, it is all the more difficult to simulate accurately a stellar disc that is embedded in a cosmological environment and therefore exposed to cosmic noise. Such experiments are essential to understand how the orbital structure of a disc may restructure on secular times. The reliability of numerical simulations over numerous dynamical times is an issue which calls for alternative probes, hence the need for analytical frameworks such as the ones presented in Chap. 2.

In the present chapter, we attempt to explain the origin of these resonant ridges in razor-thin discs while relying on two competing processes of secular diffusion, either collisionless (Sect. 2.2) for which the source of fluctuations is imposed by an external source, or collisional (Sect. 2.3) for which the source of fluctuations is self-induced and due to the system's own discreteness. Two main difficulties are encountered when implementing these diffusion equations. First, one has to explicitly construct the mapping $(\boldsymbol{x}, \boldsymbol{v}) \mapsto (\boldsymbol{\theta}, \boldsymbol{J})$, as the diffusion occurs in action space. In the context of galactic dynamics, these coordinates are now being increasingly used to construct equilibrium models of stellar systems (Binney 2010; Piffl et al. 2014) or to study the dynamics of stellar streams (Helmi and White 1999; Sellwood 2010; Eyre and Binney 2011; McMillan 2013; Sanders and Binney 2013). When considering a stellar disc, if one assumes the disc to be sufficiently tepid (i.e. the stars' orbits are not too eccentric), one can rely on the epicyclic approximation to construct such a mapping, as presented in Sect. 3.2.

The second difficulty arises when accounting for the system's self-gravity. Indeed, this requires to compute the system's response matrix $\widehat{\mathbf{M}}$ from Eq. (2.17), which asks for the introduction of potential and density basis elements as in Eq. (2.12). In order to ease the analytical inversion of $[\mathbf{I} - \widehat{\mathbf{M}}]$, one may rely on the Wentzel-Kramers-Brillouin (WKB) approximation (Liouville 1837; Toomre 1964; Kalnajs 1965; Lin and Shu 1966; Palmer et al. 1989; Binney and Tremaine 2008), which amounts to considering only the diffusion sustained by radially tightly wound spirals. This transforms Poisson's equation into a local equation and leads to a diagonal response matrix. Such an application of the WKB formalism in the context of the secular diffusion of razor-thin axisymmetric discs relies on the construction of tailored WKB basis elements presented in Sect. 3.3. As will be noted in Sect. 3.6, this WKB approximation also allows for an explicit calculation of the resonant condition

$\delta_{\mathrm{D}}(\boldsymbol{m}_1 \cdot \boldsymbol{\Omega}_1 - \boldsymbol{m}_2 \cdot \boldsymbol{\Omega}_2)$ appearing in the collisional Balescu-Lenard equation (2.67). Based on Fouvry et al. (2015a, b), Fouvry and Pichon (2015), Sect. 3.7 finally computes via the WKB approximation the collisionless and collisional diffusion fluxes to investigate radial diffusion in razor-thin axisymmetric discs, as observed in the simulations from Sellwood (2012).

3.2 Angle-Action Coordinates and Epicyclic Approximation

In order to investigate secular evolutions, a first step is to build up an explicit mapping $(\boldsymbol{x}, \boldsymbol{v}) \mapsto (\boldsymbol{\theta}, \boldsymbol{J})$ to angle-action coordinates. To do so, we assume that the disc is sufficiently cold, i.e. that the radial velocity dispersion is sufficiently small, and rely on the epicyclic approximation. Let us introduce the polar coordinates (R, ϕ), as well as their associated momenta (p_R, p_ϕ). For a razor-thin axisymmetric disc, the stationary Hamiltonian of the system takes the form

$$H_0 = \frac{1}{2}\left[p_R^2 + \frac{p_\phi^2}{R^2}\right] + \psi_0(R)\,, \tag{3.1}$$

where ψ_0 is the stationary axisymmetric background potential in the disc. The Hamiltonian being independent of ϕ, p_ϕ is a conserved quantity. This is the azimuthal action of the system, the angular momentum J_ϕ, which reads

$$J_\phi = \frac{1}{2\pi}\oint \mathrm{d}\phi\, p_\phi = p_\phi = R^2\,\dot{\phi}\,. \tag{3.2}$$

For a given value of J_ϕ, the radius R of the particle evolves according to

$$\ddot{R} = -\frac{\partial \psi_{\mathrm{eff}}}{\partial R}\,, \tag{3.3}$$

where we introduced the effective potential $\psi_{\mathrm{eff}}(R) = \psi_0(R) + J_\phi^2/(2R^2)$. Since we assume that the radial excursions of the particles are small, we may place ourselves in the vicinity of circular orbits. For a given J_ϕ, we define the guiding radius R_{g} via the implicit relation

$$0 = \frac{\partial \psi_{\mathrm{eff}}}{\partial R}\bigg|_{R_{\mathrm{g}}} = \frac{\partial \psi_0}{\partial R}\bigg|_{R_{\mathrm{g}}} - \frac{J_\phi^2}{R_{\mathrm{g}}^3}\,. \tag{3.4}$$

Here, $R_{\mathrm{g}}(J_\phi)$ corresponds to the radius of stars with an angular momentum equal to J_ϕ which are on exactly circular orbits. One should note that the mapping between R_{g} and J_ϕ is bijective and unambiguous (up to the sign of J_ϕ, i.e. whether stars are prograde or retrograde). In addition, this circular orbit is described at the azimuthal frequency Ω_ϕ given by

$$\Omega_\phi^2(R_g) = \frac{1}{R_g}\frac{\partial \psi_0}{\partial R}\bigg|_{R_g} = \frac{J_\phi^2}{R_g^4}. \tag{3.5}$$

In the vicinity of circular orbits, the Hamiltonian from Eq. (3.1) may be approximated as

$$H_0 = \frac{p_R^2}{2} + \psi_{\rm eff}(R_g, 0) + \frac{\kappa^2}{2}(R - R_g)^2, \tag{3.6}$$

where we introduced the radial epicyclic frequency κ as

$$\kappa^2(R_g) = \frac{\partial^2 \psi_{\rm eff}}{\partial R^2}\bigg|_{R_g} = \frac{\partial^2 \psi_0}{\partial R^2}\bigg|_{R_g} + 3\frac{J_\phi^2}{R_g^4}. \tag{3.7}$$

In Eq. (3.6), the radial motion takes the form of a harmonic libration. Up to an initial phase, there exists an amplitude A_R such that $R = R_g + A_R \cos(\kappa t)$. The associated radial action J_r is then given by

$$J_r = \frac{1}{2\pi}\oint \mathrm{d}R\, p_R = \frac{1}{2}\kappa A_R^2. \tag{3.8}$$

Here, $J_r = 0$ corresponds to circular orbits, and the larger J_r, the more eccentric the orbit. Let us also emphasise that the two intrinsic frequencies of motion only depend on J_ϕ, so that $\boldsymbol{\Omega}(\boldsymbol{J}) = (\Omega_\phi(J_\phi), \kappa(J_\phi))$. This will play an important role in the resonance condition appearing in the Balescu-Lenard equation (2.67). The epicyclic approximation is illustrated in Fig. 3.1.One can finally explicitly construct the mapping between (R, ϕ, p_R, p_ϕ) and $(\theta_R, \theta_\phi, J_r, J_\phi)$ (Lynden-Bell and Kalnajs 1972; Palmer 1994; Binney and Tremaine 2008). At first-order in the radial amplitude, it reads

$$R = R_g + A_R\cos(\theta_R) \quad ; \quad \phi = \theta_\phi - \frac{2\Omega_\phi}{\kappa}\frac{A_R}{R_g}\sin(\theta_R). \tag{3.9}$$

An illustration of an epicyclic orbit constructed with the mappings from Eq. (3.9) is given in Fig. 3.2. Let us note that numerous improvements of the epicyclic approximation have been proposed in the literature (Kalnajs 1979; Dehnen 1999; Lynden-Bell 2010).[1] Finally, we assume that initially the DF of the system takes the form of a quasi-isothermal DF (Binney and McMillan 2011) of the form

[1] We refer the reader to Dehnen (1999) for a detailed discussion. It especially notices that the epicyclic frequencies $(\Omega_\phi(J_\phi), \kappa(J_\phi))$ from Eqs. (3.5) and (3.7) do not satisfy the constraint from Schwarz's theorem, $\partial\Omega_\phi/\partial J_r = \partial\kappa/\partial J_\phi$, and therefore suggests to replace Ω_ϕ by $\Omega_\phi + (\mathrm{d}\kappa/\mathrm{d}J_\phi)J_r$. We do not consider such improvements in the upcoming calculations.

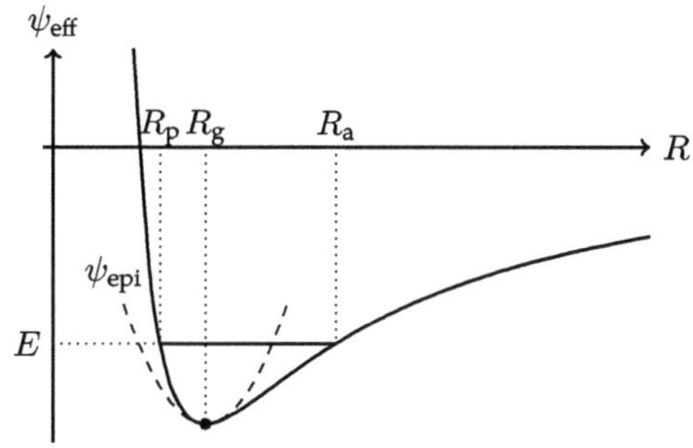

Fig. 3.1 Illustration of the epicyclic approximation in a razor-thin axisymmetric disc. The guiding radius $R_{\rm g}$ corresponds to the location of the minimum of the effective potential $\psi_{\rm eff}$. The epicyclic approximation amounts to approximating $\psi_{\rm eff}$ in the vicinity of its minimum by a harmonic potential $\psi_{\rm epi}$. In this limit, the star then undergoes radial harmonic librations between the pericentre $R_{\rm p}$ and apocentre $R_{\rm a}$ of its trajectory

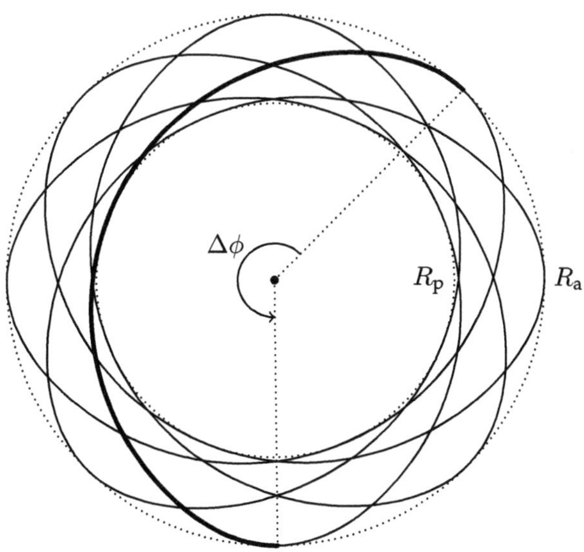

Fig. 3.2 Illustration of an epicyclic orbit for a razor-thin Mestel disc (see Sect. 3.7.1), following the angle-action mapping from Eq. (3.9). Such an orbit is the combination of an azimuthal oscillation at the mean frequency Ω_ϕ, and a harmonic libration between the star's pericentre $R_{\rm p}$ and apocentre $R_{\rm a}$ at the frequency κ. We highlighted in *bold* the azimuthal increase $\Delta\phi = 2\pi\Omega_\phi/\kappa$ during one radial oscillation. For degenerate orbits, such as the Keplerian ones (see Chap. 6), $\Delta\phi$ is a multiple of 2π, i.e. the frequencies Ω_ϕ and κ are in a rational ratio, which leads to a closed orbit (see Fig. 1.5)

$$F(R_g, J_r) = \frac{\Omega_\phi(R_g)\Sigma(R_g)}{\pi\kappa(R_g)\sigma_r^2(R_g)} \exp\left[-\frac{\kappa(R_g)J_r}{\sigma_r^2(R_g)}\right], \tag{3.10}$$

where $\Sigma(R_g)$ is the surface density of the disc and $\sigma_r^2(R_g)$ represents the local radial velocity dispersion of the stars at a given radius. Larger values of σ_r^2 correspond to hotter discs, which are therefore more stable. Such a DF becomes the Schwarzschild DF in the epicycle limit (see Eq. (4.153) in Binney and Tremaine (2008)).

3.3 The Razor-Thin WKB Basis

As we are considering the $2D$ case of a razor-thin disc, the basis elements introduced in Eq. (2.12) must be written as $\psi^{(p)}(R, \phi)$ in polar coordinates, and are associated with the surface densities $\Sigma^{(p)}(R, \phi)$. Here, as we will show, relying on the WKB approximation amounts to building up local basis elements thanks to which the response matrix $\widehat{\mathbf{M}}$ from Eq. (2.17) becomes diagonal. Let us introduce the basis elements

$$\psi^{[k_\phi, k_r, R_0]}(R, \phi) = \mathcal{A}\, \mathrm{e}^{\mathrm{i}(k_\phi \phi + k_r R)}\, \mathcal{B}_{R_0}(R)\,, \tag{3.11}$$

where the radial window function $\mathcal{B}_{R_0}(R)$ reads

$$\mathcal{B}_{R_0}(R) = \frac{1}{(\pi\sigma^2)^{1/4}} \exp\left[-\frac{(R-R_0)^2}{2\sigma^2}\right]. \tag{3.12}$$

The basis elements from Eq. (3.11) depend on three indices: k_ϕ is an azimuthal number which characterises the angular dependence of the basis elements, R_0 is the radius around which the Gaussian window $\mathcal{B}_{R_0}$ is centred, while k_r gives the radial frequency of the basis elements. One should also note the introduction of a scale-separation parameter σ, which will ensure the biorthogonality of the basis elements as detailed later on. We also introduced an amplitude $\mathcal{A}$ which will be tuned later on to correctly normalise the basis elements. The somewhat unusual normalisation of $\mathcal{B}_{R_0}$ was chosen for later convenience, to ensure that the amplitude $\mathcal{A}$ is independent of σ. Figure 3.3 illustrates the radial dependence of these basis elements, while Fig. 3.4 illustrates them in the polar $(R, \phi)-$plane.

The next step of the construction of these WKB basis elements is to compute the associated surface density basis elements $\Sigma^{[k_\phi, k_r, R_0]}$. To do so, we extend in the $z-$direction the WKB potential from Eq. (3.11) using the ansatz

$$\psi^{[k_\phi, k_r, R_0]}(R, \phi, z) = \mathcal{A}\, \mathrm{e}^{\mathrm{i}(k_\phi \phi + k_r R)}\, \mathcal{B}_{R_0}(R)\, Z(z)\,. \tag{3.13}$$

Poisson's equation in vacuum (i.e. Laplace's equation) $\Delta\psi^{[k_\phi, k_r, R_0]} = 0$ immediately leads to

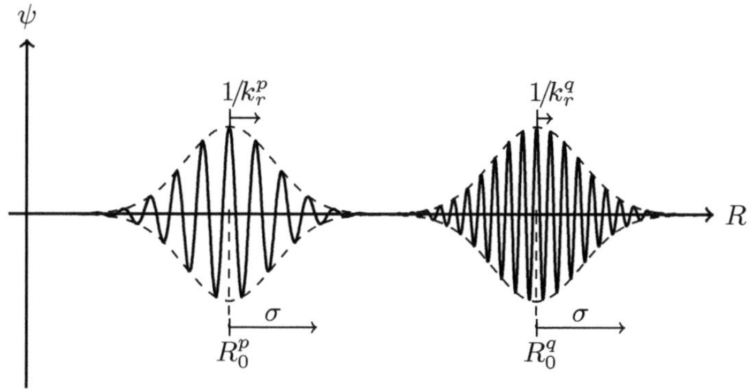

Fig. 3.3 Illustration of the radial dependence of two WKB basis elements from Eq. (3.11). Each Gaussian is centred around a radius R_0, with a typical extension given by the decoupling scale σ, and is modulated at the radial frequency k_r

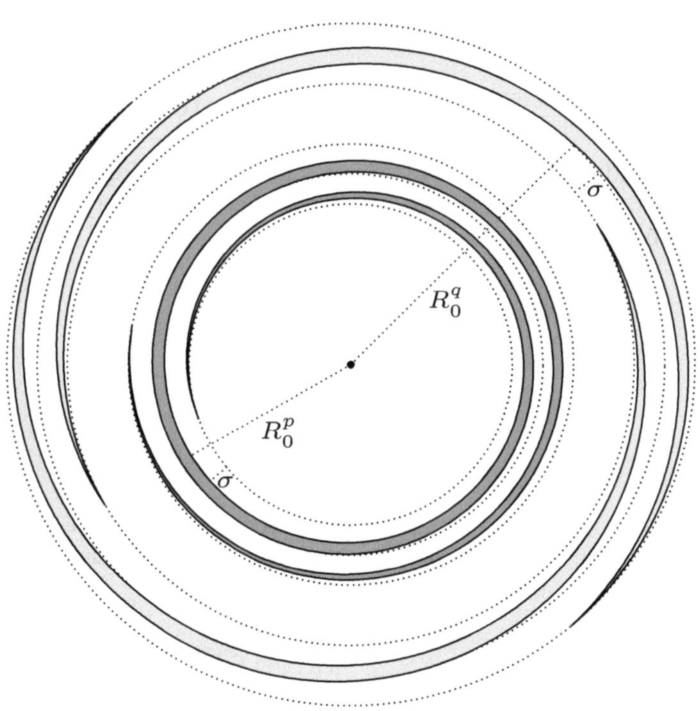

Fig. 3.4 Illustration of two WKB basis elements in the polar (R, ϕ)−plane. Each basis element is located around a central radius R_0, on a region of size σ. The winding of the spirals is governed by the radial frequency k_r, while the number of azimuthal patterns is given by the index k_ϕ, e.g., $k_\phi = 1$ for the interior *dark grey* element, and $k_\phi = 2$ for the exterior *light grey* one

$$\frac{Z''}{Z} = k_r^2\bigg[1 - \frac{\mathrm{i}}{k_r R} + 2\mathrm{i}\frac{R-R_0}{\sigma^2}\frac{1}{k_r} + \frac{R-R_0}{R}\frac{1}{(\sigma k_r)^2} + \frac{1}{(\sigma k_r)^2} + \frac{k_\phi^2}{(k_r R)^2} - \bigg[\frac{R-R_0}{\sigma^2}\frac{1}{k_r}\bigg]^2\bigg]. \tag{3.14}$$

At this stage, let us introduce explicitly our WKB assumption that all perturbations are radially tightly wound. Defining the typical size of the system by R_{sys}, we assume that

$$k_r R \gg 1 \quad ; \quad k_r \sigma \gg \frac{R_{\mathrm{sys}}}{\sigma}. \tag{3.15}$$

For azimuthal wavenumbers k_ϕ of order unity, Eq. (3.14) then becomes

$$\frac{Z''}{Z} = k_r^2. \tag{3.16}$$

As a conclusion, within the WKB limit, the extended potential from Eq. (3.13) takes the form

$$\psi^{[k_\phi,k_r,R_0]}(R,\phi,z) = \psi^{[k_\phi,k_r,R_0]}(R,\phi)\,\mathrm{e}^{-k_r|z|}, \tag{3.17}$$

where we ensured that for $z \to \pm\infty$, the potential tends to 0. Equation (3.17) introduces a discontinuity for $\partial\psi/\partial z$ in $z=0$. Gauss' theorem then gives the associated surface densities as

$$\Sigma(R,\phi) = \frac{1}{4\pi G}\bigg[\lim_{z\to 0^+}\frac{\partial\psi}{\partial z} - \lim_{z\to 0^-}\frac{\partial\psi}{\partial z}\bigg], \tag{3.18}$$

so that

$$\Sigma^{[k_\phi,k_r,R_0]}(R,\phi) = -\frac{|k_r|}{2\pi G}\,\psi^{[k_\phi,k_r,R_0]}(R,\phi). \tag{3.19}$$

As imposed by Eq. (2.12), the next step of the construction of the WKB basis elements is to ensure that the potentials and densities from Eqs. (3.11) and (3.19) form a biorthogonal basis, i.e. that one has

$$\delta_{k_\phi^p}^{k_\phi^q}\,\delta_{k_r^p}^{k_r^q}\,\delta_{R_0^p}^{R_0^q} = -\int \mathrm{d}R\,R\,\mathrm{d}\phi\,\psi^{[k_\phi^p,k_r^p,R_0^p]}(R,\phi)\,\big[\Sigma^{[k_\phi^q,k_r^q,R_0^q]}(R,\phi)\big]^*. \tag{3.20}$$

One can rewrite the r.h.s. of Eq. (3.20) as

$$(3.20) = \frac{|k_r^q|}{2\pi G}\frac{\mathcal{A}_p\mathcal{A}_q}{\sqrt{\pi\sigma^2}}\int \mathrm{d}\phi\,\mathrm{e}^{\mathrm{i}(k_\phi^p-k_\phi^q)\phi}\int \mathrm{d}R\,R\,\mathrm{e}^{\mathrm{i}(k_r^p-k_r^q)R}\exp\bigg[-\frac{(R-R_0^p)^2}{2\sigma^2}\bigg]\exp\bigg[-\frac{(R-R_0^q)^2}{2\sigma^2}\bigg]. \tag{3.21}$$

The integration on ϕ is straightforward and gives $2\pi\delta_{k_\phi^p}^{k_\phi^q}$. To perform the integration on R, we must now introduce additional assumptions to ensure the biorthogonality of the basis. The peaks of the two Gaussians in Eq. (3.21) may be assumed as separated if $\Delta R = R_0^p - R_0^q$ satisfies the separation condition

$$\Delta R \gg \sigma \quad \text{if} \quad R_0^p \neq R_0^q \,. \tag{3.22}$$

The term from Eq. (3.21) can then be assumed to be non-zero only for $R_0^p = R_0^q$. Equation (3.21) becomes

$$\delta_{k_\phi^p}^{k_\phi^q} \delta_{R_0^p}^{R_0^q} \frac{|k_r^q|}{G} \frac{\mathcal{A}_p \mathcal{A}_q}{\sqrt{\pi\sigma^2}} \int \mathrm{d}R\, R\, \mathrm{e}^{\mathrm{i}(k_r^p - k_r^q)R} \exp\!\left[-\frac{(R-R_0^p)^2}{\sigma^2}\right]. \tag{3.23}$$

The remaining integration on R now takes a form similar to the radial Fourier transform of a Gaussian of spread σ at the frequency $\Delta k_r = k_r^p - k_r^q$, and is therefore proportional to $\exp[-(\Delta k_r)^2/(4/\sigma^2)]$. As a consequence, let us assume that the frequency spread Δk_r of the WKB basis satisfies

$$\Delta k_r \gg \frac{1}{\sigma} \quad \text{if} \quad k_r^p \neq k_r^q \,. \tag{3.24}$$

With this additional assumption, Eq. (3.23) can be assumed to be non-zero only for $k_r^p = k_r^q$. Therefore, as imposed by Eqs. (3.22) and (3.24), in order to have a biorthogonal basis, one has to consider a spread σ, central radii R_0, and radial frequencies k_r such that

$$\Delta R_0 \gg \sigma \gg \frac{1}{\Delta k_r} \,. \tag{3.25}$$

With these constraints, the r.h.s. of Eq. (3.20) is non-zero only for $k_\phi^p = k_\phi^q$, $k_r^p = k_r^q$, and $R_0^p = R_0^q$. The last step of the calculation is to explicitly estimate the amplitude $\mathcal{A}$ in order to correctly normalise the basis elements. Equation (3.20) imposes

$$\frac{|k_r|}{G} \frac{\mathcal{A}^2}{\sqrt{\pi\sigma^2}} \int \mathrm{d}R\, R \exp\!\left[-\frac{(R-R_0)^2}{\sigma^2}\right] = 1 \,. \tag{3.26}$$

Thanks to the WKB assumptions from Eq. (3.15), this integration is straightforward and gives

$$\mathcal{A} = \sqrt{\frac{G}{|k_r| R_0}} \,. \tag{3.27}$$

Once the basis elements from Eq. (3.11) have been fully specified, one may compute $\psi_{\boldsymbol{m}}^{(p)}(\boldsymbol{J})$, their Fourier transform w.r.t. the angles, as defined in Eq. (2.6). Thanks to the explicit epicyclic mapping from Eq. (3.9), this takes the form

$$\psi_{\boldsymbol{m}}^{[k_\phi,k_r,R_0]}(\boldsymbol{J}) = \frac{\mathcal{A}\,\mathrm{e}^{\mathrm{i}k_r R_\mathrm{g}}}{(2\pi)^2}\int \mathrm{d}\theta_\phi \mathrm{d}\theta_R\, \mathrm{e}^{-\mathrm{i}m_\phi\theta_\phi}\, \mathrm{e}^{-\mathrm{i}m_r\theta_R}\, \mathrm{e}^{\mathrm{i}k_\phi\theta_\phi}$$
$$\times\, \mathrm{e}^{\mathrm{i}[k_r A_R \cos(\theta_R) - k_\phi \frac{2\Omega_\phi}{\kappa}\frac{A_R}{R_\mathrm{g}}\sin(\theta_R)]}\, \mathcal{B}_{R_0}(R_\mathrm{g}+A_R\cos(\theta_R))\,. \tag{3.28}$$

The integration on θ_ϕ is straightforward and gives $2\pi\delta_{m_\phi}^{k_\phi}$. Regarding the dependence on θ_R in the complex exponential, we write

$$k_r A_R \cos(\theta_R) - k_\phi \frac{2\Omega_\phi}{\kappa}\frac{A_R}{R_\mathrm{g}}\sin(\theta_R) = H_{k_\phi}(k_r)\, \sin(\theta_R+\theta_R^0)\,, \tag{3.29}$$

where we introduced the amplitude $H_{k_\phi}(k_r)$ and the phase shift θ_R^0 as

$$H_{k_\phi}(k_r) = A_R\,|k_r|\sqrt{1+\bigg[\frac{\Omega_\phi}{\kappa}\frac{2k_\phi}{k_r R_\mathrm{g}}\bigg]^2} \quad ; \quad \theta_R^0 = \tan^{-1}\bigg[-\frac{\kappa}{\Omega_\phi}\frac{k_r R_\mathrm{g}}{2k_\phi}\bigg]. \tag{3.30}$$

For typical galactic discs, one has $1/2 \leq \Omega_\phi \leq \kappa$ (Binney and Tremaine 2008). Assuming that k_ϕ is of order unity and relying on the WKB assumptions from Eq. (3.15), one can simplify Eq. (3.30) as

$$H_{k_\phi}(k_r) \simeq A_R\,|k_r| \quad ; \quad \theta_R^0 \simeq -\frac{\pi}{2}\,. \tag{3.31}$$

As we assumed the disc to be tepid, the radial oscillations of the stars are small so that $A_R \ll R_\mathrm{g}$. In Eq. (3.28), we may then get rid of the dependence on A_R in $\mathcal{B}_{R_0}(R_\mathrm{g}+A_R\cos(\theta_R))$ and replace it with $\mathcal{B}_{R_0}(R_\mathrm{g})$. The only remaining dependence with A_R in Eq. (3.28) is then in the complex exponential, and we are now in a position to explicitly perform the last integration on θ_R. To do so, let us recall the sum decomposition formula of the Bessel functions of the first kind $\mathcal{J}_\ell$, which reads

$$\mathrm{e}^{\mathrm{i}z\sin(\theta)} = \sum_{\ell\in\mathbb{Z}} \mathcal{J}_\ell[z]\,\mathrm{e}^{\mathrm{i}\ell\theta}\,. \tag{3.32}$$

The expression of the Fourier transformed WKB basis elements finally reads

$$\psi_{\boldsymbol{m}}^{[k_\phi,k_r,R_0]}(\boldsymbol{J}) = \delta_{m_\phi}^{k_\phi}\,\mathrm{e}^{\mathrm{i}k_r R_\mathrm{g}}\,\mathrm{e}^{\mathrm{i}m_r\theta_R^0}\,\mathcal{A}\,\mathcal{J}_{m_r}\big[H_{m_\phi}(k_r)\big]\,\mathcal{B}_{R_0}(R_\mathrm{g})\,. \tag{3.33}$$

3.4 WKB Razor-Thin Amplification Eigenvalues

After having explicitly constructed the WKB basis from Eq. (3.11), one may now compute the system's response matrix $\widehat{\mathbf{M}}$ from Eq. (2.17). Thanks to the Fourier transformed WKB basis elements from Eq. (3.33), one has to evaluate an expression of the form

$$\widehat{\mathbf{M}}_{[k_\phi^p,k_r^p,R_0^p],[k_\phi^q,k_r^q,R_0^q]}(\omega) = (2\pi)^2 \sum_{\boldsymbol{m}} \int \mathrm{d}\boldsymbol{J}\, \frac{\boldsymbol{m}\cdot\partial F/\partial \boldsymbol{J}}{\omega-\boldsymbol{m}\cdot\boldsymbol{\Omega}} \delta_{m_\phi}^{k_\phi^p} \delta_{m_\phi}^{k_\phi^q} \mathrm{e}^{\mathrm{i}(k_r^q-k_r^p)R_\mathrm{g}} \mathcal{A}_p \mathcal{A}_q$$
$$\times \mathcal{J}_{m_r}\!\left[\sqrt{\tfrac{2J_r}{\kappa}}k_r^p\right] \mathcal{J}_{m_r}\!\left[\sqrt{\tfrac{2J_r}{\kappa}}k_r^q\right] \mathcal{B}_{R_0^p}(R_\mathrm{g})\, \mathcal{B}_{R_0^q}(R_\mathrm{g})\,. \quad (3.34)$$

Let us now illustrate how the response matrix becomes diagonal in the WKB limit. One should first note how Eq. (3.34) is very similar to Eq. (3.21) where we discussed the biorthogonality of the WKB basis. In Eq. (3.34), the azimuthal Kronecker symbols impose $k_\phi^p = k_\phi^q$. Moreover, thanks to our assumption from Eq. (3.25) on the step distances of the basis elements, the product of the two Gaussian windows in R_g imposes $R_0^p = R_0^q$ to have a non-negligible contribution. In order to shorten temporarily the notations, let us introduce the function $h(R_\mathrm{g})$ defined as

$$h(R_\mathrm{g}) = \left|\frac{\mathrm{d}J_\phi}{\mathrm{d}R_\mathrm{g}}\right| \frac{\boldsymbol{m}\cdot\partial F/\partial \boldsymbol{J}}{\omega-\boldsymbol{m}\cdot\boldsymbol{\Omega}} \mathcal{A}_p\, \mathcal{A}_q\, \mathcal{J}_{m_r}\!\left[\sqrt{\tfrac{2J_r}{\kappa}}k_r^p\right] \mathcal{J}_{m_r}\!\left[\sqrt{\tfrac{2J_r}{\kappa}}k_r^q\right], \quad (3.35)$$

which encompasses all the additional slow radial dependences from Eq. (3.34). When estimated for $R_0^p = R_0^q$, Eq. (3.34) becomes

$$\int \mathrm{d}R_\mathrm{g}\, h(R_\mathrm{g})\, \mathrm{e}^{\mathrm{i}R_\mathrm{g}(k_r^q-k_r^p)} \exp\!\left[-\frac{(R_\mathrm{g}-R_0^p)^2}{\sigma^2}\right]. \quad (3.36)$$

This takes the form of a radial Fourier transform $\mathcal{F}_R$ at the frequency $\Delta k_r = k_r^p - k_r^q$. When rewritten as the convolution of two radial Fourier transforms, it becomes

$$(3.36) \sim \int \mathrm{d}k'\, \mathcal{F}_R[h](k')\, \exp\!\left[-\frac{(\Delta k_r - k')^2}{4/\sigma^2}\right]. \quad (3.37)$$

Because of the WKB assumption from Eq. (3.25) and the Gaussian from Eq. (3.37), one can note that if $\Delta k_r \neq 0$, the contribution from $\mathcal{F}_R[h](k')$ will come from the region $k' \sim \Delta k_r \gg 1/\sigma$. We assume that the properties of the disc are slowly varying with the radius, so that the function h has a radial Fourier transform limited to the frequency region $|k'| \lesssim 1/\sigma$. Non-negligible contributions to the response matrix can then only be obtained for $\Delta k_r = k_r^p - k_r^q = 0$. As a conclusion, we have shown that within the WKB framework, the response matrix from Eq. (2.17) can be assumed to be diagonal. To shorten the notations, let us denote the response matrix eigenvalues as

$$\lambda_{[k_\phi,k_r,R_0]}(\omega) = \widehat{\mathbf{M}}_{[k_\phi,k_r,R_0],[k_\phi,k_r,R_0]}(\omega)\,. \quad (3.38)$$

The last step of the present computation is to compute the integrals over J_ϕ and J_r in Eq. (3.34) to obtain an explicit expression of the response matrix diagonal coefficients. First, the WKB scale decoupling assumption allows us to replace the Gaussian from Eq. (3.36) by a Dirac delta $\delta_\mathrm{D}(R_\mathrm{g}-R_0^p)$ (one should pay a careful attention to the correct normalisation of the Gaussian). Equation (3.34) then becomes

$$\lambda_{[k_\phi,k_r,R_0]}(\omega) = (2\pi)^2 \mathcal{A}^2 \left| \frac{\mathrm{d}J_\phi}{\mathrm{d}R_\mathrm{g}} \right|_{R_0} \sum_{\boldsymbol{m}} \delta_{m_\phi}^{k_\phi} \int \mathrm{d}J_r \, \frac{\boldsymbol{m}\cdot\partial F/\partial \boldsymbol{J}}{\omega - \boldsymbol{m}\cdot\boldsymbol{\Omega}} \, \mathcal{J}_{m_r}^2\!\left[\sqrt{\tfrac{2J_r}{\kappa}}\,k_r\right]. \tag{3.39}$$

Here, the azimuthal Kronecker symbol allows us to execute the sum on m_ϕ. In addition, the intrinsic frequencies from Eqs. (3.5) and (3.7) immediately give

$$\left| \frac{\mathrm{d}J_\phi}{\mathrm{d}R_\mathrm{g}} \right|_{R_0} = \frac{R_0 \kappa^2}{2\Omega_\phi}. \tag{3.40}$$

As the disc is assumed to be tepid, we may assume that $|\partial F/\partial J_\phi| \ll |\partial F/\partial J_r|$, so that only the DF's gradient w.r.t. the radial action may be kept in Eq. (3.39). Using the expression of the quasi-isothermal DF from Eq. (3.10) and the basis amplitude from Eq. (3.27), Eq. (3.39) becomes

$$\lambda_{[k_\phi,k_r,R_0]}(\omega) = \frac{2\pi G\Sigma|k_r|}{\kappa^2} \frac{\kappa^4}{k_r^2\sigma_r^4} \sum_{m_r} \int \mathrm{d}J_r \, \frac{-m_r \exp[-\kappa J_r/\sigma_r^2]}{\omega - k_\phi\Omega_\phi - m_r\kappa} \, \mathcal{J}_{m_r}^2\!\left[\sqrt{\tfrac{2J_r}{\kappa}}\,k_r\right]. \tag{3.41}$$

We now rely on the integration formula (see formula (6.615) in Gradshteyn and Ryzhik (2007))

$$\int_0^{+\infty} \mathrm{d}J_r \, \mathrm{e}^{-\alpha J_r} \, \mathcal{J}_{m_r}^2\!\left[\beta\sqrt{J_r}\right] = \frac{\mathrm{e}^{-\beta^2/(2\alpha)}}{\alpha} \mathcal{I}_{m_r}\!\left[\frac{\beta^2}{2\alpha}\right], \tag{3.42}$$

where $\alpha > 0$, $\beta > 0$, $m_r \in \mathbb{Z}$, and $\mathcal{I}_{m_r}$ are modified Bessel functions of the first kind. We apply this formula with $\alpha = \kappa/\sigma_r^2$ and $\beta = \sqrt{2k_r^2/\kappa}$. We also introduce the notation

$$\chi_r = \frac{\sigma_r^2 k_r^2}{\kappa^2}, \tag{3.43}$$

so that Eq. (3.41) becomes

$$\lambda_{[k_\phi,k_r,R_0]}(\omega) = \frac{2\pi G\Sigma|k_r|}{\kappa^2} \frac{\kappa}{\chi_r} \sum_{m_r} \frac{-m_r \mathrm{e}^{-\chi_r} \mathcal{I}_{m_r}[\chi_r]}{\omega - k_\phi\Omega_\phi - m_r\kappa}. \tag{3.44}$$

Let us introduce as well the dimensionless shifted frequency s as

$$s = \frac{\omega - k_\phi\Omega_\phi}{\kappa}. \tag{3.45}$$

Using the property $\mathcal{I}_{-m_r}[\chi_r] = \mathcal{I}_{m_r}[\chi_r]$, we may finally rewrite Eq. (3.44) introducing the reduction factor (Kalnajs 1965; Lin and Shu 1966)

$$\mathcal{F}(s, \chi_r) = 2(1-s^2)\frac{\mathrm{e}^{-\chi_r}}{\chi_r}\sum_{m_r=1}^{+\infty}\frac{\mathcal{I}_{m_r}[\chi_r]}{1-[s/m_r]^2}. \tag{3.46}$$

This allows for a final rewriting of the response matrix eigenvalues in the tightly wound limit as

$$\widehat{\mathbf{M}}_{[k_\phi^p,k_r^p,R_0^p],[k_\phi^q,k_r^q,R_0^q]}(\omega) = \delta_{k_\phi^p}^{k_\phi^q}\delta_{k_r^p}^{k_r^q}\delta_{R_0^p}^{R_0^q}\frac{2\pi G\Sigma|k_r|}{\kappa^2(1-s^2)}\mathcal{F}(s, \chi_r). \tag{3.47}$$

These eigenvalues are in full agreement with the seminal results from Kalnajs (1965) and Lin and Shu (1966), which derived a WKB dispersion relation for razor-thin axisymmetric discs. In order to deal with the singularity of the previous expression when evaluated for $s = n \in \mathbb{Z}$, one should add a small imaginary part to ω, so that $s = n + \mathrm{i}\eta$. As long as η is small compared to the imaginary part of the least damped mode of the disc, adding this complex part makes a negligible contribution to $\mathrm{Re}(\lambda)$. Equation (3.47) is an important result, as it allows us to estimate straightforwardly the strength of the tightly wound self-gravitating amplification in the disc.

In order to illustrate the physical content of Eq. (3.47), let us now briefly describe how these amplification eigenvalues allow for the recovery of Toomre's stability parameter Q (Toomre 1964). This parameter characterises the local stability of an axisymmetric razor-thin stellar disc w.r.t. local tightly wound axisymmetric perturbations. We are interested in the stability w.r.t. axisymmetric modes, so that we impose $k_\phi = 0$. We place ourselves at the stability limit given by $\omega = 0$, so that Eq. (3.47) imposes $s = 0$. We now seek a criterion on the disc's parameters such that there exists no $k_r > 0$ for which $\lambda(k_r) = 1$, i.e. so that the disc is stable. Thanks to Eq. (3.47), one has

$$\lambda(k_r) = \frac{2\pi G\Sigma k_r}{\kappa^2}\mathcal{F}(0, \chi_r) = \frac{2\pi G\Sigma}{\kappa\sigma_r}K^0(\chi_r), \tag{3.48}$$

where we introduced the function $K^0(\chi_r) = \sqrt{\chi_r}\,\mathcal{F}(0, \chi_r) = (1-\mathrm{e}^{-\chi_r}\mathcal{I}_0[\chi_r])/\sqrt{\chi_r}$. The shape of the function $\chi_r \mapsto K^0(\chi_r)$ is illustrated in Fig. 3.5. From Fig. 3.5, one can note that K_0 reaches a maximum $K_{\max}^0 \simeq 0.534$ for $\chi_{\max}^0 \simeq 0.948$. As a consequence, one always has $\lambda(k_r) \leq (2\pi G\Sigma K_{\max}^0)/(\kappa\sigma_r)$. Noting that $2\pi K_{\max}^0 \simeq 3.36$, we may finally introduce the local stability parameter Q as

$$Q(J_\phi) = \frac{\sigma_r(J_\phi)\kappa(J_\phi)}{3.36\, G\, \Sigma(J_\phi)}. \tag{3.49}$$

Here, Q corresponds to the local razor-thin Toomre's parameter (Toomre 1964), which for $Q > 1$ ensures the local stability of a razor-thin stellar disc w.r.t. axisymmetric tightly wound perturbations. The straightforward derivation of this stability parameter starting from the amplification eigenvalues obtained in Eq. (3.47) illustrates how the razor-thin WKB basis introduced in Eq. (3.11) is in full agreement with previous seminal results on the WKB linear theory of razor-thin discs.

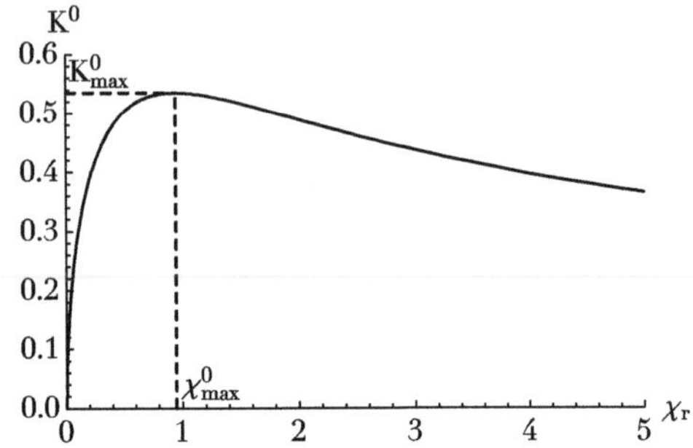

Fig. 3.5 Illustration of the behaviour of the function $\chi_r \mapsto K^0(\chi_r)$, for which one can identify the maximum amplification $K^0_{\max} \simeq 0.534$ reached for $\chi^0_{\max} \simeq 0.948$. This maximum is directly related to Toomre's Q parameter

3.5 WKB Limit for the Collisionless Diffusion

Let us now illustrate how the previous WKB calculations allow for the calculation of the secular collisionless diffusion coefficients introduced in Eq. (2.31). In order to simplify the notations, the WKB basis elements from Eq. (3.11) will be noted as

$$\psi^{(p)} = \psi^{[k_\phi^p, k_r^p, R_0^p]} \,. \tag{3.50}$$

We have shown in Eq. (3.47) that within the WKB limit, the response matrix is diagonal. Introducing its eigenvalues as λ_p, one has $\widehat{\mathbf{M}}_{pq} = \lambda_p \delta_p^q$. The matrix $[\mathbf{I} - \widehat{\mathbf{M}}]^{-1}$ is then diagonal and the diffusion coefficients from Eq. (2.32) take the form

$$D_{\boldsymbol{m}}(\boldsymbol{J}) = \frac{1}{2} \sum_{p,q} \psi_{\boldsymbol{m}}^{(p)}(\boldsymbol{J})\, \psi_{\boldsymbol{m}}^{(q)*}(\boldsymbol{J})\, \frac{1}{1-\lambda_p} \frac{1}{1-\lambda_q}\, \widehat{\mathbf{C}}_{pq}(\boldsymbol{m}\cdot\boldsymbol{\Omega}) \,, \tag{3.51}$$

where the sums on p and q run over the WKB basis elements. We recall that the basis elements $\psi_{\boldsymbol{m}}^{(p)}(\boldsymbol{J})$ as well as the matrix eigenvalues λ_p do not change from one realisation to another, so that using the definition of the perturbation autocorrelation from Eq. (2.26), we may rewrite Eq. (3.51) as

$$D_{\boldsymbol{m}}(\boldsymbol{J}) = \left\langle \frac{1}{2\pi} \int \mathrm{d}\omega' \, \frac{1}{2} \sum_{p,q} \psi_{\boldsymbol{m}}^{(p)}(\boldsymbol{J})\, \psi_{\boldsymbol{m}}^{(q)*}(\boldsymbol{J}) \frac{1}{1-\lambda_p} \frac{1}{1-\lambda_q}\, \widehat{b}_p(\boldsymbol{m}\cdot\boldsymbol{\Omega})\, \widehat{b}_q^*(\omega') \right\rangle . \tag{3.52}$$

In Eq. (3.52), note that the amplification eigenvalues λ_p, λ_q, and the basis coefficient $\widehat{b}_p$ are both evaluated at the intrinsic frequency $\boldsymbol{m}\cdot\boldsymbol{\Omega}$, while $\widehat{b}_q^*$ is evaluated at the dummy frequency ω'. In order to shorten the upcoming calculations, the frequencies of evaluation, when obvious, will not be explicitly written out. Let us now rely on the explicit expression of the Fourier transformed WKB basis elements from Eq. (3.33), so that Eq. (3.52) becomes

$$D_{\boldsymbol{m}}(\boldsymbol{J}) = \bigg\langle \frac{1}{2\pi} \!\int\! \mathrm{d}\omega' \sum_{k_r^p, k_r^q, R_0^p, R_0^q} \frac{1}{2} \frac{G}{\sqrt{R_0^p R_0^q}} \frac{1}{\sqrt{|k_r^p k_r^q|}} \mathcal{J}_{m_r}\bigg[\sqrt{\tfrac{2J_r}{\kappa}} k_r^p\bigg] \mathcal{J}_{m_r}\bigg[\sqrt{\tfrac{2J_r}{\kappa}} k_r^q\bigg] \mathrm{e}^{\mathrm{i}R_{\mathrm{g}}(k_r^p - k_r^q)}$$
$$\times \frac{1}{1-\lambda_p} \frac{1}{1-\lambda_q} \frac{1}{\sqrt{\pi\sigma^2}} \exp\bigg[-\frac{(R_{\mathrm{g}} - R_0^p)^2}{2\sigma^2}\bigg] \exp\bigg[-\frac{(R_{\mathrm{g}} - R_0^q)^2}{2\sigma^2}\bigg] \widehat{b}_p \widehat{b}_q^* \bigg\rangle . \tag{3.53}$$

Note that in Eq. (3.53), we got rid of the sum on k_ϕ^p and k_ϕ^q as Eq. (3.33) imposes $m_\phi = k_\phi^p = k_\phi^q$. The next step of the calculation is to rewrite Eq. (3.53) so as to be independent from the exact choice of the WKB basis, i.e. the precise value of σ. To do so, one should replace the basis coefficients $\widehat{b}_p$ by expressions involving only the true external potential perturbation $\delta\psi^{\mathrm{e}}$. Relying on the biorthogonality property of the basis elements imposed in Eq. (2.12), the basis coefficients $\widehat{b}_p$ are immediately given by

$$\widehat{b}_p(\omega) = -\!\int\! \mathrm{d}\boldsymbol{x}\, \Sigma^{(p)*}(\boldsymbol{x})\, \delta\widehat{\psi^{\mathrm{e}}}(\boldsymbol{x}, \omega) , \tag{3.54}$$

where the $\widehat{\cdot}$ corresponds to the temporal Fourier transform as defined in Eq. (2.9). Thanks to the explicit expression of the WKB surface density elements obtained in Eq. (3.19), some simple algebra (see Fouvry et al. (2015c) for details) easily leads to the relation

$$\widehat{b}_p(\omega) = \sqrt{\frac{|k_r^p| R_0^p}{G}} \frac{2\pi}{(\pi\sigma^2)^{1/4}} \mathrm{e}^{-\mathrm{i}R_0^p k_r^p}\, \delta\widehat{\psi^{\mathrm{e}}}_{m_\phi, k_r^p}[R_0^p, \omega] . \tag{3.55}$$

In Eq. (3.55), the exterior potential $\delta\widehat{\psi^{\mathrm{e}}}$ has been transformed according to two transformations: (i) an azimuthal Fourier transform of indice m_ϕ, (ii) a local radial Fourier transform centred around R_0^p at the frequency k_r^p. These two transformations are defined as

$$\text{(i):}\;\; f_{m_\phi} = \frac{1}{2\pi} \!\int\! \mathrm{d}\phi\, f[\phi]\, \mathrm{e}^{-\mathrm{i}m_\phi\phi} ,$$
$$\text{(ii):}\;\; f_{k_r}[R_0] = \frac{1}{2\pi} \!\int\! \mathrm{d}R\, \mathrm{e}^{-\mathrm{i}k_r(R-R_0)} \exp\bigg[-\frac{(R-R_0)^2}{2\sigma^2}\bigg] f[R] . \tag{3.56}$$

Equation (3.55) therefore allowed us to express the basis coefficients $\widehat{b}_p$ as a function of the exterior perturbing potential $\delta\widehat{\psi^{\mathrm{e}}}$. Using this relation and disentangling the sums on (k_r^p, R_0^p) and (k_r^q, R_0^q), Eq. (3.53) can be rewritten as

$$D_{\boldsymbol{m}}(\boldsymbol{J}) = \bigg\langle \frac{1}{2\pi} \!\int\! \mathrm{d}\omega'\, g(\boldsymbol{m}\cdot\boldsymbol{\Omega})\, g^*(\omega') \bigg\rangle , \tag{3.57}$$

where we defined the function $g(\omega)$ as

$$g(\omega) = 2\pi \sum_{k_r^p, R_0^p} g_s(k_r^p, R_0^p, \omega)\, \mathrm{e}^{\mathrm{i}(R_g - R_0^p)k_r^p}\, \mathcal{G}_r(R_g - R_0^p)\,. \tag{3.58}$$

In Eq. (3.58), $\mathcal{G}_r(R) = 1/\sqrt{2\pi\sigma^2}\mathrm{e}^{-R^2/(2\sigma^2)}$ is a normalised Gaussian of width σ, and g_s encompasses all the slow dependences of the diffusion coefficients w.r.t. the radial position so that

$$g_s(k_r^p, R_0^p, \omega) = \mathcal{J}_{m_r}\left[\sqrt{\tfrac{2J_r}{\kappa}}k_r^p\right] \frac{1}{1-\lambda_{k_r^p}}\, \delta\widehat{\psi^{\mathrm{e}}}_{m_\phi, k_r^p}[R_0^p, \omega]\,. \tag{3.59}$$

Next, let us replace the sums on k_r^p and R_0^p appearing in Eq. (3.58) by continuous integrals. To do so, we rely on Riemann sum formula $\sum f(x)\Delta x \simeq \int \mathrm{d}x f(x)$. One can note in the discrete sums from Eq. (3.58) that the basis elements are separated by step distances Δk_r and ΔR_0. We suppose that generically k_r^p and R_0^p are given by $k_r^p = n_k \Delta k_r$, and $R_0^p = R_g + n_r \Delta R_0$, where n_k is a strictly positive integer and n_r is an integer that can be both positive or negative. In addition, one can note in Eq. (3.58) the presence of a rapidly evolving complex exponential, which may cancel out the diffusion coefficients if the basis step distances are not chosen carefully. When summed over the basis elements, this complex exponential has the dependence

$$\exp\big[\mathrm{i}(R_g - (R_g + n_r \Delta R_0)) n_k \Delta k_r\big] = \exp\big[-\,\mathrm{i} n_r n_k \Delta R_0 \Delta k_r\big]\,. \tag{3.60}$$

As a consequence, since $n_r n_k$ is an integer, in order not to have any contributions from the complex exponential in Eq. (3.58), one should choose the step distances so that

$$\Delta R_0\, \Delta k_r = 2\pi\,. \tag{3.61}$$

Such a choice corresponds to a critical sampling condition (Gabor 1946; Daubechies 1990). As illustrated in Eq. (3.60), this allows us to leave out the complex exponential from Eq. (3.58) when performing the change to continuous expressions. This transformation is a subtle stage of the calculation, since the step distances should be simultaneously large to comply with the WKB constraints from Eq. (3.25) and small to justify the use of Riemann sum formula. In this process, as the radial Gaussian in Eq. (3.58) is sufficiently peaked and correctly normalised, it may be replaced by $\delta_{\mathrm{D}}(R_g - R_0^p)$. Equation (3.58) finally becomes

$$g(\omega) = \int \mathrm{d}k_r^p\, g_s(k_r^p, R_g, \omega)\,. \tag{3.62}$$

Let us now define the autocorrelation $\widehat{C}_{\delta\psi^{\mathrm{e}}}$ of the external perturbations as

$$\widehat{C}_{\delta\psi^{\mathrm{e}}}[m_\phi, \omega, R_g, k_r^p, k_r^q] = \frac{1}{2\pi}\int \mathrm{d}\omega' \left\langle \delta\widehat{\psi^{\mathrm{e}}}_{m_\phi, k_r^p}[R_g, \omega]\, \delta\widehat{\psi^{\mathrm{e}}}^{*}_{m_\phi, k_r^q}[R_g, \omega'] \right\rangle. \tag{3.63}$$

The expression (3.57) of the diffusion coefficients then takes the form

$$D_{\boldsymbol{m}}(\boldsymbol{J}) = \int \mathrm{d}k_r^p \, \mathcal{J}_{m_r}\!\left[\sqrt{\tfrac{2J_r}{\kappa}}\, k_r^p\right] \frac{1}{1-\lambda_{k_r^p}} \int \mathrm{d}k_r^q \, \mathcal{J}_{m_r}\!\left[\sqrt{\tfrac{2J_r}{\kappa}}\, k_r^q\right] \frac{1}{1-\lambda_{k_r^q}} \, \widehat{C}_{\delta\psi^{\mathrm{e}}}[m_\phi, \boldsymbol{m}\cdot\boldsymbol{\Omega}, R_{\mathrm{g}}, k_r^p, k_r^q] \, , \tag{3.64}$$

where the amplification eigenvalues, λ_{k_r}, are given by Eq. (3.47) and read

$$\lambda_{k_r}[R_{\mathrm{g}}, \boldsymbol{m}\cdot\boldsymbol{\Omega}] = \frac{2\pi G \Sigma |k_r|}{\kappa^2 (1-s^2)} \, \mathcal{F}(s, \chi) \, . \tag{3.65}$$

Let us now further simplify the diffusion coefficients from Eq. (3.64) by assuming some stationarity properties on the stochasticity of the external perturbations $\delta\psi^{\mathrm{e}}$. We assume that these are spatially quasi-stationary and satisfy

$$\big\langle \delta\psi^{\mathrm{e}}_{m_\phi}[R_1, t_1] \, \delta\psi^{\mathrm{e}*}_{m_\phi}[R_2, t_2] \big\rangle = C[m_\phi, t_1 - t_2, (R_1+R_2)/2, R_1 - R_2] \, , \tag{3.66}$$

where the dependence of the autocorrelation function C w.r.t. $(R_1+R_2)/2$ is supposed to be slow. Thanks to some simple algebra (see Appendix C of Fouvry et al. (2015c) for details), one can write

$$\big\langle \widehat{\delta\psi^{\mathrm{e}}}_{m_\phi, k_r^1}[R_{\mathrm{g}}, \omega_1] \, \widehat{\delta\psi^{\mathrm{e}}}^{*}_{m_\phi, k_r^2}[R_{\mathrm{g}}, \omega_2] \big\rangle = 2\pi \delta_{\mathrm{D}}(\omega_1 - \omega_2) \, \delta_{\mathrm{D}}(k_r^1 - k_r^2) \, \widehat{C}[m_\phi, \omega_1, R_{\mathrm{g}}, k_r^1] \, , \tag{3.67}$$

where $\widehat{C}[...]$ has been transformed twice, according to a temporal Fourier transform as defined in Eq. (2.9), and according to a local radial Fourier transform as in Eq. (3.56) of spread $\sqrt{2}\sigma$ w.r.t. $R_1 - R_2$ in the neighbourhood of $R_1 - R_2 = 0$ and $(R_1+R_2)/2 = R_{\mathrm{g}}$. Here, note that in Eq. (3.67), the autocorrelation was diagonalised w.r.t. ω and k_r, as can be seen from the two Dirac deltas. The diffusion coefficients from Eq. (3.64) then take the simple form

$$D_{\boldsymbol{m}}(\boldsymbol{J}) = \int \mathrm{d}k_r \, \mathcal{J}^2_{m_r}\!\left[\sqrt{\tfrac{2J_r}{\kappa}}\, k_r\right] \left[\frac{1}{1-\lambda_{k_r}}\right]^2 \widehat{C}[m_\phi, \boldsymbol{m}\cdot\boldsymbol{\Omega}, R_{\mathrm{g}}, k_r] \, . \tag{3.68}$$

This explicit expression of the collisionless diffusion coefficients is the main result of this section. Equation (3.68) is indeed a simple quadrature involving the power spectrum of the external fluctuations at the resonant frequencies boosted by the eigenvalues of the gravitational susceptibility squared.

In some situations, one may further simplify Eq. (3.68), thanks to the so-called approximation of the small denominators, which amounts to focusing on the waves that yield the maximum amplification. Indeed, let us assume that the function $k_r \mapsto \lambda(k_r)$ is a sharp function reaching a maximum value $\lambda_{\mathrm{max}}(R_{\mathrm{g}}, \omega = \boldsymbol{m}\cdot\boldsymbol{\Omega})$ for $k_r = k_r^{\mathrm{max}}(R_{\mathrm{g}}, \omega)$. One can then introduce the two frequency bounds k_r^{inf} and k_r^{sup}, such that $\lambda(k_r^{\mathrm{inf/sup}}) = \lambda_{\mathrm{max}}/2$. The characteristic spread of the region of maximum amplification is then given by $\Delta k_\lambda(R_{\mathrm{g}}, \omega) \simeq k_r^{\mathrm{sup}} - k_r^{\mathrm{inf}}$. Focusing only on this region,

Eq. (3.68) can be approximated as

$$D_{\boldsymbol{m}}(\boldsymbol{J}) = \Delta k_\lambda \, \mathcal{J}_{m_r}^2\left[\sqrt{\tfrac{2J_r}{\kappa}}\, k_r^{\max}\right]\left[\frac{1}{1-\lambda_{\max}}\right]^2 \widehat{\mathcal{C}}[m_\phi, \boldsymbol{m}\cdot\boldsymbol{\Omega}, R_{\mathrm{g}}, k_r^{\max}]\,. \tag{3.69}$$

The previous approximation can also be improved by performing the integration for $k_r \in \big[k_r^{\inf}; k_r^{\sup}\big]$. Such an approach is more numerically demanding but does not alter the conclusions drawn in the applications presented in Sect. 3.7. In Eq. (3.69), one should note that the external perturbation autocorrelation $\widehat{\mathcal{C}}$, which sources the diffusion coefficients, depends on four different parameters: the azimuthal wavenumber m_ϕ, the local intrinsic frequency of the system $\boldsymbol{m}\cdot\boldsymbol{\Omega}$, the location in the disc via R_{g}, and finally the radial frequency $k_r^{\max}$ of the most amplified tightly wound perturbation at this location. As a conclusion, thanks to the explicit WKB basis introduced in Eq. (3.11), we obtained in Eqs. (3.68) and (3.69) explicit expressions for the system's externally-induced diffusion coefficients, whose evaluations are now straightforward. In Sect. 3.7, we illustrate how this formalism may be applied to recover the important features observed in numerical simulations of the long-term evolution of stable quasi-stationary isolated and self-gravitating stellar discs.

3.6 WKB Limit for the Collisional Diffusion

In this section, let us now emphasise how the previous WKB calculations also allow for the calculation of the dressed susceptibility coefficients, as well as the collisional drift and diffusion coefficients, appearing in the inhomogeneous Balescu-Lenard equation (2.67). Here, rather than considering a situation where the disc evolves as a result of external stochastic perturbations, we consider the collisional case, where the source of secular evolution is finite$-N$ fluctuations. In this context, we will especially emphasise how the WKB approximation allows us to deal with the resonance condition present in the Balescu-Lenard equation.

A crucial property of the WKB basis from Eq. (3.11) is that the response matrix, $\widehat{\mathbf{M}}$, becomes diagonal, as shown in Eq. (3.47). Using the same shortening notations as in Eq. (3.50), one can rewrite the Balescu-Lenard susceptibility coefficients from Eq. (2.50) as

$$\frac{1}{\mathcal{D}_{\boldsymbol{m}_1,\boldsymbol{m}_2}(\boldsymbol{J}_1,\boldsymbol{J}_2,\omega)} = \sum_p \psi_{\boldsymbol{m}_1}^{(p)}(\boldsymbol{J}_1)\,\frac{1}{1-\lambda_p(\omega)}\,\psi_{\boldsymbol{m}_2}^{(p)*}(\boldsymbol{J}_2)\,. \tag{3.70}$$

Thanks to the expression of the Fourier transformed WKB basis elements from Eq. (3.33), this becomes

$$\frac{1}{\mathcal{D}_{\mathbf{m}_1,\mathbf{m}_2}(\mathbf{J}_1,\mathbf{J}_2,\omega)} = \sum_{k_\phi^p,k_r^p,R_0^p} \delta_{m_1^\phi}^{k_\phi^p} \delta_{m_2^\phi}^{k_\phi^p} \frac{G}{k_r^p R_0^p} \frac{1}{1-\lambda_p} \mathcal{J}_{m_1^r}\bigg[\sqrt{\frac{2J_r^1}{\kappa_1}}\, k_r^p\bigg] \mathcal{J}_{m_2^r}\bigg[\sqrt{\frac{2J_r^2}{\kappa_2}}\, k_r^p\bigg] \mathrm{e}^{\mathrm{i}k_r^p(R_1-R_2)} \mathrm{e}^{\mathrm{i}\theta_R^{0p}(m_1^r-m_2^r)}$$
$$\times \frac{1}{\sqrt{\pi\sigma^2}} \exp\bigg[-\frac{(R_1-R_0^p)^2}{2\sigma^2}\bigg] \exp\bigg[-\frac{(R_2-R_0^p)^2}{2\sigma^2}\bigg], \tag{3.71}$$

where we used the shortening notations $\kappa_i = \kappa(\mathbf{J}_i)$ and $R_i = R_{\mathrm{g}}(\mathbf{J}_i)$. The azimuthal Kronecker symbols immediately impose

$$m_1^\phi = m_2^\phi = k_\phi^p, \tag{3.72}$$

so that the sum on k_ϕ^p is limited to only one term. Before proceeding further with the evaluation of the susceptibility coefficients, let us first emphasise an additional consequence of the WKB assumptions, which is the restriction to local resonances.

Note that the Balescu-Lenard drift and diffusion coefficients from Eqs. (2.69) and (2.70) involve an integration over the dummy variable $\mathbf{J}_2$. For given values of $\mathbf{J}_1, \mathbf{m}_1$, and $\mathbf{m}_2$, this should be seen as a scan of the entire action space, searching for resonant regions, where the resonance constraint $\mathbf{m}_1 \cdot \mathbf{\Omega}_1 - \mathbf{m}_2 \cdot \mathbf{\Omega}_2 = 0$ is satisfied (see Fig. 2.4). As we placed ourselves within the epicyclic approximation, the intrinsic frequencies $\mathbf{\Omega} = (\Omega_\phi, \kappa)$ from Eqs. (3.5) and (3.7) only depend on the action J_ϕ. This significantly simplifies the resonance condition. For a given value of $R_1 = R_{\mathrm{g}}(\mathbf{J}_1)$, $\mathbf{m}_1$, and $\mathbf{m}_2$, one has to find the resonant radii R_2^{r} such that the resonance condition $f(R_2^{\mathrm{r}}) = 0$ is satisfied, where we defined the function $f(R_2^{\mathrm{r}})$ as

$$f(R_2^{\mathrm{r}}) = \mathbf{m}_1 \cdot \mathbf{\Omega}(R_1) - \mathbf{m}_2 \cdot \mathbf{\Omega}(R_2^{\mathrm{r}}). \tag{3.73}$$

After having identified the resonance radii R_2^{r}, one can then rely on the rule for the composition of a Dirac delta and a smooth function, which reads

$$\delta_{\mathrm{D}}(f(x)) = \sum_{y \in Z_f} \frac{\delta_{\mathrm{D}}(x-y)}{|f'(y)|}, \tag{3.74}$$

where $Z_f = \{y \mid f(y) = 0\}$ is the set of all the poles of f. Equation (3.74) also assumes that all the poles of f are simple (i.e. non-degenerate), which in our context amounts to assuming that

$$\frac{\mathrm{d}(\mathbf{m}_2 \cdot \mathbf{\Omega})}{\mathrm{d}R}\bigg|_{R_2^{\mathrm{r}}} \neq 0. \tag{3.75}$$

As long as the rates of change of the two intrinsic frequencies are not in a rational ratio, resonance poles will be simple. Note that the harmonic case, for which $\kappa = 2\Omega_\phi$, and the Keplerian case, for which $\kappa = \Omega_\phi$, are in this sense degenerate. Such dynamical degeneracies, which occur for example in the vicinity of supermassive black holes or for protoplanetary discs, require a more involved evaluation of the Balescu-Lenard collision operator, and will be considered in detail in Chap. 6.

As noted in Eq. (3.72), in order to have non-zero susceptibility coefficients, one must have $m_1^\phi = m_2^\phi$. As a consequence, the resonance requirement from Eq. (3.73) takes the form

$$m_1^\phi \Omega_\phi(R_1) + m_1^r \kappa(R_1) = m_1^\phi \Omega_\phi(R_2^{\mathrm{r}}) + m_2^r \kappa(R_2^{\mathrm{r}}) . \tag{3.76}$$

Note in Eq. (3.71) the presence of narrow radial Gaussians in R_1 and R_2. As a consequence, the relevant resonant radii R_2^{r} must necessarily be close to R_1, so that $|\Delta R| = |R_2^{\mathrm{r}} - R_1| \lesssim (\text{few})\, \sigma$. In this limit, one can rewrite Eq. (3.76) as

$$\left[m_2^\phi \frac{\mathrm{d}\Omega_\phi}{\mathrm{d}R} + m_2^r \frac{\mathrm{d}\kappa}{\mathrm{d}R} \right] \Delta R = \left[m_1^r - m_2^r \right] \kappa(R_1) . \tag{3.77}$$

On the one hand, in the l.h.s. of Eq. (3.77), the term within brackets is non-zero, because of our assumption of non-degeneracy from Eq. (3.75). Moreover, because of the WKB scale decoupling approach, the additional prefactor ΔR is small. On the other hand, the r.h.s. of Eq. (3.77) is discrete: it is either zero, or at least of the order of $\kappa(R_1)$. To be satisfied, Eq. (3.77) therefore imposes that its two sides should be equal to 0. As a consequence, within the WKB limit, only local resonances are allowed so that

$$R_2^{\mathrm{r}} = R_1 \quad ; \quad m_2^r = m_1^r . \tag{3.78}$$

This is an important consequence of the WKB approximation. This forbids distant orbits to resonate, and allows for an explicit calculation of the collision operator.

As a result of this restriction, let us proceed with the evaluation of the dressed susceptibility coefficients from Eq. (3.71), by restricting ourselves only to $\boldsymbol{m}_2 = \boldsymbol{m}_1$ and $R_2 = R_1$. Equation (3.71) becomes

$$\frac{1}{\mathcal{D}_{\boldsymbol{m}_1,\boldsymbol{m}_1}} = \sum_{k_r^p, R_0^p} \frac{G}{k_r^p R_0^p} \frac{1}{1-\lambda_p} \mathcal{J}_{m_1^r}\!\left[\sqrt{\frac{2J_r^1}{\kappa_1}} k_r^p \right] \mathcal{J}_{m_1^r}\!\left[\sqrt{\frac{2J_r^2}{\kappa_1}} k_r^p \right] \frac{1}{\sqrt{\pi\sigma^2}} \exp\!\left[-\frac{(R_1 - R_0^p)^2}{\sigma^2} \right] , \tag{3.79}$$

where we introduced the shortening notation $1/\mathcal{D}_{\boldsymbol{m}_1,\boldsymbol{m}_1} = 1/\mathcal{D}_{\boldsymbol{m}_1,\boldsymbol{m}_1}(R_1, J_r^1, R_1, J_r^2, \omega)$. As in Eq. (3.62), the next step of our calculation is to replace the discrete sums on the indices k_r^p and R_0^p in Eq. (3.79) by continuous integrals. As previously, we rely on Riemann sum formula, and assume that the step distances of the WKB basis ΔR_0 and Δk_r satisfy the critical sampling condition from Eq. (3.61), i.e. one has $\Delta R_0 \Delta k_r = 2\pi$. We also note in Eq. (3.79) the presence of a narrow radial Gaussian in $(R_1 - R_0^p)$. As it is correctly normalised, we may replace it with a Dirac delta $\delta_{\mathrm{D}}(R_1 - R_0^p)$. Equation (3.79) becomes

$$\frac{1}{\mathcal{D}_{\boldsymbol{m}_1,\boldsymbol{m}_1}} = \frac{1}{2\pi} \frac{G}{R_1} \int_{1/\sigma_k}^{+\infty} \mathrm{d}k_r \, \frac{1}{k_r} \frac{1}{1-\lambda_{k_r}(R_1, \omega)} \mathcal{J}_{m_1^r}\!\left[\sqrt{\frac{2J_r^1}{\kappa_1}} k_r \right] \mathcal{J}_{m_1^r}\!\left[\sqrt{\frac{2J_r^2}{\kappa_1}} k_r \right] , \tag{3.80}$$

where we introduced a cut-off at $1/\sigma_k$ for the integration on k_r. This bound is justified by the WKB constraint from Eq. (3.25), which imposes that the radial frequency k_r is bounded from below and avoids the divergence associated with the factor $1/k_r$. It is also important to recall that the susceptibility coefficients should only be evaluated at $R_2 = R_1$ and $\boldsymbol{m}_2 = \boldsymbol{m}_1$, as a result of the restriction to local resonances obtained in Eq. (3.78). The explicit expression of the susceptibility coefficients from Eq. (3.80) constitutes the main result of this section. Equation (3.80) also implies that only orbits with similar J_r^1 and J_r^2 contribute significantly, i.e. the resonances are local.

Finally, following Eq. (3.69), one can further simplify Eq. (3.80) by relying on the approximation of the small denominators. This amounts to assuming that the biggest contribution to the susceptibility coefficients comes from the tightly wound waves with the largest λ_{k_r}. With the same notations than in Eq. (3.69), one can write

$$\frac{1}{\mathcal{D}_{\boldsymbol{m}_1,\boldsymbol{m}_1}} = \frac{1}{2\pi} \frac{G}{R_1} \frac{\Delta k_\lambda}{k_r^{\max}} \frac{1}{1-\lambda_{\max}} \mathcal{J}_{m_1^r}\bigg[\sqrt{\frac{2J_r^1}{\kappa_1}} k_r^{\max}\bigg] \mathcal{J}_{m_1^r}\bigg[\sqrt{\frac{2J_r^2}{\kappa_1}} k_r^{\max}\bigg] . \tag{3.81}$$

While still focusing on the most amplified waves, one can improve the approximation of Eq. (3.81). Indeed, starting from Eq. (3.80), one can instead perform the k_r–integration for $k_r \in [k_r^{\mathrm{inf}}; k_r^{\mathrm{sup}}]$, where the frequency bounds are defined by $\lambda(k_r^{\mathrm{inf/sup}}) = \lambda_{\max}/2$. This approach is numerically more demanding, but allows for a more precise determination of the secular diffusion flux properties.

Once the susceptibility coefficients have been estimated, one may finally evaluate the Balescu-Lenard drift and diffusion coefficients from Eqs. (2.69) and (2.70). Thanks to the restriction to local resonances obtained in Eq. (3.78), the sum on $\boldsymbol{m}_2$ in Eqs. (2.69) and (2.70) is limited to the only term $\boldsymbol{m}_2 = \boldsymbol{m}_1$. Relying on the formula from Eq. (3.74), one can immediately perform the integration w.r.t. J_ϕ^2, which adds a prefactor of the form $1/|\partial(\boldsymbol{m}_1 \cdot \boldsymbol{\Omega}_1)/\partial J_\phi|$. Let us finally introduce the shortening notation

$$\frac{1}{(\boldsymbol{m}_1 \cdot \boldsymbol{\Omega}_1)'} = \frac{1}{\big|\frac{\partial}{\partial J_\phi}[\boldsymbol{m}_1 \cdot \boldsymbol{\Omega}_1]\big|_{J_\phi^1}} , \tag{3.82}$$

so that the drift coefficients from Eq. (2.69) become

$$A_{\boldsymbol{m}_1}(\boldsymbol{J}_1) = -\frac{4\pi^3 \mu}{(\boldsymbol{m}_1 \cdot \boldsymbol{\Omega}_1)'} \int \mathrm{d}J_r^2 \, \frac{\boldsymbol{m}_1 \cdot \partial F/\partial \boldsymbol{J}(J_\phi^1, J_r^2)}{|\mathcal{D}_{\boldsymbol{m}_1,\boldsymbol{m}_1}(J_\phi^1, J_r^1, J_\phi^1, J_r^2, \boldsymbol{m}_1 \cdot \boldsymbol{\Omega}_1)|^2} , \tag{3.83}$$

while the diffusion coefficients from Eq. (2.70) become

$$D_{\boldsymbol{m}_1}(\boldsymbol{J}_1) = \frac{4\pi^3 \mu}{(\boldsymbol{m}_1 \cdot \boldsymbol{\Omega}_1)'} \int \mathrm{d}J_r^2 \, \frac{F(J_\phi^1, J_r^2)}{|\mathcal{D}_{\boldsymbol{m}_1,\boldsymbol{m}_1}(J_\phi^1, J_r^1, J_\phi^1, J_r^2, \boldsymbol{m}_1 \cdot \boldsymbol{\Omega}_1|^2} . \tag{3.84}$$

In both Eqs. (3.83) and (3.84), the susceptibility coefficients are given by Eq. (3.80), or by Eq. (3.81) within the approximation of the small denominators. These explicit expressions of the drift and diffusion coefficients constitute an important result of this section. Let us emphasise that this WKB Balescu-Lenard formalism is self-contained and does not require any ad hoc fittings of the fluctuations occurring in the system. Except for the explicit calculation of the amplification eigenvalues in Eq. (3.47), the previous calculations are not limited to the quasi-isothermal DF from Eq. (3.10). Indeed, these drift and diffusion coefficients are valid for any tepid disc, provided one may rely on the epicyclic angle-action mapping from Eq. (3.9).

3.7 Application to Radial Diffusion

Let us now apply the previous razor-thin WKB collisionless and collisional diffusion equations to investigate how shot noise may induce radial diffusion in razor-thin axisymmetric stellar discs. In Sect. 3.7.1, we present a model of razor-thin disc model, while in Sect. 3.7.2, we investigate how the previous diffusion fluxes allow us to qualitatively understand the diffusion features observed in direct numerical simulations.

3.7.1 A Razor-Thin Disc Model

Recently, Sellwood (2012) (hereafter S12) investigated the secular evolution of an isolated, stable, quasi-stationary and self-gravitating razor-thin disc, via tailored and careful numerical simulations. After letting this disc evolve for hundreds of dynamical times, S12 observed an irreversible diffusion of the disc's DF in action space along narrow resonant ridges (see Fig. 3.10). This evolution was sustained by the spontaneous generation of transient spiral waves in the disc, as we will later detail. The disc considered by S12 is a razor-thin Mestel disc (Mestel 1963), for which the circular speed $v_\phi^2 = R\partial\psi_{\rm M}/\partial R = V_0^2$ is independent of the radius, where $\psi_{\rm M}$ is initial total potential in the system. One interest of such a simple analytical model is that it reproduces fairly well the observed flat rotation curves of galaxies. The stationary background potential $\psi_{\rm M}$ and its associated surface density $\Sigma_{\rm M}$ are given by

$$\psi_{\rm M}(R) = V_0^2 \ln\left[\frac{R}{R_{\rm max}}\right] \quad ; \quad \Sigma_{\rm M}(R) = \frac{V_0^2}{2\pi G R} , \tag{3.85}$$

where $R_{\rm max}$ is a scale parameter. Because $\psi_{\rm M}$ is scale invariant, the relationship from Eq. (3.4) between the angular momentum J_ϕ and the guiding radius $R_{\rm g}$ takes the simple form

$$J_\phi = R_{\rm g} V_0 . \tag{3.86}$$

Within the epicyclic approximation, it is also straightforward to obtain from Eq. (3.4) that the intrinsic frequencies of motion $\Omega_\phi^{\rm epi}$ and $\kappa^{\rm epi}$ take the simple form

$$\Omega_\phi^{\rm epi}(J_\phi) = \frac{V_0^2}{J_\phi} \quad ; \quad \kappa^{\rm epi}(J_\phi) = \sqrt{2}\,\Omega_\phi^{\rm epi}(J_\phi)\,. \tag{3.87}$$

Note that the Mestel disc appears as an intermediate non-degenerate disc for which $\kappa^{\rm epi}/\Omega_\phi^{\rm epi} = \sqrt{2}$, between the Keplerian case ($\kappa/\Omega_\phi = 1$) and the harmonic one ($\kappa/\Omega_\phi = 2$). Following Toomre (1977) and Binney and Tremaine (2008), a self-consistent DF for such a Mestel disc is given by

$$F_{\rm M}(E, J_\phi) = C_{\rm M}\, J_\phi^q\, \exp[-E/\sigma_r^2]\,, \tag{3.88}$$

where the exponent q and the normalisation prefactor $C_{\rm M}$ are given by

$$q = \frac{V_0^2}{\sigma_r^2} - 1 \quad ; \quad C_{\rm M} = \frac{V_0^2}{2^{1+q/2}\pi^{3/2} G \sigma_r^{q+2} \Gamma[\frac{1}{2} + \frac{q}{2}] R_{\rm max}^{q+1}}\,. \tag{3.89}$$

In Eqs. (3.88) and (3.89), we introduced σ_r as the constant radial velocity dispersion within the disc. Relying on the epicyclic approximation, the DF from Eq. (3.88) may be approximated by a quasi-isothermal DF as in Eq. (3.10), where the intrinsic frequencies are given by Eq. (3.87), the velocity dispersion σ_r is constant throughout the disc, and the surface density is given by $\Sigma_{\rm star}$, the active surface density of the disc. In order to deal with the central singularity of the Mestel disc and its infinite extent, one introduces two tapering functions $T_{\rm inner}$ and $T_{\rm outer}$ as

$$T_{\rm inner}(J_\phi) = \frac{J_\phi^{\nu_{\rm t}}}{(R_{\rm i} V_0)^{\nu_{\rm t}} + J_\phi^{\nu_{\rm t}}} \quad ; \quad T_{\rm outer}(J_\phi) = \left[1 + \left[\frac{J_\phi}{R_{\rm o} V_0}\right]^{\mu_{\rm t}}\right]^{-1}, \tag{3.90}$$

where the two power indices $\nu_{\rm t}$ and $\mu_{\rm t}$ control the sharpness of the two tapers, while the radii $R_{\rm i}$ and $R_{\rm o}$ are two scale parameters. These tapers intend to mimic the presence of a bulge and the outer truncation of the stellar disc. In addition to these tapers, we also assume that only a fraction ξ (with $0 \leq \xi \leq 1$) of the disc is indeed active, i.e. self-gravitating, while the missing component will correspond to a static contribution from the dark matter halo. As a consequence, the active DF, $F_{\rm star}$, is given by

$$F_{\rm star}(E, J_\phi) = \xi\, F_{\rm M}(E, J_\phi)\, T_{\rm inner}(J_\phi)\, T_{\rm outer}(J_\phi)\,. \tag{3.91}$$

One may also rewrite the active surface density $\Sigma_{\rm star}$ of the disc as

$$\Sigma_{\rm star}(J_\phi) = \xi\, \Sigma_{\rm M}(J_\phi)\, T_{\rm inner}(J_\phi)\, T_{\rm outer}(J_\phi)\,. \tag{3.92}$$

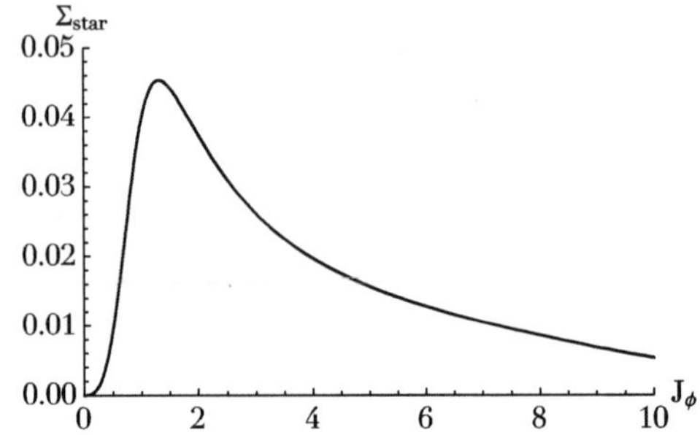

Fig. 3.6 Illustration of the active surface density $\Sigma_{\rm star}$ of the tapered Mestel disc from Eq. (3.92). Because of the two tapers from Eq. (3.90), the self-gravity of the disc is turned off in its inner and outer regions

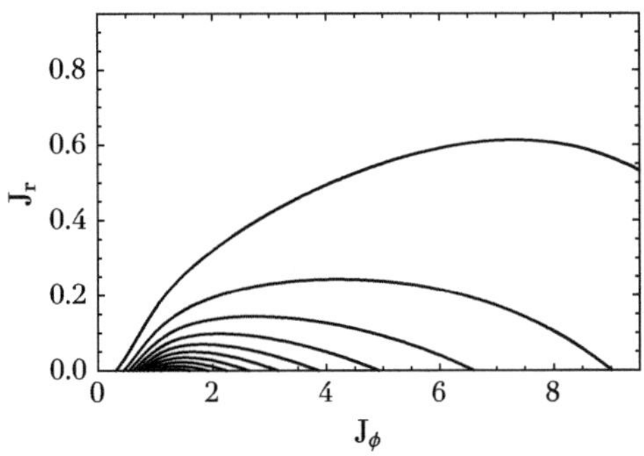

Fig. 3.7 Contours of the initial DF $F_{\rm star}$ from Eq. (3.91) in action space (J_ϕ, J_r). Contours are spaced linearly between 95 and 5% of the distribution function maximum

We place ourselves in the same unit system as S12, so that $V_0 = G = R_{\rm i} = 1$. The other numerical factors are chosen as $q = 11.4$, $\nu_{\rm t} = 4$, $\mu_{\rm t} = 5$, $\xi = 0.5$, $R_{\rm o} = 11.5$, and $R_{\rm max} = 20$. The total active mass of the disc is then straightforward to estimate and reads $M_{\rm tot} = 5.4$. The shape of the active surface density $\Sigma_{\rm star}$ is illustrated in Fig. 3.6. The initial contours of the active DF $F_{\rm star}$ are illustrated in Fig. 3.7. Because this disc is scale invariant (except for the presence of the tapering functions from Eq. (3.90)), the local Toomre's parameter Q (Toomre 1964), rederived in Eq. (3.49), becomes almost independent of the radius, especially in the intermediate regions of the disc far from the tapers. As illustrated in Fig. 3.8, one has $Q \simeq 1.5$ between the tapers and it increases strongly in the tapered regions.

At this stage, let us emphasise that S12 restricted perturbations to the sole harmonic sector $m_\phi = 2$ in order to clarify the dynamical mechanisms at play and avoid any decentring effects prohibitive for its code based on a polar grid. We note that the expressions (2.33) and (2.71) of the collisionless and collisional diffusion flux require us to sum on all the resonance vectors $\boldsymbol{m} = (m_\phi, m_r)$. Following S12's restriction, we may therefore limit ourselves to the only case $m_\phi = 2$. Throughout our calculations, in addition to this azimuthal restriction, we will more drastically restrict the resonance

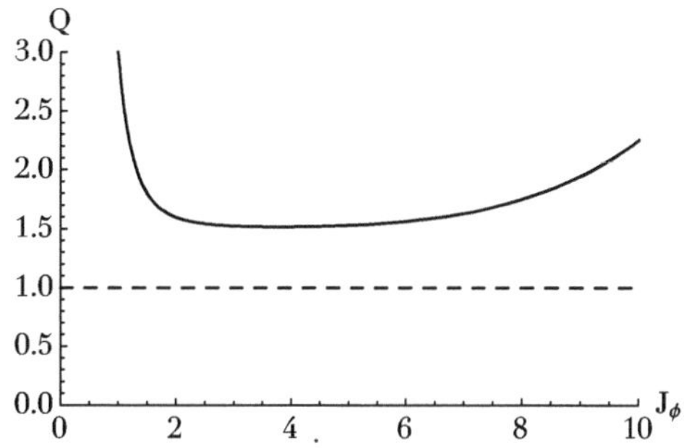

Fig. 3.8 Dependence of the local Toomre's parameter Q as a function of the angular momentum (i.e. the position within the disc) for the tapered Mestel disc from Eq. (3.92). It is scale invariant except in the inner and outer regions as a result of the presence of the tapering functions T_{inner} and T_{outer}

vectors to only three different resonances, namely the inner Lindblad resonance (ILR), $\boldsymbol{m}=(2,-1)$, the corotation resonance (COR), $\boldsymbol{m}=(2,0)$, and finally the outer Lindblad resonance, $\boldsymbol{m}=(2,1)$. Figure 3.9 illustrates how these three resonances can be interpreted when considering stars' individual orbits. All the calculations presented in the upcoming sections were also performed while accounting for the contributions from the resonances $m_r=\pm 2$, which were checked to be subdominant.

When simulating the previous razor-thin Mestel disc, S12 invariably observed sequences of transient spirals, even if the disc was specifically tailored to be isolated, stable, and quasi-stationary. On long timescales, this led to an irreversible diffusion in action space of the system's DF as illustrated in Fig. 3.10. Indeed, this figure illustrates the late time formation of a resonant ridge of particles of larger radial actions in the inner regions of disc. This narrow ridge along a very specific resonant direction is a signature of the secular evolution occurring in the system. This caused a long-term aperiodic evolution of the disc, during which small resonant and cumulative effects add up in a coherent way. It generically encompasses both processes of churning and blurring (Schönrich and Binney 2009). We will explain below in Sect. 3.7.2 the formation of this feature thanks to the previously derived razor-thin WKB diffusion coefficients. In Chap. 4, we will revisit the exact same problem while properly accounting for the system's self-gravity which can dress and strongly amplify the fluctuations within the system. Because the simulated disc was isolated, the origin of these small effects, amplified via collective effects, must come from finite$-N$ effects, i.e. be induced by the system's own discreteness.

Let us now illustrate in Fig. 3.11, the dependence of the system's response with the number of particles used to represent the disc. In this figure, the disc's evolution is characterised by the peak overdensity $\delta_{\text{max}}=\delta\Sigma_{\text{star}}/\Sigma_{\text{star}}$, which offers an estimation of how much the disc has evolved compared to its initial quasi-stationary state. Important remarks can be made from Fig. 3.11. First, because of the unavoidable Poisson shot noise in the initial conditions, the larger the number of particles, the smaller the initial value of δ_{max}, with an expected scaling given by $\delta_{\text{max}}\propto 1/\sqrt{N}$. See

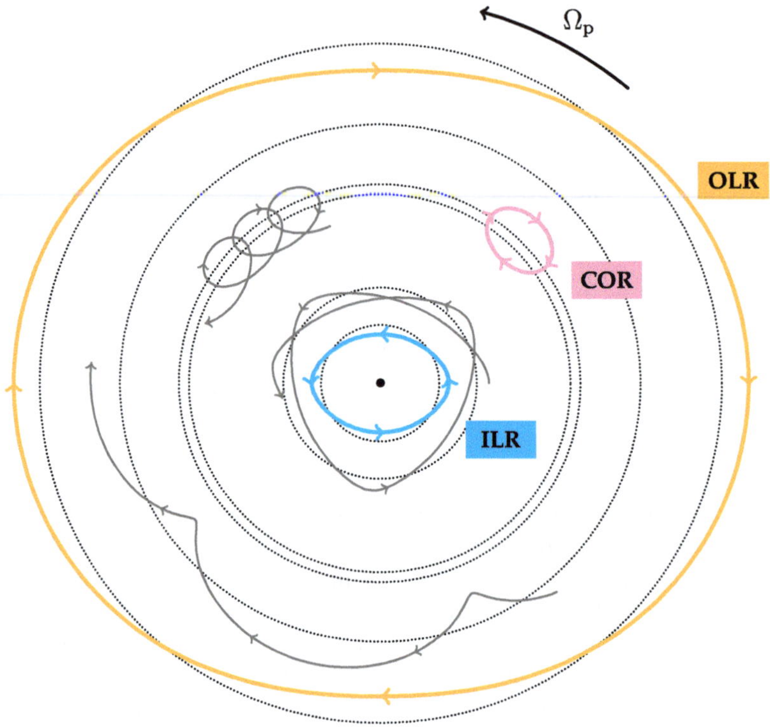

Fig. 3.9 Inspired from Fig. 1.10 of Kormendy (2013). Illustration of stellar orbits–within the epicyclic approximation–and some associated resonances as seen in the rotating frame attached to a $m_p = 2$ pattern rotating anticlockwise at the pattern frequency Ω_p (see top arrow). In this rotating frame, the pattern remains fixed, while, because of differential shearing (i.e. the fact the orbital frequencies decay as stars move outwards), stars will drift w.r.t. it. Inside the corotation, stars drift forward (anticlockwise) w.r.t. the pattern, i.e. they have an azimuthal frequency larger than the one of the pattern. Outside the corotation, stars drift backwards (clockwise) w.r.t. the pattern, as they have a smaller azimuthal frequency. In addition to their azimuthal oscillations at the frequency Ω_ϕ, stars also undergo a harmonic libration at the frequency κ. At corotation (COR, *pink* orbit), for which $\Omega_p = \Omega_\phi$, because of the radial motion, stars move clockwise along a closed ellipse. At the inner Lindblad resonance (ILR, *blue* orbit), defined as $\Omega_p = \Omega_\phi - \kappa/2$, the stellar epicyclic orbit in the rotating frame is a closed ellipse: the star executes two radial oscillations for every forward (anticlockwise) azimuthal revolution around the centre. At the outer Lindblad resonance (OLR, *orange* orbit), for which $\Omega_p = \Omega_\phi + \kappa/2$, the orbit is also closed in the rotating frame: the star executes two radial oscillations for every backwards (clockwise) azimuthal revolution around the centre. For other guiding radii, illustrated with grey orbits, the stellar orbits are not closed: the stars are not at resonance with the pattern

the left panel of Fig. 4.9 for a detailed confirmation of this prediction. One can also note an initial systematic steep rise in $\delta_{\max}$ in the very first dynamical times. This corresponds to the initial swing amplification (see Fig. 3.19) of the initial Poisson shot noise. The quieter the initial sampling (see Sellwood (1983) for a presentation of the quiet start sampling procedure used in S12), i.e. the closer the system is from

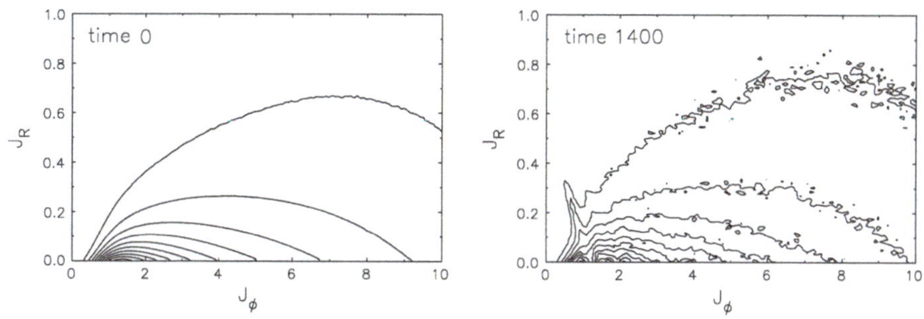

Fig. 3.10 Extracted from Sellwood (2012). Illustration of the evolution of the active stellar DF $F_{\rm star}$ from Eq. (3.91) in action space (J_ϕ, J_r). *Left panel* Initial contours of $F_{\rm star}(J_\phi, J_r)$ for $t=0$. Contours are spaced linearly between 95 and 5% of the function maximum. One can note how the inner taper from Eq. (3.90) suppresses the system's density for low angular momentum. *Right panel* Same as in the *left panel* but at a much later stage of the evolution $t=1400$. In the inner regions of the disc, one can note the formation on secular timescales of a narrow ridge of enhanced radial actions J_r. This is a signature of secular evolution

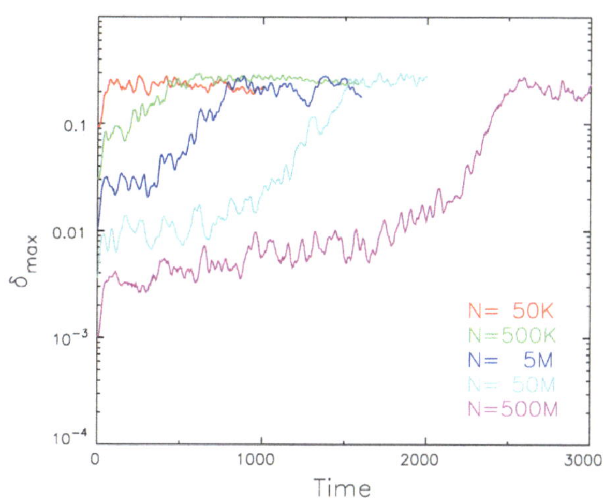

Fig. 3.11 Extracted from Sellwood (2012). Illustration in S12's simulation of the dependence of the peak overdensity $\delta_{\rm max}=\delta\Sigma_{\rm star}/\Sigma_{\rm star}$ as a function of time and the number of particles in the simulation (represented by different colours). Because of the initial Poisson shot noise in the sampling, the larger the number of particles, the smaller the initial value of $\delta_{\rm max}$, which decreases like $1/\sqrt{N}$. The initial systematic steep rises in $\delta_{\rm max}$ in the very first dynamical times correspond to the swing amplification of the system's initial Poisson shot noise. Two phases can then be identified in the growth of $\delta_{\rm max}$. The first slow phase, up to $\delta_{\rm max}\lesssim 0.02$, corresponds to a slow collisional dynamics driven by finite$-N$ effects, which gets slower as the number of particles increases. The second faster phase, for $\delta_{\rm max}\gtrsim 0.02$, corresponds to an unstable collisionless evolution whose growth rate is independent of the number of particles used. See Sect. 4.4 for a detailed discussion on these various dependences

equilibrium, the weaker this initial phase. See Sect. 4.4 and especially Fig. 4.9 for a more thorough investigation of the dependence of the system's response w.r.t. the number of particles. Right after this initial amplification, the system undergoes two successive dynamical regimes. The first stage is a stage of slow evolution, during which $\delta_{\max}$ slowly increases. Then, at a later stage, for $\delta_{\max} \gtrsim 0.02$, the growth of $\delta_{\max}$ becomes much steeper and eventually reaches a saturation. As discussed in detail in Sect. 4.4.4 and the associated Fig. 4.13, the first stage of slow evolution corresponds to a regime of secular collisional dynamics during which the system evolves as a result of dressed finite$-N$ effects. The larger the number of particles, the slower this evolution: the dynamics is sourced by discreteness effects. As for the second regime of fast growth, it corresponds to an unstable collisionless dynamics. This can hardly be seen in Fig. 3.11, but one expects the growth rate of $\delta_{\max}$ in the first initial slow phase to decrease as N gets larger, while the growth rate of $\delta_{\max}$ in the second regime is independent of the values of N used in the simulation. All the various properties inferred from Fig. 3.11 will be discussed and recovered in detail in Sect. 4.4. Let us finally emphasise that Fig. 5 of S12 also shows that after redistributing randomly the azimuthal phases of the particles at some stage of the evolution, the ridge would still appear on the same timescale. This resonant ridge is therefore not a phase-dependent feature: it only depends on the system's mean orbital structure, i.e. its mean DF $F = F(\boldsymbol{J}, t)$.

Having detailed the main results from S12's long-term simulations of razor-thin stable discs, let us now investigate in Sect. 3.7.2 how the razor-thin WKB limits of the collisionless and collisional diffusion equations, obtained in Sects. 3.5 and 3.6, allow us to explain the formation of the narrow ridge of resonant orbits observed in the direct $N-$body simulations from Fig. 3.10.

3.7.2 *Shot-Noise-Driven Radial Diffusion*

In order to compute the diffusion fluxes associated with the collisionless and collisional razor-thin WKB diffusion equations, let us first investigate the disc's self-gravitating amplification. This is captured by the razor-thin WKB amplification eigenvalues $\lambda(k_r)$ obtained in Eq. (3.47). The behaviour of this amplification is indeed essential to implement the approximation of the small denominators needed to estimate the disc's diffusion properties as in Eqs. (3.69) and (3.81). For a given position J_ϕ and a given resonance $\boldsymbol{m}$, Fig. 3.12 illustrates the behaviour of the function $k_r \mapsto \lambda(k_r)$. This figure allows us to determine what are the wave frequencies that yield locally the maximum amplification. Such waves sustain the system's WKB diffusion. Note that because Eq. (3.47) only depends on s^2, the ILR and OLR resonances will always have the same amplification eigenvalues. Thanks to Fig. 3.12, one can determine the frequency of maximum amplification $k_r^{\max}$ such that $\lambda(k_r^{\max}) = \lambda_{\max}$. We also define the domain of maximum amplification $k_r \in [k_r^{\inf}, k_r^{\sup}]$ such that $\lambda_{\max}/2 \leq \lambda(k_r)$, over which the integrations on k_r may be performed in Eqs. (3.69) and (3.81). Let us note that because of the scale-invariance of the razor-thin Mestel

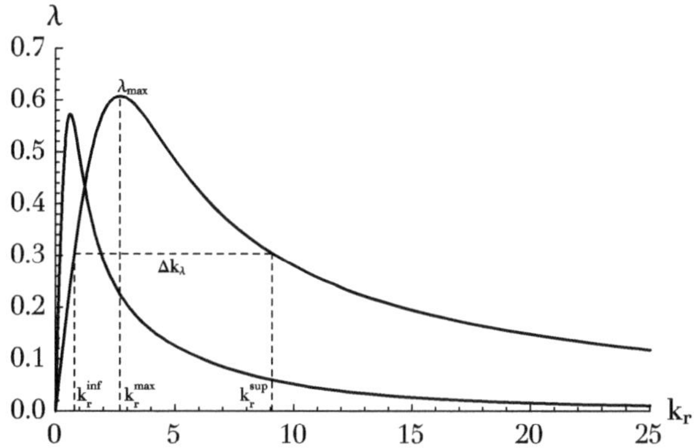

Fig. 3.12 Variations of the response matrix amplification eigenvalues λ from Eq. (3.47) with the WKB radial frequency k_r, for $\boldsymbol{m} = (m_\phi, m_r) = (2, 0)$ and two different values of J_ϕ. The curve that peaks at large k_r is for the smaller value of J_ϕ. We illustrated as well the domain of maximum amplification given by $k_r \in [k_r^{\mathrm{inf}}; k_r^{\mathrm{sup}}]$ for which one has $\lambda_{\max}/2 \leq \lambda(k_r) \leq \lambda_{\max}$, where $\lambda_{\max} = \lambda(k_r^{\max})$. This domain corresponds to the regions over which the integration for the approximation of the small denominators will be performed in Eqs. (3.69) and (3.81)

disc, it is straightforward to show that $k_r^{\max} \propto 1/J_\phi$, as well as $k_r^{\mathrm{inf/sup}} \propto 1/J_\phi$. Figure 3.13 illustrates the behaviour of the amplification factor $1 \mapsto 1/(1 - \lambda_{\max}(\boldsymbol{m}, J_\phi))$ for the different resonances $\boldsymbol{m}$. We note that the COR resonance is always more amplified than the ILR and OLR resonances (see Fig. 3.15 for a discussion of one consequence of such an ordering), but the overall maximum WKB amplification ($\sim$3 for the COR and $\sim$1.5 for the ILR and OLR) remains sufficiently small, for the system's diffusion not to be dictated only by the properties of the disc's self-gravity. Having estimated the system's susceptibility, one can now estimate in turn the collisionless diffusion flux (Sect. 3.7.2.1) as well as the collisional one (Sect. 3.7.2.2) to recover the formation of the radial resonant ridge observed in Fig. 3.10.

3.7.2.1 Collisionless Forced Radial Diffusion

The aim of this section is to understand the formation of the radial ridge observed in Fig. 3.10. A first approach is to rely on the razor-thin WKB limit of the collisionless diffusion equation obtained in Sect. 3.5. Let us already emphasise that S12's simulations modelled an isolated stellar disc in the absence of any external perturbations. As a consequence, in order to rely on our collisionless formalism, one has to assume some form for the perturbation power spectrum $\widehat{\mathcal{C}}[m_\phi, \omega, R_{\mathrm{g}}, k_r]$ appearing in the diffusion coefficients from Eq. (3.68). Here, we will assume that the "exterior" potential felt by the system represents the inevitable source of noise caused by the finite number of stars in the disc. Of course, such perturbations originate from the disc itself, but could also mimic the effects that massive compact gas clouds have on the disc. Because of Poisson shot noise, potential fluctuations scale like $\delta\psi^{\mathrm{e}} \propto \sqrt{\Sigma_{\mathrm{star}}}$, so

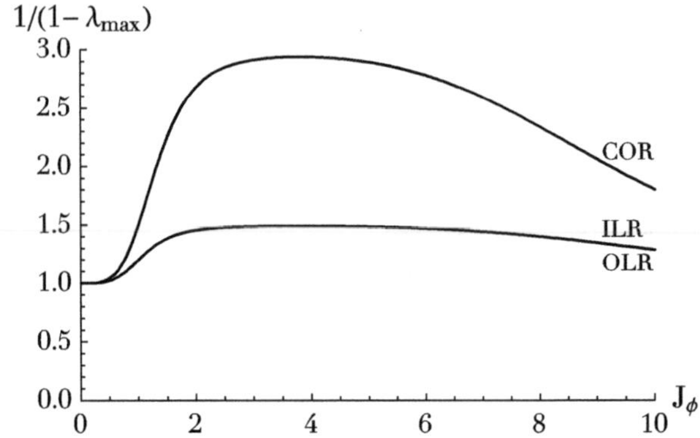

Fig. 3.13 Dependence of the maximum amplification factor $1/(1-\lambda_{\max})$ for various resonances as a function of the position J_ϕ within the disc. The amplification associated with the COR is always larger than the one associated with the ILR or OLR. Self-gravity is turned off in the inner and outer regions as a result of the tapering functions from Eq. (3.90)

that we may assume that, up to a normalisation, the system undergoes perturbations following a power spectrum given by

$$\widehat{\mathcal{C}}[m_\phi, \omega, R_{\mathrm{g}}, k_r] = \delta_{m_\phi}^2 \, \Sigma_{\mathrm{star}}(R_{\mathrm{g}}) \, . \tag{3.93}$$

Such an approximation is relatively crude as we only accounted for the dependence of the noise with J_ϕ, and neglected any dependence w.r.t. ω and k_r. In Eq. (3.93), we also added an azimuthal Kronecker symbol to account for the fact that the perturbing forces in the system were restricted to the sole harmonic sector $m_\phi = 2$. For a system perturbed by a more realistic exterior source, one expects the spectrum of the external perturbations to be more coloured and to depend on the full statistical properties of the perturbers. We also note that the absence of dependence w.r.t. ω in Eq. (3.93) implies that the three resonances ILR, COR, and OLR will undergo the same perturbations when considered for the same location in the disc, even if they are associated with different local frequencies $\boldsymbol{m}\cdot\boldsymbol{\Omega}$. Looking at the shape of the active surface density Σ_{star} in Fig. 3.6, one can note that the inner region of the disc ($J_\phi \simeq 1.5$) in the vicinity of the inner taper will be the most perturbed region. Let us also emphasise that the disc's self-induced shot noise fluctuations are not external perturbations, so that one should rely on the Balescu-Lenard formalism from Sect. 3.6 to account self-consistently for the system internal graininess. This will be the focus of Sect. 3.7.2.2.

Having estimated in Fig. 3.13 the characteristics of the disc's WKB amplification eigenvalues, and having specified in Eq. (3.93) our model to describe the spectral properties of the system's internal shot noise fluctuations, we may now compute the system's razor-thin WKB collisionless diffusion coefficients given by Eq. (3.69), and then the associated collisionless diffusion flux $\boldsymbol{\mathcal{F}}_{\mathrm{tot}}$. Figure 3.14 illustrates the initial behaviour of the diffusion flux norm $|\boldsymbol{\mathcal{F}}_{\mathrm{tot}}|$. In Fig. 3.14, the dark contours show the magnitude of the collisionless diffusion flux $|\boldsymbol{\mathcal{F}}_{\mathrm{tot}}|$ generated by the contributions

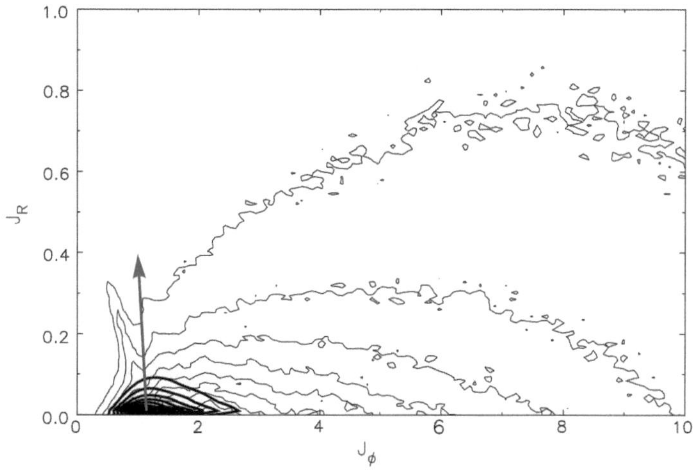

Fig. 3.14 Illustration of the norm of the collisionless diffusion flux $|\mathcal{F}_{\rm tot}|$ summed over the three resonances (ILR, COR and OLR) in *bold* lines. The contours are spaced linearly between 95 and 5% of the maximum norm. The *grey* vector gives the direction of the particle's diffusion vector associated with the norm maximum (arbitrary length). The background *thin* lines correspond to the diffused distribution from S12, which exhibits a narrow resonant ridge of diffusion

from the two Lindblad resonances and the corotation resonance. The grey arrow shows the direction of particles' individual diffusion at the location of the peak flux (the direction is similar at neighbouring points). In Fig. 3.14, one can note the presence of only one maximum peak of diffusion located in $(J_\phi, J_r) \simeq (1, 0.01)$. Note that the position of the peak of diffusion is slightly offset from the one observed in S12's simulations illustrated with the background contours of Fig. 3.14. This difference is due both to the crude noise model of Eq. (3.93), as well as to the intrinsic limitations of the explicit razor-thin WKB formalism which prevents us from correctly describing the diffusion regime associated with loosely wound transient spirals. See Sect. 4.3.3 for a full justification of why these contributions are indeed essential for the secular formation of the resonant ridge. However, our analytical results remain in good qualitative agreement with the numerical experiments from S12. We also note in Fig. 3.14 that the dominant net flux makes an angle of $111°$ with the J_ϕ−axis, while the diffusion associated with the ILR resonance is inclined by 153° w.r.t the J_ϕ−axis. This corresponds to the direction associated with the resonance vector $\boldsymbol{m} = (-2, 1)$. These two similar inclinations illustrate the dominant role played by waves at the ILR resonance in the inner region of the disc, where the DF peaks. Finally, we note quite surprisingly that despite having assumed in Eq. (3.93) that the driving fluctuations are white noise, we recovered in Fig. 3.14 that the norm of the diffusion flux is sharply peaked in action space. This is a clear illustration of the localisation of the disc's inner taper (which can be seen in Fig. 3.6). Because of this sharp taper, one expects the DF's gradients $\partial F/\partial \boldsymbol{J}$ to be significant in these regions, which naturally enhances the collisionless diffusion flux $\mathcal{F}_{\rm tot}$ from Eq. (2.33).

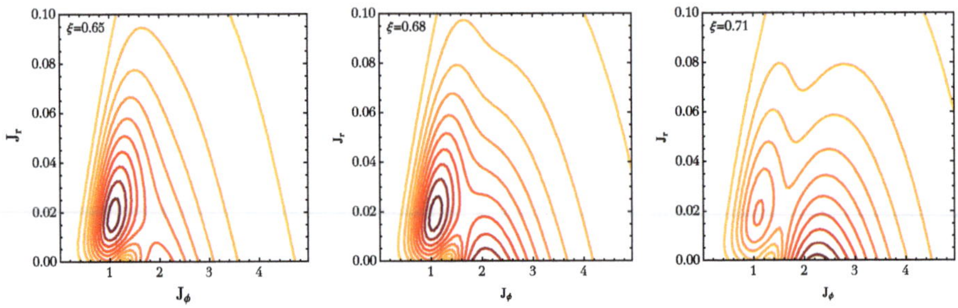

Fig. 3.15 Illustration of the norm of the collisionless diffusion flux $|\mathcal{F}_{\mathrm{tot}}|$ summed over the three resonances (ILR, COR and OLR), as one varies the disc's active fraction ξ. From *left* to *right*: $\xi = 0.65,\, 0.68,\, 0.71$. Such values of ξ still comply with the stability constraint $Q(R_{\mathrm{g}}) > 1$ everywhere in the disc. The contours are spaced linearly between 95 and 5% of the norm maximum. As the active fraction ξ increases, the disc's gravitational susceptibility gets stronger. This leads to a transition in the diffusion regime of the disc from a regime of heating in the inner regions through the ILR resonance for small values of ξ, to a regime of radial migration of quasi-circular orbits through the corotation resonance in more external regions of the disc for larger values of ξ

Following our characterisation of the collisionless diffusion sourced by the Poisson shot noise from Eq. (3.93), let us now explore how the disc's gravitational susceptibility may impact its secular evolution. This is illustrated in Fig. 3.15, where we investigate the effect of changing the disc's active fraction ξ, as introduced in Eq. (3.91). In Fig. 3.15, one can note that as the disc's active fraction increases, the disc's susceptibility gets larger, so that the diffusion gets hastened because perturbations are more amplified. In addition to this acceleration, one can also note that the contours of the norm of the collisionless diffusion flux also change qualitatively of behaviour. Indeed, as ξ increases, one observes a transition between an ILR-dominated heating in the inner region of the disc ($J_{\phi} \simeq 1$), to a regime a radial migration of quasi-circular orbits in more intermediate regions of the disc ($J_{\phi} \simeq 2$). Such a transition can be understood from Fig. 3.13, where we note that the COR resonance is always more amplified than the ILR and OLR resonances. As can be seen from Eq. (3.47), increasing the active fraction ξ immediately leads to an increase of the amplification eigenvalues λ. As a consequence, as ξ increases, both $\lambda_{\mathrm{max}}^{\mathrm{ILR}}$ and $\lambda_{\mathrm{max}}^{\mathrm{COR}}$ increase. However, since one has $\lambda_{\mathrm{max}}^{\mathrm{ILR}} < \lambda_{\mathrm{max}}^{\mathrm{COR}} < 1$, for $\lambda_{\mathrm{max}}^{\mathrm{COR}}$ close to 1, because the effective amplification is given by the factor $1/(1 - \lambda_{\mathrm{max}})$, the COR resonance gets much more amplified than the ILR and OLR resonances. This is the reason for the transition of diffusion regime observed in Fig. 3.15. As ξ increases, the COR resonance ends up being dominant leading to the transition observed in Fig. 3.15. With such higher active fractions of the disc, the system's diffusion regime is dictated by the higher intrinsic susceptibility of the disc, whose effect is indeed captured by the WKB collisionless diffusion coefficients from Eq. (3.69).

3.7.2.2 Collisional Radial Diffusion

In the previous section, we applied the razor-thin WKB collisionless formalism from Sect. 3.5 to understand the spontaneous formation of the radial resonant ridge observed in Fig. 3.10. The main assumption of the previous section was to treat the disc's internal Poisson shot noise as an exterior perturbation and model it according to Eq. (3.93). The self-induced fluctuations associated with the finite number of particles in the disc, because they are created by the disc itself, should ideally not be treated as imposed by an external perturber, but should be accounted for self-consistently. This is the main purpose of the Balescu-Lenard equation (2.67), whose razor-thin WKB limit was obtained in Sect. 3.6. Given the knowledge of the disc's WKB amplification eigenvalues obtained in Fig. 3.13, one can straightforwardly proceed to the evaluation of the razor-thin WKB susceptibility coefficients derived in Eq. (3.81). The associated collisional drift and diffusion coefficients from Eqs. (3.83) and (3.84) may then be estimated. This allows us to estimate the system's total collisional diffusion flux $\boldsymbol{\mathcal{F}}_{\text{tot}}$ from Eq. (2.71) and its associated divergence $\text{div}(\boldsymbol{\mathcal{F}}_{\text{tot}})$. As previously discussed, we restrict here the sums on the resonance vectors $\boldsymbol{m}$ only to the ILR, COR and OLR resonances. Since the Balescu-Lenard formalism self-consistently accounts for the system's internal graininess, these calculations do not require any ad hoc fittings or assumptions on the spectral properties of the system's internal fluctuations. Because the individual mass of the particles scales like $\mu = M_{\text{tot}}/N$, let us rather consider the quantity $N\text{div}(\boldsymbol{\mathcal{F}}_{\text{tot}})$ which is independent of the number of particles. This allows for a quantitative comparison with the results obtained in Fig. 3.10 via S12's numerical simulations.

Figure 3.16 illustrates the initial contours of $N\text{div}(\boldsymbol{\mathcal{F}}_{\text{tot}})$ in action space as predicted by the razor-thin WKB limit of the Balescu-Lenard equation. In Fig. 3.16, red contours are associated with regions for which $N\text{div}(\boldsymbol{\mathcal{F}}_{\text{tot}}) < 0$, so that following the

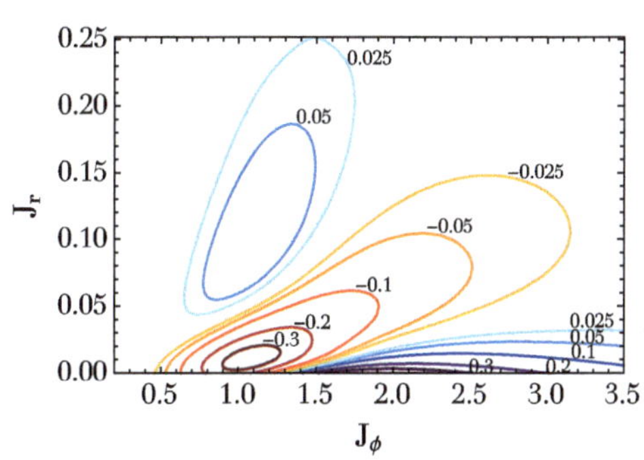

Fig. 3.16 Map of $N\text{div}(\boldsymbol{\mathcal{F}}_{\text{tot}})$ where the total collisional flux $\boldsymbol{\mathcal{F}}_{\text{tot}}$ has been summed over the three resonances (ILR, COR and OLR). *Red* contours, for which $N\text{div}(\boldsymbol{\mathcal{F}}_{\text{tot}}) < 0$, correspond to regions from which the orbits will be depleted, while *blue* contours, for which $N\text{div}(\boldsymbol{\mathcal{F}}_{\text{tot}}) > 0$, correspond to regions where the secular diffusion will tend to increase the value of the DF. The net fluxes involve both heating near $(J_\phi, J_r) \simeq (1, 0.1)$, and radial migration near $(J_\phi, J_r) \simeq (1.8, 0)$

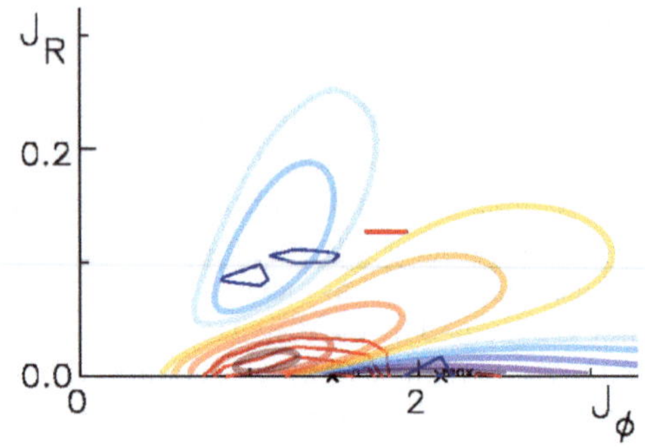

Fig. 3.17 Overlay of the WKB predictions for the divergence of the diffusion flux $N\text{div}(\mathcal{F}_{\text{tot}})$ and the differences between the initial and the evolved DF in S12's simulation. The opaque contours correspond to the differences in action space for the DF in S12 between the late time $t_{\text{S12}}=1000$ and the initial time $t_{\text{S12}}=0$ (see the upper panel of S12's figure 10). The *red* opaque contours correspond to negative differences, so that these regions are emptied from their orbits, while *blue* opaque contours are associated with positive differences, i.e. regions where the value of the DF has increased as a result of secular diffusion. The transparent contours correspond to the predicted values of $N\text{div}(\mathcal{F}_{\text{tot}})$ from the WKB limit of the Balescu-Lenard equation, using the same conventions as in Fig. 3.16. One can note the overlap between the predicted transparent contours and the measured opaque ones

convention from Eq. (2.72), they correspond to regions where the razor-thin WKB Balescu-Lenard equation predicts a decrement of the disc's DF. In contrast, blue contours are associated with regions for which $N\text{div}(\mathcal{F}_{\text{tot}})>0$, i.e. regions where the DF will increase. The overall diffusion obtained in Fig. 3.16 involves two simultaneous diffusions, namely the beginning of the formation of a resonant ridge towards larger radial actions in the vicinity of $(J_\phi, J_r)\simeq(1.0, 0.1)$, and the formation of an overdensity along the J_ϕ−axis around $(J_\phi, J_r)\simeq(1.8, 0)$. This first diffusion feature is in fact consistent with S12's early time measurements, as shown in Fig. 3.17, and with the similar late time ones, as shown in Fig. 3.18. The qualitative agreements observed in Figs. 3.17 and 3.18 are in fact surprisingly good, given all the approximations needed in the derivation of the WKB theory, and the fact that the collisional diffusion flux was only computed at the initial time $t=0^+$. Interestingly, note that the early time measurement reported in Fig. 3.17 also displays the hint of the formation of an overdensity along the $J_r=0$ axis, in agreement with the second diffusion process discussed previously. We note that the late time measurement from Fig. 3.18 suggests that this overdensity has split, with the hint of the formation of a second ridge. The disappearance of this overdensity is most likely to be explained by the integration forward in time of the Balescu-Lenard equation, while here we limited ourselves to the sole computation of the diffusion flux at the initial time.

Thanks to the explicit computation of $N\text{div}(\mathcal{F}_{\text{tot}})$ in Fig. 3.16, let us now study the typical timescale of diffusion associated with the collisional diffusion predicted by the razor-thin WKB Balescu-Lenard equation.

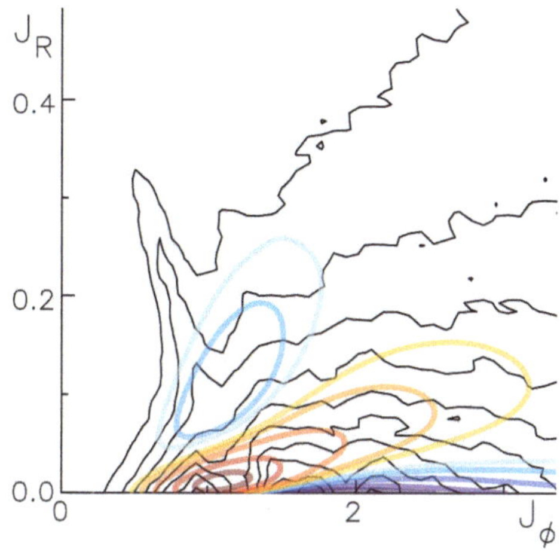

Fig. 3.18 Overlay of the WKB predictions for the divergence of the diffusion flux $N\mathrm{div}(\mathcal{F}_{\mathrm{tot}})$ on top of the contours of the DF in action space measured in S12's simulation. The *black* background contours are the level contours of the DF at time $t_{\mathrm{S12}}=1400$ (see the lower panel of Fig. 7 of S12). These contours are spaced linearly between 95 and 5% and clearly exhibit the appearance of a resonant ridge. The coloured transparent contours correspond to the predicted values of $N\mathrm{div}(\mathcal{F}_{\mathrm{tot}})$ within the WKB approximation, using the same conventions as in Fig. 3.16. One can note that the late time developed ridge is consistent with the predicted depletion (*red*) and enrichment (*blue*) of orbits

3.7.3 Diffusion Timescale

The previous estimation of the collisional diffusion flux $N\mathrm{div}(\mathcal{F}_{\mathrm{tot}})$ allows us to compare explicitly the timescale of appearance of finite$-N$ effects captured by the Balescu-Lenard equation with the duration of S12's simulations. One can note that the Balescu-Lenard equation (2.67) depends on the total number N of particles only through the particles' individual mass $\mu=M_{\mathrm{tot}}/N$. As a consequence, let us rewrite the Balescu-Lenard equation (2.67) as

$$\frac{\partial F}{\partial t}=\frac{1}{N}C_{\mathrm{BL}}[F], \tag{3.94}$$

where $C_{\mathrm{BL}}[F]=N\mathrm{div}(\mathcal{F}_{\mathrm{tot}})$ is the $N-$independent Balescu-Lenard collisional operator, i.e. the r.h.s. of Eq. (2.67) multiplied by $N=M_{\mathrm{tot}}/\mu$. Of course, this rewriting immediately illustrates that the larger the number of particles, the slower the secular evolution. One also recovers the fact that the Balescu-Lenard equation was obtained thanks to a kinetic development at first-order in the small parameter $1/N\ll 1$. Let us therefore introduce the rescaled time

$$\tau=\frac{t}{N}, \tag{3.95}$$

so that Eq. (3.94) becomes

$$\frac{\partial F}{\partial \tau} = C_{\mathrm{BL}}[F] \,, \tag{3.96}$$

so as to rewrite the Balescu-Lenard equation without any explicit appearance of N. This allows us to quantitatively compare the time during which S12's simulation was performed to the diffusion timescale predicted by the Balescu-Lenard formalism.

The ridge observed in Fig. 3.10 was obtained after letting a disc made of $N = 50 \times 10^6$ particles evolve for a time $\Delta t_{\mathrm{S12}} = 1400$. Following Eq. (3.95), the ridge was therefore observed in S12's simulation after a rescaled time $\Delta \tau_{\mathrm{S12}} = \Delta t_{\mathrm{S12}}/N \simeq 3 \times 10^{-5}$. One may then compare this time with the typical time required to form a resonant ridge within the WKB Balescu-Lenard formalism. Starting from the map of $N \mathrm{div}(\boldsymbol{\mathcal{F}}_{\mathrm{tot}})$ obtained in Fig. 3.16, one can estimate the typical time needed for such a flux to lead to the features observed in S12's simulations. The contours presented in Fig. 3.10 are separated by a value $0.1 \times F_0^{\max}$, where $F_0^{\max} \simeq 0.12$ corresponds to the maximum of the initial normalised DF from Eq. (3.91). In order to observe the resonant ridge, the DF should therefore change by an amount of the order of $\Delta F_0 \simeq 0.1 \times F_0^{\max}$. In Fig. 3.16, we obtained that the maximum of the norm of the divergence of the collisional diffusion flux is given by $|N \mathrm{div}(\boldsymbol{\mathcal{F}}_{\mathrm{tot}})|_{\max} \simeq 0.4$. Thanks to Eq. (3.96), one can then write $\Delta F_0 \simeq \Delta \tau_{\mathrm{WKB}} |N \mathrm{div}(\boldsymbol{\mathcal{F}}_{\mathrm{tot}})|_{\max}$, where $\Delta \tau_{\mathrm{WKB}}$ is the typical time during which the WKB Balescu-Lenard should be considered in order to allow for the development of a ridge. With the previous numerical values, we obtain $\Delta \tau_{\mathrm{WKB}} \simeq 3 \times 10^{-2}$. Comparing the timescale $\Delta \tau_{\mathrm{S12}}$ measured in $N-$body simulations and the timescale $\Delta \tau_{\mathrm{WKB}}$ predicted by the razor-thin WKB Balescu-Lenard equation, one gets

$$\frac{\Delta \tau_{\mathrm{S12}}}{\Delta \tau_{\mathrm{WKB}}} \simeq 10^{-3} \,. \tag{3.97}$$

In Eq. (3.97), the direct application of the razor-thin WKB Balescu-Lenard equation does not allow us to predict and recover the observed timescale of appearance of diffusion features in the numerical simulations. Indeed, the timescale of collisional diffusion predicted by our present WKB formalism appears as much larger than the time during which the numerical simulation was effectively performed. This discrepancy is also strengthened by the use of a softening length in the numerical simulations, which induces an effective thickening of the disc and therefore a slowdown of the collisional relaxation. Let us now discuss the origin of this discrepancy.

3.7.4 Interpretation

In order to interpret S12's simulation in the light of a collisional diffusion equation such as the Balescu-Lenard equation (2.67), let us first emphasise the undisputed presence of collisional effects in the simulations. This is especially noticeable in

Fig. 3.11, where we note that as the number of particles increases, the growth of the density fluctuations is delayed. The larger the number of particles, the later the effects of the secular evolution. Such a dependence indubitably underlines the role played by discreteness as the seed for the appearance of the diffusion features observed in Fig. 3.10.

Sellwood and Kahn (1991) have argued that sequences of causally connected transient waves in the disc could occur subject to a (possibly non-local) resonant condition between successive spirals. The Balescu-Lenard equation captures precisely such sequences, in the sense that it integrates over dressed correlated potential fluctuations subject to relative resonant conditions, but does not preserve causality nor resolve correlations on dynamical timescales. As emphasised in the derivation of the Balescu-Lenard equation, the system's exact initial phases are not relevant in this formalism, which focuses on describing the system's mean orbit-averaged secular evolution. This independence on the initial phases was already emphasised in Fig. 5 of S12, where it is shown that even after redistributing randomly the stars' azimuthal phases at a given time of the simulation, the growth trends observed in Fig. 3.11 remain the same, and the disc still develops the resonant ridge of Fig. 3.10. As a conclusion, the resonant secular features observed in S12's simulation correspond to a process induced by the finite number of particles and independent of the disc's particular initial phases. This corresponds exactly to the grounds on which the Balescu-Lenard equation (2.67) was derived, so that it should be the master equation to understand and capture the features observed in Fig. 3.10.

While we had qualitatively recovered in Sects. 3.7.2.1 and 3.7.2.2 the formation of the resonant ridge, we noted in Eq. (3.97) a timescale discrepancy, whose origin remains to be understood. The main assumption in the application of the Balescu-Lenard equation was the use of the razor-thin WKB basis from Eq. (3.11). We therefore argue that the timescale discrepancy observed in Eq. (3.97) is caused by the incompleteness of this basis. Indeed, the WKB basis elements, thanks to which the dressed susceptibility coefficients were estimated in Eq. (3.81), do not form a complete set as they can only represent correctly tightly wound spirals. As emphasised in Eq. (3.78), they also enforce local resonances, so that they do not allow for remote orbits to resonate, or wave packets to propagate between such non-local resonances. The seminal works from Goldreich and Lynden-Bell (1965); Julian and Toomre (1966) showed that any leading spiral wave during its unwinding to a trailing wave undergoes a significant amplification, coined swing amplification and illustrated in Fig. 3.19. Because it involves open spirals, this linear amplifying mechanism is not captured by the WKB razor-thin formalism. This important additional dressing is expected to increase the susceptibility of the disc and therefore accelerate the system's long-term diffusion, so that the timescale discrepancy from Eq. (3.97) should become less restrictive. Following the notations from Toomre (1981), the razor-thin tapered disc presented in Sect. 3.7.1 is such that $Q \simeq 1.5$ and $X \simeq 2$, so that Fig. 7 from Toomre (1981) shows that significant swing amplification of around a factor 10 or more is to be expected. We note in the Balescu-Lenard equation (2.67) that the dressed susceptibility coefficients come squared, so that the amplification associated

Fig. 3.19 From dust to ashes–extracted from Fig. 8 of Toomre (1981). Illustration of the swing amplification process occurring in a razor-thin Mestel disc. Here, one observes the transient strong amplification of a leading perturbation as it unwinds due to the disc's differential shearing. Such a linear process involves loosely wound perturbations and cannot therefore be captured by the WKB formalism from Sect. 3.3

with swing amplification will be even larger, hastening even more the disc's secular evolution.

We showed in Sect. 3.7.2.2 that the WKB Balescu-Lenard equation captures qualitatively the main features of the disc's diffusion process. In order to reconcile the timescale discrepancy from Eq. (3.97), one should get rid of the WKB approximation, and evaluate the Balescu-Lenard diffusion flux while fully accounting for the disc's susceptibility, to capture the missing mechanism of swing amplification. This is the purpose of the next chapter, where we show that the Balescu-Lenard equation associated with a complete evaluation of the disc's self-gravity recovers the resonant ridge (Sect. 4.3.1), matches the diffusion timescale (Sect. 4.3.2), and confirms that swing amplification is indeed the main driver of the disc's evolution (Sect. 4.3.3).

3.8 Conclusion

In this chapter we implemented the collisionless and collisional diffusion equations in the context of razor-thin stellar discs. In order to seek straightforward estimations of the diffusion fluxes, we relied on the epicyclic approximation to construct angle-action coordinates (Sect. 3.2) and on a tailored WKB basis to deal with the system's self-gravity (Sects. 3.3 and 3.4). Following this approach, we obtained simple quadratures for both the collisionless diffusion equation (Sect. 3.5) and the collisional one (Sect. 3.6). In particular, these simple WKB expressions yield, to our knowledge, the first non-trivial explicit expressions of the Balescu-Lenard drift and diffusion coefficients in the astrophysical context. They are therefore certainly useful to provide insight into the physical processes at play during the secular diffusion of self-gravitating razor-thin stellar discs.

In Sect. 3.7, we applied these two formalisms to describe the shot-noise-driven radial diffusion occurring spontaneously in razor-thin stellar discs when considered on secular timescales. We illustrated how the calculation in the WKB limit of the full diffusion flux recovered most of the secular features observed in the direct simulations from Sellwood (2012), especially the hints for the formation of a resonant ridge, i.e. the depletion and enrichment of orbits along a narrow preferred direction in action space. We also noted that as $Q \rightarrow 1$, the corotation resonance of waves becomes more important, as self-gravity amplifies perturbations at corotation very strongly. This leads to a transition between an ILR-dominated diffusion to a regime of radial migration in more external regions of the disc (see Fig. 3.14). These various qualitative agreements are impressive given the level of approximation involved in the WKB limit.

The timescale comparison proposed in Eq. (3.97) highlighted however a significant quantitative overestimation w.r.t. the numerical observations. In Sect. 3.7.4, we interpreted this discrepancy as being due to the intrinsic limitations of the WKB formalism, which cannot account for swing amplification, during which unwinding transient spirals get strongly amplified. This additional amplification, which involves non-local waves absorption and emission, appears therefore as the missing contribution required to reconcile quantitatively our predictions and the numerical ones. One venue is to compute numerically the Balescu-Lenard equation (2.67) in action space, without assuming tightly wound spirals or epicyclic orbits. This is the topic of the next chapter.

3.8.1 Future Works

In the light of the upcoming GAIA data, our previous description of the secular dynamics of razor-thin stellar discs could be significantly developed by extending for example the system's DF with an additional degree of freedom, namely the metallicity Z of the stars. The previous formalisms could then straightforwardly be tailored to describe the diffusion of such extended DFs $F(\boldsymbol{J}, Z)$. In order to account for the disc stellar history, one would also add a source term in the diffusion equation associated with the time dependent birth of new stars throughout the lifetime of the galaxy, while keeping track of the time and radial evolution of the gas metallicity. Such extended DFs have recently been considered in, e.g., Schönrich and Binney (2009), Binney and Sanders (2014), Sanders and Binney (2015), in the context of Galactic archeology (Binney 2013). Let us briefly detail here how one can proceed.

We introduce an extended DF $F_{\mathrm{Z}}(\mathrm{Z}, \boldsymbol{J}, t)$, so that $F_{\mathrm{Z}}\mathrm{dZ}\mathrm{d}\boldsymbol{J}$ is proportional to the mass of stars with a metallicity in the range $[\mathrm{Z}, \mathrm{Z}+\mathrm{dZ}]$ and an action $\boldsymbol{J}$ in the volume $\mathrm{d}\boldsymbol{J}$. Let us also introduce the traditional reduced DF, F, as

$$F(\boldsymbol{J}, t) = \int \mathrm{dZ}\, F_{\mathrm{Z}}(\mathrm{Z}, \boldsymbol{J}, t)\,. \tag{3.98}$$

We assume that at a given time t and position $R_{\mathrm{g}} = R_{\mathrm{g}}(J_\phi)$ in the disc, the metallicity of the interstellar medium (ISM) is known and characterised by the function $\mathrm{Z}_{\mathrm{g}}(J_\phi, t)$. See, e.g., Sanders and Binney (2015) for an example of Z_{g} for the Milky Way. We also assume that the star formation rate (SFR) is a known function of position and time written as $\mathrm{SFR}(J_\phi, t)$. When a new star is formed, it satisfies two conditions. First, stars are like time capsules and preserve the state of the ISM at the remote epoch of their formation. Moreover, stars are born on the cold orbits of the gas, so that they initially have $J_r \simeq 0$ (and $J_z \simeq 0$ in the case of thickened discs, see Chap. 5). Up to a normalisation, the source term describing the birth of new stars, F_{s}, then reads

$$\frac{\partial F_{\mathrm{s}}(\mathrm{Z}, \boldsymbol{J}, t)}{\partial t} = \mathrm{SFR}(J_\phi, t)\, \delta_{\mathrm{D}}(\mathrm{Z} - \mathrm{Z}_{\mathrm{g}}(J_\phi, t))\, \delta_{\mathrm{D}}(J_r)\,. \tag{3.99}$$

Here, $\partial F_{\mathrm{s}}/\partial t$ quantifies the amount of new stars created per unit time. The collisionless and collisional diffusion equations (2.31) and (2.67) describe the dynamics of the system's reduced DF, F, and can be written under the shortened form

$$\frac{\partial F}{\partial t} = \mathrm{Diff}\big[F, F\big]\,, \tag{3.100}$$

where we wrote the diffusion operator as $\mathrm{Diff}[F, F]$, for both the collisionless and collisional diffusion equation. In this operator, the first occurence of "F" stands for the bath DF, i.e. the DF which secularly sources the drift and diffusion coefficients. The second occurence of "F" stands for the diffusing DF, i.e. the DF whose time and action gradients appear in Eq. (3.100). See, e.g., Chavanis (2012) for a discussion on the distinction between these two DFs. Because the collisionless and collisional diffusion equations are self-consistent, these two DFs are the same. Adding the source term from Eq. (3.99), the generic diffusion equation (3.100) can be written as

$$\int \mathrm{dZ}\, \frac{\partial F_{\mathrm{Z}}}{\partial t} = \mathrm{Diff}\bigg[F, \int \mathrm{dZ}\, F_{\mathrm{Z}}\bigg] + \int \mathrm{dZ}\, \frac{\partial F_{\mathrm{s}}}{\partial t}\,. \tag{3.101}$$

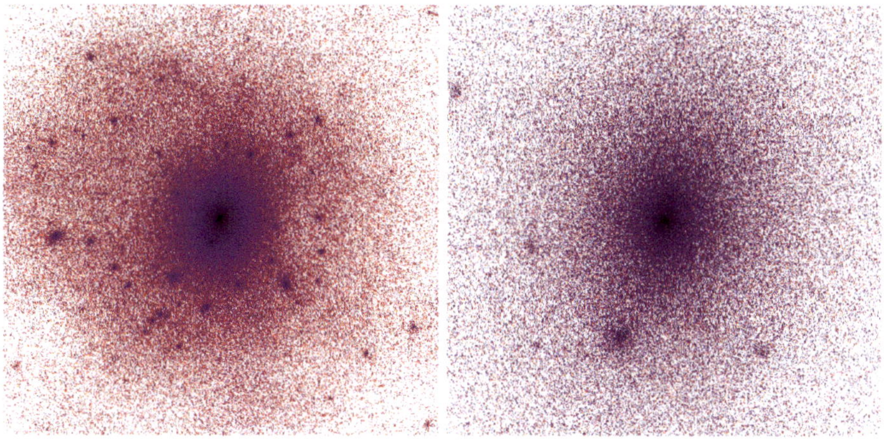

Fig. 3.20 Illustration of the dark matter density in a zoomed dark matter only simulation run with the AMR code Ramses (Teyssier 2002). The two snapshots were taken at the same time and are centred on the same dark matter halo. The *left panel* corresponds to a cubic region of extension 500 kpc, while the *right panel* only extends up to 100 kpc. The halo was chosen to be quiet, i.e. did not undergo any recent major mergers. On large scales, one can note the presence of various clumps in the dark matter density, which get much fainter as one gets closer to the centre of the halo. Here, any infalling clump gets rapidly dissolved by dynamical friction (see Fig. 1.9). On the scale of the inner galactic disc (approximately 10 kpc), these clumps are therefore expected to be screened by the dark matter halo, and the disc shielded from them. Such simulations seem to indicate that the perturbations induced by the dark matter halo are weak and will not trigger a strong diffusion in the disc

In Eq. (3.101), one can then invert the integration w.r.t. Z and the derivatives w.r.t. the actions occurring in the diffusion operator. By considering each $Z-$slice independently, one can untangle this equation to obtain a sourced diffusion equation for the extended DF F_Z reading

$$\frac{\partial F_Z}{\partial t} = \text{Diff}[F, F_Z] + \frac{\partial F_s}{\partial t} . \tag{3.102}$$

Equation (3.102) describes the dynamics of each $Z-$slice of the extended DF with its specific source term. Each $Z-$slice follows an independent diffusion equation, except for the fact that the drift and diffusion coefficients appearing in Diff[F, .] are sourced by the same reduced DF, F, integrated over all metallicities. The $Z-$slices therefore only see each other through these shared coefficients, i.e. the diffusions are self-consistent and simultaneous.

These considerations allow us to describe more precisely how the radial migration of stars interplays with the disc's chemical structure and leads for example to the appearance of metallicity gradients within the disc. One crucial strength of the previous formalisms is that they allow for detailed comparisons of the relative strengths of different diffusion mechanisms, i.e. different characteristics for the diffusion operator Diff[F, .]. One can for example characterise the statistical properties of the dark matter overdensities (i.e. clumps) and investigate how the potential fluctuations they induce may lead to a secular diffusion in the stellar disc. Figure 3.20 illustrates such clumps in thc vicinity of a simulated quiet dark matter halo, which did not undergo any recent major mergers. Once these fluctuations have been characterised, the associated collisionless diffusion can be computed and so can its effects on the disc's orbital structure. Characterising the incoming cosmic flux of perturbations could therefore in turn constrain the ΛCDM scenario on galactic scales.

References

Aumer, M., & Binney, J. (2009). *MNRAS, 397*, 1286.
Binney, J., & Sanders, J. L. (2014). *Setting the scene for Gaia and LAMOST*. In *IAU Symposium* (vol. 298, pp. 117–129).
Binney, J., & Tremaine, S. (2008). *Galactic dynamics* (2nd ed.). Princeton University Press.
Binney, J. (2010). *MNRAS, 401*, 2318.
Binney, J. (2013). *New Astron. Rev., 57*, 29.
Binney, J., & McMillan, P. J. (2011). *MNRAS, 413*, 1889.
Chavanis, P.-H. (2012). *Physica A, 391*, 3680.
Daubechies, I. (1990). *IEEE Transactions on Information Theory, 36*, 961.
Dehnen, W. (1998). *AJ, 115*, 2384.
Dehnen, W. (1999). *AJ, 118*, 1190.
Eyre, A., & Binney, J. (2011). *MNRAS, 413*, 1852.
Famaey, B., Jorissen, A., Luri, X., et al. (2005). *A&A, 430*, 165.
Fouvry, J.-B., & Pichon, C. (2015). *MNRAS, 449*, 1982.
Fouvry, J.-B., Binney, J., & Pichon, C. (2015a). *ApJ, 806*, 117.
Fouvry, J.-B., Pichon, C., & Chavanis, P.-H. (2015b). *A&A, 581*, A139.
Fouvry, J.-B., Pichon, C., & Prunet, S. (2015c). *MNRAS, 449*, 1967.
Gabor, D. (1946). *Electrical Engineers, 93*, 429.
Goldreich, P., & Lynden-Bell, D. (1965). *MNRAS, 130*, 125.
Gradshteyn, I. S., & Ryzhik, I. M. (2007). *Table of integrals, series, and products*. Elsevier Academic Press.
Helmi, A., & White, S. D. M. (1999). *MNRAS, 307*, 495.
Julian, W. H., & Toomre, A. (1966). *ApJ, 146*, 810.
Kalnajs, A. J. (1965). Ph.D. thesis, Harvard University.
Kalnajs, A. J. (1979). *AJ, 84*, 1697.
Kormendy, J. (2013). *Secular evolution in disk galaxies* (p. 1). Cambridge University Press.
Lin, C. C., & Shu, F. H. (1966). *Proc. Natl. Acad. Sci. USA, 55*, 229.
Liouville, J. (1837). *J. Math. Pures. Appl., 1*, 16.
Lynden-Bell, D. (2010). *MNRAS, 402*, 1937.
Lynden-Bell, D., & Kalnajs, A. J. (1972). *MNRAS, 157*, 1.
McMillan, P. J. (2013). *MNRAS, 430*, 3276.
Mestel, L. (1963). *MNRAS, 126*, 553.
Palmer, P. (1994). *Stability of collisionless stellar systems*. Netherlands: Springer.
Palmer, P. L., Papaloizou, J., & Allen, A. J. (1989). *MNRAS, 238*, 1281.
Piffl, T., Binney, J., McMillan, P. J., et al. (2014). *MNRAS, 445*, 3133.
Planck Collaboration, Ade, P. A. R., Aghanim, N., et al. (2014). *A&A, 571*, A16.
Sanders, J. L., & Binney, J. (2013). *MNRAS, 433*, 1826.
Sanders, J. L., & Binney, J. (2015). *MNRAS, 449*, 3479.

Schönrich, R., & Binney, J. (2009). *MNRAS*, *396*, 203.
Sellwood, J. A., & Carlberg, R. G. (2014). *ApJ*, *785*, 137.
Sellwood, J. A. (1983). *J. Comput. Phys.*, *50*, 337.
Sellwood, J. A. (2010). *MNRAS*, *409*, 145.
Sellwood, J. A. (2012). *ApJ*, *751*, 44.
Sellwood, J. A., & Kahn, F. D. (1991). *MNRAS*, *250*, 278.
Teyssier, R. (2002). *A&A*, *385*, 337.
Toomre, A. (1981). *Structure and evolution of normal galaxies*, (pp. 111–136). Cambridge University Press.
Toomre, A. (1964). *ApJ*, *139*, 1217.
Toomre, A. (1977). *ARA&A*, *15*, 437.
Wielen, R. (1977). *A&A*, *60*, 263.

Chapter 4
Razor-Thin Discs and Swing Amplification

4.1 Introduction

In Chap. 3, we investigated the formation of a narrow ridge of resonant orbits in action space appearing spontaneously on secular timescales in razor-thin stable isolated self-gravitating stellar discs. These ridges are the orbital counterparts of the processes of churning and blurring (Schönrich and Binney 2009), when considered in idealised N−body simulations. In order to understand the origin of this feature, we considered two possible approaches, either based on the collisionless diffusion equation (introduced in Sect. 2.2) or the collisional Balescu–Lenard equation (introduced in Sect. 2.3). In addition, in order to obtain simple and tractable quadratures for the associated diffusion fluxes, we relied on the assumptions that the disc's transient response could be described with tightly wound spirals via the WKB and epicyclic approximations. These simple expressions provided insight into the physical processes at work during the secular diffusion of self-gravitating discrete discs. We also reached a qualitative agreement with the results from numerical simulations, recovering the presence of an enhanced diffusion in the inner regions of the disc.

However, the WKB approximation is quantitatively questionable to capture the phase during which transient spirals unwind and undergo a strong amplification. This discrepancy was quantified in Eq. (3.97) when comparing the timescale of diffusion predicted by the WKB Balescu–Lenard formalism and the timescale inferred from numerical simulations. In Sect. 3.7.4, we blamed it on the incompleteness of the WKB basis, as it can only represent correctly tightly wound perturbations. The WKB approximation also led us to consider only resonances between orbits that are close one to another in radius. This prevented remote orbits to resonate, or wave packets to propagate between such non-local resonances. As illustrated in Fig. 3.19, the seminal works from Goldreich and Lynden-Bell (1965), Julian and Toomre (1966), Toomre (1981), have shown that any leading spiral wave undergoes a significant amplification

The work presented in this chapter is based on Fouvry et al. (2015).

J.-B. Fouvry, *Secular Evolution of Self-Gravitating Systems Over Cosmic Age*,
Springer Theses, DOI 10.1007/978-3-319-66760-7_4

as it unwinds to become a trailing wave, and this is not captured by the WKB approximation investigated in Chap. 3.

In the present chapter, we avoid such approximations by relying on the matrix method (Kalnajs 1976), in order to estimate the whole self-gravitating amplification of the disc as well as to account for the roles possibly played by non-local resonances. Once the disc's susceptibility is estimated, one can compute numerically the drift and diffusion coefficients appearing in the collisional Balescu–Lenard equation. The associated diffusion predictions can then be compared to crafted sets of numerical simulations, allowing us to estimate ensemble averaged secular responses of a sizeable number of simulations, from which we extract robust predictions on the scalings of the disc's response w.r.t. the number of particles or the disc to halo mass fraction.

In Sect. 4.2, we detail one implementation of the matrix method to compute the Balescu–Lenard diffusion flux in a razor-thin disc. In Appendix 4.D, we specify how this same approach may straightforwardly be generalised to $3D$ spherical systems, whose secular dynamics can also be probed in the same manner. Section 4.3 computes numerically the collisional drift and diffusion coefficients in action space for a razor-thin truncated Mestel disc, and compares the divergence of the corresponding flux to the results obtained in the direct $N-$body simulations from Sellwood (2012). It also illustrates how the correct timescales of diffusion are recovered. In the same section, we also emphasise how the strong self-gravitating amplification of loosely wound perturbations is indeed the main driver of the disc's secular evolution. Section 4.4 presents our $N-$body simulations of the same setting, compares the scalings of the flux w.r.t. the number of particles and the active fraction of the disc, and illustrates the late-time unstable phase transition induced in the disc as a result of the slow collisional evolution.

4.2 Calculating the Balescu–Lenard Diffusion Flux

In order to compute the Balescu–Lenard diffusion flux, three main difficulties have to be overcome. First, one has to construct explicitly the mapping $(\boldsymbol{x}, \boldsymbol{v}) \mapsto (\boldsymbol{\theta}, \boldsymbol{J})$, as the collisional drift and diffusion coefficients are associated with a diffusion in action space. Fortunately, for a razor-thin axisymmetric system, the integrability of the potential is imposed by symmetry and angle-action coordinates can be determined as shown in Sect. 4.2.1. The second difficulty arises from the non-locality of Poisson's equation and the associated complexity of the system's response matrix $\widehat{\mathbf{M}}$. As already noted in Eq. (2.12), this requires the use of a biorthogonal basis of potentials $\psi^{(p)}$, which must then be integrated over the whole action space along with functions possessing a pole like $1/(\omega - \boldsymbol{m}\cdot\boldsymbol{\Omega})$ as in Eq. (2.17). This cumbersome evaluation has to be performed numerically, along with the inversion of $[\mathbf{I} - \widehat{\mathbf{M}}]$ required to estimate the collisional dressed susceptibility coefficients from Eq. (2.50). We will show in Sects. 4.2.2, 4.2.3 and 4.2.4 how these various numerical evaluations may be performed. Finally, a third difficulty in the Balescu–Lenard equation comes from the resonance condition $\delta_{\mathrm{D}}(\boldsymbol{m}_1\cdot\boldsymbol{\Omega}_1 - \boldsymbol{m}_2\cdot\boldsymbol{\Omega}_2)$, which requires to determine how orbits resonate one with another. Once the intrinsic orbital frequencies of the stars have been determined, we will show in Sect. 4.2.5 how these resonances can be dealt with.

4.2.1 Calculating the Actions

Razor-thin axisymmetric potentials are guaranteed by symmetry to be integrable. Following Lynden-Bell and Kalnajs (1972), Tremaine and Weinberg (1984), the two natural actions $\boldsymbol{J} = (J_1, J_2)$ of the system are given by

$$J_1 = J_r = \frac{1}{\pi} \int_{r_\mathrm{p}}^{r_\mathrm{a}} \mathrm{d}r \, \sqrt{2(E - \psi_0(r)) - L^2/r^2} \quad ; \quad J_2 = J_\phi = L \, , \tag{4.1}$$

where we introduced as r_p and r_a the pericentre and apocentre of the trajectory, while E and L are the energy and angular momentum of the considered star. Here, the action J_r encodes the amount of radial energy of the star, so that $J_r = 0$ corresponds to exactly circular orbits. The two intrinsic frequencies of motion can then be introduced as $\Omega_1 = \kappa$ associated with the radial libration, and $\Omega_2 = \Omega_\phi$, associated with the azimuthal rotation. The radial frequency Ω_1 is given by

$$\frac{2\pi}{\Omega_1} = 2 \int_{r_\mathrm{p}}^{r_\mathrm{a}} \frac{\mathrm{d}r}{\sqrt{2(E - \psi_0(r)) - J_2^2/r^2}} \, , \tag{4.2}$$

while the azimuthal frequency Ω_2 satisfies

$$\frac{\Omega_2}{\Omega_1} = \frac{J_2}{\pi} \int_{r_\mathrm{p}}^{r_\mathrm{a}} \frac{\mathrm{d}r}{r^2 \sqrt{2(E - \psi_0(r)) - J_2^2/r^2}} \, . \tag{4.3}$$

At this stage, one can note that various coordinates may be used to represent the $2D$ action space. Indeed, for a given background potential ψ_0, any orbit can equivalently be represented by the pairs $(r_\mathrm{p}, r_\mathrm{a}) \leftrightarrow (E, L) \leftrightarrow (J_r, J_\phi)$. However, determining the actions associated with one set $(r_\mathrm{p}, r_\mathrm{a})$ only requires to compute an integral as in Eq. (4.1), while determining the pericentre and apocentre associated with a set of actions (J_1, J_2) requires the inversion of the same non-trivial implicit relation. In addition, as the peri/apocentres are the two roots of the equation $2(E - \psi_0(r)) - L^2/r^2 = 0$, one gets that for a given value of r_p and r_a, the energy E and the angular momentum L are straightforwardly obtained as

$$E = \frac{r_\mathrm{a}^2 \psi_\mathrm{a} - r_\mathrm{p}^2 \psi_\mathrm{p}}{r_\mathrm{a}^2 - r_\mathrm{p}^2} \quad ; \quad L = \sqrt{\frac{2(\psi_\mathrm{a} - \psi_\mathrm{p})}{r_\mathrm{p}^{-2} - r_\mathrm{a}^{-2}}} \, , \tag{4.4}$$

where we used the shortening notations $\psi_{\mathrm{p/a}} = \psi_0(r_{\mathrm{p/a}})$. As a consequence, in the upcoming applications, we use $(r_{\mathrm{p}}, r_{\mathrm{a}})$ as the representative variables of the $2D$ action space.

4.2.2 The Basis Elements

We assume that the considered $2D$ basis elements depend on two indices spanning the two degrees of freedom of a razor-thin disc. Let us therefore define

$$\psi^{(p)}(R, \phi) = \psi_n^{\ell}(R\phi) = \mathrm{e}^{\mathrm{i}\ell\phi}\, \mathcal{U}_n^{\ell}(R), \tag{4.5}$$

where $\mathcal{U}_n^{\ell}$ is a real radial function and (R, ϕ) are the usual polar coordinates. Similarly, the associated surface densities are of the form

$$\Sigma^{(p)}(R, \phi) = \Sigma_n^{\ell}(R, \phi) = \mathrm{e}^{\mathrm{i}\ell\phi}\, \mathcal{D}_n^{\ell}(R)\,, \tag{4.6}$$

where $\mathcal{D}_n^{\ell}$ is a real radial function. In Eqs. (4.5) and (4.6), the basis elements depend on two indices $\ell \geq 0$ and $n \geq 0$. In all the upcoming numerical calculations, we use the explicit radial functions from Kalnajs (1976) presented in Appendix 4.A. Once the basis elements have been specified, one may compute their Fourier transform w.r.t. the angles $\boldsymbol{\theta} = (\theta_1, \theta_2)$. Indeed, the expression of the response matrix from Eq. (2.17) requires the use of $\psi_{\boldsymbol{m}}^{(p)}(\boldsymbol{J})$ computed for the resonance vector $\boldsymbol{m} = (m_1, m_2)$. Following the convention from Eq. (2.6), one has

$$\psi_{\boldsymbol{m}}^{(p)}(\boldsymbol{J}) = \frac{1}{(2\pi)^2} \int \mathrm{d}\theta_1 \mathrm{d}\theta_2\, \psi^{(p)}(R, \phi)\, \mathrm{e}^{-\mathrm{i}m_1\theta_1}\, \mathrm{e}^{-\mathrm{i}m_2\theta_2}\,. \tag{4.7}$$

Lynden-Bell and Kalnajs (1972) gives us that the angles θ_1 and θ_2 associated with the actions from Eq. (4.1) read

$$\theta_1 = \Omega_1 \int_{C_1} \mathrm{d}r \frac{1}{\sqrt{2(E - \psi_0(r)) - J_2^2/r^2}} \quad ; \quad \theta_2 = \phi + \int_{C_1} \mathrm{d}r \frac{\Omega_2 - J_2/r^2}{\sqrt{2(E - \psi_0(r)) - J_2^2/r^2}}\,, \tag{4.8}$$

where C_1 is a contour starting from the pericentre r_{p} and going up to the current position $r = r(\theta_1)$ along the radial oscillation. Following the notations from Tremaine and Weinberg (1984), one can straightforwardly show that Eq. (4.7) becomes

$$\psi_{\boldsymbol{m}}^{(p)}(\boldsymbol{J}) = \delta_{m_2}^{\ell^p}\, \mathcal{W}_{\ell^p m_2 n^p}^{m_1}(\boldsymbol{J})\,, \tag{4.9}$$

where $\mathcal{W}^{m_1}_{\ell^p m_2 n^p}(\boldsymbol{J})$ is given by

$$\mathcal{W}^{m_1}_{\ell^p m_2 n^p}(\boldsymbol{J}) = \frac{1}{\pi}\int_{r_\mathrm{p}}^{r_\mathrm{a}} \mathrm{d}r \, \frac{\mathrm{d}\theta_1}{\mathrm{d}r} \, \mathcal{U}^{\ell^p}_{n^p}(r) \, \cos\big[m_1\theta_1[r]+m_2(\theta_2-\phi)[r]\big] . \tag{4.10}$$

One should note that the integration boundaries in Eq. (4.10) are given by the peri/apocentre $r_{\mathrm{p/a}}$ associated with the actions $\boldsymbol{J}$. Such a property illustrates once more why $(r_\mathrm{p}, r_\mathrm{a})$ appear as natural coordinates to parametrise the $2D$ action space. Equation (4.10) was written as an integral over r, thanks to the change of variables $\theta_1 \mapsto r$, which satisfies

$$\frac{\mathrm{d}\theta_1}{\mathrm{d}r} = \frac{\Omega_1}{\sqrt{2(E-\psi_0(r))-J_2^2/r^2}} . \tag{4.11}$$

Finally, in Eq. (4.10), one should note that $\theta_1[r]$ and $(\theta_2-\phi)[r]$ only depend on r, thanks to the mappings from Eq. (4.8). If $\mathcal{U}^{\ell^p}_{n^p}$ is a real function, then the coefficients $\mathcal{W}^{m_1}_{\ell^p m_2 n^p}$ are real as well. Because these coefficients involve two intricate integrals, they are numerically expensive to compute. However, they satisfy by parity the symmetry relation $\mathcal{W}^{(-m_1)}_{\ell^p (-m_2) n^p} = \mathcal{W}^{m_1}_{\ell^p m_2 n^p}$, which offers a significant reduction of the effective number of coefficients to compute.

4.2.3 *Computing the Response Matrix*

Thanks to the computation of the Fourier transformed basis elements in Eq. (4.9), one may now evaluate the response matrix from Eq. (2.17). One should note that its definition involves an integration over the dummy variable $\boldsymbol{J}$, which as discussed previously, will be performed in the $2D$ $(r_\mathrm{p}, r_\mathrm{a})$–space. The first step is to perform the change of variables $\boldsymbol{J} = (J_1, J_2) \mapsto (E, L)$, whose Jacobian is given by

$$\frac{\partial(E, L)}{\partial(J_1, J_2)} = \begin{vmatrix} \dfrac{\partial E}{\partial J_1} & \dfrac{\partial E}{\partial J_2} \\ \dfrac{\partial L}{\partial J_1} & \dfrac{\partial L}{\partial J_2} \end{vmatrix} = \begin{vmatrix} \Omega_1 & \Omega_2 \\ 0 & 1 \end{vmatrix} = \Omega_1 , \tag{4.12}$$

so that one has $\mathrm{d}J_1\mathrm{d}J_2 = \mathrm{d}E\mathrm{d}L/\Omega_1$. Thanks to the expression (4.9), the response matrix may then be rewritten as

$$\widehat{\mathbf{M}}_{pq}(\omega) = (2\pi)^2 \delta_{\ell^p}^{\ell^q} \sum_{m_1} \int \mathrm{d}E\mathrm{d}L \, \frac{1}{\Omega_1} \frac{(m_1, \ell^p)\cdot\partial F/\partial \boldsymbol{J}}{\omega-(m_1, \ell^p)\cdot\boldsymbol{\Omega}} \, \mathcal{W}^{m_1}_{\ell^p \ell^p n^p}(\boldsymbol{J}) \, \mathcal{W}^{m_1}_{\ell^p \ell^p n^q}(\boldsymbol{J}) , \tag{4.13}$$

where the sum on m_2 has been executed thanks to the Kronecker delta from Eq. (4.9). In addition, we also dropped the conjugate on $\mathcal{W}^{m_1}_{\ell^p \ell^p n^p}$ as it is real. Performing an additional change of variables $(E, L) \mapsto (r_{\mathrm{p}}, r_{\mathrm{a}})$, one can finally rewrite Eq. (4.13) as

$$\widehat{\mathbf{M}}_{pq}(\omega) = \delta_{\ell^p}^{\ell^q} \sum_{m_1} \int \mathrm{d}r_{\mathrm{p}} \mathrm{d}r_{\mathrm{a}} \frac{g_{m_1}^{\ell^p n^p n^q}(r_{\mathrm{p}}, r_{\mathrm{a}})}{h^{\omega}_{m_1 \ell^p}(r_{\mathrm{p}}, r_{\mathrm{a}})} , \tag{4.14}$$

where the functions $g_{m_1}^{\ell^p n^p n^q}(r_{\mathrm{p}}, r_{\mathrm{a}})$ and $h^{\omega}_{m_1 \ell^p}(r_{\mathrm{p}}, r_{\mathrm{a}})$ are defined as

$$g_{m_1}^{\ell^p n^p n^q}(r_{\mathrm{p}}, r_{\mathrm{a}}) = (2\pi)^2 \left| \frac{\partial(E, L)}{\partial(r_{\mathrm{p}}, r_{\mathrm{a}})} \right| \frac{1}{\Omega_1} \left[(m_1, \ell^p) \cdot \frac{\partial F}{\partial \boldsymbol{J}} \right] \mathcal{W}^{m_1}_{\ell^p \ell^p n^p}(\boldsymbol{J}) \, \mathcal{W}^{m_1}_{\ell^p \ell^p n^q}(\boldsymbol{J}) , \tag{4.15}$$

and

$$h^{\omega}_{m_1 \ell^p}(r_{\mathrm{p}}, r_{\mathrm{a}}) = \omega - (m_1, \ell^p) \cdot \boldsymbol{\Omega} . \tag{4.16}$$

The Jacobian $\partial(E, L)/\partial(r_{\mathrm{p}}, r_{\mathrm{a}})$ of the transformation $(E, L) \mapsto (r_{\mathrm{p}}, r_{\mathrm{a}})$ appearing in Eq. (4.15) can straightforwardly be computed from the expressions (4.4) of $E(r_{\mathrm{p}}, r_{\mathrm{a}})$ and $L(r_{\mathrm{p}}, r_{\mathrm{a}})$. Finally, if ever the system's DF was not defined as $F = F(\boldsymbol{J})$, but rather as $F = F(E, L)$, its gradients are immediately given by

$$\boldsymbol{m} \cdot \frac{\partial F}{\partial \boldsymbol{J}} = m_1 \Omega_1 \left(\frac{\partial F}{\partial E} \right)_L + m_2 \left[\Omega_2 \left(\frac{\partial F}{\partial E} \right)_L + \left(\frac{\partial F}{\partial L} \right)_E \right] . \tag{4.17}$$

4.2.4 Sub-region Integration

The next step of the calculation is to perform the remaining integrations over $(r_{\mathrm{p}}, r_{\mathrm{a}})$ in Eq. (4.14). However, because of the presence of the pole $1/h^{\omega}_{m_1 \ell^p}$, such integrations have to be performed carefully. To do so, we cut out the integration domain $(r_{\mathrm{p}}, r_{\mathrm{a}})$ in various subregions indexed by i. The ith region is centred around the position $(r^i_{\mathrm{p}}, r^i_{\mathrm{a}})$ and corresponds to the square domain $r_{\mathrm{p}} \in [r^i_{\mathrm{p}} - \Delta r/2; r^i_{\mathrm{p}} + \Delta r/2]$, and $r_{\mathrm{a}} \in [r^i_{\mathrm{a}} - \Delta r/2; r^i_{\mathrm{a}} + \Delta r/2]$, where Δr characterises the extension of the subregions. Such a truncation is illustrated in Fig. 4.1.

The smaller Δr, the more accurate the estimation of the response matrix. Within the ith region, one may perform limited developments of the functions g and h from Eqs. (4.15) and (4.16) around the centre $(r^i_{\mathrm{p}}, r^i_{\mathrm{a}})$, so as to write

$$g(r^i_{\mathrm{p}} + \Delta r_{\mathrm{p}}, r^i_{\mathrm{a}} + \Delta r_{\mathrm{a}}) \simeq a^i_g + b^i_g \Delta r_{\mathrm{p}} + c^i_g \Delta r_{\mathrm{a}} \quad ; \quad h(r^i_{\mathrm{p}} + \Delta r_{\mathrm{p}}, r^i_{\mathrm{a}} + \Delta r_{\mathrm{a}}) \simeq a^i_h + b^i_h \Delta r_{\mathrm{p}} + c^i_h \Delta r_{\mathrm{a}} , \tag{4.18}$$

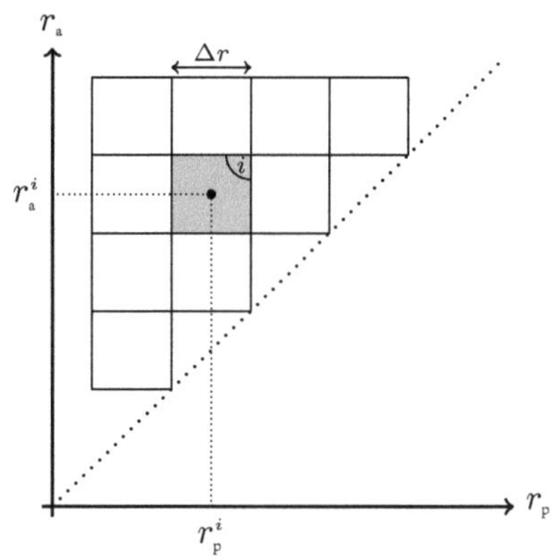

Fig. 4.1 Illustration of truncation of the $(r_{\rm p}, r_{\rm a})-$domain in small subregions to allow for the calculation of the response matrix. Each region is centred on the position $(r_{\rm p}^i, r_{\rm a}^i)$ with an extension characterised by Δr

where we dropped the indices to simplify the notations. The coefficients a_g^i, b_g^i and c_g^i (similarly for h) are given by

$$a_g^i = g(r_{\rm p}^i, r_{\rm a}^i) \;\; ; \;\; b_g^i = \frac{\partial g}{\partial r_{\rm p}}\bigg|_{(r_{\rm p}^i, r_{\rm a}^i)} \;\; ; \;\; c_g^i = \frac{\partial g}{\partial r_{\rm a}}\bigg|_{(r_{\rm p}^i, r_{\rm a}^i)} . \tag{4.19}$$

In order to minimise the number of evaluations of g required in the numerical implementation, the coefficients involving partial derivatives are estimated by finite differences, so that one has for instance

$$b_g(r_{\rm p}^i, r_{\rm a}^i) = \frac{g(r_{\rm p}^i + \Delta r, r_{\rm a}^i) - g(r_{\rm p}^i - \Delta r, r_{\rm a}^i)}{2\Delta r} . \tag{4.20}$$

One can now perform an approximated integration on each subregion. It takes the form

$$\begin{aligned}\iint_i \mathrm{d}r_{\rm p}\mathrm{d}r_{\rm a} \, \frac{g(r_{\rm p}, r_{\rm a})}{h(r_{\rm p}, r_{\rm a})} &\simeq \int_{-\frac{\Delta r}{2}}^{\frac{\Delta r}{2}} \int_{-\frac{\Delta r}{2}}^{\frac{\Delta r}{2}} \mathrm{d}x_{\rm p}\mathrm{d}x_{\rm a} \, \frac{a_g^i + b_g^i x_{\rm p} + c_g^i x_{\rm a}}{a_h^i + b_h^i x_{\rm p} + c_h^i x_{\rm a} + \mathrm{i}\eta} \\ &= \aleph(a_g^i, b_g^i, c_g^i, a_h^i, b_h^i, c_h^i, \eta, \Delta r) ,\end{aligned} \tag{4.21}$$

where $\aleph$ is an analytic function depending only on the coefficients obtained in the limited development of Eq. (4.18). It is important to note here that in order to have a well-defined integral, we added an imaginary part $\eta > 0$ to the temporal frequency ω, so that $\omega = \omega_0 + \mathrm{i}\eta$. When investigating unstable modes in discs, this imaginary part η corresponds to the growth rate of the unstable modes (see Appendix 4.C). Finally, let us note that one always has $a_g^i, b_g^i, c_g^i \in \mathbb{R}$, as well as $a_h^i, b_h^i, c_h^i \in \mathbb{R}$. The effective calculation of the function $\aleph$ is briefly presented in Appendix 4.B. Thanks to the approximation from Eq. (4.21), the response matrix from Eq. (4.14) finally becomes

$$\widehat{\mathbf{M}}_{pq}(\omega) = \delta_{\ell p}^{\ell q} \sum_{m_1} \sum_{i} \aleph(a_g^i, b_g^i, c_g^i, a_h^i, b_h^i, c_h^i, \eta, \Delta r) \,. \tag{4.22}$$

To effectively compute Eq. (4.22), one has to introduce a bound $m_1^{\max}$, so that the sum on m_1 is only limited to $|m_1| \leq m_1^{\max}$. As it requires to truncate the action space in various subregions, the calculation of the response matrix remains a daunting task, in particular to ensure appropriate numerical convergence. One natural way to validate this calculation is to recover known unstable modes of razor-thin disc. A seminal example is given by truncated Mestel discs (Zang 1976; Evans and Read 1998b; Sellwood and Evans 2001). An illustration of the validation of the present method with such discs is presented in Appendix 4.C. Once the response matrix $\widehat{\mathbf{M}}$ has been estimated, the calculation of the dressed susceptibility coefficients $1/|\mathcal{D}|^2$ from Eq. (2.50) is immediate and only involves summations over the basis elements, whose Fourier transforms in angles have already been computed.[1]

4.2.5 Critical Resonant Lines

Once the response matrix and the dressed susceptibility coefficients have been estimated, one may proceed to the evaluation of the collisional drift and diffusion coefficients from Eqs. (2.69) and (2.70). However, the resonance condition in the Dirac delta $\delta_{\mathrm{D}}(\boldsymbol{m}_1 \cdot \boldsymbol{\Omega}_1 - \boldsymbol{m}_2 \cdot \boldsymbol{\Omega}_2)$ generates an additional difficulty. Let us recall the definition of the composition of a Dirac delta and a smooth function (Hörmander 2003), which in a d-dimensional setup takes the form

$$\int_{\mathbb{R}^d} \mathrm{d}\boldsymbol{x}\, f(\boldsymbol{x})\, \delta_{\mathrm{D}}(g(\boldsymbol{x})) = \int_{g^{-1}(0)} \mathrm{d}\sigma(\boldsymbol{x})\, \frac{f(\boldsymbol{x})}{|\nabla g(\boldsymbol{x})|} \,. \tag{4.23}$$

In Eq. (4.23), we introduced as $g^{-1}(0) = \{\boldsymbol{x} \,|\, g(\boldsymbol{x}) = 0\}$ the hyper-surface of (generically) dimension $(d-1)$ defined by the constraint $g(\boldsymbol{x}) = 0$, along with $\mathrm{d}\sigma(\boldsymbol{x})$ the surface measure on $g^{-1}(0)$. The euclidean norm of the gradient of g is also naturally defined as $|\nabla g(\boldsymbol{x})| = \sqrt{|\partial g/\partial x_1|^2 + ... + |\partial g/\partial x_d|^2}$. Finally, we also assume that the resonance condition associated with the function $g(\boldsymbol{J}_2) = \boldsymbol{m}_1 \cdot \boldsymbol{\Omega}_1 - \boldsymbol{m}_2 \cdot \boldsymbol{\Omega}_2$ is non-degenerate, so that $\forall \boldsymbol{x} \in g^{-1}(0)$, $|\nabla g(\boldsymbol{x})| > 0$. This also ensures that the dimension of the set $g^{-1}(0)$ is $(d-1)$. When one considers a degenerate potential such as the harmonic or Keplerian potentials, the resonant domain fills space. Dealing with such a degeneracy requires a more involved evaluation of the Balescu–Lenard collision operator, and will be the subject of Chap. 6, where we investigate in detail the secular evolution of quasi-Keplerian systems. Here, we consider the case of a razor-thin disc, so that $d = 2$. As a consequence, $g^{-1}(0)$ is of dimension 1 and takes the form a curve

[1] As the basis effectively used may be significantly truncated, one could need to regularise the inversion of $[\mathbf{I} - \widehat{\mathbf{M}}]$ to avoid Gibbs rigging. This was not needed in the numerical applications presented here.

γ, called the critical resonant line. Generically, γ can be represented as an application of the form

$$\gamma \,:\, u \mapsto \gamma(u) = (\gamma_1(u), \gamma_2(u)) \,, \tag{4.24}$$

so that the r.h.s. of Eq. (4.23) can be rewritten as

$$\int_\gamma \!\mathrm{d}\sigma(\boldsymbol{x}) \, \frac{f(\boldsymbol{x})}{|\nabla g(\boldsymbol{x})|} = \int \!\mathrm{d}u \, \frac{f(\gamma(u))}{|\nabla(g)(\gamma(u))|} \, |\gamma'(u)| \,, \tag{4.25}$$

where we naturally introduced $|\gamma'(u)| = \sqrt{|\mathrm{d}\gamma_1/\mathrm{d}u|^2 + |\mathrm{d}\gamma_2/\mathrm{d}u|^2}$.

Using once again $(r_\mathrm{p}, r_\mathrm{a})$ as the representative variables of the action space, one can rewrite the drift and diffusion coefficients from Eqs. (2.69) and (2.70) as

$$\begin{aligned} A_{\boldsymbol{m}_1}(\boldsymbol{J}_1) &= \sum_{\boldsymbol{m}_2} \int \!\mathrm{d}r_\mathrm{p}\mathrm{d}r_\mathrm{a} \, \delta_\mathrm{D}(\boldsymbol{m}_1 \cdot \boldsymbol{\Omega}_1 - \boldsymbol{m}_2 \cdot \boldsymbol{\Omega}_2) \, G^A_{\boldsymbol{m}_1,\boldsymbol{m}_2}(r_\mathrm{p}, r_\mathrm{a}) \,, \\ D_{\boldsymbol{m}_1}(\boldsymbol{J}_1) &= \sum_{\boldsymbol{m}_2} \int \!\mathrm{d}r_\mathrm{p}\mathrm{d}r_\mathrm{a} \, \delta_\mathrm{D}(\boldsymbol{m}_1 \cdot \boldsymbol{\Omega}_1 - \boldsymbol{m}_2 \cdot \boldsymbol{\Omega}_2) \, G^D_{\boldsymbol{m}_1,\boldsymbol{m}_2}(r_\mathrm{p}, r_\mathrm{a}) \,. \end{aligned} \tag{4.26}$$

In Eq. (4.26), we respectively introduced the function $G^A_{\boldsymbol{m}_1,\boldsymbol{m}_2}(r_\mathrm{p}, r_\mathrm{a})$ and $G^D_{\boldsymbol{m}_1,\boldsymbol{m}_2}$ $(r_\mathrm{p}, r_\mathrm{a})$ as

$$\begin{aligned} G^A_{\boldsymbol{m}_1,\boldsymbol{m}_2}(r_\mathrm{p}, r_\mathrm{a}) &= -\frac{1}{\Omega_1} \left| \frac{\partial(E, L)}{\partial(r_\mathrm{p}, r_\mathrm{a})} \right| \frac{4\pi^3 \mu \, \boldsymbol{m}_2 \cdot \partial F / \partial \boldsymbol{J}_2}{|\mathcal{D}_{\boldsymbol{m}_1,\boldsymbol{m}_2}(\boldsymbol{J}_1, \boldsymbol{J}_2, \boldsymbol{m}_1 \cdot \boldsymbol{\Omega}_1)|^2} \,, \\ G^D_{\boldsymbol{m}_1,\boldsymbol{m}_2}(r_\mathrm{p}, r_\mathrm{a}) &= \frac{1}{\Omega_1} \left| \frac{\partial(E, L)}{\partial(r_\mathrm{p}, r_\mathrm{a})} \right| \frac{4\pi^3 \mu \, F(\boldsymbol{J}_2)}{|\mathcal{D}_{\boldsymbol{m}_1,\boldsymbol{m}_2}(\boldsymbol{J}_1, \boldsymbol{J}_2, \boldsymbol{m}_1 \cdot \boldsymbol{\Omega}_1)|^2} \,. \end{aligned} \tag{4.27}$$

For a given value of $\boldsymbol{J}_1$, $\boldsymbol{m}_1$, and $\boldsymbol{m}_2$, and defining $\omega_1 = \boldsymbol{m}_1 \cdot \boldsymbol{\Omega}_1$, we introduce the critical curve $\gamma_{\boldsymbol{m}_2}(\omega_1)$ as

$$\gamma_{\boldsymbol{m}_2}(\omega_1) = \left\{ (r_\mathrm{p}, r_\mathrm{a}) \,\middle|\, \boldsymbol{m}_2 \cdot \boldsymbol{\Omega}(r_\mathrm{p}, r_\mathrm{a}) = \omega_1 \right\} . \tag{4.28}$$

Relying on the formula from Eq. (4.23), one can rewrite Eq. (4.26) as

$$A_{\boldsymbol{m}_1}(\boldsymbol{J}_1) = \sum_{\boldsymbol{m}_2} \int_{\gamma_{\boldsymbol{m}_2}(\omega_1)} \!\mathrm{d}\sigma \, \frac{G^A_{\boldsymbol{m}_1,\boldsymbol{m}_2}}{|\nabla(\boldsymbol{m}_2 \cdot \boldsymbol{\Omega}_2)|} \quad ; \quad D_{\boldsymbol{m}_1}(\boldsymbol{J}_1) = \sum_{\boldsymbol{m}_2} \int_{\gamma_{\boldsymbol{m}_2}(\omega_1)} \!\mathrm{d}\sigma \, \frac{G^D_{\boldsymbol{m}_1,\boldsymbol{m}_2}}{|\nabla(\boldsymbol{m}_2 \cdot \boldsymbol{\Omega}_2)|} \,, \tag{4.29}$$

where the resonant contribution $|\nabla(\boldsymbol{m}_2 \cdot \boldsymbol{\Omega}_2)|$ is given by

$$|\nabla(\boldsymbol{m}_2 \cdot \boldsymbol{\Omega}_2)| = \sqrt{\left[\boldsymbol{m}_2 \cdot \frac{\partial \boldsymbol{\Omega}_2}{\partial r_\mathrm{p}} \right]^2 + \left[\boldsymbol{m}_2 \cdot \frac{\partial \boldsymbol{\Omega}_2}{\partial r_\mathrm{a}} \right]^2} \,. \tag{4.30}$$

In Eq. (4.30), the derivatives of the intrinsic frequencies w.r.t. r_p and r_a may be computed using finite differences, as was done in Eq. (4.20). Once the critical lines of resonance have been determined, the computation of the drift and diffusion coefficients from Eq. (4.29) is straightforward, and the secular diffusion flux $\mathcal{F}_{tot}$ from Eq. (2.71) follows immediately.

4.3 Application to Self-induced Radial Diffusion

Let us now illustrate how the previous computations of the response matrix and the Balescu–Lenard drift and diffusion coefficients may be used to interpret the diffusion features observed in the simulation of Sellwood (2012) (hereafter S12), already presented in detail in Sect. 3.7.1. Our aim here is to recover the formation of a narrow resonant ridge, as observed in Fig. 3.10. Our general motivation is churning and blurring (Schönrich and Binney 2009), which are the astrophysically relevant underlying processes. Following the WKB results from Sect. 3.7, we also aim to resolve the diffusion timescale discrepancy obtained in Eq. (3.97). Indeed, when considering the non-WKB Balescu–Lenard equation (2.67), we expect that the use of a non-local basis such as in Eq. (4.5), as well as the numerical computation of the response matrix will allow us to account for the contributions previously ignored in the WKB approximation.

Let us consider the same razor-thin disc as considered in S12's simulation and already presented in Sect. 3.7.1. Here, we recall that S12 restricted potential perturbations to the harmonic sector $m_\phi = 2$, so that we consider the same restriction on the azimuthal number m_ϕ. In the double sum on the resonance vectors $\boldsymbol{m}_1$ and $\boldsymbol{m}_2$ present in the Balescu–Lenard equation (2.67), we also assume that $\boldsymbol{m}_1$ and $\boldsymbol{m}_2$ belong to the restricted set $(m_\phi, m_r) \in \{(2,-1), (2,0), (2,1)\}$. As previously, $(2,-1)$ corresponds to the inner Lindblad resonance (ILR), $(2,0)$ to the corotation resonance (COR), and $(2,1)$ to the outer Lindblad resonance (OLR). See Fig. 3.9 for an illustration of these resonances. All the upcoming calculations were also performed while considering the contributions associated with $m_r = \pm 2$, which were checked to be subdominant.

4.3.1 *Initial Diffusion Flux*

In the previous section, we detailed how one could compute both the system's response matrix as well as the Balescu–Lenard diffusion flux in razor-thin axisymmetric discs. The calculation of the response matrix especially requires to build up a grid in the (r_p, r_a)−space. Here, we consider a grid such that $r_p^{min} = 0.08$, $r_a^{max} = 4.92$, with a grid spacing given by $\Delta r = 0.05$. When computing the response matrix in Eq. (4.22), the sum on m_1 was reduced to $|m_1| \leq m_1^{max} = 7$. The basis elements were taken following Kalnajs $2D$ basis, as presented in Appendix 4.A, with the parameters

$k_{\mathrm{Ka}}=7$ and a truncation radius given by $R_{\mathrm{Ka}}=5$. Let us note that even though the disc considered extends up to $R_{\mathrm{max}}=20$, one can still safely consider a basis truncated at such a small radius to efficiently capture the system's diffusion properties. In addition, we restricted the basis elements to $0 \leq n \leq 8$. As emphasised in Eq. (4.21), in order to evaluate the response matrix, one has to add a small imaginary part η to the frequency to regularise the resonant denominator. Throughout the upcoming calculations, we considered $\eta=10^{-4}$ and checked that this choice had no impact on our results.

Here, the total potential ψ_{M} is known analytically from Eq. (3.85), so that, following Sect. 4.2.1, the mapping to the angle-action coordinates is straightforward to obtain. As given by Eqs. (4.2) and (4.3), one can compute the disc's intrinsic frequencies Ω_{ϕ} and κ on the $(r_{\mathrm{p}}, r_{\mathrm{a}})$−grid. Once these frequencies are known, one can determine the system's critical resonant lines, introduced in Eq. (4.28). It is along these curves that one has to perform the integrations present in the expression (4.29) of the Balescu–Lenard drift and diffusion coefficients. Figure 4.2 illustrates these critical resonant lines. In this figure, one can note that by getting rid of the WKB approximation, we allow for non-local resonances between distant orbits.

Following Eq. (4.29), one can then compute the disc's drift and diffusion coefficients, and finally the collisional diffusion flux $\boldsymbol{\mathcal{F}}_{\mathrm{tot}}$ introduced in Eq. (2.71). As already made in Sect. 3.7.2.2, because the mass of the particles scales like $\mu=M_{\mathrm{tot}}/N$, it is natural to consider the quantity $N\boldsymbol{\mathcal{F}}_{\mathrm{tot}}$. Following the convention from Eq. (2.72), the direction along which individual orbits diffuse is given by the vector field $-N\boldsymbol{\mathcal{F}}_{\mathrm{tot}}=-(N\mathcal{F}^{\phi}_{\mathrm{tot}}, N\mathcal{F}^{r}_{\mathrm{tot}})$ defined over the action space (J_{ϕ}, J_r). Figure 4.3 illustrates this diffusion flux. In Fig. 4.3, one can already note how the diffusion vector

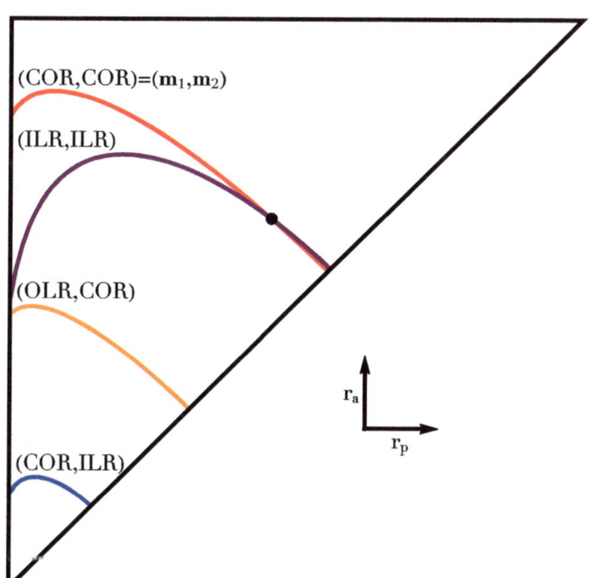

Fig. 4.2 Illustration of four different resonant critical lines in the $(r_{\mathrm{p}}, r_{\mathrm{a}})$−space. As introduced in Eq. (4.28), a critical line is characterised by two resonance vectors $\boldsymbol{m}_1, \boldsymbol{m}_2$, and a location $\boldsymbol{J}_1=(r^1_{\mathrm{p}}, r^1_{\mathrm{a}})$ in action space. Each of the four plotted critical lines is associated with the same location $(r^1_{\mathrm{p}}, r^1_{\mathrm{a}})$ represented by the *black* dot, but with a different choice for the resonance vectors $\boldsymbol{m}_1$ and $\boldsymbol{m}_2$, among the three resonances ILR, COR and OLR. One can note that for $\boldsymbol{m}_1=\boldsymbol{m}_2$, the critical line goes through the point $(r^1_{\mathrm{p}}, r^1_{\mathrm{a}})$

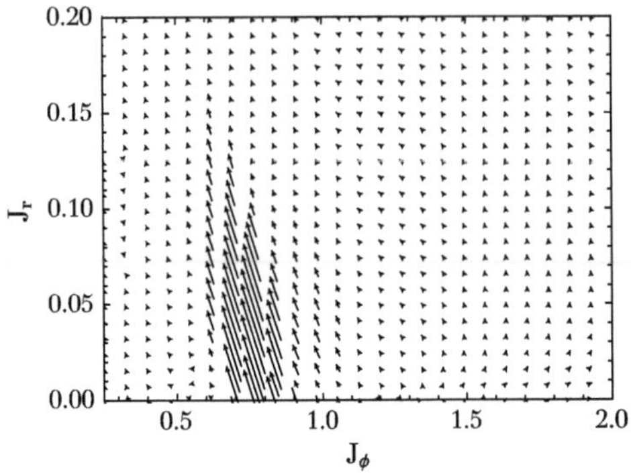

Fig. 4.3 Map of the diffusion flux $-N\boldsymbol{\mathcal{F}}_{\mathrm{tot}}$ computed for $\boldsymbol{m}_1, \boldsymbol{m}_2 \in \{\boldsymbol{m}_{\mathrm{ILR}}, \boldsymbol{m}_{\mathrm{COR}}, \boldsymbol{m}_{\mathrm{OLR}}\}$. As defined in the rewriting from Eq. (2.72), $-N\boldsymbol{\mathcal{F}}_{\mathrm{tot}}$ corresponds to the direction along which individual particles diffuse in action space

field is concentrated in the inner region of the disc and aligned with a narrow resonant direction. Along this ridge of diffusion, one typically has $\mathcal{F}^{\phi}_{\mathrm{tot}} = -2\mathcal{F}^{r}_{\mathrm{tot}}$, so that the diffusion is aligned with the direction of the ILR resonance vector given by $\boldsymbol{m}_{\mathrm{ILR}} = (2, -1)$.

After having determined the collisional diffusion flux $N\boldsymbol{\mathcal{F}}_{\mathrm{tot}}$, one can compute its divergence, to characterise the regions in action space where the disc's DF is expected to change as a result of diffusion. This is illustrated in Fig. 4.4 which represents the initial contours of $N\mathrm{div}(\boldsymbol{\mathcal{F}}_{\mathrm{tot}})$. Figure 4.4 is the main result of this section. In this figure, we note that the Balescu–Lenard formalism predicts the formation of a narrow resonant ridge in the inner regions of the disc, aligned the direction of the ILR resonance. One also recovers that the stars which will populate the ridge originate from the base of the ridge and diffuse along the ILR direction. It is most likely that the slight shift in the position of the ridge w.r.t. S12's numerical measurement is due to the fact that the Balescu–Lenard diffusion flux was only estimated for $t = 0^{+}$, while S12's measurement was made at $t = 1400$. Other possible origins for this small difference could be the use of a softening length in the numerical simulations, which modifies the two-body interaction potential. This could as well be due to the difference between the ensemble averaged evolution, predicted by the Balescu–Lenard formalism, and one specific realisation, as probed in S12. Indeed, our own simulations (see Sect. 4.4) suggest some variations in the position of the ridge between different runs. Having determined explicitly the value of $N\mathrm{div}(\boldsymbol{\mathcal{F}}_{\mathrm{tot}})$, let us now compare the typical timescale of diffusion predicted by the Balescu–Lenard equation to what was measured numerically in S12's simulation. This is the purpose of the next section. Finally, in order to get a better a grasp of the driving mechanisms of this secular diffusion, we will investigate in Sect. 4.3.3 the respective roles of the self-gravitating amplification and the limitation to the tightly wound basis elements, to emphasise the crucial role played by swing amplification (illustrated in Fig. 3.19).

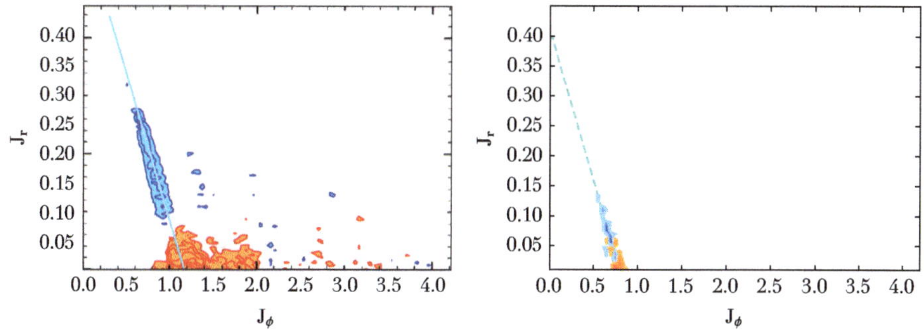

Fig. 4.4 *Left panel* Extracted from Sellwood (2012), Fig. 7. Illustration of the contours of the changes in the DF between the time $t_{S12}=1400$ and $t_{S12}=0$, for a run with $N=50\times10^6$ particles. As in the *right panel*, *red* contours correspond to negative differences, i.e. regions emptied from their orbits, while *blue* contours correspond to positive differences, i.e. regions where the system's DF increases during the diffusion. *Right panel* Map of $N\mathrm{div}(\boldsymbol{\mathcal{F}}_{\mathrm{tot}})$, where the total flux has been computed with $\boldsymbol{m}_1, \boldsymbol{m}_2 \in \{\boldsymbol{m}_{\mathrm{ILR}}, \boldsymbol{m}_{\mathrm{COR}}, \boldsymbol{m}_{\mathrm{OLR}}\}$. *Red* contours, for which $N\mathrm{div}(\boldsymbol{\mathcal{F}}_{\mathrm{tot}})<0$, correspond to regions from which the orbits will be depleted during the diffusion, while *blue* contours, for which $N\mathrm{div}(\boldsymbol{\mathcal{F}}_{\mathrm{tot}})>0$, are associated with regions for which the value of the DF will increase as a result of diffusion. The contours are spaced linearly between the minimum and maximum of $N\mathrm{div}(\boldsymbol{\mathcal{F}}_{\mathrm{tot}})$. The maximum value for the positive *blue* contours is given by $N\mathrm{div}(\boldsymbol{\mathcal{F}}_{\mathrm{tot}})\simeq 350$, while the minimum value for the negative *red* contours is $N\mathrm{div}(\boldsymbol{\mathcal{F}}_{\mathrm{tot}})\simeq -250$. The contours in both panels are aligned with the ILR direction $\boldsymbol{m}_{\mathrm{ILR}}=(2,-1)$ in the (J_ϕ, J_r)−plane, as illustrated with the *cyan* line

4.3.2 *Diffusion Timescale*

In Sect. 3.7.3, when relying on the WKB approximation, the main disagreement was obtained in Eq. (3.97) when comparing the timescales of diffusion. We noted that the timescale of collisional diffusion predicted by the WKB Balescu–Lenard equation was about a factor 10^3 too slow compared to what was effectively measured in S12. This is because the WKB approximation cannot account for the swing amplification of loosely wound perturbations, which significantly boosts and hastens the diffusion in cold dynamical systems such as razor-thin stellar discs. Thanks to our explicit and quantitative estimation of the collisional diffusion flux $N\mathrm{div}(\boldsymbol{\mathcal{F}}_{\mathrm{tot}})$, let us now perform the same analysis, by comparing the rescaled times of diffusion $\Delta\tau$ as defined in Eq. (3.95).

Section 3.7.3 showed that the time $\Delta\tau_{S12}$ required to observe the numerical ridge in S12's numerical simulation was $\Delta\tau_{S12}\simeq 3\times10^{-5}$. In Fig. 3.16 obtained with the WKB approximation, we noted that the maximum of the norm of the diffusion flux was given by $|N\mathrm{div}(\boldsymbol{\mathcal{F}}_{\mathrm{tot}})|_{\max}\simeq 0.4$, which led to a WKB rescaled time of diffusion given by $\Delta\tau_{\mathrm{WKB}}\simeq 3\times10^{-2}$. When fully accounting for the system's self-gravity, we obtained in Fig. 4.4 that the maximum of the norm of the divergence of the diffusion flux was given by $|N\mathrm{div}(\boldsymbol{\mathcal{F}}_{\mathrm{tot}})|_{\max}\simeq 350$, which corresponds to an enhancing of the diffusion flux of a factor 10^3 compared to the WKB case. Hence, one can write $\Delta\tau_{\mathrm{BL}}\simeq\Delta\tau_{\mathrm{WKB}}/10^3\simeq 3\times10^{-5}$, where $\Delta\tau_{\mathrm{BL}}$ stands for the time during which the

Balescu–Lenard diffusion flux from Fig. 4.4 has to be considered to allow for the formation of the resonant ridge. Comparing the numerically measured rescaled time $\Delta\tau_{\mathrm{S12}}$ and the time $\Delta\tau_{\mathrm{BL}}$ predicted by the Balescu–Lenard equation, one gets

$$\frac{\Delta\tau_{\mathrm{S12}}}{\Delta\tau_{\mathrm{BL}}} \simeq 1\,. \tag{4.31}$$

As a consequence, the projection of the disc's response over an unbiased basis led to over a hundredfold increase of the disc's susceptibility and therefore to a very significant acceleration of the disc's secular evolution. Thanks to this mechanism, we reached a very good agreement between the diffusion timescale observed in numerical simulations and the prediction from the Balescu–Lenard formalism. This quantitative match is rewarding, both from the point of view of the accuracy of the $N-$body integrator (symplecticity, timestep size, softening, etc.) and from the point of view of the relevance of the Balescu–Lenard formalism and the various approximations on which it relies (timescale decoupling, $1/N$ truncation of the BBGKY hierarchy, neglect of the collision term, see, e.g., Appendix 2.B for a discussion).

In the next section, we will emphasise how the main source of secular collisional evolution in S12's simulation is indeed the strong self-gravitating amplification of loosely wound perturbations, i.e. sequences of uncorrelated swing amplified spirals sourced by finite$-N$ effects.

4.3.3 *Why Swing Amplification Matters*

Let us now briefly show the importance of both collective effects and the completeness of the basis to capture accurately the swing amplification of loosely wound perturbations, and how it is indeed driving the formation of the diffusion features recovered in Fig. 4.4.

4.3.3.1 Turning Off Collective Effects

In order to assess the importance of self-gravity, one could proceed to the same evaluation of the diffusion flux as in Fig. 4.4, while however neglecting collective effects. This amounts to assuming that the system's response matrix becomes $\widehat{\mathbf{M}}=0$. The Balescu–Lenard equation (2.67) then becomes the Landau equation (2.73), where the dressed susceptibility coefficients $1/|\mathcal{D}_{\boldsymbol{m}_1,\boldsymbol{m}_2}|^2$ from Eq. (2.50) are replaced by their bare analogs $|A_{\boldsymbol{m}_1,\boldsymbol{m}_2}|^2$ from Eq. (2.74). In this context, the computation of the diffusion flux does not require the calculation of the response matrix. However, one must still perform the integrations along the resonant lines, as presented in Sect. 4.2.5. We finally rely on the same numerical parameters as the ones detailed in Sect. 4.3.1. This allows us to compute the associated bare diffusion flux $N\mathrm{div}(\boldsymbol{\mathcal{F}}_{\mathrm{tot}}^{\mathrm{bare}})$, whose initial contours are illustrated in Fig. 4.5. One should compare the maps of the dressed dif-

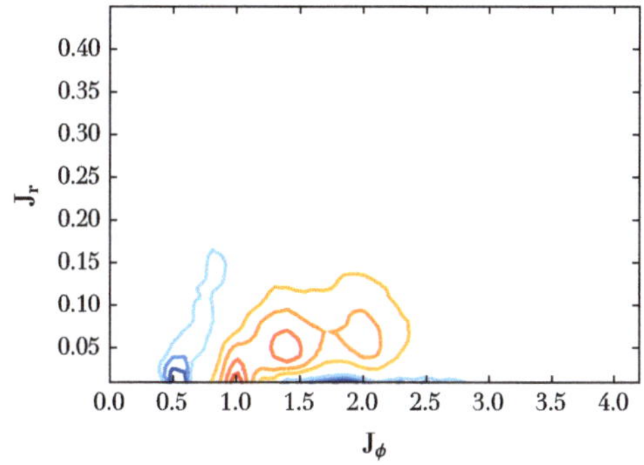

Fig. 4.5 Map of the $N\mathrm{div}(\mathcal{F}_{\mathrm{tot}}^{\mathrm{bare}})$ corresponding to the bare diffusion flux (i.e. without accounting for collective effects), following the same conventions as in Fig. 4.4. The contours are spaced linearly between the minimum and maximum of $N\mathrm{div}(\mathcal{F}_{\mathrm{tot}}^{\mathrm{bare}})$. The maximum value for the positive *blue* contours is given by $N\mathrm{div}(\mathcal{F}_{\mathrm{tot}}^{\mathrm{bare}}) \simeq 0.30$, while the minimum value for the negative *red* contours reads $N\mathrm{div}(\mathcal{F}_{\mathrm{tot}}^{\mathrm{bare}}) \simeq -0.50$. One should note that turning off collective effects led to the disappearance of the strong narrow radial ridge obtained in Fig. 4.4. This figure is qualitatively similar to the results presented in Fig. 3.16, obtained via the razor-thin WKB limit of the Balescu–Lenard equation

fusion flux $N\mathrm{div}(\mathcal{F}_{\mathrm{tot}})$ obtained in Fig. 4.4 to the present map of the bare diffusion flux $N\mathrm{div}(\mathcal{F}_{\mathrm{tot}}^{\mathrm{bare}})$, thanks to which one may assess the importance and the strength of the system's self-gravitating amplification. As expected for dynamically cold systems such as razor-thin stellar discs, turning off self-gravity reduces significantly the system's susceptibility and slows down its secular evolution by a factor of about 1000. One can also note that while the secular appearance of the narrow resonant ridge was obvious in the dressed diffusion from Fig. 4.4, this is much less clear in the bare case from Fig. 4.5. Let us also note that the overall shape observed in Fig. 4.5 is somewhat similar to what was obtained in Fig. 3.16, when relying on the razor-thin WKB limit of the Balescu–Lenard equation. The amplitudes of the bare divergence contours are also similar to the WKB values obtained in Fig. 3.16.

As a conclusion, the comparison of Figs. 4.4 and 4.5 strongly emphasises how the self-gravitating amplification of loosely wound perturbations is indeed responsible for the appearance of a narrow ridge, while also drastically hastening the system's diffusion to ensure a rapid appearance of the ridge, as seen in the timescales comparison from Eq. (4.31).

4.3.3.2 Turning Off Loosely Wound Contributions

In order to emphasise once again the role played by loosely wound perturbations, let us try to reproduce the results presented in Sect. 3.7.2.2, which relied on the razor-thin WKB limit of the Balescu–Lenard equation. Indeed, using the generic numerical methods presented in Sect. 4.2, one can mimic these WKB results by carefully choosing the basis elements introduced in Eq. (4.5). Recall that the basis

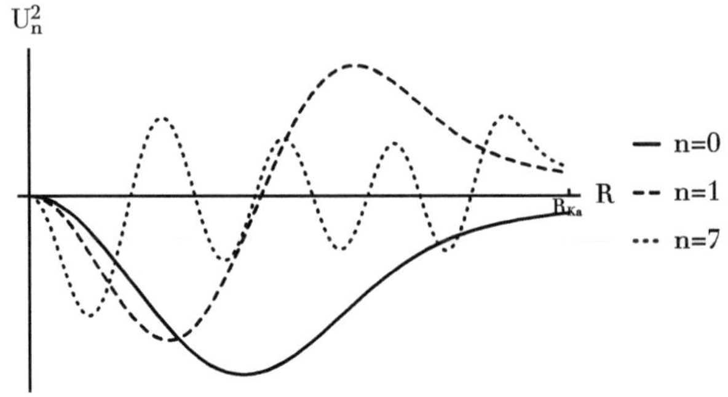

Fig. 4.6 Illustration of the radial dependence of Kalnajs basis elements for $\ell = 2$ and $k_{\mathrm{Ka}} = 7$, as defined in Appendix 4.A. These basis elements are the ones used to estimate the Balescu–Lenard diffusion flux in Sect. 4.3.1. One can note that as the radial index n increases, the basis elements get more and more radially wound

elements depend on two indices: an azimuthal index ℓ and a radial one n. Because S12's simulation was restricted to the harmonic sector $m_\phi = 2$, we only consider basis elements associated with $\ell = 2$. Figure 4.6 illustrates the radial dependence of the basis elements as one changes the radial index n. One can note from Fig. 4.6 that the larger n, the faster the radial variation of the basis elements, i.e. the more tightly wound the basis elements. As a consequence, in order to get rid of the loosely wound basis elements which can get swing amplified, let us perform a truncation on the radial indices considered. Let us define the diffusion flux $N\mathrm{div}(\boldsymbol{\mathcal{F}}_{\mathrm{tot}}^{\mathrm{WKB}})$ computed in the same manner as the total dressed flux $N\mathrm{div}(\boldsymbol{\mathcal{F}}_{\mathrm{tot}})$, except that here we restrict ourselves to basis elements such that $n_{\mathrm{cut}} \leq n \leq n_{\mathrm{max}}$, with $n_{\mathrm{cut}} = 2$ and $n_{\mathrm{max}} = 8$. By keeping only the sufficiently wound basis elements, our aim is to consider the same contributions as the ones captured by the razor-thin WKB limit. Figure 4.7 illustrates the initial contours of $N\mathrm{div}(\boldsymbol{\mathcal{F}}_{\mathrm{tot}}^{\mathrm{WKB}})$. One can first note that the amplitudes of the contours in Fig. 4.7 are of the same order than the WKB contours from Fig. 3.16. The presence in Fig. 4.4 of positive blue contours is also in qualitative agreement with a secular heating of the disc leading to an increase in the radial action J_r. However, these contours do not exhibit the formation of a narrow resonant ridge as was predicted in Fig. 4.4, when accounting as well for loosely wound contributions.

As a conclusion, Figs. 4.5 and 4.7 illustrate how the strong self-gravitating amplification of loosely wound perturbations (i.e. swing amplification) is indeed responsible for both the appearance of a narrow ridge of resonant orbits as well as for the associated rapid timescale of growth. Having emphasised the relevance of the Balescu–Lenard formalism to describe the secular dynamics of razor-thin discs, let us investigate in the next section some additional properties of these long-term evolutions (already advertised in Fig. 3.11), by relying on our own $N-$body simulations.

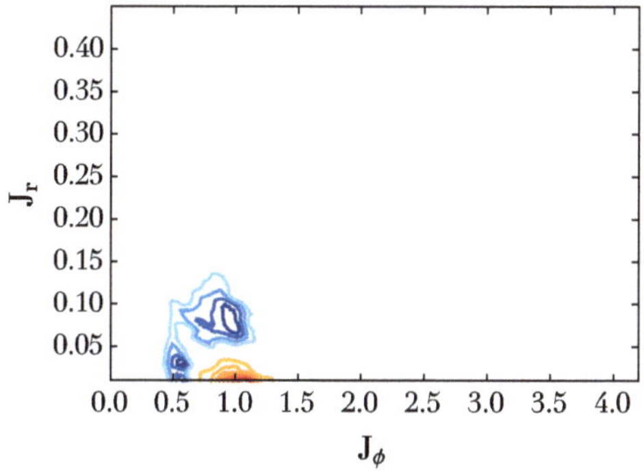

Fig. 4.7 Map of $N\mathrm{div}(\mathcal{F}_{\mathrm{tot}}^{\mathrm{WKB}})$ corresponding to the dressed diffusion flux when loosely wound contributions are not accounted for, following the same conventions as in Fig. 4.4. In order to restrict ourselves only to tightly wound contributions, we did not consider the contributions associated with the basis elements for the radial index $n=0, 1$, as these elements are loosely wound (see Fig. 4.6). The contours are spaced linearly between the minimum and the maximum of $N\mathrm{div}(\mathcal{F}_{\mathrm{tot}}^{\mathrm{WKB}})$. The maximum value for the positive *blue* contours is given by $N\mathrm{div}(\mathcal{F}_{\mathrm{tot}}^{\mathrm{WKB}})\simeq 0.7$, while the minimum value for the negative *red* contours reads $N\mathrm{div}(\mathcal{F}_{\mathrm{tot}}^{\mathrm{WKB}})\simeq -4.5$. This figure should be compared to Fig. 3.16, which was obtained by relying on the razor-thin WKB limit of the Balescu–Lenard equation

4.4 Comparisons with N−Body Simulations

In order to investigate in detail the scalings of the system's evolution w.r.t. the number of particles or the active fraction of the disc, we now resort to our own N−body simulations. We present in Sect. 4.4.1 the characteristics of the N−body code that was used, while Sects. 4.4.2 and 4.4.3 focus on the respective dependences of the system's response with the number of particles and its active fraction. Finally, in Sect. 4.4.4, we illustrate how the very late evolution of the system exhibits an out-of-equilibrium phase transition.

4.4.1 A N−Body Implementation

When simulating the evolution of self-gravitating discs, one should pay a particular attention to the sampling of the initial conditions in order to ensure that the disc is initially in a state of collisionless equilibrium. In the present context, this requires to be able to sample particles' positions and velocities from the DF of Eq. (3.19). We do not repeat here the sampling strategy that was used here, whose details can be found in Appendix E of Fouvry et al. (2015). One should note that we relied on a random sampling procedure of the DF. This does not correspond to a quiet start sampling (Sellwood 1983), which would have allowed for a reduction of the initial shot noise within the disc.

Once particles have been sampled, their positions and velocities are evolved using a straightforward particle-mesh $N-$body code with a single-timestep leapfrog integrator (see Binney and Tremaine (2008), Sect. 3.4.1.). As was done in S12, the potential in which particles evolve is decomposed in two components: (i) an axisymmetric static contribution $\psi_{\mathrm{M}}(R)$ from the unperturbed Mestel disc, as introduced in Eq. (3.85), (ii) a non-axisymmetric contribution $\delta\psi(R, \phi)$ which grows as perturbations develop in the disc. Thanks to this splitting, we avoid difficulties associated with dealing with the rigid part of the potential associated with the tapering functions and the disc's active fraction, and not due to the DF. Here, we calculate $\delta\psi$ using a cloud-in-cell interpolation (see Binney and Tremaine (2008), Sect. 2.9.3.) of the particles' masses onto a $N_{\mathrm{mesh}} \times N_{\mathrm{mesh}}$ mesh of square cells of size Δx_{mesh}. We then filter the resulting density field $\rho(x, y)$ to isolate the disc's response, as discussed below. We then rely on the traditional "doubling-up" procedure to determine the potential $\delta\psi$ at the cell vertices. The contribution of $\delta\psi$ to each particles' acceleration is finally obtained using the same cloud-in-cell interpolation scheme.

Similarly to S12's restriction, when computing the density, we add a filtering scheme to account only for the $m_\phi = 2$ response. This contribution is obtained by calculating

$$\rho_2(r) = \frac{1}{2\pi} \int \mathrm{d}\phi \, \rho(r \cos(\phi), r \sin(\phi)) \, \mathrm{e}^{-2\mathrm{i}\phi} \, . \tag{4.32}$$

This calculation is performed at each timestep immediately after the cloud-in-cell assignment of mass to the mesh. We then impose to the mesh the new mass distribution

$$\rho(x_k, y_k) = \rho_2(r_k) \, \mathrm{e}^{2\mathrm{i}\phi_k} \, , \tag{4.33}$$

where (r_k, ϕ_k) are given by $(x_k, y_k) = (r_k \cos(\phi_k), r_k \sin(\phi_k))$. To compute $\rho_2(r)$, we rely on a brute-force computation of Eq. (4.32) on a series of N_{ring} radial rings with a spacing $\Delta r_{\mathrm{ring}} \ll \Delta x_{\mathrm{mesh}}$, with the trapezium rule with $N_\phi = 720$ points in ϕ for the angular dependence. While being very simple, this $N-$body code aims at reproducing as closely as possible the details of S12's implementation. One should still underline two important differences: S12 relies on a polar mesh to compute $\delta\psi$, while we use here a cartesian mesh with a $m_\phi = 2$ filtering of the density field; S12 uses a block timestep scheme, while we use here a simpler single-timestep scheme.

The results presented thereafter were obtained with a timestep $\Delta t = 10^{-3} R_{\mathrm{i}} / V_0$ (where R_{i} and V_0 were introduced in Sect. 3.7.1), with a mesh that extends up to $\pm R_{\max} = 20 R_{\mathrm{i}}$ and $N_{\mathrm{mesh}} = 120$ cells, so that $\Delta x_{\mathrm{mesh}} = R_{\mathrm{i}}/3$. To filter the potential to the $m_\phi = 2$ harmonic sector, we used $N_{\mathrm{ring}} = 1000$ radial rings and $N_\phi = 720$ points in the azimuthal direction, so that $\Delta r_{\mathrm{ring}} = 2R_{\mathrm{i}}/100$. The computation of the potential from the density was performed via a Fourier transform on the mesh, with a softening length $\varepsilon = R_{\mathrm{i}}/6$, comparable to the Plummer softening $\varepsilon = R_{\mathrm{i}}/8$ used in S12. We checked that the results are not significantly changed when the timestep or the mesh size are divided by 2. We detail in Appendix 4.C, how this $N-$body implementation was validated by recovering known unstable modes of truncated Mestel discs (Zang 1976; Evans and Read 1998b; Sellwood and Evans 2001).

4.4.2 Scaling with N

The Balescu–Lenard equation (2.67) describes the system's mean secular collisional evolution, in the sense of it being averaged over different realisations. Therefore, in order to investigate such dynamics via N−body simulations, we run multiple simulations for the same number of particles, which only differ initially by the sampling of the initial conditions. For a given number of particles, we then perform an ensemble average of the different evolution realisations. Having estimated this mean evolution, we are in a position to compare it with the Balescu–Lenard predictions.

In order to study the scaling w.r.t. the number of particles of these various numerical simulations, one should define a function quantifying the "amount of diffusion" undergone by the system and compare it with the predictions of the Balescu–Lenard formalism. An additional difficulty also comes from the statistical nature of the initial sampling. Indeed, as only N stars are being sampled, the system's initial DF will necessarily fluctuate w.r.t. the smooth underlying DF as a result of the unavoidable Poisson shot noise. Let us insist on the fact that these fluctuations directly originate from the initial sampling and are not as such specific to the collisional diffusion process described by the Balescu–Lenard equation. It is therefore important to disentangle these two effects. To do so, let us define the function $\widetilde{h}(t, N)$ as

$$\widetilde{h}(t, N) = \big\langle h_i(t, N) \big\rangle , \tag{4.34}$$

where we introduced the operator $\big\langle \,.\, \big\rangle$ as the ensemble average over different realisations for the same number of particles indexed by i. In the upcoming applications, it is approximated as the arithmetic average of $p=32$ different realisations, so that $\big\langle \,.\, \big\rangle = 1/(p) \sum_i (\,.\,)$. In Eq. (4.34), we introduced the distance function $h_i(t, N)$ as

$$h_i(t, N) = \!\int\! \mathrm{d}\boldsymbol{J} \, \big[F_i(t, \boldsymbol{J}, N) - \big\langle F(t=0, \boldsymbol{J}, N) \big\rangle \big]^2 . \tag{4.35}$$

In Eq. (4.35), we noted as $F_i(t, \boldsymbol{J}, N)$ the normalised DF at time t of the ith simulation with a number N of particles, while $\big\langle F(t=0) \big\rangle$ stands for the averaged system's DF at the initial time. Defined in such a way, the function h_i quantifies the "distance" between the initial mean DF $\big\langle F(t=0) \big\rangle$ and the evolved DF $F_i(t)$. When interested in the early time behaviour of the function $\widetilde{h}$, one can perform a limited development as

$$\widetilde{h}(t, N) \simeq \widetilde{h}_0(N) + \widetilde{h}_1(N)\, t + \widetilde{h}_2(N) \, \frac{t^2}{2} , \tag{4.36}$$

where the coefficients $\widetilde{h}_0$, $\widetilde{h}_1$ and $\widetilde{h}_2$ only depend on N. Starting from Eq. (4.35), these coefficients can easily be estimated. One has

$$\widetilde{h}_0(N) = \!\int\! \mathrm{d}\boldsymbol{J} \, \Big\langle \big[F \quad \big\langle F_0 \big\rangle \big]^2 \Big\rangle , \tag{4.37}$$

where we used the shortening notations $\langle F_0\rangle = \langle F(t=0, \boldsymbol{J}, N)\rangle$ and $F = F(t=0, \boldsymbol{J}, N)$. One can note that this coefficient only depends on the properties of the initial sampling of the DF, and not on its dynamics. As discrete sampling obeys Poisson statistics, one can immediately rewrite Eq. (4.37) as

$$\widetilde{h}_0(N) = \frac{\alpha_0}{N}\,, \tag{4.38}$$

where the constant α_0 is independent of N. One can similarly compute the coefficient $\widetilde{h}_1(N)$, which reads

$$\widetilde{h}_1(N) = 2\int \mathrm{d}\boldsymbol{J}\, \langle [F - \langle F_0\rangle]\, F'\rangle\,, \tag{4.39}$$

where we used the notation $F' = \big[\partial F/\partial t\big](t=0)$. The two terms appearing in the crossed term from Eq. (4.39) have two different physical origins. The first one, $\big[F - \langle F_0\rangle\big]$, is associated with the initial sampling of the DF, while the second, F', is driven by the system's dynamics. Assuming that the sampling and the system's dynamics are uncorrelated, one can write $\langle [F - \langle F_0\rangle] F'\rangle = \langle F - \langle F_0\rangle\rangle \langle F'\rangle = 0$, i.e. one has $\widetilde{h}_1(N) = 0$. Finally, one can compute the coefficient $\widetilde{h}_2(N)$ given by

$$\widetilde{h}_2(N) = 2\int \mathrm{d}\boldsymbol{J} \Big\langle \big[F'\big]^2 + \big[F - \langle F_0\rangle\big] F''\Big\rangle\,, \tag{4.40}$$

where we used the notation $F'' = \big[\partial^2 F/\partial t^2\big](t=0)$. Using the same argument of uncorrelation as in Eq. (4.39), one can get rid of the second term in the r.h.s. of Eq. (4.40). In addition, let us assume that the variance of $\big[F'\big]^2$ is small compared to its expectation, so that $\langle \big[F'\big]^2\rangle = \big[\langle F'\rangle\big]^2$. Equation (4.40) becomes

$$\widetilde{h}_2(N) = 2\int \mathrm{d}\boldsymbol{J}\, \big[\langle F'\rangle\big]^2\,. \tag{4.41}$$

The dependence of the term $\langle F'\rangle$ with N is directly given by the Balescu–Lenard equation (see Eq. (3.94)), which gives the scaling

$$\widetilde{h}_2(N) = \frac{\alpha_2}{N^2}\,, \tag{4.42}$$

where the amplitude α_2 is independent of N. Let us insist on the fact that such a scaling is a prediction of the Balescu–Lenard formalism. Should the secular evolution observed in S12 be a Vlasov-only evolution, i.e. a collisionless evolution, one would not have such a scaling.

Let us now compare the scalings from Eqs. (4.38) and (4.42) with the ones measured in $N-$body simulations. We consider number of particles given by $N \in \big\{8, 12, 16, 24, 32, 48, 64\big\} \times 10^5$, and for each number of particles, we perform

32 different simulations with different initial conditions, using the $N-$body implementation presented in Sect. 4.4.1. For each value of N, one may first study the behaviour of the function $t \mapsto \widetilde{h}(t, N)$, as illustrated in Fig. 4.8. Once these functions are computed, one can fit parabolas to them, following Eq. (4.36), to determine the behaviours of $N \mapsto \widetilde{h}_0(N)$, $\widetilde{h}_2(N)$. The dependence with N of these two coefficients is illustrated in Fig. 4.9. In the left panel of Fig. 4.9, we recover the scaling $\widetilde{h}_0(N) \propto 1/N$ obtained in Eq. (4.38). Such a dependence is fully due to the initial Poisson shot noise in the initial conditions. From the right panel of Fig. 4.9, we obtain the scaling $\widetilde{h}_2(N) \propto N^{-1.91}$. This has to be compared to the prediction $\widetilde{h}_2(N) \propto N^{-2}$

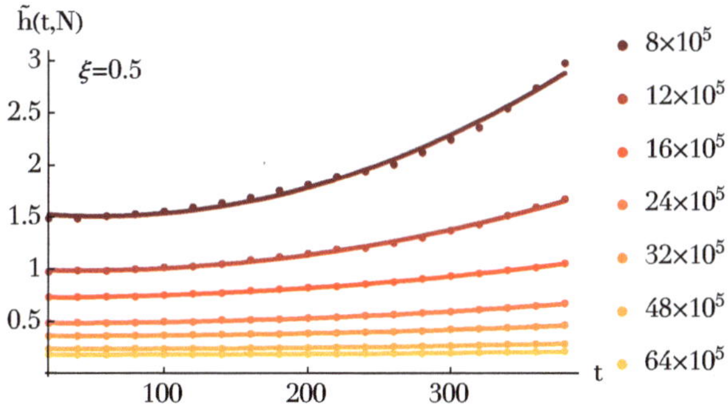

Fig. 4.8 Illustration of the behaviour of the function $t \mapsto \widetilde{h}(t, N)$ from Eq. (4.34), for an active fraction $\xi = 0.5$. The function is averaged over 32 different realisations with particle numbers $N \in \{8, 12, 16, 24, 32, 48, 64\} \times 10^5$. To compute $\widetilde{h}(t, N)$, the action space domain $(J_\phi, J_r) = [0; 2.5] \times [0; 0.2]$ was binned in 100×50 regions. The values of $\widetilde{h}(t, N)$ have also been uniformly renormalised to clarify this representation. Dots correspond to the snapshots of the simulations for which $\widetilde{h}(t, N)$ was effectively computed, while the lines correspond to second-order fits, following Eq. (4.36). As expected, the larger the number of particles, the less noisy the simulation and the smaller $\widetilde{h}(t, N)$

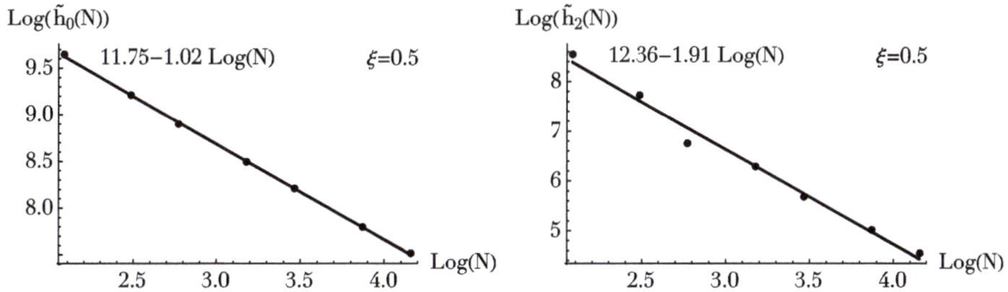

Fig. 4.9 *Left panel* Illustration of the behaviour of the function $\log(N) \mapsto \log(\widetilde{h}_0(N))$, where N was rescaled by a factor 10^{-5} to clarify the representation. *Dots* are associated with the values computed thanks to Fig. 4.8, while the line corresponds to a linear fit reading $\log(\widetilde{h}_0(N)) \simeq 11.75 - 1.02 \log(N)$. The coefficients $\widetilde{h}_0(N)$ have been uniformly renormalised to clarify the representation. *Right panel* Same conventions as in the *left panel*. Illustration of the behaviour of the function $\log(N) \mapsto \log(\widetilde{h}_2(N))$, whose linear fit takes the form $\log(\widetilde{h}_2(N)) \simeq 12.36 - 1.91 \log(N)$

from Eq. (4.42) derived from the Balescu–Lenard equation. Given the small number of realisations considered here and the various uncertainties in the fits, the measurements and the predictions appear to be in good agreement. Such a scaling of $\widetilde{h}_2(N)$ with N therefore confirms the relevance of the Balescu–Lenard framework. It demonstrates that the secular evolution of S12's stable Mestel disc is the result of a collisional diffusion seeded by the system's discrete nature and the effects of amplified distant resonant encounters.

In order to investigate in more detail this collisional scaling, let us now describe another measurement, which allows us to get rid of the pollution by the Poisson shot noise present in Eq. (4.38). Indeed, one of the difficulty of the previous measurements was to disentangle the contributions from the Poisson shot noise, as in $\widetilde{h}_0(N)$ from Eq. (4.38), from the ones associated with the collisional scaling of the Balescu–Lenard equation, as in $\widetilde{h}_2(N)$ from Eq. (4.42). In order not to be sensitive to fluctuations around the mean DF associated with Poisson shot noise, let us restrict ourselves only to sufficiently large fluctuations, i.e. fluctuations effectively induced by the secular evolution rather than by the inevitable Poisson shot noise. Let us therefore define the function $\widetilde{V}(t, N)$ as

$$\widetilde{V}(t, N) = \int \mathrm{d}\boldsymbol{J}\, \chi\big[\langle F(t, \boldsymbol{J}, N)\rangle - \langle F(t{=}0, \boldsymbol{J}, N)\rangle < C_{\widetilde{V}}\big], \tag{4.43}$$

where we introduced a threshold $C_{\widetilde{V}} < 0$, as well as $\chi\big[x < C_{\widetilde{V}}\big]$ a characteristic function equal to 1 if $x < C_{\widetilde{V}}$ and 0 otherwise. The function $\widetilde{V}(t, N)$ therefore measures the volume in action space of the regions depleted from particles (as $C_{\widetilde{V}} < 0$), for which the mean DF value has changed by more than $C_{\widetilde{V}}$. Provided that $C_{\widetilde{V}}$ is chosen to be sufficiently large, such a definition allows us not to be polluted anymore by the Poisson sampling shot noise. The scaling of $\widetilde{V}(t, N)$ for the initial times is straightforward to obtain (see Eq. (3.94)) and reads

$$\widetilde{V}(t, N) \simeq \frac{t}{N} \widetilde{V}_0 , \tag{4.44}$$

where $\widetilde{V}_0$ is a constant independent of N. As a consequence, for a fixed value of N, one expects a linear time dependence of the function $t \mapsto \widetilde{V}(t, N)$, as illustrated in the left panel of Fig. 4.10.

Finally, in order to test the scaling w.r.t. N predicted by Eq. (4.44), one may introduce a threshold value $\widetilde{V}_{\text{thold}}$, and, for a given value of N, define the associated threshold time $t_{\text{thold}}(N)$ such that

$$\widetilde{V}(t_{\text{thold}}(N), N) = \widetilde{V}_{\text{thold}} . \tag{4.45}$$

Equation (4.44) immediately gives the scaling

$$t_{\text{thold}}(N) \simeq N \, \frac{\widetilde{V}_{\text{thold}}}{\widetilde{V}_0} . \tag{4.46}$$

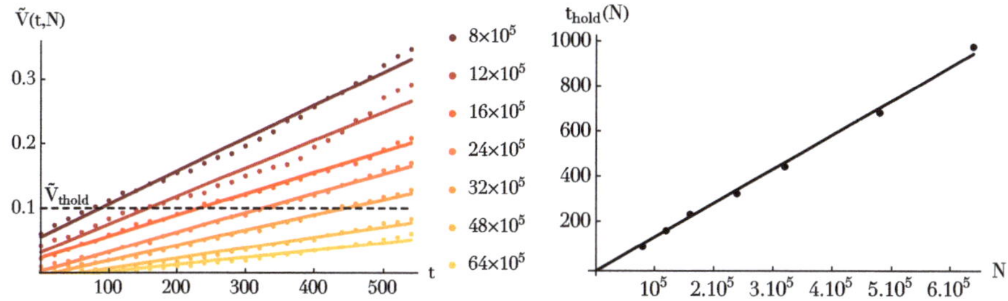

Fig. 4.10 *Left panel* Illustration of the behaviour of the function $t \mapsto \widetilde{V}(t, N)$ defined in Eq. (4.43), when averaged over 32 different realisations for particle numbers $N \in \{8, 12, 16, 24, 32, 48, 64\} \times 10^5$, along with the associated linear fits. To effectively compute $\widetilde{V}(t, N)$, we used the same binning of action space as in Fig. 4.8. As predicted in Eq. (4.44), one recovers that for a fixed value of N, the function $t \mapsto \widetilde{V}(t, N)$ is linear. The horizontal *dashed* line illustrates the threshold value $\widetilde{V}_{\mathrm{thold}}$ which was used to determine the threshold time t_{thold} as defined in Eq. (4.45). *Right panel* Illustration of the behaviour of the function $N \mapsto t_{\mathrm{thold}}(N)$ and its associated linear fit. As predicted by the Balescu–Lenard equation in Eq. (4.46), one recovers a linear dependence of $t_{\mathrm{thold}}(N)$ with N

This linear scaling of $t_{\mathrm{thold}}(N)$ with N is a prediction from the Balescu–Lenard formalism and the associated collisional dynamics, and is nicely recovered in the right panel of Fig. 4.10.

4.4.3 Scaling with ξ

Following the previous study of the scaling of the system's response w.r.t. the number of particles, let us now study the impact of the disc's active fraction ξ on the properties of the diffusion. Indeed, one of the strength of the Balescu–Lenard formalism is to capture the effect of gravitational polarisation via the response matrix and the dressed susceptibility coefficients. Following S12, the simulations considered in the previous sections were all performed with an active fraction $\xi = 0.5$. Only one half of the total potential was effectively generated self-consistently by the stars, while the rest was associated with the contributions from a static and rigid halo. By increasing the disc's active fraction, one can strengthen the self-gravitating amplification, and therefore hasten the diffusion, while still remaining in a collisional regime of evolution. If one keeps increasing even more ξ, the system will eventually become dynamically unstable and its dynamics will be driven by the collisionless Vlasov equation. See Sect. 4.4.4 for a detailed discussion on the transition between the two regimes of diffusion: slow collisional vs. unstable collisionless. Provided that ξ is not too large, the dynamics of the the disc is still driven by the Balescu–Lenard equation, and the scaling on $\widetilde{h}_2$ with N obtained in Eq. (4.42) remains the same. However, because of the increased self-gravity, the prefactor $\alpha_2(\xi)$ from Eq. (4.42) will increase as the system becomes more responsive. Let us now investigate the dependence of α_2 with ξ,

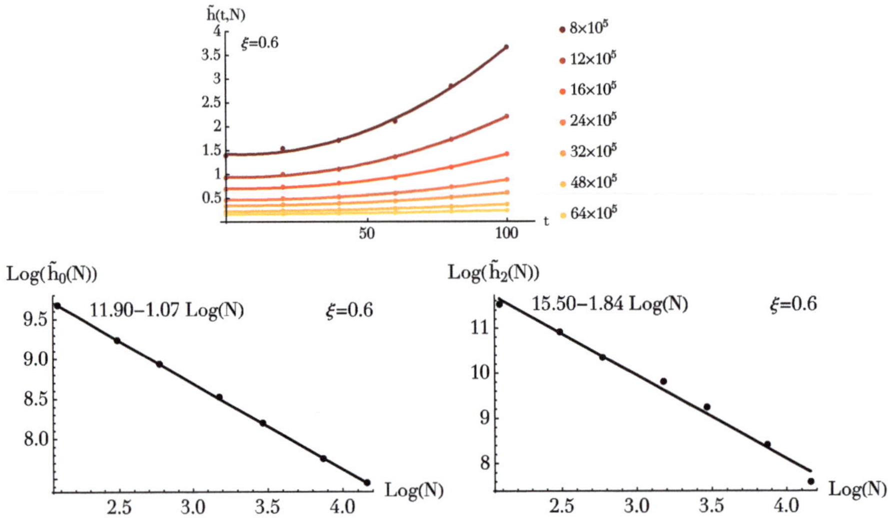

Fig. 4.11 *Top panel* Illustration of the behaviour of the function $t \mapsto \widetilde{h}(t, N)$ for an active fraction $\xi = 0.6$, following the same conventions as in Fig. 4.8. As expected, increasing the active fraction hastens the secular diffusion and therefore hastens the growth of the function $\widetilde{h}(t, N)$. *Bottom-left* panel Behaviour of the function $\log(N) \mapsto \log(\widetilde{h}_0(N))$ for an active fraction $\xi = 0.6$, following the conventions from Fig. 4.9. The associated linear fit reads $\log(\widetilde{h}_0(N)) \simeq 11.90 - 1.07 \log(N)$. One recovers the expected scaling of the initial Poisson shot noise sampling as obtained in Eq. (4.38). *Bottom-right* panel Illustration of the behaviour of the function $\log(N) \mapsto \log(\widetilde{h}_2(N))$ for an active fraction $\xi = 0.6$, following the conventions from Fig. 4.9. The associated linear fit takes the form $\log(\widetilde{h}_2(N)) \simeq 15.50 - 1.84 \log(N)$. One recovers the expected Balescu–Lenard collisional scaling obtained in Eq. (4.42)

which can be both measured via direct $N-$body simulations following Sect. 4.4.2, as well as predicted by the Balescu–Lenard formalism, following Sect. 4.3.1.

Let us consider the same set of simulations as previously, with $N \in \{8, 12, 16, 24, 32, 48, 64\} \times 10^5$ and 32 different realisations for each value of N, while increasing the value of ξ to $\xi = 0.6$. The equivalents of Figs. 4.8 and 4.9 are illustrated for $\xi = 0.6$ in Fig. 4.11. Even in this more susceptible regime, one recovers the fact that the function $t \mapsto \widetilde{h}(t, N)$ behaves initially like a parabola as obtained in Eq. (4.36). One also recovers the expected scalings of the functions $N \mapsto \widetilde{h}_0(N), \widetilde{h}_2(N)$, associated respectively with the initial Poisson shot noise and the collisional scaling of the Balescu–Lenard equation. Thanks to these fits, let us now study the ratio $\alpha_2(\xi = 0.6)/\alpha_2(\xi = 0.5)$ to estimate the amplitude of the associated polarisations. The fits $N \mapsto \widetilde{h}_2(N)$ from Figs. 4.9 and 4.11 allow us to write $\log(\widetilde{h}_2(N)) \simeq 6.40 - 1.91(\log(N) - 3.12)$ for $\xi = 0.5$, and $\log(\widetilde{h}_2(N)) \simeq 9.76 - 1.84(\log(N) - 3.12)$ for $\xi = 0.6$, where we shifted the intercept of the fits to correspond to the centre of the considered domain $\log(N) \in [\log(8); \log(64)]$. As a consequence, from the $N-$body realisations, one obtains the ratio

$$\left.\frac{\alpha_2(0.6)}{\alpha_2(0.5)}\right|_{\mathrm{NB}} \simeq \exp\big[9.76-6.40\big] \simeq 29\,. \tag{4.47}$$

Let us now compare this measurement from numerical simulations to the same measurement estimated via the Balescu–Lenard formalism. Equation (4.41) immediately gives us this ratio as

$$\frac{\alpha_2(\xi_1)}{\alpha_2(\xi_2)} = \frac{\displaystyle\int \mathrm{d}\boldsymbol{J}\,\big[\mathrm{div}(\boldsymbol{\mathcal{F}}_{\mathrm{tot}}^{\xi_1})\big]^2}{\displaystyle\int \mathrm{d}\boldsymbol{J}\,\big[\mathrm{div}(\boldsymbol{\mathcal{F}}_{\mathrm{tot}}^{\xi_2})\big]^2}\,, \tag{4.48}$$

where $\boldsymbol{\mathcal{F}}_{\mathrm{tot}}^{\xi}$ stands for the initial Balescu–Lenard diffusion flux from Eq. (2.72) with an active fraction ξ. The value of $\alpha_2(\xi=0.5)$ can be determined from Fig. 4.4, while we illustrate in Fig. 4.12 the secular diffusion flux predicted for $\xi=0.6$.

Comparing Figs. 4.4 and 4.12, we note that they both exhibit similar diffusion features, but the one associated with the larger value of ξ predicts a faster diffusion. Thanks to the contours from both Figs. 4.4 and 4.12, we may estimate the ratio of the coefficients α_2. In order to focus on the regions associated with the resonant ridge, the integrals on $\boldsymbol{J}$ in Eq. (4.48) are performed for $J_\phi \in \big[0.5;\,1.2\big]$ and $J_r \in \big[0.06;\,0.15\big]$. These Balescu–Lenard predictions lead to the measurement

$$\left.\frac{\alpha_2(0.6)}{\alpha_2(0.5)}\right|_{\mathrm{BL}} \simeq 42\,. \tag{4.49}$$

Despite the noise associated with considering a much more sensitive disc with $\xi=0.6$, the ratios of α_2 measured either via direct $N-$body simulations as in

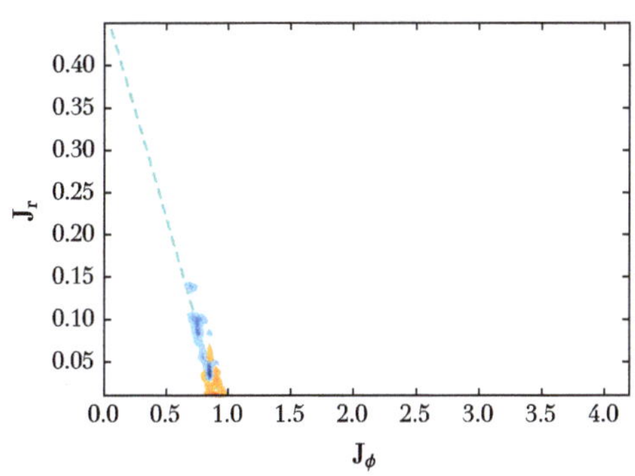

Fig. 4.12 Map of $N\mathrm{div}(\boldsymbol{\mathcal{F}}_{\mathrm{tot}})$ with an active fraction $\xi=0.6$, following the same conventions as in Fig. 4.4. The contours are spaced linearly between the minimum and the maximum of $N\mathrm{div}(\boldsymbol{\mathcal{F}}_{\mathrm{tot}})$. The maximum value for the positive *blue* contours corresponds to $N\mathrm{div}(\boldsymbol{\mathcal{F}}_{\mathrm{tot}})\simeq 4200$, while the minimum value for the negative *red* contours reads $N\mathrm{div}(\boldsymbol{\mathcal{F}}_{\mathrm{tot}})\simeq -3200$. Increasing the active fraction of the disc increases its susceptibility, so that the norm of $N\mathrm{div}(\boldsymbol{\mathcal{F}}_{\mathrm{tot}})$ gets larger and the secular diffusion is hastened

Eq. (4.47) or via the Balescu–Lenard formalism as in Eq. (4.49) are within the same order of magnitude. We recovered here a crucial strength of the Balescu–Lenard formalism which, because it accounts for collective effects, captures the relative effects of the disc's susceptibility on the characteristics of the secular collisional diffusion. This is essential for dynamically cold systems such as razor-thin discs.

4.4.4 Secular Phase Transitions

In Sect. 4.3.1, we computed the Balescu–Lenard predictions for the initial diffusion flux $\mathcal{F}_{\text{tot}}(t=0^{+})$. The direct $N-$body simulations presented in Sect. 4.4.2 also allowed us to check the appropriate scaling of the system's diffusion w.r.t. the number of particles and its active fraction. In order to probe the late secular evolution of the system via the Balescu–Lenard formalism, one has to integrate forward in time the Balescu–Lenard equation. In order to account for the fact that this diffusion is self-induced, i.e. the fact the drift and diffusion coefficients are self-consistent with the system's DF, i.e. the fact that the Balescu–Lenard equation is an integro-differential equation, this integration forward in time has to be made iteratively by updating the drift and diffusion coefficients as the system's DF changes. Such difficult iterations are beyond the scope of the present chapter, but we refer to Appendix 6.C for an illustration of how the Balescu–Lenard diffusion equation may be rewritten as a stochastic Langevin equation, for which numerical integrations appear simpler. Now that they have been validated at $t=0^{+}$, the direct $N-$body simulations may be used to investigate the late times evolution of the system.

The Balescu–Lenard equation (2.67) describes the long-term evolution of a discrete stable quasi-stationary self-gravitating inhomogeneous system. As already underlined in the derivation of this kinetic equation (see, e.g., Eq. (2.128)), for such slow evolutions to occur, it is mandatory for the system to be dynamically stable w.r.t. the collisionless Vlasov dynamics. This also has to remain valid as the system diffuses through a series of quasi-stationary stable equilibria. The Balescu–Lenard equation being associated with a kinetic development at order $1/N$, let us note that such an equation is valid for secular timescales of order $N t_{\text{D}}$, where t_{D} is the dynamical time.

When considering long-term evolutions, the Balescu–Lenard dynamics may lead to two distinct outcomes. On the one hand, if the system remains dynamically stable during its entire evolution, the Balescu–Lenard equation will drive the system towards a $1/N-$stationary state.[2] After having reached such a stationary state, the

---

[2]Boltzmann's DFs of the form $F(\boldsymbol{J}) \propto \exp[-\beta H(\boldsymbol{J})]$, when physically reachable, are obvious stationary states of the Balescu–Lenard equation. Let us emphasise that self-gravitating systems cannot in the strict sense reach statistical equilibrium, as entropy is not bounded from above (Padmanabhan 1990; Chavanis 2006). Indeed, for a self-gravitating system, it only takes two particles to satisfy the conservation of energy (by bringing them arbitrarily close to each other) and another two to satisfy the conservation of angular momentum (by sending one of them arbitrarily far from the cluster). Lynden-Bell and Kalnajs (1972) have shown that, when given the opportunity, waves within the

$1/N$ effects vanish, and the system's dynamics is then driven by $1/N^2$ effects, which are not accounted for in the Balescu–Lenard equation. On the other hand, the Balescu–Lenard equation may also lead on secular timescales to a dynamical destabilisation of the system. As a result of the long-term resonant effects of the collisional diffusion, the irreversible changes induced in the system's DF may make the system unstable. After a slow, stable, and quasi-stationary evolution sourced by collisional $1/N-$effects, the system may at some point become dynamically unstable w.r.t. the collisionless dynamics, which then becomes the main driver of the system's later time evolution. This was already suggested in Sellwood (2012), which observed an out-of-equilibrium phase transition between the $1/N$ Balescu–Lenard collisional evolution and the collisionless Vlasov evolution.

Relying on the $N-$body simulations presented in Sect. 4.4.2, let us now illustrate in detail this phase transition. In order to capture the change of evolution regime within the disc (collisional vs. collisionless), let us define for a given number N of particles the quantity $\Sigma_2(t, N)$ as

$$\Sigma_2(t, N) = \left\langle \int_{R_{\rm inf}}^{R_{\rm sup}} \mathrm{d}R\, R\, \mathrm{d}\phi\, \Sigma_{\rm star}(t, N, R, \phi)\, \mathrm{e}^{-\mathrm{i}2\phi} \right\rangle = \left\langle \mu \sum_n \mathrm{e}^{-\mathrm{i}2\phi_n} \right\rangle, \tag{4.50}$$

where, similarly to Eq. (4.34), $\langle\, \cdot \,\rangle$ stands for the ensemble average over the 32 different realisations with the same number of particles. The radii considered here are restricted to the range $R \in [R_{\rm inf};\, R_{\rm sup}] = [1.2;\, 5]$, where the active surface density of the disc is only weakly affected by the inner and outer tapers. Finally, to obtain the second equality in Eq. (4.50), we replaced the active surface density, $\Sigma_{\rm star}$, by a discrete sum over all the particles of the system. Here, the sum on n is restricted only to particles whose radius lies between $R_{\rm inf}$ and $R_{\rm sup}$, and we noted their azimuthal phase as ϕ_n. The function Σ_2 aims at quantifying the strength of non-axisymmetric features within the disc and should therefore be seen as a way to estimate how much the disc has evolved. During the initial Balescu–Lenard collisional evolution of the system, one expects low values of Σ_2, as such an evolution is an orbit-averaged evolution, i.e. we assumed that $F = F(\boldsymbol{J}, t)$, so that the mean system's DF should not depend on the angles $\boldsymbol{\theta}$. During this first slow collisional phase, Σ_2 still remains non-zero, because of both unavoidable Poisson sampling shot noise and the fact that the disc sustains transient spiral waves driving its secular evolution. On the long-term, this collisional evolution leads to a destabilisation of the system. The dynamical drivers of the system's evolution are no more discrete distant resonant collisional encounters, but exponentially growing collisionless dynamical instabilities. Because of the appearance of strong non-axisymmetric features, in this collisionless regime, one

(Footnote 2 continued)
system will reshuffle orbits so that mass flows inwards and angular momentum flows outwards, which leads to an increase in entropy.

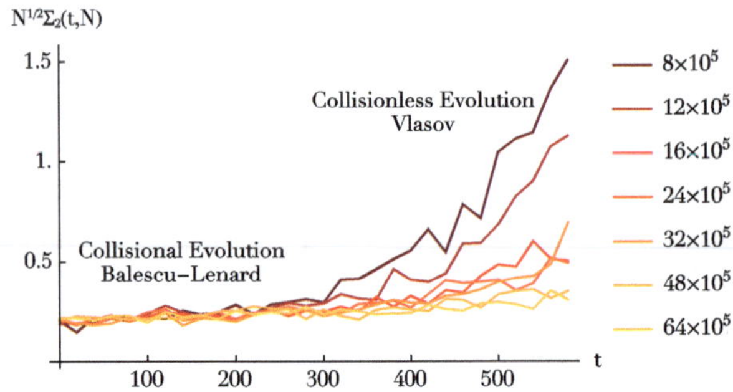

Fig. 4.13 Illustration of the behaviour of the function $t \mapsto \sqrt{N}\Sigma_2(t, N)$, as introduced in Eq. (4.50), as one varies the number of particles. The prefactor $\sqrt{N}$ was added to mask Poisson shot noise allowing for the initial values of $\sqrt{N}\Sigma_2$ to be independent of N. This illustrates the out-of-equilibrium transition between the initial Balescu–Lenard collisional evolution, for which low values of Σ_2 are expected, and the collisionless Vlasov evolution, for which the system loses its mean axisymmetry and larger values of Σ_2 are reached. As expected, the larger the number of particles, the later the transition

expects much larger values of Σ_2. Figure 4.13 illustrates this transition between the two regimes of diffusion, thanks to the behaviour of the function $t \mapsto \sqrt{N}\Sigma_2(t, N)$.[3] This phase transition can also easily be seen by directly looking at the disc's active surface density Σ_{star} during these two regimes. This is illustrated in Fig. 4.14, where one notices that during the late time collisionless evolution, the disc becomes strongly non-axisymmetric.

This phase transition can also easily be seen by directly looking at the disc's active surface density Σ_{star} during these two regimes. This is illustrated in Fig. 4.14, where one notices that during the late time collisionless evolution, the disc becomes strongly non-axisymmetric.

In order to illustrate this change of dynamical regime, Sellwood (2012) fitted unstable growing modes to the disc in this regime to effectively recover the presence of a dynamical instability. Right after the instability settles in, S12 noted that the pattern speed of the spiral response is consistent with the ILR frequency associated with the ridge. At this stage, one could also rely on the matrix method from Appendix 4.C, to show that a perturbed DF with a sufficiently large ridge is indeed associated with an unstable configuration (De Rijcke and Voulis 2016). In conclusion, let us emphasise that an isolated stellar disc, fully stable in the mean sense, will, given time, drive itself through two-point resonant correlations towards dynamical instability.

[3] Similar dynamical phase transitions have been observed in the long-range interacting HMF (Hamiltonian Mean Field) toy model (Campa et al. 2008). During the slow collisional evolution, finite−N effects get the system's DF to change. In some situations, the system may then become (dynamically) unstable and undergoes a rapid phase transition from a homogeneous phase to an inhomogeneous one. This transition can be monitored by the magnetisation (see Fig. 1 in Campa et al. (2008)), which is an order parameter playing a role similar to Σ_2 here. See also Benetti and Marcos (2017) for a recent application of the Balescu–Lenard equation to the HMF model.

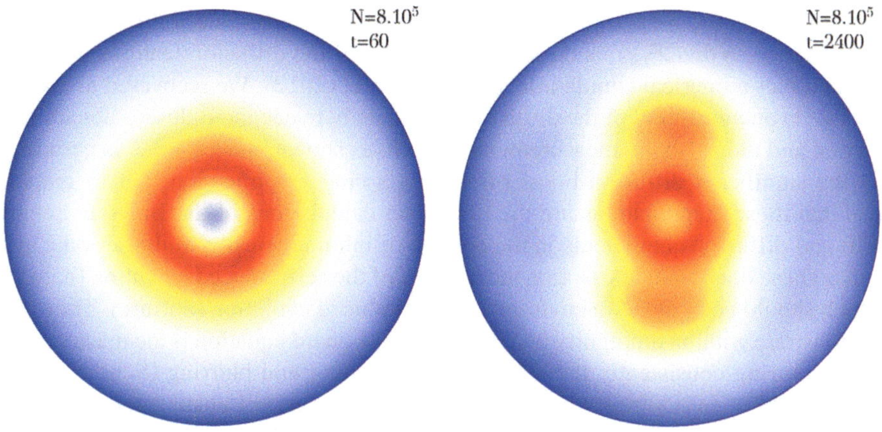

Fig. 4.14 Illustration of the disc's active surface density $\Sigma_{\rm star}$ for a N−body run with $N = 8 \times 10^5$ particles and restricted to the radial range $R \leq 6$. *Left panel* At an early time $t = 60$, for which the mean disc remains axisymmetric. In this regime, the dynamics of the disc is collisional and governed by the Balescu–Lenard equation (2.67). *Right panel* At a much later time $t = 2400$, for which the disc is strongly non-axisymmetric. In this regime, the dynamics of the disc is collisionless and governed by Vlasov equation

This illustrates the extent to which cold quasi-stationary systems such as stellar discs are truly secularly metastable.

4.5 Conclusion

Most astrophysical discs were formed through dissipative processes and typically evolved over many dynamical times. When isolated, long-range gravitational interactions allow their components to interact effectively through resonances, which may secularly drive discs toward more likely equilibria. These processes are captured by recent extensions of the kinetic theory of self-gravitating systems rewritten in angle-action variables and captured by the inhomogeneous Balescu–Lenard equation (2.67). Solving these equations provides astronomers with a new opportunity to quantify in detail the secular dynamics of these systems. While numerically challenging, the computation of the diffusion fluxes predicted by these kinetic equations is, as demonstrated in this chapter, within reach of an extension of the matrix method (Kalnajs 1976), allowing for an estimation of the strength of the self-gravitating orbital response.

In this chapter, we estimated the drift and diffusion coefficients of the inhomogeneous Balescu–Lenard equation in the context of razor-thin stellar discs. The details of the disc's self-gravity were taken into account via the matrix method, as detailed in Sect. 4.2.3. This method was validated on unstable Mestel discs in Appendix C. In Sect. 4.3, we computed the divergence of the self-induced diffusion flux in action space, $N\mathrm{div}(\mathcal{F}_{\mathrm{tot}})$, and recovered in Fig. 4.4 the diffusion features observed in the direct numerical simulations from Sellwood (2012). We recovered as well in Eq. (4.31) an agreement of the diffusion timescale between the Balescu–Lenard prediction and the $N-$body measurements, which, as shown in Sect. 4.3.3, is permitted by the significant diffusion boost offered by swing amplification. Let us emphasise that these computations are the first exact calculations of the Balescu–Lenard drift and diffusion coefficients in the context of inhomogeneous multi-periodic systems. These computations capture the essence of the self-induced evolution (nature), which should compete with environmentally-induced evolution (nurture). They also demonstrate without ambiguity that the Balescu–Lenard equation is the master equation capturing consistently the self-induced churning and blurring (Schönrich and Binney 2009). In addition, the multi-component Balescu–Lenard equation (2.76) can also account for the joint evolutions of multiple populations, e.g., stars and giant molecular clouds. The presence of a spectrum of masses can have a significant effect on the system's secular dynamics, as detailed in Sect. 5.7.6 for thickened discs.

In Sect. 4.4, we compared these predictions to idealised numerical simulations of stable razor-thin Mestel discs sampled by pointwise particles and evolved for hundreds of dynamical times. Relying on ensemble averages of these $N-$body runs, we identified a clear signature of the Balescu–Lenard process in the scaling of the diffusion features with N and ξ, the active fraction of mass within the disc. We also emphasised how, for late times, the collisional diffusion features slowly appearing in the disc's DF eventually lead to a destabilisation of the disc. As originally identified in Goldreich and Lynden-Bell (1965), Julian and Toomre (1966) by studying their linear response, the susceptibility of cold self-gravitating discs plays a crucial role in their secular evolution as it appears squared in the Balescu–Lenard equation, which significantly boosts the effects induced by the system's discreteness. In these early works, the relevance of the susceptibility was shown via the study of discs' linear response. Here, we have shown how central this susceptibility is to discs' secular response.

Figure 4.15 summarises the main stages in the evolution of the razor-thin stellar disc considered in Sellwood (2012). Let us now briefly recall how the inhomogeneous Balescu–Lenard equation allows for the description of such a complex dynamics. First, as a direct consequence of the unavoidable Poisson shot noise in the initial sampling, the larger the number of particles, the smaller the initial value of δ_{max}, as can be seen in Fig. 4.15. We recovered explicitly this scaling in Fig. 4.9, via the study of the function $N \mapsto \widetilde{h}_0(N)$. Then, in the first few dynamical times, the initial perturbations in the disc are strongly amplified by the disc's self-gravity, via the mechanism of swing amplification (see Fig. 3.20). In order to account for such collective effects, one has to rely on the matrix method, as detailed in Appendix 4.C. Because of the finite number of particles and the discreteness noise it induces, the disc begins subsequently its slow secular evolution. This corresponds to a collisional dynamics captured by the inhomogeneous Balescu–Lenard equation (2.67). This dynamics is sourced by $1/N$ effects, so that the larger the number of particles, the slower the evolution. This can be seen in Fig. 4.15, where one notes that the initial slope of the magenta line is smaller than the initial slope of the cyan line. We emphasised this scaling of the collisional dynamics in Fig. 4.9, thanks to the function $N \mapsto \widetilde{h}_2(N)$. During this first stage of evolution, the system's slow orbital distortion leads to the appearance of a narrow ridge in the disc's DF in action space. We recovered this ridge explicitly in Fig. 4.4, thanks to the calculation of the initial diffusion flux of the inhomogeneous Balescu–Lenard equation. When the resonant ridge gets too large, the disc becomes linearly unstable and undergoes an out-of-equilibrium dynamical phase transition, as illustrated in Figs. 4.13 and 4.14. Following this destabilisation, the disc's dynamics is governed by Vlasov equation (1.9). The evolution is then collisionless, i.e. independent of the number of particles. This can be seen in Fig. 4.15, where one notes that the final slopes of the magenta and cyan lines are the same. The disc finally reaches a saturation, where its orbital structure has been drastically reshuffled. Figure 4.15 illustrates therefore how the kinetic theory of long-range interacting systems allows for the detailed description of the long-term dynamics of self-gravitating systems such as razor-thin stellar discs.

The various illustrations presented in this chapter offer a qualitative and quantitative understanding of the secular diffusion processes induced by discreteness effects occurring in galactic discs. Our qualitative agreements in both amplitude, position, width and scaling of the induced orbital signatures suggest therefore that the secular evolution of such razor-thin stellar discs is indeed driven by discrete resonances, as captured by the Balescu–Lenard equation. Let us finally emphasise that such an evolution does not depend on the initial phases of the disc's constituents, since the matching Balescu–Lenard fluxes are phase averaged. The Balescu–Lenard collisional equation therefore reproduces the initial orbital evolution of self-gravitating discs driven by discrete two-point correlations beyond the mean field approximation.

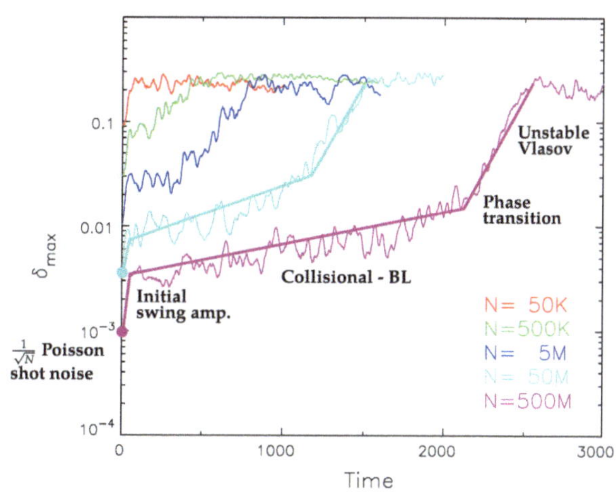

Fig. 4.15 Background extracted from Sellwood (2012). Illustration of the different dynamical regimes successively at play in the evolution of a razor-thin stellar disc. The present chapter illustrated how these different regimes can be understood via the inhomogeneous Balescu–Lenard equation. See the main text for a detailed discussion

4.5.1 Future Works

We have seen in this chapter how the inhomogeneous Balescu–Lenard equation was indeed able to accurately capture the diffusion features observed in direct $N-$body simulations of razor-thin discs, once one accounts correctly for the system's self-gravity. It appears as particularly important in this context, as razor-thin stellar discs are cold dynamical systems, within which the swing amplification of loosely wound fluctuations can be very large.

A first direct follow-up of this work would be to integrate forward in time the Balescu–Lenard equation, in order to estimate the system's diffusion flux at some later time. Because of the self-consistency of the Balescu–Lenard equation, such an integration remains technically difficult. One possible approach would be to rely on its Langevin rewriting (see Appendix 6.C), which describes the diffusion of individual particles rather than of the system's DF as a whole. This integration would especially allow for a more detailed investigation of the process of secular phase transition during which the system becomes at some point unstable for the collisionless dynamics.

As was emphasised in Appendix 4.C, $N-$body simulations require the use of an additional parameter, the softening length ε, which leads to a modification of the pairwise interaction potential. As can be seen in Fig. 1 of Sellwood and Evans (2001), this parameter has a strong impact on the characteristics of the unstable modes recovered via $N-$body simulations. It would be of particular interest to investigate in a systematic manner the various influences of this parameter. This would for example involve looking at the effects of ε on unstable modes measurements in numerical simulations, update the linear theory to account for a modified softened interaction potential, and investigate how such a softening may also impact the Balescu–Lenard predictions for the secular diffusion features.

We described in Appendix 4.D a possible follow-up of this work which would be to consider the secular dynamics of $3D$ spherical systems, whose dynamics is very similar to that presented in this chapter. Accounting for potential fluctuations induced by supernova feedback would allow us to investigate one possible mechanism of softening of dark matter haloes' profiles on cosmic times. By characterising from hydrodynamical simulations the typical feedback-induced perturbations from a galactic disc onto its dark matter halo, one should be in a position to quantitatively estimate the amplitude of the subsequent secular diffusion in the halo. Such mechanisms are for now beyond the reach of current simulations, so that such a precise theoretical framework appears as a necessary first step to probe these processes.

Similarly, the secular dynamics of $3D$ spherical globular clusters should also be re-investigated within the Balescu–Lenard formalism, following the approach presented in Appendix 4.D. Indeed, though it has long been known that for such systems, their discreteness is the main driver of their secular evolution, it has up to now only be described as local encounters (see Heggie and Hut (2003) for a review). The Balescu–Lenard equation, complemented with the estimation of the system's self-gravity from linear theory, would allow us to account for non-local resonances, assess the importance of the cluster's self-gravity, and recover the observed dependence of the system's response with the number of particles and the fraction of radial orbits.

Appendix

A Kalnajs $2D$ Basis

Let us detail the $2D$ basis introduced in Kalnajs (1976) and used in Sect. 4.3 to compute the diffusion flux. A similar rewriting of the basis normalisations can also be found in Earn and Sellwood (1995). This basis depends on two parameters, namely $k_{\mathrm{Ka}} \in \mathbb{N}$ and a scale radius $R_{\mathrm{Ka}} > 0$. In order to shorten the notations in the upcoming expressions, let us write r for the dimensionless quantity r/R_{Ka}. As introduced in Eqs. (4.5) and (4.6), the $2D$ basis elements depend on two indices: the azimuthal number $\ell \geq 0$ and the radial number $n \geq 0$. The radial component of the potential elements is given by

$$\mathcal{U}_n^{\ell}(r) = -\frac{\sqrt{G}}{R_{\mathrm{Ka}}^{1/2}} \mathcal{P}(k_{\mathrm{Ka}}, \ell, n)\, r^{\ell} \sum_{i=0}^{k} \sum_{j=0}^{n} \alpha_{\mathrm{Ka}}(k_{\mathrm{Ka}}, \ell, n, i, j)\, r^{2i+2j} \,, \tag{4.51}$$

while the radial component of the density elements reads

$$\mathcal{D}_n^{\ell}(r) = \frac{(-1)^n}{\sqrt{G}\, R_{\mathrm{Ka}}^{3/2}} \mathcal{S}(k_{\mathrm{Ka}}, \ell, n)\, (1-r^2)^{k_{\mathrm{Ka}}-1/2}\, r^{\ell} \sum_{j=0}^{n} \beta_{\mathrm{Ka}}(k_{\mathrm{Ka}}, \ell, n, j)\, (1-r^2)^j \,. \tag{4.52}$$

In Eqs. (4.51) and (4.52), we introduced the coefficients $\mathcal{P}(k, \ell, n)$ and $\mathcal{S}(k, \ell, n)$ as

$$\begin{aligned}
\mathcal{P}(k, \ell, n) &= \left\{ \frac{[2k+\ell+2n+(1/2)]\Gamma[2k+\ell+n+(1/2)]}{\Gamma[2k+n+1]\,\Gamma^2[\ell+1]\,\Gamma[n+1]} \Gamma[\ell+n+(1/2)] \right\}^{1/2} , \\
\mathcal{S}(k, \ell, n) &= \frac{\Gamma[k+1]}{\pi\,\Gamma[2k+1]\,\Gamma[k+(1/2)]} \left\{ \frac{[2k+\ell+2n+(1/2)]\,\Gamma[2k+n+1]\,\Gamma[2k+\ell+n+(1/2)]}{\Gamma[\ell+n+(1/2)]\,\Gamma[n+1]} \right\}^{1/2} .
\end{aligned} \tag{4.53}$$

In Eqs. (4.51) and (4.52), we also introduced the coefficients α_{Ka} and β_{Ka} as

$$\begin{aligned}
\alpha_{\mathrm{Ka}}(k, \ell, n, i, j) &= \frac{[-k]_i\, [\ell+(1/2)]_i\, [2k+\ell+n+(1/2)]_j [i+\ell+(1/2)]_j\, [-n]_j}{[\ell+1]_i\, [1]_i\, [\ell+i+1]_j [\ell+(1/2)]_j [1]_j} , \\
\beta_{\mathrm{Ka}}(k, \ell, n, j) &= \frac{[2k+\ell+n+(1/2)]_j\, [k+1]_j\, [-n]_j}{[2k+1]_j\, [k+(1/2)]_j\, [1]_j} ,
\end{aligned} \tag{4.54}$$

where the two previous expressions relied on the rising Pochhammer symbol $[a]_i$ defined as

$$[a]_i = \begin{cases} 1 & \text{if} \quad i = 0 \,, \\ a\,(a+1) \ldots (a+n-1) & \text{if} \quad i > 0 \,. \end{cases} \tag{4.55}$$

B Calculation of ℵ

In this Appendix, we briefly detail how the analytic function ℵ, introduced in Eq. (4.21) to compute the response matrix, may be estimated. In order to ease the effective numerical implementation of this calculation, let us first rewrite ℵ in a dimensionless fashion so that

$$
\begin{aligned}
\aleph(a_g,b_g,c_g,a_h,b_h,c_h,\eta,\Delta r) &= \int_{-\frac{\Delta r}{2}}^{\frac{\Delta r}{2}}\!\int_{-\frac{\Delta r}{2}}^{\frac{\Delta r}{2}} \mathrm{d}x_\mathrm{p}\mathrm{d}x_\mathrm{a}\, \frac{a_g+b_g x_\mathrm{p}+c_g x_\mathrm{a}}{a_h+b_h x_\mathrm{p}+c_h x_\mathrm{a}+\mathrm{i}\eta} \\
&= \frac{a_g}{a_h}(\Delta r)^2 \int_{-\frac{1}{2}}^{\frac{1}{2}}\!\int_{-\frac{1}{2}}^{\frac{1}{2}} \mathrm{d}x\mathrm{d}y\, \frac{1+\frac{b_g\Delta r}{a_g}x+\frac{c_g\Delta r}{a_g}y}{1+\frac{b_h\Delta r}{a_h}x+\frac{c_h\Delta r}{a_h}y+\mathrm{i}\frac{\eta}{a_h}} \\
&= \frac{a_g}{a_h}(\Delta r)^2\, \aleph_\mathrm{D}\left[\frac{b_g\Delta r}{a_g},\frac{c_g\Delta r}{a_g},\frac{b_h\Delta r}{a_h},\frac{c_h\Delta r}{a_h},\frac{\eta}{a_h}\right],
\end{aligned}
\tag{4.56}
$$

where we assumed that $a_g, a_h \neq 0$ and used the change of variables $x = x_\mathrm{p}/\Delta r$ and $y = x_\mathrm{a}/\Delta r$. Finally, we introduced the dimensionless function $\aleph_\mathrm{D}$ as

$$
\aleph_\mathrm{D}(b,c,e,f,\eta) = \int_{-\frac{1}{2}}^{\frac{1}{2}}\!\int_{-\frac{1}{2}}^{\frac{1}{2}} \mathrm{d}x\mathrm{d}y\, \frac{1+bx+cy}{1+ex+fy+\mathrm{i}\eta}. \tag{4.57}
$$

To effectively compute this integral, it only remains to exhibit a function $G(x,y)$ so that

$$
\frac{\partial^2 G}{\partial x \partial y} = \frac{1+bx+cy}{1+ex+fy+\mathrm{i}\eta}. \tag{4.58}
$$

One possible choice for G is given by

$$
\begin{aligned}
G(x,y) &= \frac{1}{4e^2 f^2}\log[e^2x^2+2e(fxy+x)+f^2y^2+2fy+\eta^2+1] \\
&\times\Big\{bf(e^2x^2-(fy+\mathrm{i}\eta+1)^2)+2ef(ex+\mathrm{i}\eta+1)-ce(ex+\mathrm{i}\eta+1)^2\Big\} \\
&+\frac{\mathrm{i}}{2e^2f^2}\left\{\frac{\pi}{2}-\tan^{-1}\!\left[\frac{ex+fy+1}{\eta}\right]\right\}\Big\{bf(e^2x^2-(fy+\mathrm{i}\eta+1)^2)+2ef(ex+\mathrm{i}\eta+1)-ce(ex+\mathrm{i}\eta+1)^2\Big\} \\
&+\frac{y}{4e^2 f}\Big\{f(-4e+b(2ex+fy+2\mathrm{i}\eta+2))+ce(2ex-fy+2\mathrm{i}\eta+2)+2ef(cy+2)\log[ex+fy+\mathrm{i}\eta+1]\Big\}.
\end{aligned}
\tag{4.59}
$$

In the previous expression, one should be careful with the presence of a complex logarithm and a $\tan^{-1}$. Fortunately, because $e, f, \eta \in \mathbb{R}$, and $\eta \neq 0$, one can straightforwardly show that the arguments of both of these functions never cross the usual branch-cut of these functions chosen to be $\{\mathrm{Im}(z)=0;\ \mathrm{Re}(z)\leq 0\}$. Equation (4.57) can then be computed as

$$
\aleph_\mathrm{D} = G[\tfrac{1}{2},\tfrac{1}{2}]-G[\tfrac{1}{2},-\tfrac{1}{2}]-G[-\tfrac{1}{2},\tfrac{1}{2}]+G[-\tfrac{1}{2},-\tfrac{1}{2}]. \tag{4.60}
$$

C Recovering Unstable Modes

Let us detail how the matrix code presented in Sect. 4.2.3 as well as the $N-$body code described in Sect. 4.4.1 can be validated by recovering known unstable modes of razor-thin discs. The direct numerical calculation of modes in a galactic disc is a complex task, which has only been made for a small number of discs models (Zang 1976; Kalnajs 1977; Vauterin and Dejonghe 1996; Pichon and Cannon 1997; Evans and Read 1998b; Jalali and Hunter 2005; Polyachenko 2005; Jalali 2007, 2010; De Rijcke and Voulis 2016). Here, we will recover the results of the pioneer work of Zang (1976), extended in Evans and Read (1998), Evans and Read (1998b), and recovered numerically in Sellwood and Evans (2001). These three works were interested in recovering the precession rate $\omega_0 = m_{\mathrm{p}} \Omega_{\mathrm{p}}$ and growth rate $\eta = s$ of the unstable modes a truncated Mestel disc very similar to the stable one presented in Sect. 3.7.1. The unstable discs considered thereafter are fully active discs, so that $\xi = 1$, and their radial velocity dispersion is given by $q = (V_0/\sigma_r)^2 - 1 = 6$. Finally, we consider different models of disc by varying the power index ν_{t} of the inner taper function as defined in Eq. (3.90). Here, we will look for $m_{\mathrm{p}} = 2$ modes, and will consider three different truncation indices given by $\nu_{\mathrm{t}} = 4, 6, 8$. In Sect. C.1, we first recover the associated unstable modes by computing the system's response matrix, following Sect. 4.2.3, while in Sect. C.2, we recover these modes via direct $N-$body simulations using the $N-$body implementation presented in Sect. 4.4.1.

C.1 The Response Matrix Validation

In order to compute the system's response matrix, we follow the method presented in Sects. 4.2.3 and 4.2.4. In addition, we use the same numerical parameters as the ones used in Sect. 4.3.1. The grid in the $(r_{\mathrm{p}}, r_{\mathrm{a}})-$space is characterised by $r_{\mathrm{p}}^{\mathrm{min}} = 0.08$, $r_{\mathrm{a}}^{\mathrm{max}} = 4.92$ and $\Delta r = 0.05$. The sum on the resonant index m_1 is limited to $|m_1| \leq m_1^{\mathrm{max}} = 7$. Finally, we consider basis elements given by Appendix 4.A, with the parameters $k_{\mathrm{Ka}} = 7$ and $R_{\mathrm{Ka}} = 5$, with a restriction of the radial basis elements to $0 \leq n \leq 8$. One should note that despite having a disc that extends up to $R_{\mathrm{max}} = 20$, one can still safely consider a basis truncated at such a small radius R_{Ka}, which allows us to efficiently capture the self-gravitating properties of the disc in the inner regions.

In order to search for unstable modes in a disc, one has to look for complex frequencies $\omega = \omega_0 + \mathrm{i}\eta$, such that the complex response matrix $\widehat{\mathbf{M}}(\omega_0, \eta)$ from Eq. (4.13) possesses an eigenvalue equal to 1. This complex frequency is then associated with an unstable mode of pattern speed $\omega_0 = m_{\mathrm{p}} \Omega_{\mathrm{p}}$ and growth rate η. In order to effectively determine the characteristics of the unstable modes, we follow an approach based on Nyquist contours, as presented in Pichon and Cannon (1997). For a fixed value of η, one studies the behaviour of the function $\omega_0 \mapsto \det\big[\mathbf{I} - \widehat{\mathbf{M}}(\omega_0, \eta)\big]$, which takes the form a continuous curve in the complex plane, called a Nyquist contour. For $\eta \to +\infty$, one has $\widehat{\mathbf{M}}(\omega, \eta) \to 0$, so that the contour will shrink around the point $(1, 0)$.

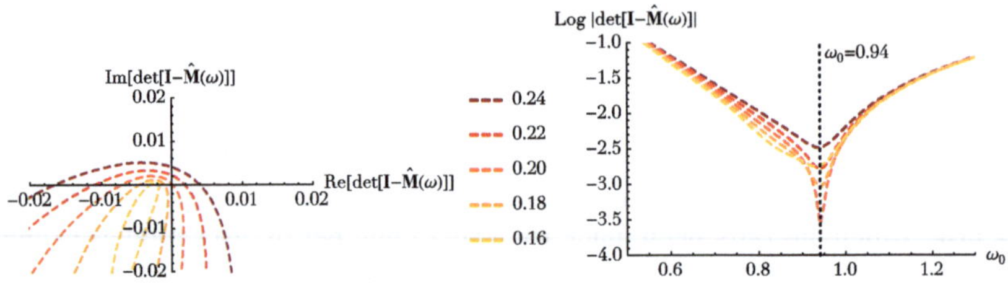

Fig. 4.16 *Left panel* Zoomed-in Nyquist contours in the complex plane of the function $\omega_0 \mapsto \det\big[\mathbf{I}-\widehat{\mathbf{M}}(\omega_0,\eta)\big]$ for various fixed values of η illustrated with different colours. These contours were obtained via the matrix method for a truncated Mestel disc with $\nu_t = 6, q = (V_0/\sigma_r)^2 - 1 = 6$, and looking for $m_p = 2$ modes. One can note that for $\eta = 0.20$, the contour crosses the origin, which corresponds to the presence of an unstable mode. *Right panel* Illustration of the behaviour of the function $\omega_0 \mapsto \log|\det\big[\mathbf{I}\cdot\widehat{\mathbf{M}}(\omega_0,\eta)\big]|$, when considering the same truncated Mestel disc as in the *left panel*. Each coloured curve corresponds to a different fixed value for η. This representation allows us to determine the pattern speed $\omega_0 = m_p\Omega_p \simeq 0.94$ of the system's unstable mode

As a consequence, for a given value of η, the number of windings of the Nyquist contour around the origin of the complex plane gives a lower bound of the number of unstable modes with a growth rate superior to η. By varying the value of η, one can then determine the largest value of η which admits an unstable mode, and this is the growth rate of the most unstable mode of the disc. Figure 4.16 illustrates these Nyquist contours for an unstable Mestel disc with the truncation index $\nu_t = 6$. We gathered in Table 4.1 the results of the measurements for the three discs considered.

After having determined the characteristics (ω_0, η) of the unstable modes, one can then study their shapes in the physical space. To do so, one can compute $\widehat{\mathbf{M}}(\omega_0, \eta)$ and numerically diagonalise this matrix of size $n_{\max} \times n_{\max}$, where $n_{\max}$ is the number of basis elements considered. One of the matrix eigenvalues is then almost equal to 1, and one can consider its associated eigenvector $\boldsymbol{X}_{\text{mode}}$. The shape of the mode is then given by

$$\Sigma_{\text{mode}}(R, \phi) = \text{Re}\bigg[\sum_p \boldsymbol{X}^p_{\text{mode}}\, \Sigma^{(p)}(R, \phi)\bigg], \tag{4.61}$$

where we wrote the considered surface density basis elements as $\Sigma^{(p)}$. Figure 4.17 illustrates the shape of the recovered unstable mode for the truncated $\nu_t = 4$ Mestel disc.

C.2 The N−Body Code Validation

Let us now investigate the same unstable modes via direct $N-$body simulations, in order to validate the $N-$body implementation on which Sect. 4.4 is based. We do not detail here the initial sampling of the particles required to setup the simulations,

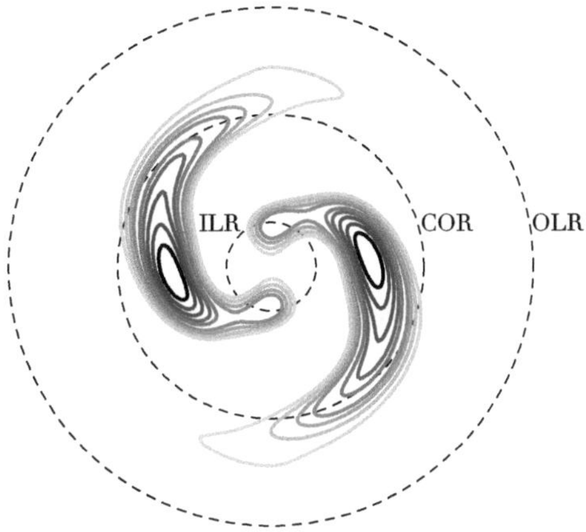

Fig. 4.17 Illustration of the dominant $m_\mathrm{p}=2$ unstable mode for a truncated $\nu_\mathrm{t}=4$ Mestel disc as recovered via the matrix method presented in Sect. 4.2.3. Only positive level contours are shown and they are spaced linearly between 10 and 90% of the maximum norm. The three resonance radii, associated with the resonance ILR, COR, and OLR have been represented, as defined by $\omega_0=m_\mathrm{p}\Omega_\mathrm{p}=\boldsymbol{m}\cdot\boldsymbol{\Omega}(R_{\boldsymbol{m}})$, where the intrinsic frequencies $\boldsymbol{\Omega}(R)=(\Omega_\phi(R),\kappa(R))$ are computed within the epicyclic approximation, as in Eq. (3.87). See Fig. 3.9 for an illustration of the signification of these resonance radii

and details can be found in Appendix E of Fouvry et al. (2015). In order not to be significantly impacted by the initial Poisson shot noise and the lack of a quiet start sampling (Sellwood 1983), for each value of the truncation power ν_t, the simulations were performed with $N=20\times10^6$ particles. As can be observed in Fig. 1 of Sellwood and Evans (2001), in order to recover correctly the characteristics of the disc's unstable modes, an appropriate setting of the $N-$body code parameters is crucial. Following the description from Sect. 4.4.1, we consider a cartesian grid made of $N_\mathrm{mesh}=120$ grid cells, while using a softening length given by $\varepsilon=R_\mathrm{i}/60$. We also restrict the perturbing forces only to the harmonic sector $m_\phi=2$, thanks to $N_\mathrm{ring}=2400$ radial rings with $N_\phi=720$ azimuthal points.

In order to extract the characteristics of the unstable modes from $N-$body realisations, one may proceed as follows. For each snapshot of the simulation, one can estimate the disc's surface density as

$$\begin{aligned}\Sigma_\mathrm{star}(\boldsymbol{x},t) &= \mu\sum_{i=1}^{N}\delta_\mathrm{D}(\boldsymbol{x}-\boldsymbol{x}_i(t))\\ &= \sum_p b_p(t)\,\Sigma^{(p)}(\boldsymbol{x})\,,\end{aligned}\tag{4.62}$$

where the sum on i in the first line is made on all particles in the simulation, and $\boldsymbol{x}_i(t)$ stands for the position of the ith particle at time t. In the second line of Eq. (4.62), the sum on p is made on all the basis elements considered. Here, we consider the same basis elements as the ones considered previously in Sect. C.1. The basis coefficients $b_p(t)$ are straightforward to estimate thanks to the biorthogonality property from Eq. (2.12). They read

$$b_p(t) = -\int \mathrm{d}\boldsymbol{x}\, \Sigma_{\mathrm{star}}(\boldsymbol{x}, t)\, \psi^{(p)*}(\boldsymbol{x}) = -\mu \sum_i \psi^{(p)}(\boldsymbol{x}_i(t)) \,. \tag{4.63}$$

Because we are looking for unstable modes, we expect the coefficients $b_p(t)$ to have a temporal dependence of the form $b_p(t) \propto \exp[-\mathrm{i}(\omega_0 + \mathrm{i}\eta)t]$, where $\omega_0 = m_{\mathrm{p}}\Omega_{\mathrm{p}}$ is the pattern speed of the mode and η its growth rate. If an unstable mode is present in the disc, one therefore expects the relations

$$\frac{\mathrm{d}\,\mathrm{Re}\big[\log(b_p(t))\big]}{\mathrm{d}t} = \eta \;\; ; \;\; \frac{\mathrm{d}\,\mathrm{Im}\big[\log(b_p(t))\big]}{\mathrm{d}t} = -\omega_0 \,, \tag{4.64}$$

provided one pays a careful attention to the branch-cut of the complex logarithm. These linear time dependences appear therefore as appropriate measurements to estimate the growth rate and pattern speed of unstable modes. Let us note that Eq. (4.64) does not hold anymore if more than one unstable mode of similar strength are present in the disc. Figure 4.18 illustrates such measurements for different values of the truncation index ν_{t}.

Following the determination of the basis coefficients $b_p(t)$, one can consequently study the shape of the recovered unstable mode in the physical space. Indeed, similarly to Eq. (4.61), the shape of the mode is given by

$$\Sigma_{\mathrm{mode}}(R, \phi, t) = \mathrm{Re}\bigg[\sum_p b_p(t)\, \Sigma^{(p)}(R, \phi)\bigg] \,. \tag{4.65}$$

Similarly to Fig. 4.17, we illustrate in Fig. 4.19 the unstable mode of the same truncated $\nu_{\mathrm{t}} = 4$ Mestel disc, as recovered from N-body simulations.

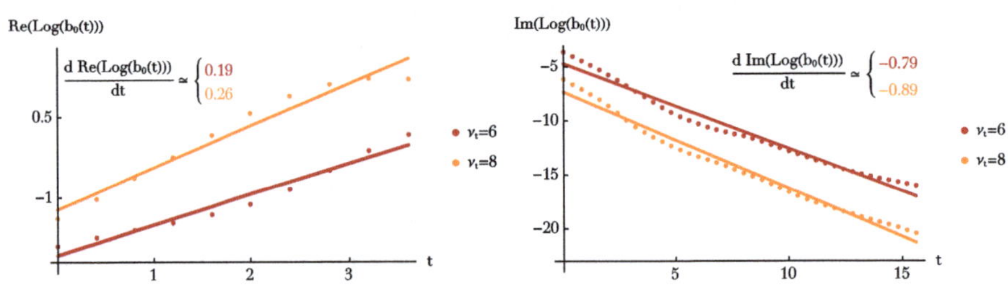

Fig. 4.18 Illustration of the measurements of the growth rates η (*left panel*) and pattern speeds ω_0 (*right panel*) of the $m_{\mathrm{p}} = 2$ unstable modes of truncated Mestel discs, with a radial velocity dispersion given by $q = (V_0/\sigma_r)^2 - 1 = 6$, and truncation indices given by $\nu_{\mathrm{t}} = 6, 8$. The basis coefficient plotted corresponds to the indices $(\ell, n) = (2, 0)$

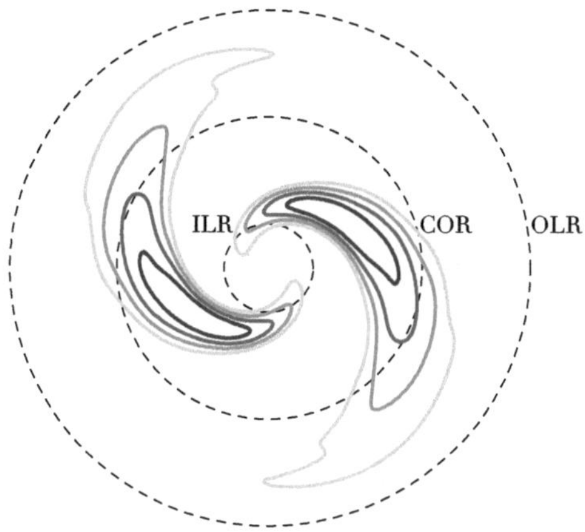

Fig. 4.19 Illustration of the dominant $m_p = 2$ unstable mode for a truncated $\nu_t = 4$ Mestel disc as recovered via direct N−body simulations. Only positive level contours are shown and they are spaced linearly between 20 and 80% of the maximum norm. Similarly to Fig. 4.17, the radii associated with the resonances ILR, COR, and OLR are represented

As a conclusion, we gathered in Table 4.1 the growth rates and pattern speeds obtained either via the matrix method or via direct N−body simulations. As already noted in Sellwood and Evans (2001) when considering truncated Mestel discs, the recovery of the characteristics of unstable modes from direct N−body simulations remains a difficult task, for which the convergence to the values predicted through linear theory can be delicate.

Table 4.1 Measurements of the pattern speed $\omega_0 = m_p \Omega_p$ and growth rate η for unstable $m_p = 2$ modes in truncated Mestel discs. The velocity dispersion is characterised by $q = (V_0/\sigma_r)^2 - 1 = 6$, and the power indices of the inner taper are given by $\nu_t = 4$, 6, 8. The theoretical values were obtained from a tailored linear theory in Evans and Read (1998b). Our measurements were performed either via the response matrix as in Sect. C.1, or via direct N−body simulations as in Sect. C.2

Unstable $m_p = 2$ modes of truncated Mestel discs, $q = 6$.						
Method	$\nu_t = 4$		$\nu_t = 6$		$\nu_t = 8$	
	ω_0	η	ω_0	η	ω_0	η
Linear Theory	0.88	0.13	0.90	0.22	0.92	0.27
Matrix Method	0.93	0.11	0.94	0.20	0.95	0.24
N−body	0.99	0.13	0.79	0.19	0.89	0.26

D The Case of Self-Gravitating Spheres

In this Appendix, let us show how the previous calculations of the system's response matrix and the associated diffusion flux presented for razor-thin discs, can straightforwardly be extended to $3D$ spherical systems. Analytical studies of the linear collective response of spherical self-gravitating systems have been considered by a number of authors (Tremaine and Weinberg 1984; Weinberg 1989; Seguin and Dupraz 1994; Murali and Tremaine 1998; Murali 1999; Weinberg 2001a; Pichon and Aubert 2006). Such calculations are of interest if one wants to describe the long-term evolution of spherical systems such as dark matter haloes while accounting for self-gravity. In Sect. D.1, we show how the main text calculations can straightforwardly be extended to such systems, while in Sect. D.2, we illustrate how such a formalism may be applied to the study of the cusp-core problem in the context of the long-term evolution of dark matter haloes.

D.1 The 3D calculation

As in the case of razor-thin axisymmetric potentials, $3D$ spherically symmetric potentials are also guaranteed by symmetry to be integrable. The three natural actions are given by

$$J_1 = J_r = \frac{1}{\pi}\int_{r_{\mathrm{p}}}^{r_{\mathrm{a}}} \mathrm{d}r\, \sqrt{2(E-\psi_0(r)) - L^2/r^2} \;\; ; \;\; J_2 = L \;\; ; \;\; J_3 = L_z \, , \tag{4.66}$$

where the radial action J_r was already introduced in Eq. (4.1), $L>0$ stands for the magnitude of the particle's angular momentum, and L_z its projection along the $z-$axis. Here, as previously, the first action J_r encodes the amount of radial energy of the star, L encodes the typical distance of the star to the centre, while finally L_z characterises the vertical orientation of the orbital plane to which the particle motion is restricted. The intrinsic frequencies of motion of the associated angles are given by $\boldsymbol{\Omega}=(\Omega_1, \Omega_2, \Omega_3)$. For spherical systems, one should see the third action J_3 as a mute variable, so that $\Omega_3=0$. Therefore, one has an additional conserved quantity, namely $\theta_3=\mathrm{cst.}$, which corresponds to the longitude of the ascending node. As for razor-thin discs, the two additional frequencies of motion $\Omega_1=\kappa$ and $\Omega_2=\Omega_\phi$ are given by Eqs. (4.2) and (4.3). In this context, we may also use the pericentre and apocentre $(r_{\mathrm{p}}, r_{\mathrm{a}})$ to represent the two actions (J_1, J_2).

Let us now introduce the $3D$ spherical coordinates (R, θ, ϕ). For $3D$ systems, the generic basis element can be decomposed as

$$\psi^{(p)}(R,\theta,\phi) = \psi_n^{\ell m}(R,\theta,\phi) = Y_\ell^m(\theta,\phi)\,\mathcal{U}_n^\ell(R) \, , \tag{4.67}$$

where Y_ℓ^m are the usual spherical harmonics and $\mathcal{U}_\ell^n$ is a real radial function. Equation (4.67) is the direct spherical equivalent of the $2D$ decomposition from Eq. (4.5), and allows us to separate the angular dependences from the radial one. Similarly, the

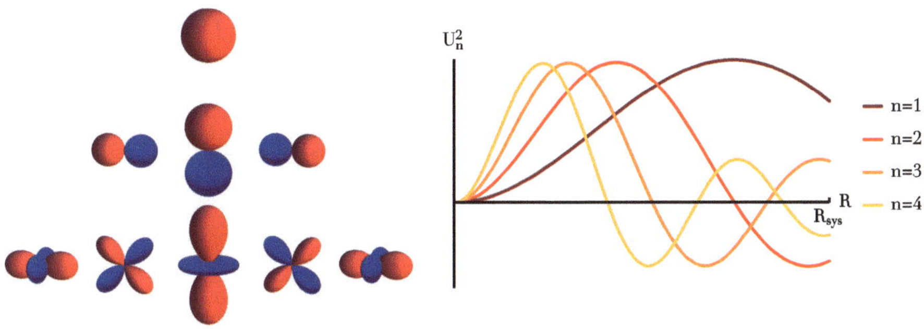

Fig. 4.20 *Left panel* Illustration of the spherical harmonics Y_ℓ^m used to construct the $3D$ basis elements from Eq. (4.67). From top to bottom, the lines are associated with $\ell = 0, 1, 2$, and on a given line, the harmonics are represented for $-\ell \leq m \leq \ell$. *Right panel* Illustration of the radial dependence of the basis function $\mathcal{U}_n^{\ell=2}$, based on spherical Bessel functions and introduced in Weinberg (1989), for various values of the radial index $n \geq 1$. Here, the basis elements are defined on a finite radial range $R \leq R_{\mathrm{sys}}$

associated density elements are given by

$$\rho^{(p)}(R, \theta, \phi) = \rho_n^{\ell m}(R, \theta, \phi) = Y_\ell^m(\theta, \phi)\, \mathcal{D}_n^\ell(R)\,, \tag{4.68}$$

where $\mathcal{D}_n^\ell$ is a real radial function. Explicit $3D$ basis of potentials and densities elements can for example be built from spherical Bessel functions (Fridman and Poliachenko 1984; Weinberg 1989; Rahmati 2009) or from ultraspherical polynomials (Hernquist and Ostriker 1992). The spherical basis elements suggested in Weinberg (1989) are illustrated in Fig. 4.20.

In $3D$, the Fourier transformed basis elements from Eq. (4.7) become

$$\psi_{\boldsymbol{m}}(\boldsymbol{J}) = \frac{1}{(2\pi)^3} \int \mathrm{d}\theta_1 \mathrm{d}\theta_2 \mathrm{d}\theta_2\, \psi^{(p)}(\boldsymbol{x}(\boldsymbol{\theta}, \boldsymbol{J}))\, \mathrm{e}^{-\mathrm{i}\boldsymbol{m}\cdot\boldsymbol{\theta}}\,, \tag{4.69}$$

while the angle mapping from Eq. (4.8) still holds. In order to describe the orientation of the orbital plane, let us define the orbit's inclination, $\beta = \beta(\boldsymbol{J})$, as

$$J_3 = J_2 \cos(\beta) \;\; \text{with} \;\; 0 \leq \beta \leq \pi\,. \tag{4.70}$$

Following Tremaine and Weinberg (1984), Weinberg (1989), the Fourier transform in angles of the basis element $\psi^{(p)} = \psi_{n^p}^{\ell^p m^p}$ w.r.t. the resonance vector $\boldsymbol{m} = (m_1, m_2, m_3)$ takes the form

$$\psi_{\boldsymbol{m}}^{(p)}(\boldsymbol{J}) = \delta_{m^p}^{m_3}\, \mathcal{V}_{\ell^p m_2 m^p}(\beta)\, \mathcal{W}_{\ell^p m_2 n^p}^{m_1}(\boldsymbol{J})\,. \tag{4.71}$$

In the previous equation, one should pay attention to the difference between the index m^p, which is the second index of the basis element from Eq. (4.67), and $\boldsymbol{m} = (m_1, m_2, m_3)$ corresponding to the three indices of the Fourier transform w.r.t.

the angles. In Eq. (4.71), we introduced the coefficient $\mathcal{W}^{m_1}_{\ell^p m_2 n^p}(\boldsymbol{J})$, whose expression was already obtained in Eq. (4.10). We also introduced the coefficient $\mathcal{V}_{\ell^p m_2 m^p}(\beta)$, specific to the $3D$ basis, which reads

$$\mathcal{V}_{\ell^p m_2 m^p}(\beta) = \mathrm{i}^{m^p - m_2} \, Y^{m_2}_{\ell^p}(\pi/2, 0) \, \mathcal{R}^{\ell^p}_{m_2 m^p}(\beta) \,, \tag{4.72}$$

where we introduced the rotation matrix for spherical harmonics, given by

$$\begin{aligned} \mathcal{R}^{\ell}_{m_2 m}(\beta) = \sum_{t} (-1)^t \frac{\sqrt{(\ell+m_2)!\,(\ell-m_2)!\,(\ell+m)!\,(\ell-m)!}}{(\ell-m-t)!\,(\ell+m_2-t)!\,t!\,(t+m-m_2)!} \\ \times \, [\cos(\beta/2)]^{2\ell+m_2-m-2t} [\sin(\beta/2)]^{2t+m-m_2} \,. \end{aligned} \tag{4.73}$$

In Eq. (4.73), the sum is to be made on all the "t" such that the arguments of the factorials are either zero of positive. It corresponds to $t_{\mathrm{min}} \leq t \leq t_{\mathrm{max}}$, with $t_{\mathrm{min}} = \mathrm{Max}[0, m_2 - m]$ and $t_{\mathrm{max}} = \mathrm{Min}[\ell+m, \ell+m_2]$.

Having computed the Fourier transformed basis elements, one may then proceed to the evaluation of the system's response matrix. As already detailed in Sect. 4.2.3, we perform the estimation of the response matrix by using $(r_{\mathrm{p}}, r_{\mathrm{a}})$ as our variables. To do so, the first step of the calculation is to go from $\boldsymbol{J} = (J_1, J_2, J_3)$ to $(E, L, \cos(\beta))$. Similarly to Eq. (4.12), the Jacobian of this transformation is given by

$$\frac{\partial(E, L, \cos(\beta))}{\partial(J_1, J_2, J_3)} = \begin{vmatrix} \Omega_1 & \Omega_2 & 0 \\ 0 & 1 & 0 \\ 0 & -L_z/L^2 & 1/L \end{vmatrix} = \frac{\Omega_1}{L} \,. \tag{4.74}$$

One may now perform the integration w.r.t. the inclination β. Let us therefore assume that the system's DF is such that $F = F(J_1, J_2) = F(E, L)$. In addition, we noted previously that the system's intrinsic frequencies $\boldsymbol{\Omega} = (\Omega_1, \Omega_2, \Omega_3)$ are independent of J_3, so that in the expression (2.17) of the response matrix, the only remaining dependences w.r.t. β are in the Fourier transformed basis elements from Eq. (4.71) through the rotation matrix from Eq. (4.73). Following Edmonds (1996), the rotation matrices satisfy the orthogonality relation

$$\int_{-1}^{1} \mathrm{d}\cos(\beta) \, \mathcal{R}^{\ell^p}_{m_2 m_3}(\beta) \, \mathcal{R}^{\ell^q}_{m_2 m_3}(\beta) = \delta^{\ell^q}_{\ell^p} \frac{2}{2\ell^p + 1} \,. \tag{4.75}$$

Equation (4.72) then gives

$$\int_{-1}^{1} \mathrm{d}\cos(\beta) \, \mathcal{V}^*_{\ell^p m_2 m_3}(\beta) \, \mathcal{V}_{\ell^q m_2 m_3}(\beta) = \delta^{\ell^q}_{\ell^p} \, \mathcal{C}_{\ell^p m_2} \,, \tag{4.76}$$

where we introduced the coefficient $\mathcal{C}_{\ell^p m_2}$ as

$$\mathcal{C}_{\ell^p m_2} = \frac{2}{2\ell^p+1} \left| Y_{\ell^p}^{m_2}(\pi/2, 0) \right|^2 . \tag{4.77}$$

The expression (4.71) of the Fourier transformed basis elements allows us then to rewrite the response matrix from Eq. (2.17) as

$$\widehat{\mathbf{M}}_{pq}(\omega) = (2\pi)^3 \delta_{\ell^p}^{\ell^q} \delta_{m^p}^{m^q} \sum_{\substack{m_1 \\ |m_2| \leq \ell^p \\ (\ell^p - m_2)\text{even}}} \mathcal{C}_{\ell^p m_2} \int \mathrm{d}E \mathrm{d}L \, \frac{L}{\Omega_1} \frac{\boldsymbol{m} \cdot \partial F / \partial \boldsymbol{J}}{\omega - \boldsymbol{m} \cdot \boldsymbol{\Omega}} \mathcal{W}_{\ell^p m_2 n^p}^{m_1}(\boldsymbol{J}) \, \mathcal{W}_{\ell^p m_2 n^q}^{m_1}(\boldsymbol{J}) , \tag{4.78}$$

where one may note that the sum on m_3 was dropped thanks to the Kronecker symbol from Eq. (4.71). In addition, expression (4.77) also imposes the additional constraints $|m_2| \leq \ell^p$ and $(\ell^p - m_2)$ even, so that the sum on m_2 may also be reduced. Finally, we also relied on the fact that the coefficient $\mathcal{W}_{\ell^p m_2 n^p}^{m_1}(\boldsymbol{J})$ from Eq. (4.10) is real, so that no conjugates are present in Eq. (4.78). Let us now perform the change of variables $(E, L) \mapsto (r_\mathrm{p}, r_\mathrm{a})$ to rewrite Eq. (4.78) as

$$\widehat{\mathbf{M}}_{pq}(\omega) = \delta_{\ell^p}^{\ell^q} \delta_{m^p}^{m^q} \sum_{\substack{m_1 \\ |m_2| \leq \ell^p \\ (\ell^p - m_2)\text{ even}}} \int \mathrm{d}r_\mathrm{p} \mathrm{d}r_\mathrm{a} \, \frac{g_{m_1 m_2}^{\ell^p n^p n^q}(r_\mathrm{p}, r_\mathrm{a})}{h_{m_1 m_2}^{\omega}(r_\mathrm{p}, r_\mathrm{a})} , \tag{4.79}$$

where the functions $g_{m_1 m_2}^{\ell^p n^p n^q}(r_\mathrm{p}, r_\mathrm{a})$ and $h_{m_1 m_2}^{\omega}(r_\mathrm{p}, r_\mathrm{a})$ are respectively given by

$$g_{m_1 m_2}^{\ell^p n^p n^q}(r_\mathrm{p}, r_\mathrm{a}) = (2\pi)^3 \, \mathcal{C}_{\ell^p m_2} \left| \frac{\partial(E, L)}{\partial(r_\mathrm{p}, r_\mathrm{a})} \right| \frac{L}{\Omega_1} \left[\boldsymbol{m} \cdot \frac{\partial F}{\partial \boldsymbol{J}} \right] \mathcal{W}_{\ell^p m_2 n^p}^{m_1}(\boldsymbol{J}) \, \mathcal{W}_{\ell^p m_2 n^q}^{m_1}(\boldsymbol{J}) , \tag{4.80}$$

and

$$h_{m_1 m_2}^{\omega}(r_\mathrm{p}, r_\mathrm{a}) = \omega - (m_1, m_2) \cdot (\Omega_1, \Omega_2) . \tag{4.81}$$

In Eq. (4.80), if the system's DF is such that $F = F(E, L)$, the gradient $\boldsymbol{m} \cdot \partial F / \partial \boldsymbol{J}$ w.r.t. the actions may be computed following Eq. (4.17). Let us finally note the very strong analogies that exist between Eqs. (4.79) and (4.14) obtained for razor-thin discs. This allows us to apply to Eq. (4.79) the exact same method as described in Sect. 4.2.4 by truncating the $(r_\mathrm{p}, r_\mathrm{a})-$space in small regions on which linear approximations may be performed. We do not repeat here these calculations. As the calculation of the response matrix can be a cumbersome numerical calculation, it is important to validate its implementation by recovering known unstable modes for $3D$ spherical systems, e.g., in Polyachenko and Shukhman (1981), Saha (1991), Weinberg (1991). Following this approach, one is therefore able to compute the response matrix of a $3D$ spherical system. In addition, the computation of the Fourier transformed basis elements in Eq. (4.71) allows us to subsequently compute the associated collisionless and collisional fluxes. Let us now illustrate in the upcoming section one possible application of such an approach to describe the secular evolution of dark matter haloes.

D.2 An Exemple of Application: The Cusp-Core Problem

Dark matter (DM) only simulations favour the formation of a cusp in the inner region of DM haloes (Dubinski and Carlberg 1991; Navarro et al. 1997), following what appears to be an universal profile, the NFW profile. However, observations tend to recover profiles more consistent with a shallower core profile (Moore 1994; de Blok and McGaugh 1997; de Blok et al. 2001; de Blok and Bosma 2002; Kuzio de Naray et al. 2008). This discrepancy between the cuspy profile predicted by direct DM only simulations and the core profile inferred from observations is one current important challenge in astrophysics, coined the cusp-core problem.

Various solutions have been proposed to resolve this discrepancy. A first set of solutions relies on modifying the dynamical properties of the collisionless dark matter preventing it from indeed collapsing into cuspy profile. Examples include the possibility of warm DM (Kuzio de Naray et al. 2010) or of self-interacting dark matter (Spergel and Steinhardt 2000; Rocha et al. 2013). Another set of solutions rely on the remark that accounting self-consistently for the baryonic physics and its back-reactions on the DM may also be at the origin of the discrepancy. These mechanisms can be divided into three broad categories. The first one relies on dynamical friction from infalling baryonic clumps and disc instabilities (El-Zant et al. 2001; Weinberg and Katz 2002; Tonini et al. 2006; Romano-Diz et al. 2008; Goerdt et al. 2010; Cole et al. 2011; Del Popolo and Pace 2016). A second possible mechanism is associated with AGN-driven feedback (Peirani et al. 2008; Martizzi et al. 2012; Dubois et al. 2016). Finally, a third possible mechanism relies on the long-term effects associated with supernova-driven feedback (Binney et al. 2001; Gnedin and Zhao 2002; Read and Gilmore 2005; Mashchenko et al. 2006, 2008; Governato et al. 2010; Teyssier et al. 2013; Pontzen and Governato 2012; El-Zant et al. 2016).

The collisionless diffusion equation (2.31), and its associated customisation to $3D$ systems presented in this Appendix is the appropriate framework to investigate in detail the role that supernova feedback may have on the secular evolution of DM haloes. Can the presence of an inner stellar disc, because of the potential fluctuations it induces, lead to the secular diffusion of a cuspy DM halo to a core one?

The first step of such an analysis is to characterise these fluctuations. To do so, we rely on hydrodynamical simulations. In order to decouple the source of perturbations, i.e. the disc, from the perturbed system, i.e. the halo, these simulations are performed while using a static and inert halo. Therefore, during the numerical simulations, feedback, while still present, cannot lead to a secular evolution of the halo profile. Similarly, any back-reaction from the halo onto the disc cannot be accounted for. Such a setup allows us to measure and characterise the statistical properties of the fluctuations induced by the disc directly from simulations. Because the DM halo is analytical, this also prevents any shot noise associated with the use of a finite number of DM particles. Once these fluctuations have been estimated, their effects on the DM halo may then be quantified using the secular collisionless diffusion equation (2.31).

In order to characterise these fluctuations, we consider an analytic NFW halo profile, and embed within it a gaseous and stellar disc, paying attention to preparing the

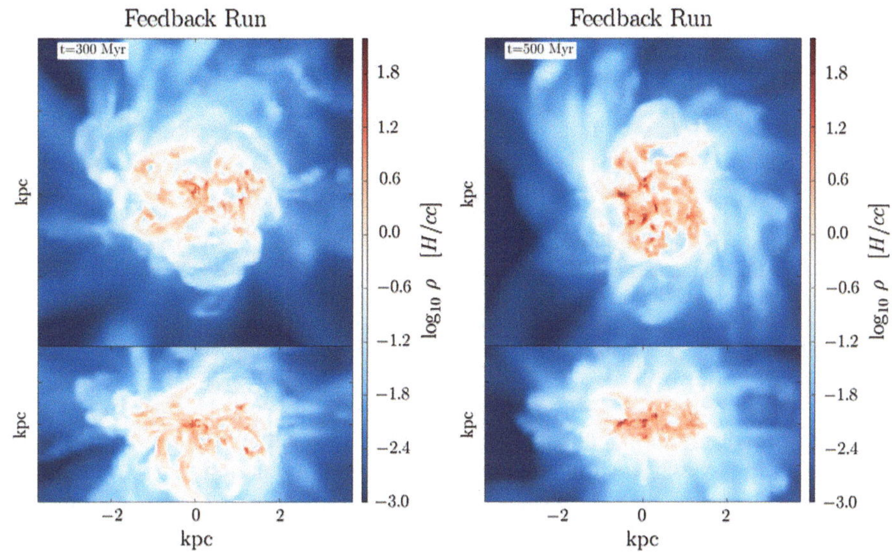

Fig. 4.21 Simulations run by Rebekka Bieri. Illustration of the gas density in a hydrodynamical simulation performed with the AMR code Ramses (Teyssier 2002). The stellar and gaseous discs are embedded in a static NFW DM halo and are seen from the top (*top panel*) and the edge (*bottom panel*). A supernova feedback recipe was implemented following Kimm et al. (2005). As can be seen in these two snapshots, this leads to fluctuations in the gas density, which resonantly couple to the DM halo and induce therein a resonant diffusion

system in a quasi-stationary state. In addition, we implement a supernova feedback recipe allowing for the release of energy from the supernovae into the interstellar medium. Figure 4.21 illustrates two successive snapshots of such a hydrodynamical simulation. In Fig. 4.21, one can note that because of supernova feedback, the gas density undergoes some fluctuations. These fluctuations in the potential due to the gas will be felt by the DM halo and may therefore be the driver of a resonant forced secular diffusion in the DM halo. Because we are interested in the ensemble average autocorrelation of the feedback fluctuations, various realisations are run for the same physical setup.

Once these simulations are performed, we characterise their statistics by computing the autocorrelation matrix $\widehat{\mathbf{C}}$ introduced in Eq. (2.25). This first requires to compute the basis coefficients $b_p(t)$, so that the perturbing potential $\delta\psi^{\mathrm{e}}$ may be written as $\delta\psi^{\mathrm{e}}(\boldsymbol{x}, t) = \sum_p b_p(t)\psi^{(p)}(\boldsymbol{x})$. Thanks to the biorthogonality property from Eq. (2.12), these basis coefficients are immediately given by

$$b_p(t) = -\int \mathrm{d}\boldsymbol{x}\, \psi^{(p)*}(\boldsymbol{x})\, \delta\rho^{\mathrm{e}}(\boldsymbol{x}, t)\,, \tag{4.82}$$

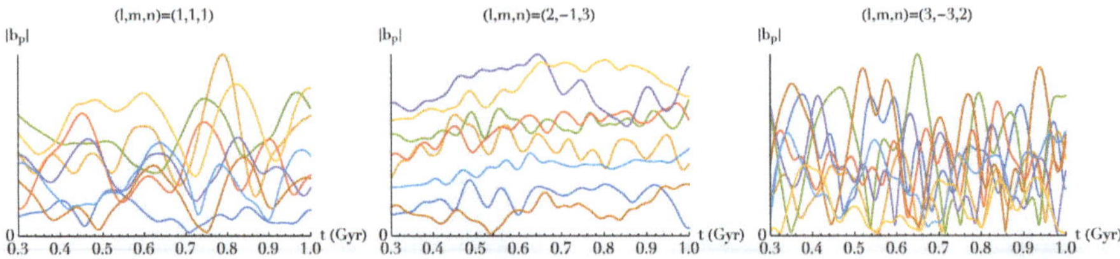

Fig. 4.22 Illustration of the behaviour of the basis coefficients $t \mapsto b_p(t)$, for three different basis elements, i.e. different values of the basis indices (ℓ, m, n). Each colour corresponds to a different realisation. One can note the presence of potential fluctuations associated with supernovae feedback. The autocorrelation of these fluctuations, captured by the matrix $\widehat{\mathbf{C}}$, will drive a feedback-induced secular diffusion in the DM halo

where $\delta\rho^{\mathrm{e}}$ stands for the star and gas density fluctuations within the disc, as illustrated in Fig. 4.21. In Fig. 4.22, we illustrate the behaviour of the function $t \mapsto b_p(t)$ for three different basis elements and ten different realisations. In Fig. 4.22, one can note, as expected, that the fluctuations history vary from one realisation to another. Similarly, the typical frequencies of the fluctuations also depend on the considered basis elements. Once the perturbations history $t \mapsto b_p(t)$ is extracted from the numerical simulations, one may follow Eq. (2.25) to compute their ensemble averaged autocorrelation matrix $\widehat{\mathbf{C}}_{pq}(\omega)$. This fully characterises the stochastic external source term, which sources the collisionless diffusion coefficients $D_{\boldsymbol{m}}(\boldsymbol{J})$ from Eq. (2.32). When the characteristics of the initial DM halo (namely its potential and associated self-consistent DF) have been specified, one may follow Appendix 4.D.1 to compute the halo's response matrix $\widehat{\mathbf{M}}$. This allows finally for the calculation of the diffusion coefficients $D_{\boldsymbol{m}}(\boldsymbol{J})$, and for the estimation of the collisionless diffusion flux $\mathcal{F}_{\mathrm{tot}}$ from Eq. (2.33). This diffusion flux characterises the initial orbital restructuration undergone by the DM halo's DF.

In this context, a final difficulty arises from the integration forward in time of the collisionless diffusion equation (2.34). Indeed, because the diffusion is self-consistent, the diffusion coefficients are slow functions of the halo's DF. Similarly, the halo's potential (initially cuspy) also secularly depends on the halo's DF. The integration of the diffusion equation therefore requires a self-consistent update of the halo's diffusion coefficients and the halo's potential. One possible approach to solve this self-consistency problem relies on an iterative approach (Prendergast and Tomer 1970; Weinberg 2001b) that we do not detail further here.

The method described previously is expected to allow for a detailed description of the resonant collisionless diffusion occurring in a DM halo as a result of stochastic external fluctuations induced by its inner galactic disc. The same approach could also allow us to investigate how much this diffusion mechanism depends on the strength of the feedback mechanisms, by for example changing the feedback recipes used in the hydrodynamical simulations. One could determine the typical power spectrum of the perturbations induced by the feedback, or give quantitative bounds on the feedback strengths required to induce a softening of the DM halo's profile. Similarly, the dependence of the diffusion efficiency w.r.t. the disc and halo masses could also be investigated. Finally, the efficiency of AGN feedback to induce a secular diffusion in the DM halo could also be studied within the same framework by characterising the associated potential fluctuations.

References

Benetti, F. P. C., & Marcos, B. (2017). *Phys. Rev. E, 95*, 022111.
Binney, J., & Tremaine, S. (2008). *Galactic dynamics* (2nd ed.). Princeton University Press.
Binney, J., Gerhard, O., & Silk, J. (2001). *MNRAS, 321*, 471.
Campa, A., Chavanis, P.-H., Giansanti, A., & Morelli, G. (2008). *Phys. Rev. E, 78*, 040102.
Chavanis, P.-H. (2006). *Int. J. Mod. Phys. B, 20*, 3113.
Cole, D. R., Dehnen, W., & Wilkinson, M. I. (2011). *MNRAS, 416*, 1118.
de Blok, W. J. G., & Bosma, A. (2002). *A&A, 385*, 816.
de Blok, W. J. G., & McGaugh, S. S. (1997). *MNRAS, 290*, 533.
de Blok, W. J. G., McGaugh, S. S., Bosma, A., & Rubin, V. C. (2001). *ApJ, 552*, L23.
De Rijcke, S., & Voulis, I. (2016). *MNRAS, 456*, 2024.
Del Popolo, A., & Pace, F. (2016). *Ap&SS, 361*, 162.
Dubinski, J., & Carlberg, R. G. (1991). *ApJ, 378*, 496.
Dubois, Y., Peirani, S., Pichon, C., et al. (2016). *MNRAS, 463*, 3948.
Earn, D. J. D., & Sellwood, J. A. (1995). *ApJ, 451*, 533.
Edmonds, A. (1996). *Angular momentum in quantum mechanics*. Princeton University Press.
El-Zant, A., Freundlich, J., & Combes, F. (2016). *MNRAS, 461*, 1745.
El-Zant, A., Shlosman, I., & Hoffman, Y. (2001). *ApJ, 560*, 636.
Evans, N. W., & Read, J. C. A. (1998a). *MNRAS, 300*, 83.
Evans, N. W., & Read, J. C. A. (1998b). *MNRAS, 300*, 106.
Fouvry, J.-B., Pichon, C., Magorrian, J., & Chavanis, P.-H. (2015). *A&A, 584*, A129.
Fridman, A. M., & Poliachenko, V. L. (1984). *Physics of gravitating systems. II*. Berlin Heidelberg: Springer.
Gnedin, O. Y., & Zhao, H. (2002). *MNRAS, 333*, 299.
Goerdt, T., Moore, B., Read, J. I., & Stadel, J. (2010). *ApJ, 725*, 1707.
Goldreich, P., & Lynden-Bell, D. (1965). *MNRAS, 130*, 125.
Governato, F., Brook, C., Mayer, L., et al. (2010). *Nature, 463*, 203.
Heggie, D., & Hut, P. (2003). *The gravitational million-body problem*. Cambridge University Press.
Hernquist, L., & Ostriker, J. P. (1992). *ApJ, 386*, 375.
Hörmander, L. (2003). *The analysis of linear partial differential operators I*. Berlin: Springer.
Jalali, M. A. (2007). *ApJ, 669*, 218.
Jalali, M. A. (2010). *MNRAS, 404*, 1519.
Jalali, M. A., & Hunter, C. (2005). *ApJ, 630*, 804.
Julian, W. H., & Toomre, A. (1966). *ApJ, 146*, 810.
Kalnajs, A. J. (1976). *ApJ, 205*, 745.
Kalnajs, A. J. (1977). *ApJ, 212*, 637.
Kimm, T., Cen, R., Devriendt, J., Dubois, Y., & Slyz, A. (2015). *MNRAS, 451*, 2900.
Kuzio de Naray, R., Martinez, G. D., Bullock, J. S., & Kaplinghat, M. (2010). *ApJ, 710*, L161
Kuzio de Naray, R., McGaugh, S. S., & de Blok, W. J. G. (2008). *ApJ, 676*, 920
Lynden-Bell, D., & Kalnajs, A. J. (1972). *MNRAS, 157*, 1.
Martizzi, D., Teyssier, R., Moore, B., & Wentz, T. (2012). *MNRAS, 422*, 3081.
Mashchenko, S., Couchman, H. M. P., & Wadsley, J. (2006). *Nature, 442*, 539.
Mashchenko, S., Wadsley, J., & Couchman, H. M. P. (2008). *Science, 319*, 174.
Moore, B. (1994). *Nature, 370*, 629.
Murali, C. (1999). *ApJ, 519*, 580.

Murali, C., & Tremaine, S. (1998). *MNRAS, 296*, 749.
Navarro, J. F., Frenk, C. S., & White, S. D. M. (1997). *ApJ, 490*, 493.
Padmanabhan, T. (1990). *Phys. Rep., 188*, 285.
Peirani, S., Kay, S., & Silk, J. (2008). *A&A, 479*, 123.
Pichon, C., & Aubert, D. (2006). *MNRAS, 368*, 1657.
Pichon, C., & Cannon, R. C. (1997). *MNRAS, 291*, 616.
Polyachenko, E. V. (2005). *MNRAS, 357*, 559.
Polyachenko, V. L., & Shukhman, I. G. (1981). *Soviet Ast., 25*, 533.
Pontzen, A., & Governato, F. (2012). *MNRAS, 421*, 3464.
Prendergast, K. H., & Tomer, E. (1970). *AJ, 75*, 674.
Rahmati, A., & Jalali, M. A. (2009). *MNRAS, 393*, 1459.
Read, J. I., & Gilmore, G. (2005). *MNRAS, 356*, 107.
Rocha, M., Peter, A. H. G., Bullock, J. S., et al. (2013). *MNRAS, 430*, 81.
Romano-Díaz, E., Shlosman, I., Hoffman, Y., & Heller, C. (2008). *ApJ, 685*, L105.
Saha, P. (1991). *MNRAS, 248*, 494.
Schönrich, R., & Binney, J. (2009). *MNRAS, 396*, 203.
Seguin, P., & Dupraz, C. (1994). *A&A, 290*, 709.
Sellwood, J. A. (1983). *J. Comput. Phys., 50*, 337.
Sellwood, J. A. (2012). *ApJ, 751*, 44.
Sellwood, J. A., & Evans, N. W. (2001). *ApJ, 546*, 176.
Spergel, D. N., & Steinhardt, P. J. (2000). *Phys. Rev. Lett., 84*, 3760.
Teyssier, R. (2002). *A&A, 385*, 337.
Teyssier, R., Pontzen, A., Dubois, Y., & Read, J. I. (2013). *MNRAS, 429*, 3068.
Tonini, C., Lapi, A., & Salucci, P. (2006). *ApJ, 649*, 591.
Toomre, A. (1981). *Structure and evolution of normal galaxies*, (pp. 111–136). Cambridge University Press.
Tremaine, S., & Weinberg, M. D. (1984). *MNRAS, 209*, 729.
Vauterin, P., & Dejonghe, H. (1996). *A&A, 313*, 465.
Weinberg, M. D. (1989). *MNRAS, 239*, 549.
Weinberg, M. D. (1991). *ApJ, 368*, 66.
Weinberg, M. D. (2001a). *MNRAS, 328*, 311.
Weinberg, M. D. (2001b). *MNRAS, 328*, 321.
Weinberg, M. D., & Katz, N. (2002). *ApJ, 580*, 627.
Zang, T. A. (1976). Ph.D. thesis, Massachusetts Institute of Technology.

Chapter 5
Thickened Discs

5.1 Introduction

The problem of explaining the origin of thick discs in our Galaxy has been around for some time (e.g., Gilmore and Reid 1983; Freeman 1987). The interest for this dynamical question has been particularly revived recently in the light of the current APOGEE survey (Eisenstein et al. 2011) and the upcoming data collected by the GAIA mission. Star formation within stellar discs typically occurs on the circular orbits of the gas, so that young stars should form a very thin disc (Wielen 1977). However, chemo-kinematic observations of old stars within our Milky Way (Jurić et al. 2008; Ivezić et al. 2008; Bovy et al. 2012), or in other galactic discs (Burstein 1979; Mould 2005; Yoachim and Dalcanton 2006; Comerón et al. 2011) have all shown that thick components are very common. Yet the formation and the origin of thickened stellar discs remains a significant puzzle for galactic formation theory.

Various dynamical mechanisms, either internal or external, have been proposed to explain this observed thickening, but their respective impacts and roles remain to be clarified. First, some violent major events could be at the origin of the vertically extended distribution of stars in disc galaxies. These could possibly be due to the accretion of galaxy satellites (Meza et al. 2005; Abadi et al. 2003), major mergers of gas-rich systems (Brook et al. 2004), or even gravitational instabilities in gas-rich turbulent clumpy discs (Noguchi 1998; Bournaud et al. 2009). While such violent mergers definitely have a strong impact on galactic structure, these extreme events may not be required to form a thickened stellar disc, which could also originate from a slow, secular and continuous heating of a pre-existing thin disc.

The work presented in this chapter is based on Fouvry et al. (2017).

J.-B. Fouvry, *Secular Evolution of Self-Gravitating Systems Over Cosmic Age*,
Springer Theses, DOI 10.1007/978-3-319-66760-7_5

Numerous smooth thickening mechanisms have been investigated in detail. Galactic discs could be thickened as a result of galactic infall of cosmic origin leading to multiple minor mergers (Toth and Ostriker 1992; Quinn et al. 1993; Villalobos and Helmi 2008; Matteo et al. 2011), and evidences for such events have been found in the phase space structure of the Milky Way (e.g., Purcell et al. 2011). Spiral density waves (Sellwood and Carlberg 1984; Minchev and Quillen 2006; Monari et al. 2016) are also possible candidates to increase the velocity dispersion within the disc, which can in turn be converted into vertical motion through deflections from giant molecular clouds (GMCs) (Spitzer and Schwarzschild 1953; Lacey 1984; Hänninen and Flynn 2002). In addition, radial migration (Lynden-Bell and Kalnajs 1972; Sellwood and Binney 2002), which describes the change of angular momentum of a star with no increase in its radial energy, could also play an important role in the secular evolution of stellar discs. Radial migration could be induced by spiral-bar coupling (Minchev and Famaey 2010), transient spiral structures (Barbanis and Woltjer 1967; Carlberg and Sellwood 1985; Solway et al. 2012, or perturbations induced by minor mergers (Quillen et al. 2009; Bird et al. 2012). An analytical model of radial migration was extensively used in Schönrich and Binney (2009b) to investigate in detail its impact on vertical heating, and recovered the main features of the Milky Way thin and thick discs. Recent $N-$body simulations also investigated the role played by radial migration (Haywood 2008; Loebman et al. 2011; Minchev et al. 2014), but the efficiency of this mechanism was recently shown to be limited (Minchev et al. 2012). Finally, owing to the increase of computing power, large numerical simulations are now in a position to investigate these processes in a self-consistent cosmological setup (Minchev et al. 2015; Grand et al. 2016). The developments of these global approaches are expected to offer new clues on the effective interplay between these various competing thickening mechanisms. As discussed in Chap. 2, recall that all these investigations can be broadly characterised as induced by an external (nurture) or internal (nature) source to trigger a vertical orbital restructuration in the disc.

In the present chapter, we intend to write down, in the context of tepid stellar discs of finite thickness, the two equations corresponding to collisionless or collisional diffusion derived in Chap. 2. As already discussed, on the one hand, the first formalism, presented in Sect. 2.2, assumes the system to be collisionless and considers the secular effects induced by external perturbations. The second collisional formalism, presented in Sect. 2.3, focuses on the role played by the system's intrinsic graininess. Here, both diffusion processes should be investigated, as it is not a priori known which will be the most efficient at thickening discs.

Following Chap. 3, implementing Eqs. (2.31) and (2.67) raises two main difficulties. These are respectively the explicit construction of the angle-action mapping $(\boldsymbol{x}, \boldsymbol{v}) \mapsto (\boldsymbol{\theta}, \boldsymbol{J})$ as well as the computation of the response matrix from Eq. (2.17), which requires the introduction of a basis of potentials and densities. Both problems are significantly more challenging in the thickened regime. For thickened discs, we will solve the first difficulty by introducing in Sect. 5.2 the thickened epicyclic approximation offering in the limit of sufficiently cold discs explicit angle-action coordinates. We will deal with the second difficulty in Sects. 5.3 and 5.4 by generalising the razor-thin WKB basis (see Chap. 3) to the thickened geometry, which will offer an analytical expression of the disc's amplification eigenvalues thanks to which we will account for the disc's gravitational susceptibility. Once these two difficulties are addressed, we will show in Sect. 5.5 (resp. sect. 5.6) how one can estimate the diffusion fluxes associated with the collisionless diffusion equation (2.31) (resp. the collisional diffusion equation (2.67)). Finally, Sect. 5.7 will be dedicated to applications of both formalisms to investigate the dynamical mechanisms at play in the secular thickening of stellar discs. These applications will be compared in particular to the numerical experiments from Solway et al. (2012).

5.2 Angle-Action Coordinates and Epicyclic Approximation

A first step towards the secular dynamics of inhomogeneous systems is to construct a set of angle-action coordinates. Let us follow the same method as what was presented in Sect. 3.2 in the razor-thin case. Assuming that the disc is sufficiently cold, one can decouple the vertical motion and treat it as a harmonic libration. Let us introduce the cylindrical coordinates (R, ϕ, z) along with their associated momenta (p_R, p_ϕ, p_z). We also assume that the axisymmetric potential of the disc is symmetric w.r.t. the equatorial plane $z=0$. In the vicinity of circular orbits, the stationary Hamiltonian from Eq. (3.6) becomes here

$$H_0 = \frac{1}{2}\Big[p_R^2 + p_z^2\Big] + \psi_{\text{eff}}(R_g, 0) + \frac{\kappa^2}{2}(R - R_g)^2 + \frac{\nu^2}{2}z^2 , \tag{5.1}$$

where we introduced the vertical epicyclic frequency ν as

$$\nu^2(R_g) = \frac{\partial^2 \psi_{\text{eff}}}{\partial z^2}\bigg|_{(R_g,0)} . \tag{5.2}$$

Of course, for a thickened disc, the azimuthal and radial frequencies Ω_ϕ and κ from Eqs. (3.5) and (3.7) should be computed in the equatorial plane $z=0$. We also note here that with the epicyclic approximation, ν depends only on J_ϕ. In Eq. (5.1), the radial and vertical motions have been decoupled, and, up to initial phases, there exists then a vertical amplitude A_z such that $z = A_z \cos(\nu t)$. Similarly to Eq. (3.8), the associated vertical action J_z is immediately given by

$$J_z = \frac{1}{2\pi}\oint \mathrm{d}z\, p_z = \frac{1}{2}\nu A_z^2 . \tag{5.3}$$

In this context, $(J_r, J_z) = (0, 0)$ corresponds to circular orbits. Increasing J_r (resp. J_z) amounts therefore to increasing the amplitude of the radial (resp. vertical) oscillations of the stars, so that the orbits get hotter. This is illustrated in Fig. 5.1. It

Fig. 5.1 Illustration of an epicyclic orbit (*left panel*) as one respectively increases its radial action J_r (*middle panel*) or its vertical action J_z (*right panel*)

Fig. 5.2 Illustration of the epicyclic approximation in a thickened disc. *Left panel* Three different orbits characterised by different sets of actions $\boldsymbol{J} = (J_\phi, J_r, J_z)$. *Middle panel* The same orbits represented in the plane (R, v_r) of radius and radial velocity. In this space, the epicyclic trajectories are closed. The centre of the ellipse represents the guiding radius R_g of the orbit, i.e. the action J_ϕ. The size of the ellipse directly relates to the radial action J_r: the wider the ellipse, the larger J_r. The frequency of motion along this ellipse is given by the radial frequency $\kappa(J_\phi)$. *Right panel* The same orbits in the plane (z, v_z) of height and vertical velocity. In this space, epicyclic trajectories are closed as well. The potential being symmetric w.r.t. the equatorial plane $z = 0$, the closed ellipses are always centred on the origin. The size of the ellipse directly relates to the vertical action J_z: the wider the ellipse, the larger J_z. The frequency of motion along this ellipse is given by the vertical frequency $\nu(J_\phi)$

is straightforward to complete Eq. (3.9) to obtain an explicit relation between the physical phase space coordinates and the angle-action ones. Indeed, one has

$$R = R_g + A_R \cos(\theta_R) \;\; ; \;\; \phi = \theta_\phi - \frac{2\Omega_\phi}{\kappa} \frac{A_R}{R_g} \sin(\theta_R) \;\; ; \;\; z = A_z \cos(\theta_z) \, . \tag{5.4}$$

In Fig. 5.2, we illustrate epicyclic orbits described by this angle-action mapping.

Finally, in the thickened context, the razor-thin quasi-isothermal DF from Eq. (3.10) becomes

$$F(R_g, J_r, J_z) = \frac{\Omega_\phi \Sigma}{\pi \kappa \sigma_r^2} \exp\!\left[- \frac{\kappa J_r}{\sigma_r^2} \right] \frac{\nu}{2\pi \sigma_z^2} \exp\!\left[- \frac{\nu J_z}{\sigma_z^2} \right] , \tag{5.5}$$

where Σ stands for the projected surface density of the disc, so that $\Sigma(R) = \int \mathrm{d}z \, \rho(R, z)$, where ρ is the disc's density. In Eq. (5.5), σ_z represents the vertical velocity dispersion of the stars at a given radius, and only depends on the position in the disc.

5.3 The Thickened WKB Basis

In Sect. 3.3, we presented in detail how one could construct a biorthogonal basis of tightly wound spirals in the context of razor-thin axisymmetric discs. Let us now generalise this construction to thickened stellar discs by specifying the vertical components of these basis elements. As the in-plane dependence of our WKB basis elements will be the same than the one presented in Sect. 3.3, we will here focus on the additional degree of freedom associated with the vertical dimension. The cylindrical coordinates are noted as (R, ϕ, z), and we introduce the $3D$ WKB basis elements as

$$\psi^{[k_\phi, k_r, R_0, n]}(R, \phi, z) = \mathcal{A}\, \psi_r^{[k_\phi, k_r, R_0]}(R, \phi)\, \psi_z^{[k_r, n]}(z)\,, \tag{5.6}$$

where $\mathcal{A}$ is an amplitude, which will be tuned later on to ensure the correct normalisation of the basis elements. In Eq. (5.6), we introduced as $\psi_r^{[k_\phi, k_r, R_0]}(R, \phi)$ the same in-plane dependence as for the razor-thin WKB basis elements from Eq. (3.11), so that

$$\psi_r^{[k_\phi, k_r, R_0]}(R, \phi) = \mathrm{e}^{\mathrm{i}(k_\phi \phi + k_r R)}\, \mathcal{B}_{R_0}(R)\,, \tag{5.7}$$

where the radial window function $\mathcal{B}_{R_0}$ was introduced in Eq. (3.12). In Eq. (5.6), one can note that the basis elements depend on four indices. Here, (k_ϕ, k_r, R_0) are the same indices as in the razor-thin case, so that k_ϕ characterises the angular dependence of the basis elements, k_r is the radial frequency of the elements, and R_0 the position in the disc around which the window $\mathcal{B}_{R_0}$ is centred. Finally, we introduced the index $n \geq 1$, specific to the thick case, which numbers the considered vertical dependences. We also recall that the window function from Eq. (3.12) involves a decoupling scale σ, which ensures the biorthogonality of the basis elements. The radial dependence of the basis elements is illustrated in Fig. 3.1, while their dependence in the equatorial plane $z = 0$ is given by Fig. 3.2. The thickened basis elements from Eq. (5.6) are therefore constructed by multiplying the in-plane razor-thin WKB basis elements by a vertical component $\psi_z^{[k_r, n]}(z)$, which we now specify.

The construction of the basis elements requires us to satisfy Poisson's equation (2.12), which characterises the associated density elements. Relying on the same tight winding assumptions as in Eq. (3.15), Poisson's equation becomes

$$-k_r^2 \mathcal{A}\, \psi_r \psi_z + \mathcal{A}\, \psi_r \frac{\mathrm{d}^2 \psi_z}{\mathrm{d}z^2} = 4\pi G \rho\,, \tag{5.8}$$

where we dropped the superscripts $[k_\phi, k_r, R_0, n]$ to shorten the notations and introduced the associated density ρ. At this stage, we now assume that the density elements satisfy an ansatz of separability of the form

$$\rho(R, \phi, z) = \frac{\lambda_\rho}{4\pi G}\, \mathcal{A}\, \psi_r(R, \phi)\, \psi_z(z)\, w(z)\,, \tag{5.9}$$

where $\lambda_p = \lambda_p^{[k_r,n]}$ is a proportionality constant, while $w(z)$ is a cavity function, which is chosen to be independent of the basis' indices. Equation (5.8) immediately becomes

$$\frac{d^2\psi_z}{dz^2} - k_r^2 \psi_z = \lambda_p \, w(z) \, \psi_z \, . \tag{5.10}$$

Equation (5.10) is a self-consistent relation that the vertical component ψ_z has to satisfy. It takes the form of a Sturm-Liouville equation (Courant and Hilbert 1953), which requires us to determine the eigenfunctions $\psi_z^{[k_r,n]}$ as well as the associated eigenvalues $\lambda_p^{[k_r,n]}$. Assuming a sufficient regularity for the functions involved, the Sturm-Liouville theory ensures that there exists a discrete spectrum of real eigenvalues $\lambda_1 < \lambda_2 < \ldots < \lambda_n \to +\infty$, with their associated eigenfunctions ψ_z^1, ..., ψ_z^n. In addition, when correctly normalised, these eigenfunctions form a biorthogonal basis so that $\int dz \, w(z) \, \psi_z^p(z) \, \psi_z^q(z) = \delta_p^q$.

By explicitly specifying the considered cavity function $w(z)$, one can get explicit expressions for these eigenfunctions. We assume that the density elements vanish out of a sharp cavity, so that they are zero for $|z| > h$. This corresponds to the choice

$$w(z) = \Theta(z/h) \, , \tag{5.11}$$

where $\Theta(x)$ is a door function, equal to 1 for $x \in [-1; 1]$ and 0 elsewhere. See Griv and Gedalin (2012) for a similar vertical ansatz. Let us now specify what has to be chosen for the height parameter h. The WKB basis being local, one may adapt $h = h(R_0)$ a function of the location within the disc, to allow for a better representation of the disc's mean vertical density profile. Equation (5.11) approximates the mean physical profile of the disc by a sharp cavity. As illustrated in Fig. 5.3, we choose h to match the volume of the physical and sharp cavities, i.e. we impose $\int dz \, \rho_{\mathrm{tot}}(R_0, z) = 2h \rho_{\mathrm{tot}}(R_0, 0)$.

If one assumes that the vertical density profile of the disc takes the form of a Spitzer profile, see Eq. (5.71), h is then given by $h(R_0) = 2z_0(R_0)$. One should

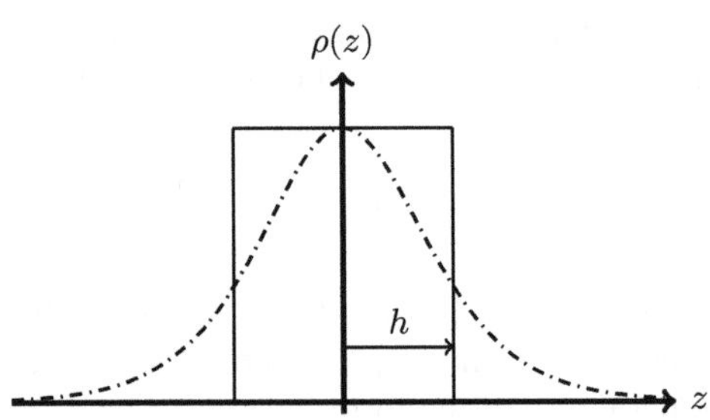

Fig. 5.3 Illustration of the sharp cavity (*solid* lines), introduced in Eq. (5.11), consistent with the mean underlying vertical density (*dotted-dashed* lines). The cavity is constructed to match the total volume of the vertical density profile. In this figure, the mean density profile corresponds to a Spitzer profile as defined in Eq. (5.71)

therefore consider the cavity scale h from Eq. (5.11) not as a free parameter of the formalism, but as imposed by the mean density profile of the considered disc. Once the cavity function from Eq. (5.11) has been specified, one may explicitly solve Poisson's equation (5.10) to determine the density basis elements. It takes the form of a wave equation, and we therefore assume that ψ_z follows the ansatz

$$\psi_z(z) = \begin{cases} A\,\mathrm{e}^{-k_r z} & \text{if } \; z > h\,, \\ B\,\mathrm{e}^{\mathrm{i}k_z z} + C\,\mathrm{e}^{-\mathrm{i}k_z z} & \text{if } \; |z| \leq h\,, \\ D\,\mathrm{e}^{k_r z} & \text{if } \; z < -h\,, \end{cases} \tag{5.12}$$

where the frequency k_z remains to be determined. The eigenvalue λ_ρ immediately reads $\lambda_\rho = -(k_r^2 + k_z^2)$. In addition to the ansatz from Eq. (5.12), one also has to impose for ψ_z and $\mathrm{d}\psi_z/\mathrm{d}z$ to be continuous at $z = \pm h$. Let us now restrict ourselves to symmetric perturbations, $\psi_z(-z) = \psi_z(z)$, while the very similar antisymmetric case will be presented in Appendix 5.A. Symmetric perturbations immediately lead to $A = D$ and $B = C$, while the continuity requirements impose

$$\begin{cases} A\,\mathrm{e}^{-k_r h} = 2B\,\cos(k_z h)\,, \\ k_r\,A\,\mathrm{e}^{-k_r h} = 2k_z B\,\sin(k_z h)\,. \end{cases} \tag{5.13}$$

To be non-trivial, the vertical frequency k_z must therefore satisfy the quantisation relation

$$\tan(k_z h) = \frac{k_r}{k_z}\,. \tag{5.14}$$

Once k_r and h have been specified, Eq. (5.14) restricts the allowed values of k_z. Following the definition of the basis elements from Eq. (5.6), let us introduce the index $n \geq 1$, such that k_z^n is the nth solution of Eq. (5.14), i.e. such that

$$k_z^1 < k_z^2 < ... < k_z^n < ... \quad \text{and} \quad \tan(k_z^n h) = \frac{k_r}{k_z^n}\,. \tag{5.15}$$

At this stage, we note that for a sufficiently thin disc such that $k_z^1 h, k_r h \lesssim 1$, the first quantised symmetric frequency k_z^1 may be approximated as

$$k_z^1 \simeq \sqrt{\frac{k_r}{h}}\,. \tag{5.16}$$

In Fig. 5.4, we illustrate the quantisation relation from Eq. (5.14), as well as its antisymmetric analog from Eq. (5.100). Two important properties should already be noted. First, the fundamental symmetric frequency k_z^1 is the only quantised frequency such that $k_z h < \pi/2$. Following the approximated expression from Eq. (5.16), in the limit of a razor-thin disc, for which $h \to 0$, this is the only frequency for which

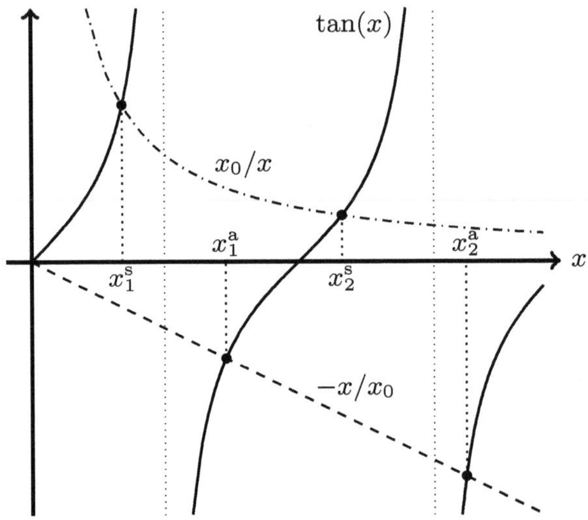

Fig. 5.4 Illustration of the quantisation relations for the vertical frequency imposed by the sharp cavity from Eq. (5.11). Dimensionless quantities are represented using the notations $x = k_z h$ and $x_0 = k_r h$. The top *dotted-dashed* curve is associated with the symmetric quantisation relation from Eq. (5.14), which imposes the quantised dimensionless frequencies x_1^{s}, x_2^{s}, ... The bottom *dashed* curve is associated with the antisymmetric quantisation relation from Eq. (5.100) imposing the frequencies x_1^{a}, x_2^{a}, ... One can already note the specific role played by the fundamental symmetric frequency x_1^{s}, which is the only dimensionless frequency inferior to $\pi/2$

$k_z^1 h \to 0$, while all the other quantised frequencies are such that $k_z h > \pi/2$. This already emphasises why the fundamental symmetric frequency k_z^1 will play a crucial role in the razor-thin limit. This will especially become clear in Appendix 5.C, where we recover the razor-thin limits of the two diffusion equations. Moreover, the periodicity of the "tan" function ensures that in the limit of sufficiently thick disc for which $k_r h \gtrsim \pi$, one can assume that both symmetric and antisymmetric frequencies satisfy

$$\Delta k_z = k_z^{n+1} - k_z^n \simeq \frac{\pi}{h} . \tag{5.17}$$

Straightforward calculations finally lead to the complete expression of the symmetric potential elements which read

$$\psi^{[k_\phi, k_r, R_0, n]}(R, \phi, z) = \mathcal{A}\, \psi_r^{[k_\phi, k_r, R_0]}(R, \phi) \begin{cases} \cos(k_z^n z) & \text{if } |z| \leq h , \\ \mathrm{e}^{k_r h} \cos(k_z^n h)\, \mathrm{e}^{-k_r |z|} & \text{if } |z| \geq h , \end{cases} \tag{5.18}$$

while the associated density elements read

$$\rho^{[k_\phi, k_r, R_0, n]}(R, \phi, z) = -\frac{k_r^2 + (k_z^n)^2}{4\pi G}\, \psi^{[k_\phi, k_r, R_0, n]}(R, \phi, z)\, \Theta\!\left[\frac{z}{h}\right] . \tag{5.19}$$

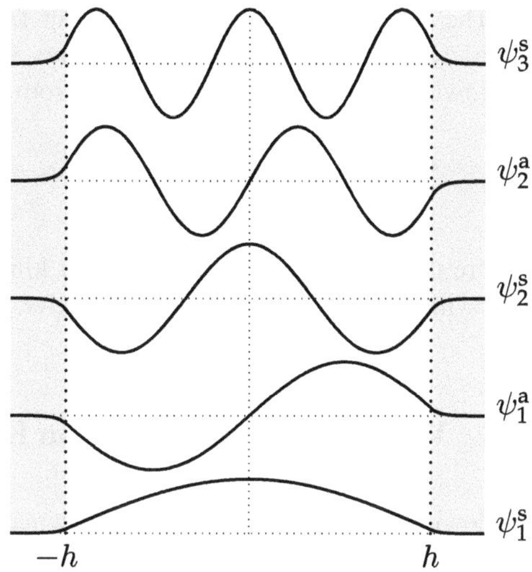

Fig. 5.5 Illustration of the vertical dependence of the first WKB potential basis elements. Here, ψ^{s} stands for the symmetric elements from Eq. (5.18), while ψ^{a} is associated with the antisymmetric ones from Eq. (5.101). The basis elements can also be ordered thanks to their number of nodes within the cavity, as expected from the Sturm-Liouville theory

The equivalent expressions for the antisymmetric basis elements are given in Eqs. (5.101) and (5.102). The vertical components of these basis elements are illustrated in Fig. 5.5.

As imposed by Eq. (2.12), the final step of the construction of the thickened WKB basis elements is to ensure that the basis is biorthogonal. We already showed in Eq. (3.20), when constructing the razor-thin WKB basis elements, that for $(k_\phi^p, k_r^p, R_0^p) \neq (k_\phi^q, k_r^q, R_0^q)$ the orthogonality was satisfied, provided that the decoupling assumptions from Eq. (3.25) were satisfied. Regarding the vertical component, we also underlined in Eq. (5.10), that the Sturm-Liouville theory enforces the orthogonality w.r.t. the n_p and n_q indices, even when considering both symmetric and antisymmetric basis elements. As a consequence, the basis elements from Eqs. (5.18) and (5.19) form a biorthogonal basis. Our final step is to specify the amplitude $\mathcal{A}$ to ensure a correct normalisation. Following the razor-thin calculations from Eq. (3.27), one straightforwardly gets

$$\mathcal{A} = \sqrt{\frac{G}{R_0 h (k_r^2 + (k_z^n)^2)}}\, \alpha_n \,, \tag{5.20}$$

where we introduced the prefactor $1 \leq \alpha_n \leq 1.6$ as

$$\alpha_n = \sqrt{\frac{2}{1 + \sin(2k_z^n h)/(2k_z^n h)}} \,. \tag{5.21}$$

The epicyclic angle-action mapping from Eq. (5.4) allows us to compute the Fourier transform of the basis elements w.r.t. the angles, as defined in Eq. (2.6). Following the razor-thin calculations from Eq. (3.33), one gets

$$\psi_{\boldsymbol{m}}^{[k_\phi,k_r,R_0,n]}(\boldsymbol{J}) = \delta_{m_\phi}^{k_\phi}\,\delta_{m_z}^{\mathrm{even}}\,\mathcal{A}\,\mathrm{e}^{\mathrm{i}k_r R_\mathrm{g}}\,\mathrm{i}^{m_z-m_r}\,\mathcal{B}_{R_0}(R_\mathrm{g})\,\mathcal{J}_{m_r}\!\left[\sqrt{\tfrac{2J_r}{\kappa}}\,k_r\right]\mathcal{J}_{m_z}\!\left[\sqrt{\tfrac{2J_z}{\nu}}\,k_z^n\right], \tag{5.22}$$

where the Bessel functions of the first kind $\mathcal{J}_\ell$ were introduced in Eq. (3.32). The antisymmetric analog of Eq. (5.22) is given in Eq. (5.105).

5.4 WKB Thick Amplification Eigenvalues

Following the construction of our thickened WKB basis elements, let us evaluate the system's response matrix, as given by Eq. (2.17).

5.4.1 WKB Response Matrix

A key result of Sect. 3.4 was to show that in the razor-thin limit, the response matrix computed with WKB basis elements was diagonal. This was an essential result of the derivation which allowed us to obtain in Sects. 3.5 and 3.6 explicit analytical expressions for the collisionless and collisional diffusion fluxes. As illustrated in the previous section, the thick WKB basis elements have the same in-plane dependence as the razor-thin ones. However, they may in principle interact one with another via their vertical components. In Appendix 5.B, we show that even for a thick disc, with our thick WKB basis elements, one may still assume that the response matrix is diagonal so that

$$\widehat{\mathbf{M}}_{[k_\phi^p,k_r^p,R_0^p,n_p],[k_\phi^q,k_r^q,R_0^q,n_q]}(\omega) = \delta_{k_\phi^p}^{k_\phi^q}\,\delta_{k_r^p}^{k_r^q}\,\delta_{R_0^p}^{R_0^q}\,\delta_{n_p}^{n_q}\,\lambda_{[k_\phi^p,k_r^p,R_0^p,n_p]}(\omega)\,. \tag{5.23}$$

Such a property is a crucial step of the present derivation, which allows us to account analytically for the system's self-gravity.

Following the calculations presented in Sect. 3.4, one can straightforwardly compute the diagonal elements of the response matrix, using the Fourier transformed WKB basis elements from Eq. (5.22). The additional integral w.r.t. J_z can also be computed using the integration formula from Eq. (3.42). As in the razor-thin case, we assume the disc to be sufficiently cold, so that one may neglect the contributions from $\partial F/\partial J_\phi$ w.r.t. the ones associated with $\partial F/\partial J_r$ and $\partial F/\partial J_z$. After some simple algebra, the symmetric amplification eigenvalues read

$$\lambda^{\text{sym}}_{[k_\phi,k_r,R_0,n]}(\omega) = \frac{2\pi G\Sigma\alpha_n^2}{h\kappa^2(1+(k_z/k_r)^2)} \sum_{\ell_z \text{even}} \frac{\mathrm{e}^{-\chi_z}\mathcal{I}_{\ell_z}[\chi_z]}{(1-s_{\ell_z}^2)} \bigg\{ \mathcal{F}(s_{\ell_z},\chi_r) - \ell_z \frac{\nu}{\sigma_z^2}\frac{\sigma_r^2}{\kappa}\mathcal{G}(s_{\ell_z},\chi_r) \bigg\} . \tag{5.24}$$

Similarly to Eqs. (3.43) and (3.45), we introduced the dimensionless quantities χ_r and χ_z as

$$\chi_r = \frac{\sigma_r^2 k_r^2}{\kappa^2} \quad ; \quad \chi_z = \frac{\sigma_z^2 k_z^2}{\nu^2} , \tag{5.25}$$

and the shifted dimensionless frequency s_{ℓ_z} as

$$s_{\ell_z} = \frac{\omega - k_\phi\Omega_\phi - \ell_z\nu}{\kappa} . \tag{5.26}$$

Finally, in Eq. (5.24), as already made in Eq. (3.46), we introduced the reduction functions $\mathcal{F}$ and $\mathcal{G}$ as

$$\mathcal{F}(s,\chi) = 2(1-s^2)\frac{\mathrm{e}^{-\chi}}{\chi}\sum_{\ell=1}^{+\infty}\frac{\mathcal{I}_\ell[\chi]}{1-(s/\ell)^2} \quad ; \quad \mathcal{G}(s,\chi) = 2(1-s^2)\frac{\mathrm{e}^{-\chi}}{\chi}\bigg[\frac{1}{2}\frac{\mathcal{I}_0[\chi]}{s} + \frac{1}{s}\sum_{\ell=1}^{+\infty}\frac{\mathcal{I}_\ell[\chi]}{1-(\ell/s)^2}\bigg] . \tag{5.27}$$

These two reduction functions are illustrated in Fig. 5.23. Relying on the antisymmetric basis elements from Appendix 5.A, one also obtains the associated antisymmetric amplification eigenvalues as

$$\lambda^{\text{anti}}_{[k_\phi,k_r,R_0,n]}(\omega) = \frac{2\pi G\Sigma\beta_n^2}{h\kappa^2(1+(k_z/k_r)^2)} \sum_{\ell_z \text{odd}} \frac{\mathrm{e}^{-\chi_z}\mathcal{I}_{\ell_z}[\chi_z]}{(1-s_{\ell_z}^2)} \bigg\{ \mathcal{F}(s_{\ell_z},\chi_r) - \ell_z \frac{\nu}{\sigma_z^2}\frac{\sigma_r^2}{\kappa}\mathcal{G}(s_{\ell_z},\chi_r) \bigg\} , \tag{5.28}$$

where the prefactor β_n was introduced in Eq. (5.104).

Equations (5.24) and (5.28) are important results as they allow us to easily estimate the strength of the self-gravitating amplification in a thick disc. We will show in Sect. 5.4.2 how these amplification eigenvalues are in full agreement with the razor-thin ones obtained in Sect. 3.4. We will emphasise in particular how these amplification eigenvalues allow us to generalise the razor-thin Toomre's parameter Q from Eq. (3.49) to the thickened geometry.

When effectively computing the thick amplification eigenvalues from Eqs. (5.24) and (5.28), one has to enforce two additional restrictions. These amount to neglecting the contributions from the vertical action gradients w.r.t. the radial ones, and restricting the sum on resonance vectors only to closed orbits on resonance. Let us now motivate these two assumptions.

The general expression of the response matrix from Eq. (2.17) involves the gradients $\partial F/\partial \boldsymbol{J}$ of the system's DF w.r.t. all the actions coordinates. As in the razor-thin case, we may assume the disc to be sufficiently cold, so that one may neglect the contributions from $\partial F/\partial J_\phi$ w.r.t. $\partial F/\partial J_r$ and $\partial F/\partial J_z$. In addition, let us now also neglect the contributions from the vertical gradients w.r.t. the radial ones, the radial ones being the only gradients which remain in the razor-thin limit. In Eqs. (5.24) and (5.28), this amounts to neglecting the reduction function $\mathcal{G}$ and conserving only the contributions from the reduction function $\mathcal{F}$.

One should also note that in the diffusion equations, the response matrix eigenvalues always have to be evaluated at resonance. As a consequence, for the collisionless diffusion coefficients $D_{\boldsymbol{m}}(\boldsymbol{J})$ from Eq. (2.32) and the collisional drift and diffusion coefficients $A_{\boldsymbol{m}}(\boldsymbol{J})$ and $D_{\boldsymbol{m}}(\boldsymbol{J})$ from Eqs. (2.69) and (2.70), the amplification eigenvalues have to be evaluated at the resonant frequency $\omega = \boldsymbol{m} \cdot \boldsymbol{\Omega}$. Following Eq. (5.26), the shifted dimensionless frequency $s_{\ell_z}^{\boldsymbol{m}}$ associated with a given resonance $\boldsymbol{m}$ reads

$$s_{\ell_z}^{\boldsymbol{m}} = m_r + (m_z - \ell_z)\frac{\nu}{\kappa} + \mathrm{i}\eta \,, \tag{5.29}$$

where a small imaginary part η was added. The potential is assumed to be dynamically non-degenerate (see Eq. (5.60)), so that ν/κ is not a rational number. Consequently, $s_{\ell_z}^{\boldsymbol{m}}$, when evaluated for a resonance $\boldsymbol{m}$, is an integer only for $m_z = \ell_z$. Here, $s_{\ell_z}^{\boldsymbol{m}}$ being an integer, means that there exists a rotating frame in which the star's orbit is closed, i.e. in which the considered star is exactly on resonance. In the razor-thin case, this was always possible, but this is no more guaranteed in the thickened case. As illustrated in Fig. 5.23, the reduction function $\mathcal{F}$ diverges in the neighbourhood of integers, but is well defined when evaluated for exactly integer values, as long as one adds a small imaginary part η as in Eq. (5.29). In order never to probe the diverging branches of this reduction function, one should therefore always evaluate this function for exactly integer values of s. As $s_{\ell_z}^{\boldsymbol{m}}$ is an integer only for $\ell_z = m_z$, we may restrict the sum on ℓ_z to this only term in the generic expressions (5.24) and (5.28) of the thickened amplification eigenvalues.

When accounting for the two previous approximations, the amplification eigenvalues from Eqs. (5.24) and (5.28), when computed for a resonance $\boldsymbol{m}$, take the form

$$\lambda_{\boldsymbol{m}}(J_\phi, k_r, k_z) = \frac{2\pi G \Sigma \gamma_{\boldsymbol{m}}^2}{h\kappa^2(1+(k_z/k_r)^2)} \frac{\mathrm{e}^{-\chi_z} \mathcal{I}_{m_z}[\chi_z]}{(1-m_r^2)} \mathcal{F}(m_r, \chi_r) \,, \tag{5.30}$$

where we introduced the numerical prefactor $\gamma_{\boldsymbol{m}}$ as

$$\gamma_{\boldsymbol{m}}(J_\phi, k_r, k_z) = \begin{cases} \alpha(J_\phi, k_r, k_z) & \text{if} \quad m_z \text{ even} \,, \\ \beta(J_\phi, k_r, k_z) & \text{if} \quad m_z \text{ odd} \,. \end{cases} \tag{5.31}$$

Equation (5.30), because of its generic writing, applies to both symmetric and anti-symmetric vertical resonances. Let us emphasise that these reduced expressions are fully compatible with the discussions from Appendix 5.B showing that the thickened response matrix is diagonal. These expressions are also consistent with the upcoming section, where we recover the razor-thin WKB amplification eigenvalues and generalise Toomre's Q parameter to thick discs. The applications of the WKB formalism presented in Sect. 5.7 all rely on the simplified expression of the amplification eigenvalues obtained in Eq. (5.30).

5.4.2 A Thickened Q Factor

Following our calculation of the amplification eigenvalues, let us now show how these are in full agreement with the razor-thin results obtained in Sect. 3.4, and also offer a generalisation of Toomre's Q parameter to thick discs.

In the razor-thin limit, only resonances with $m_z = 0$ are allowed, so that only the symmetric basis elements may play a role. Let us note that the symmetric quantisation relation from Eq. (5.14) is such that except for the fundamental symmetric frequency $k_{z,\mathrm{s}}^1$, one always has $k_{z,\mathrm{s}}^n > \pi/(2h)$. Because in the razor-thin limit one has $h \to 0$, only the fundamental symmetric mode contributes to the amplification eigenvalue in this limit. In the same infinitely thin limit, one can then get rid of the dependence of λ w.r.t. k_z and evaluate the amplification eigenvalues in $(k_r, k_z^1(k_r, h)) = (k_r, \sqrt{k_r/h})$, thanks to Eq. (5.16). Equation (5.30) then becomes

$$\lambda(\omega, k_\phi, k_r, h) = \frac{2\pi G \alpha_1^2 \Sigma k_r}{\kappa^2 (1+k_r h)} \frac{\mathrm{e}^{-\chi_z} \mathcal{I}_0[\chi_z]}{(1-s^2)} \mathcal{F}(s, \chi_r) , \tag{5.32}$$

where the dimensionless frequency s was introduced in Eq. (3.45), while the prefactor α_1 was defined in Eq. (5.21) and is a function of $k_{z,\mathrm{s}}^1 h = \sqrt{k_r h}$. One immediately has $\lim_{\mathrm{thin}} \alpha_1 = 1$. Similarly, χ_z, introduced in Eq. (5.25), should also be seen as a function of k_r and h, and reads $\chi_z = (\sigma_z^2 k_r)/(\nu^2 h)$. When considering the razor-thin limit, one should keep in mind that the kinematic height of the mean disc σ_z/ν and the size of the sharp WKB cavity h are directly related. Indeed, as detailed in Eq. (5.74), Jeans equation imposes a relation of the form

$$\frac{\sigma_z}{\nu} = c_2 \, h , \tag{5.33}$$

where c_2 is a dimensionless constant. For a Spitzer profile, as defined in Eq. (5.71), this constant reads $c_2 = 1/\sqrt{2}$. One then immediately has $\chi_z = c_2^2 k_r h$, so that $\lim_{\mathrm{thin}} \chi_z = 0$. As a consequence, in the razor-thin limit, Eq. (5.32) gives

$$\lim_{\mathrm{thin}} \lambda_{\mathrm{sym}} = \frac{2\pi G \Sigma |k_r|}{\kappa^2 (1-s^2)} \mathcal{F}(s, \chi_r) , \tag{5.34}$$

as already obtained in Eq. (3.47). This result underlines how the thick WKB basis elements constructed in Sect. 5.3 are fully consistent with the razor-thin WKB results from Sect. 3.3. Using the thickened disc model considered in the applications of Sect. 5.7, we illustrate in Fig. 5.6 how the thickened WKB amplification eigenvalues tend to the razor-thin ones, as one reduces the thickness of the disc.

Starting from the asymptotic expression (5.32) of the symmetric amplification eigenvalues in the limit of a thinner disc, let us now study how it impacts the value of the razor-thin Toomre's Q parameter from Eq. (3.49), as one accounts for the disc finite thickness, i.e. for a non-zero value of h. We are interested in the system's stability w.r.t. to axisymmetric tightly wound perturbations, so that we impose $k_\phi = 0$.

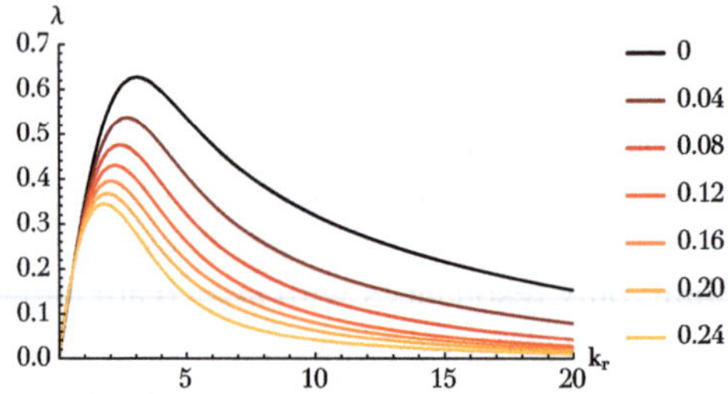

Fig. 5.6 Illustration of the effect of the disc thickness on the WKB amplification eigenvalues for the thickened Mestel disc presented in Sect. 5.7.1, at the location $J_\phi = 2$ and for the corotation resonance $\boldsymbol{m} = \boldsymbol{m}_{\mathrm{COR}} = (2, 0, 0)$. The different curves are associated with different values of the disc scale height z_0 from Eq. (5.71). For $z_0 \neq 0$, we computed $\lambda(k_r, k_z^{\mathrm{min}}(k_r))$, following Eq. (5.30). For $z_0 = 0$, i.e. for the razor-thin case, we computed $\lambda_{\mathrm{thin}}(k_r)$ from Eq. (3.47). As expected, the thickening of the disc tends to reduce its gravitational susceptibility

Placing ourselves at the stability limit $\omega = 0$ (so that $s = 0$), we seek a criterion on the disc's parameters so that there exists no $k_r > 0$ for which $\lambda(k_r, h) = 1$, i.e. such that the disc is stable. Let us then rewrite Eq. (5.32) as

$$\begin{aligned} \lambda(k_r, h) &= \frac{2\pi G \Sigma k_r}{\kappa^2} \, \mathcal{F}(0, \chi_r) \bigg\{ \frac{\alpha_1^2}{1 + k_r h} \mathrm{e}^{-\chi_z} \mathcal{I}_0[\chi_z] \bigg\} \\ &= \frac{2\pi G \Sigma k_r}{\kappa^2} \, \mathcal{F}(0, \chi_r) \bigg\{ 1 - \bigg[\frac{2}{3} + c_2^2 \bigg] k_r h \bigg\} \\ &= \frac{2\pi G \Sigma}{\kappa \sigma_r} \, K(\chi_r, \gamma) \, , \end{aligned} \tag{5.35}$$

where to obtain the second line of Eq. (5.35), we performed a limited development at first-order in $k_r h$ of α_1 and χ_z. In the third line, in order to shorten the notations, we introduced $\gamma = \big[\frac{2}{3} + c_2^2\big] h / (\sigma_r / \kappa)$, as well as the structure function $K(\chi_r, \gamma)$ as

$$K(\chi_r, \gamma) = \frac{1}{\sqrt{\chi_r}} \bigg[1 - \mathrm{e}^{-\chi_r} \mathcal{I}_0[\chi_r] \bigg] \bigg[1 - \gamma \sqrt{\chi_r} \bigg] . \tag{5.36}$$

This function is the direct thickened analog of the razor-thin function $K^0(\chi_r)$ introduced in Eq. (3.48) to derive the razor-thin Q parameter. Figure 5.7 illustrates the shape of the function $\chi_r \mapsto K(\chi_r, \gamma)$.

In order to obtain a simple expression for the thickened stability parameter, our next step is to study the behaviour of $K_{\mathrm{max}}(\gamma)$, the maximum of the function $\chi_r \mapsto K(\chi_r, \gamma)$, as one varies γ. As already obtained in Eq. (3.48), for $\gamma = 0$, one has $K_{\mathrm{max}}^0 \simeq 0.534$ reached for $\chi_{\mathrm{max}}^0 \simeq 0.948$. A first-order expansion in γ then allows us to write

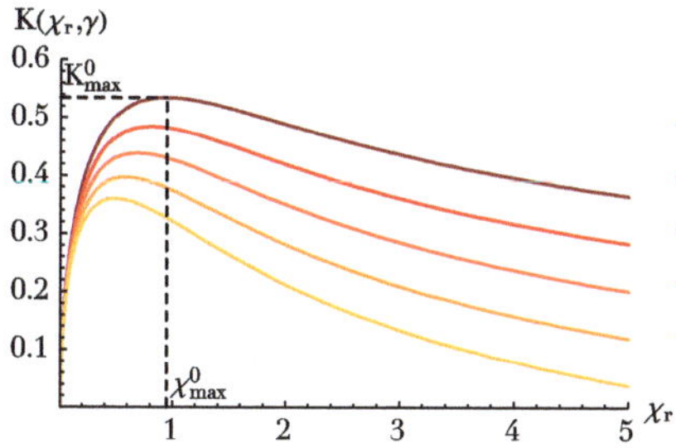

Fig. 5.7 Illustration of the function $\chi_r \mapsto K(\chi_r, \gamma)$, introduced in Eq. (5.36), for various values of γ. The razor-thin case corresponds to $\gamma=0$ and was already illustrated in Fig. 3.1. As expected, accounting for the finite thickness of the disc reduces the amplification eigenvalues

$$
\begin{aligned}
K_{\max}(\gamma) &\simeq K_{\max}^{0}\left[1-\gamma\sqrt{\chi_{\max}^{0}}\right] \\
&\simeq K_{\max}^{0}\,\mathrm{e}^{-\gamma\sqrt{\chi_{\max}^{0}}} = K_{\max}^{\mathrm{approx.}}(\gamma)\,,
\end{aligned}
\tag{5.37}
$$

where, in the second line, we replaced the linear approximation by an exponential, which offers a better fit. The shapes of the function $\gamma \mapsto K_{\max}(\gamma)$, $K_{\max}^{\mathrm{approx.}}(\gamma)$ are illustrated in Fig. 5.8.

Once one has estimated the maximum $K_{\max}^{\mathrm{approx.}}(\gamma)$, Eq. (5.35) immediately gives us the expression of the thickened Q_{thick} factor which reads

$$
Q_{\mathrm{thick}} = Q_{\mathrm{thin}}\,\mathrm{e}^{-\gamma\sqrt{\chi_{\max}^{0}}} = Q_{\mathrm{thin}}\,\exp\!\left[1.61\,\frac{\sigma_z/\nu}{\sigma_r/\kappa}\right],
\tag{5.38}
$$

where we used Eq. (5.33) to rewrite h as a function of σ_z/ν, given the value $c_2=1/\sqrt{2}$ for a Spitzer profile. We also wrote Q_{thin} for the razor-thin Toomre's parameter from Eq. (3.49). One can note that Eq. (5.38) was obtained through a rather general procedure allowing for the computation of the response matrix eigenvalues using the thickened WKB basis elements. Let us emphasise that this calculation is not specific to the Spitzer profile from Eq. (5.71). Should one consider a different mean vertical

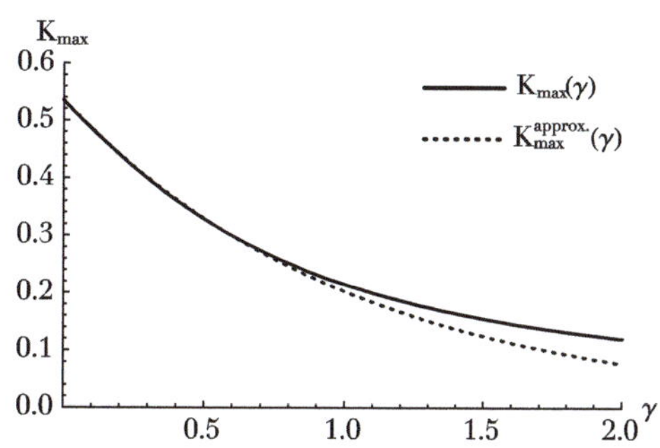

Fig. 5.8 Illustration of the behaviour of the function $\gamma \mapsto K_{\max}(\gamma)$ along with its approximation $K_{\max}^{\mathrm{approx.}}(\gamma)$, introduced in Eq. (5.37), as one varies the disc thickness characterised by γ

density profile, one would only have to change accordingly the value of the constant c_2 from Eq. (5.33), which relates the thickness of the mean density profile to the size of the sharp cavity from Eq. (5.11). A follow-up work of the present derivation will be to investigate via numerical simulations the relevance and quality of this new stability parameter to characterise instabilities in thickened stellar discs.

Let us finally discuss how this new Q_{thick} parameter compares to previous results. Vandervoort (1970) tackled in particular a similar issue of characterising tightly wound density waves in thickened stellar discs. See also Romeo (1992) for another generalisation of Q to the thickened geometry. The approach of Vandervoort (1970) relied on the collisionless Boltzmann equation limited to even perturbations. It also relied on the assumption of the existence of the adiabatic invariant J_z, thanks to which the vertical motion of the stars may be described. With our current notation, Eq. (77) of Vandervoort (1970) gives amplification eigenvalues of the form

$$\lambda_{\text{V}} = \frac{2\pi G\Sigma|k_r|}{\kappa^2(1-s^2)}\mathcal{F}(s,\chi_r)Q_{\text{V}}^{-1}(k_r h)\,, \tag{5.39}$$

where we used Eq. (5.11) to relate h and z_0. In Eq. (5.39), $Q_{\text{V}}(k_r h)$ is a non-trivial function, which can be computed via implicit variational principles. Similarly, in our present thickened WKB formalism, the analog of Eq. (5.39) is given by Eq. (5.32) and takes the form

$$\lambda_{\text{F}} = \frac{2\pi G\Sigma|k_r|}{\kappa^2(1-s^2)}\mathcal{F}(s,\chi_r)Q_{\text{F}}^{-1}(k_r h)\,, \tag{5.40}$$

where the function $Q_{\text{F}}(k_r h)$ is an explicit function reading

$$Q_{\text{F}}(k_r h) = \frac{1+k_r h}{\alpha_1^2 \text{e}^{-\chi_z}\mathcal{I}_0[\chi_z]}\,. \tag{5.41}$$

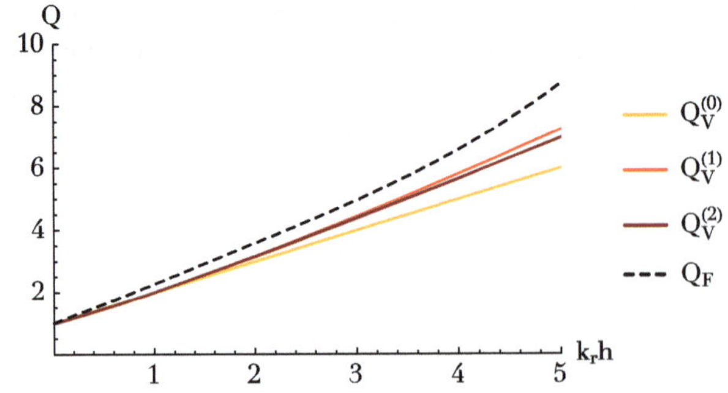

Fig. 5.9 Comparisons of the correction functions Q_{V} from Eq. (5.39) obtained in Vandervoort (1970) with the explicit function Q_{F} from Eq. (5.40) obtained thanks to the thickened WKB approximation. The behaviour of Q_{V} was obtained from Table 1 in Vandervoort (1970), which provides various approximations of increasing order: $Q_{\text{V}}^{(0)}$, $Q_{\text{V}}^{(1)}$, and $Q_{\text{V}}^{(2)}$. Despite being obtained from significantly different methods, these two approaches lead to similar results

Let us emphasise here that the expression of the correction function Q_{F} is explicit and was obtained by directly estimating the eigenvalues of the response matrix from Eq. (5.23) using the thickened WKB basis elements from Eq. (5.6). Thanks to Table 1 in Vandervoort (1970), which offers approximate values for the function $x \mapsto Q_{\mathrm{V}}(x)$, we may compare the functions Q_{V} and Q_{F}, as illustrated in Fig. 5.9. We note that the behaviours of the two functions are very similar on the considered range $0 \leq k_r h \leq 5$, even if they were obtained through different approaches.

5.5 WKB Limit for the Collisionless Diffusion

Having characterised the WKB self-gravitating amplification in thickened discs, let us now proceed to the evaluation of the diffusion coefficients involved in the collisionless diffusion equation (2.31). We follow an approach similar to Sect. 3.5. Let us first write the thick WKB basis elements from Eq. (5.6) as

$$\psi^{(p)} = \psi^{[k_\phi^p, k_r^p, R_0^p, n_p]} . \tag{5.42}$$

Relying on Eq. (5.23) to write the response matrix as $\widehat{\mathbf{M}}_{pq} = \lambda_p \delta_p^q$, the collisionless diffusion coefficients from Eq. (2.32) become

$$D_{\boldsymbol{m}}(\boldsymbol{J}) = \frac{1}{2} \sum_{p,q} \psi_{\boldsymbol{m}}^{(p)}(\boldsymbol{J})\, \psi_{\boldsymbol{m}}^{(q)*}(\boldsymbol{J})\, \frac{1}{1-\lambda_p} \frac{1}{1-\lambda_q} \widehat{\mathbf{C}}_{pq}(\boldsymbol{m}\cdot\boldsymbol{\Omega}) , \tag{5.43}$$

where $\widehat{\mathbf{C}}_{pq}$ was introduced in Eq. (2.26) and stands for the autocorrelation of the external perturbations. At this stage, let us note that the Fourier transformed basis elements from Eq. (5.22) (resp. (5.105)) involve a $\delta_{m_z}^{\mathrm{even}}$ (resp. $\delta_{m_z}^{\mathrm{odd}}$) for the symmetric (resp. antisymmetric) elements. As a consequence, in Eq. (5.43), as $\psi_{\boldsymbol{m}}^{(p)}$ and $\psi_{\boldsymbol{m}}^{(q)}$ are evaluated for the same resonance vector $\boldsymbol{m}$, the diffusion coefficients do not couple symmetric and antisymmetric basis elements. Therefore, in order to estimate $D_{\boldsymbol{m}}(\boldsymbol{J})$, depending on whether m_z is even (resp. odd), one only has to consider the symmetric (resp. antisymmetric) basis elements. As was done in Sect. 5.3, we restrict ourselves to the symmetric case, while the antisymmetric case will be straightforward to obtain by direct analogy.

Following Eq. (3.55), let us first express the basis coefficients $\widehat{b}_p$ as a function of the external perturbation $\delta\psi^{\mathrm{e}}$. After some calculation, one gets

$$\widehat{b}_p(\omega) = \frac{(k_r^p)^2 + (k_z^p)^2}{4\pi G} \frac{\mathcal{A}_p R_0^p}{(\pi\sigma)^{1/4}} (2\pi)^2\, \delta\widehat{\psi^{\mathrm{e}}}_{m_\phi, k_r^p, k_z^p}[R_0^p, \omega] , \tag{5.44}$$

where we used the shortened notation $k_z^p = k_z^{n_p}$. Following Eq. (3.55), here in Eq. (5.44), $\delta\widehat{\psi^{\mathrm{e}}}$ has undergone three transformations. (i) an azimuthal Fourier transform of indice m_ϕ, (ii) a local radial Fourier transform centred around R_0^p at the frequency k_r^p, and (iii) an even restricted vertical Fourier transform on the scale h at the frequency k_z^p. These three transformations are defined as

$$
\begin{aligned}
&\text{(i):}\;\; f_{m_\phi} = \frac{1}{2\pi}\!\int\! \mathrm{d}\phi\, f[\phi]\, \mathrm{e}^{-\mathrm{i}m_\phi \phi}\,, \\
&\text{(ii):}\;\; f_{k_r}[R_0] = \frac{1}{2\pi}\!\int\! \mathrm{d}R\, \mathrm{e}^{-\mathrm{i}k_r(R-R_0)}\, \exp\!\bigg[-\frac{(R-R_0)^2}{2\sigma^2}\bigg]\, f[R]\,, \\
&\text{(iii):}\;\; f_{k_z} = \!\int_{-h}^{+h}\! \mathrm{d}z\, \cos(k_z z)\, f[z]\,.
\end{aligned}
\tag{5.45}
$$

Following again Eq. (3.57), one may now disentangle the sums on p and q in Eq. (5.43), so that the collisionless diffusion coefficients become

$$
D_{\boldsymbol{m}}^{\mathrm{sym}}(\boldsymbol{J}) = \delta_{m_z}^{\mathrm{even}} \bigg\langle \frac{1}{2\pi}\!\int\! \mathrm{d}\omega'\, g(\boldsymbol{m}\cdot\boldsymbol{\Omega})\, g^*(\omega') \bigg\rangle\,,
\tag{5.46}
$$

where we introduced the function $g(\omega)$ as

$$
g(\omega) = \frac{2\pi}{2h} \sum_{k_r^p, R_0^p, n_p} g_{\mathrm{s}}(k_r^p, R_0^p, k_z^p, \omega)\, \mathrm{e}^{\mathrm{i}k_r^p (R_{\mathrm{g}}-R_0^p)}\, \mathcal{G}_r(R_{\mathrm{g}}-R_0^p)\,.
\tag{5.47}
$$

In Eq. (5.47), we executed the sum on k_ϕ^p thanks to the azimuthal Kronecker delta from Eq. (5.22). Here, as in Eq. (3.58), we also introduced $\mathcal{G}_r(R) = 1/\sqrt{2\pi\sigma^2}\mathrm{e}^{-R^2/(2\sigma^2)}$ a normalised radial Gaussian, and g_{s} encompasses all the slow dependences of the diffusion coefficients w.r.t. the radial position so that

$$
g_{\mathrm{s}}(k_r^p, R_0^p, k_z^p, \omega) = \mathcal{J}_{m_r}\!\bigg[\sqrt{\tfrac{2J_r}{\kappa}}k_r^p\bigg]\, \mathcal{J}_{m_z}\!\bigg[\sqrt{\tfrac{2J_z}{\nu}}k_z^p\bigg]\, \frac{\alpha_p^2}{1-\lambda_p}\, \delta\widehat{\psi}^{\mathrm{e}}{}_{m_\phi, k_r^p, k_z^p}[R_0^p, \omega]\,.
\tag{5.48}
$$

Let us emphasise the strong similarities between Eq. (5.48) and its razor-thin analog from Eq. (3.59). Following the same method as in the razor-thin case, we rely on Riemann sum formula to rewrite Eq. (5.47) with continuous integrals w.r.t. R_0^p and k_r^p using the critical step distances from Eq. (3.61). Equation (5.47) becomes

$$
g(\omega) = \frac{1}{2h} \sum_{n_p} \!\int\! \mathrm{d}k_r^p\, g_{\mathrm{s}}(k_r^p, R_{\mathrm{g}}, k_z^p, \omega)\,,
\tag{5.49}
$$

where one can note that there still remains a discrete sum on the index n_p. At this stage, to make further progress in the calculations, two strategies are possible. On the one hand, one may assume that the disc is sufficiently thick so that one can replace the sum on n_p in Eq. (5.49) by a continuous integral over k_z. On the other hand, in the limit of a thinner disc, as the quantised frequencies k_z tend to be further apart (see Fig. 5.4), one should keep the discrete sum from Eq. (5.49). In the upcoming calculations, we stick to the first approach and aim for continuous expressions. In Appendix 5.C, we will follow the second approach, show that these two approaches are in full agreement and fully consistent with the razor-thin results obtained in Sect. 3.5.

As noted in Eq. (5.17), for a sufficiently thick disc, one can assume that the distance between two successive quantised k_z frequencies is of order $\Delta k_z \simeq \pi/h$. Assuming that Δk_z is sufficiently small compared to the scale of variation of the function $k_z \mapsto g_s(k_z)$, let us rely once again on Riemann sum formula to rewrite Eq. (5.49) as

$$g(\omega) = \frac{1}{2\pi} \int \mathrm{d}k_r^p \mathrm{d}k_z^p \, g_s(k_r^p, R_g, k_z^p, \omega) . \tag{5.50}$$

Following Eq. (3.63), we introduce $\widehat{C}_{\delta\psi^e}$ the autocorrelation of the external perturbations as

$$\widehat{C}_{\delta\psi^e}[m_\phi, \omega, R_g, k_r^p, k_r^q, k_z^p, k_z^q] = \frac{1}{2\pi} \int \mathrm{d}\omega' \left\langle \delta\widehat{\psi^e}_{m_\phi, k_r^p, k_z^p}[R_g, \omega] \, \delta\widehat{\psi^e}^*_{m_\phi, k_r^q, k_z^q}[R_g, \omega'] \right\rangle , \tag{5.51}$$

so that the diffusion coefficients from Eq. (5.46) become

$$\begin{aligned} D_{\boldsymbol{m}}^{\mathrm{sym}}(\boldsymbol{J}) = \delta_{m_z}^{\mathrm{even}} \frac{1}{(2\pi)^2} & \int \mathrm{d}k_r^p \mathrm{d}k_z^p \, \mathcal{J}_{m_r}\!\left[\sqrt{\tfrac{2J_r}{\kappa}} k_r^p\right] \mathcal{J}_{m_z}\!\left[\sqrt{\tfrac{2J_z}{\nu}} k_z^p\right] \frac{\alpha_p^2}{1-\lambda_p} \\ \times & \int \mathrm{d}k_r^q \mathrm{d}k_z^q \, \mathcal{J}_{m_r}\!\left[\sqrt{\tfrac{2J_r}{\kappa}} k_r^q\right] \mathcal{J}_{m_z}\!\left[\sqrt{\tfrac{2J_z}{\nu}} k_z^q\right] \frac{\alpha_q^2}{1-\lambda_q} \, \widehat{C}_{\delta\psi^e}[m_\phi, \boldsymbol{m}\cdot\boldsymbol{\Omega}, R_g, k_r^p, k_r^q, k_z^p, k_z^q] . \end{aligned} \tag{5.52}$$

The antisymmetric equivalent of Eq. (5.52) is straightforward to obtain via the substitutions $\delta_{m_z}^{\mathrm{even}} \to \delta_{m_z}^{\mathrm{odd}}$ and $\alpha_{p/q} \to \beta_{p/q}$. In addition, one should also pay attention to the fact that the autocorrelation $\widehat{C}_{\delta\psi^e}$ should be computed slightly differently in the antisymmetric context. Indeed, as the antisymmetric basis elements from Eq. (5.105) possess an odd vertical dependence, the even restricted Fourier transform from Eq. (5.45) should be replaced by an odd restricted vertical Fourier transform defined as

$$\text{(iii):} \;\; f_{k_z} = \int_{-h}^{+h} \mathrm{d}z \, \sin(k_z z) \, f[z] . \tag{5.53}$$

Finally, in Eq. (5.52), notice that the integrations on k_z^p and k_z^q should only be performed for $k_z \geq k_z^1$, i.e. for k_z larger than the associated fundamental mode, as illustrated in Fig. 5.4.

Following Eq. (3.66), let us now further simplify Eq. (5.52) by assuming some additional properties on the stochasticity of the external perturbations. In analogy with Eq. (3.66), we suppose that the external perturbations, $\delta\psi^e$, are spatially quasi-stationary so that

$$\big\langle \delta\psi^{\mathrm{e}}_{m_\phi}[R_1,z_1,t_1]\,\delta\psi^{\mathrm{e}*}_{m_\phi}[R_2,z_2,t_2]\big\rangle = \mathcal{C}[m_\phi,t_1-t_2,(R_1+R_2)/2,R_1-R_2,z_1+z_2,z_1-z_2]\,, \tag{5.54}$$

where the dependences w.r.t. $(R_1+R_2)/2$ and z_1+z_2 are supposed to be slow. Thanks to some simple algebra (see Appendix E of Fouvry et al. (2017)), one can show that

$$\big\langle \delta\widehat{\psi}^{\mathrm{e}}_{m_\phi,k_r^1,k_z^1}[R_\mathrm{g},\omega_1]\,\delta\widehat{\psi}^{\mathrm{e}*}_{m_\phi,k_r^2,k_z^2}[R_\mathrm{g},\omega_2]\big\rangle = 2\pi^2\,\delta_\mathrm{D}(\omega_1-\omega_2)\,\delta_\mathrm{D}(k_r^1-k_r^2)\,\delta_\mathrm{D}(k_z^1-k_z^2)\,\widehat{\mathcal{C}}[m_\phi,\omega_1,R_\mathrm{g},k_r^1,k_z^1]\,, \tag{5.55}$$

where in analogy with Eq. (3.67), $\widehat{\mathcal{C}}[...]$ has been transformed three times, according to a temporal Fourier transform as defined in Eq. (2.9), according to a local radial Fourier transform as in Eq. (5.45) of spread $\sqrt{2}\sigma$ w.r.t. R_1-R_2 in the neighbourhood of $R_1-R_2=0$ and $(R_1+R_2)/2=R_\mathrm{g}$, and finally according to an even restricted vertical Fourier transform as in Eq. (5.45) w.r.t. z_1-z_2 in the neighbourhood of $z_1-z_2=0$ and $z_1+z_2=0$. In Eq. (5.55), the autocorrelation of the external perturbation was therefore diagonalised w.r.t. ω, k_r, and k_z, so that the diffusion coefficients from Eq. (5.52) become

$$D^{\mathrm{sym}}_{\boldsymbol{m}}(\boldsymbol{J}) = \delta^{\mathrm{even}}_{m_z}\frac{\pi}{(2\pi)^2}\int \mathrm{d}k_r^p\,\mathrm{d}k_z^p\,\mathcal{J}^2_{m_r}\bigg[\sqrt{\frac{2J_r}{\kappa}}k_r^p\bigg]\,\mathcal{J}^2_{m_z}\bigg[\sqrt{\frac{2J_z}{\nu}}k_z^p\bigg]\bigg[\frac{\alpha_p^2}{1-\lambda_p}\bigg]^2\,\widehat{\mathcal{C}}[m_\phi,\boldsymbol{m}\cdot\boldsymbol{\Omega},R_\mathrm{g},k_r^p,k_z^p]\,. \tag{5.56}$$

The antisymmetric analog of Eq. (5.56) is straightforward to obtain thanks to the substitutions $\delta^{\mathrm{even}}_{m_z}\to\delta^{\mathrm{odd}}_{m_z}$ and $\alpha_p\to\beta_p$. Finally, in this case despite the fact that one is considering antisymmetric diffusion coefficients, it is important to note that $\widehat{\mathcal{C}}$ still has to be transformed according to an even-restricted vertical Fourier transform, see Appendix E of Fouvry et al. (2017) for details. The explicit expression (5.56) of the collisionless diffusion coefficients is the main result of this section. It presents close similarities with Eq. (3.68) short of an extra integral along vertical k_z modes modulated by an extra Bessel function.

As in Eq. (3.69), one may further simplify Eq. (5.56) by relying on the approximation of the small denominators, for which one focuses on the tightly wound waves which yield the maximum self-gravitating amplification. Let us therefore assume that the function $(k_r,k_z)\mapsto\lambda(k_r,k_z)$ reaches in its domain a well-defined maximum $\lambda_{\mathrm{max}}(R_\mathrm{g},\omega)$ for $(k_r,k_z)=(k_r^{\mathrm{max}},k_z^{\mathrm{max}})$. Let us then define the domain of maximum amplification $\mathcal{V}_{\mathrm{max}}=\big\{(k_r,k_z)\big|\lambda(k_r,k_z)\geq\lambda_{\mathrm{max}}/2\big\}$ and its associated area $|\mathcal{V}_{\mathrm{max}}|$. Equation (5.56) can then be approximated as

$$D^{\mathrm{sym}}_{\boldsymbol{m}}(\boldsymbol{J}) = \delta^{\mathrm{even}}_{m_z}\frac{\pi|\mathcal{V}_{\mathrm{max}}|}{(2\pi)^2}\,\mathcal{J}^2_{m_r}\bigg[\sqrt{\frac{2J_r}{\kappa}}k_r^{\mathrm{max}}\bigg]\,\mathcal{J}^2_{m_z}\bigg[\sqrt{\frac{2J_z}{\nu}}k_z^{\mathrm{max}}\bigg]\bigg[\frac{\alpha_{\mathrm{max}}^2}{1-\lambda_{\mathrm{max}}}\bigg]^2\,\widehat{\mathcal{C}}[m_\phi,\boldsymbol{m}\cdot\boldsymbol{\Omega},R_\mathrm{g},k_r^{\mathrm{max}},k_z^{\mathrm{max}}]\,, \tag{5.57}$$

while the associated antisymmetric diffusion coefficients are straightforward to obtain by direct analogy. One can also improve the previous approximation by performing the integrations from Eq. (5.56) for $(k_r,k_z)\in\mathcal{V}_{\mathrm{max}}$. Such a calculation ensures a better estimation of the diffusion flux while being more numerically demanding. This does not alter the results obtained in the applications presented in Sect. 5.7.

5.6 WKB Limit for the Collisional Diffusion

Relying on the thick WKB amplification eigenvalues obtained in Eq. (5.30) and following the same approach as in the previous section, let us now evaluate the collisional drift and diffusion coefficients of the Balescu-Lenard equation (2.67).

Following the notations from Eq. (5.42) and separating the contributions from the symmetric and antisymmetric basis elements, the dressed susceptibility coefficients from Eq. (2.50) take the form

$$\frac{1}{\mathcal{D}_{\boldsymbol{m}_1,\boldsymbol{m}_2}(\boldsymbol{J}_1,\boldsymbol{J}_2,\omega)} = \sum_p \left[\frac{\psi_{\boldsymbol{m}_1}^{\mathrm{s},(p)}(\boldsymbol{J}_1)\,\psi_{\boldsymbol{m}_2}^{\mathrm{s},(p)*}(\boldsymbol{J}_2)}{1-\lambda_p^{\mathrm{s}}(\omega)} + \frac{\psi_{\boldsymbol{m}_1}^{\mathrm{a},(p)}(\boldsymbol{J}_1)\,\psi_{\boldsymbol{m}_2}^{\mathrm{a},(p)*}(\boldsymbol{J}_2)}{1-\lambda_p^{\mathrm{a}}(\omega)} \right], \tag{5.58}$$

where the superscripts "s" and "a" respectively correspond to the symmetric and antisymmetric basis elements. We showed in Eq. (5.22) (resp. (5.105)) that the Fourier transformed WKB basis elements involve an azimuthal Kronecker symbol $\delta_{m_\phi}^{k_\phi^p}$, as well as a $\delta_{m_z}^{\mathrm{even}}$ (resp. $\delta_{m_z}^{\mathrm{odd}}$) for the symmetric (resp. antisymmetric) basis elements. As a consequence, in Eq. (5.58) in order to have non-vanishing susceptibility coefficients, one must necessarily have

$$m_1^\phi = m_2^\phi = k_\phi \quad \text{and} \quad (m_1^z - m_2^z) \;\text{even}\,. \tag{5.59}$$

Because m_1^z and m_2^z must be of the same parity, one concludes that the susceptibility coefficients do not mix up symmetric and antisymmetric basis elements. As a consequence, depending on the parity of m_1^z, one can restrict oneself only to the symmetric elements or only to the antisymmetric ones.

Let us now focus on one crucial consequence of the WKB approximation, which is the restriction to local resonances. As already noted in Eq. (3.73), one technical difficulty of the Balescu-Lenard equation is to deal with the resonance condition $\boldsymbol{m}_1\cdot\boldsymbol{\Omega}_1 - \boldsymbol{m}_2\cdot\boldsymbol{\Omega}_2 = 0$. For a given value of $\boldsymbol{J}_1$, $\boldsymbol{m}_1$, and $\boldsymbol{m}_2$, one has to identify the resonant radii R_2^{r} for which the resonance condition is satisfied. In our case, one important simplification comes from the thickened epicyclic approximation, thanks to which the orbital frequencies $\boldsymbol{\Omega} = (\Omega_\phi, \kappa, \nu)$ depend only on J_ϕ. As in Eq. (3.75), we also assume here that the disc's mean potential is dynamically non-degenerate so that

$$\left. \frac{\mathrm{d}(\boldsymbol{m}_2\cdot\boldsymbol{\Omega})}{\mathrm{d}R} \right|_{R_2^{\mathrm{r}}} \neq 0\,. \tag{5.60}$$

Following the notations from Eq. (3.76), the resonance condition takes the form

$$m_1^\phi \Omega_\phi^1 + m_1^r \kappa^1 + m_1^z \nu^1 = m_1^\phi \Omega_\phi^{\mathrm{r}} + m_2^{\mathrm{r}} \kappa^{\mathrm{r}} + m_2^z \nu^{\mathrm{r}} , \tag{5.61}$$

where we used the notation $\Omega_\phi^1 = \Omega_\phi^1(R_1)$ and $\Omega_\phi^{\mathrm{r}} = \Omega_\phi(R_2^{\mathrm{r}})$. We also relied on Eq. (5.59) to impose $m_1^\phi = m_2^\phi$. Because the Fourier transformed basis elements from Eqs. (5.22) and (5.105) involve the narrow radial Gaussian $\mathcal{B}_{R_0}$, the resonant radii R_2^{r} are necessarily close to R_1 so that $|\Delta R| = |R_2^{\mathrm{r}} - R_1| \lesssim$ (few) σ. Similarly to Eq. (3.77), the resonance condition from Eq. (5.61) may be rewritten as

$$\left[m_2^\phi \frac{\mathrm{d}\Omega_\phi}{\mathrm{d}R} + m_2^r \frac{\mathrm{d}\kappa}{\mathrm{d}R} + m_2^z \frac{\mathrm{d}\nu}{\mathrm{d}R} \right] \Delta R = \left[m_1^r - m_2^r \right] \kappa^1 + \left[m_1^z - m_2^z \right] \nu^1 . \tag{5.62}$$

In the l.h.s. of Eq. (5.62), the terms within brackets is non-zero as a result of our assumption from Eq. (5.60) that the disc's mean potential is dynamically non-degenerate, while ΔR is small because of the scale decoupling approach used in the construction of the WKB basis elements. The r.h.s. of Eq. (5.62) should be seen as discrete in the sense that is the sum of a multiple of κ and of ν. For a disc not too thick, one expects to have $\nu \gg \kappa$. In addition, we showed in Eq. (5.59), that $(m_1^z - m_2^z)$ has to be an even number. As a consequence, for $(m_1^z - m_2^z) \neq 0$, one has

$$\left| (m_1^z - m_2^z)\, \nu^1 \right| \geq \left| 2\nu^1 \right| \gg \left| m_1^r - m_2^r \right| \kappa^1 , \tag{5.63}$$

provided that the resonance vectors $\boldsymbol{m}_1$ and $\boldsymbol{m}_2$ are of small order. In this situation, the l.h.s. of Eq. (5.62) is therefore small, while its r.h.s. is of order ν^1. Equation (5.62) therefore imposes $m_1^z = m_2^z$. Equation (5.62) then takes the exact same form as the razor-thin equation (3.77). We follow the same argument and therefore conclude that the thick WKB basis elements impose that only local resonances are allowed so that

$$R_2^{\mathrm{r}} = R_1 \quad ; \quad m_1^r = m_2^r \quad ; \quad m_1^z = m_2^z . \tag{5.64}$$

This an essential result of the WKB approximation, which enables us to pursue the analytical evaluation of the dressed susceptibility coefficients.

Restricting ourselves to the cases $R_2 = R_1$ and $\boldsymbol{m}_2 = \boldsymbol{m}_1$, and using the expression of the Fourier transformed basis elements from Eq. (5.22), the symmetric susceptibility coefficients from Eq. (5.58) now read

$$\begin{aligned} \frac{1}{\mathcal{D}_{\boldsymbol{m}_1,\boldsymbol{m}_1}} = \sum_{k_r^p, R_0^p, n_p} & \frac{G}{R_0^p h} \frac{1}{(k_r^p)^2 + (k_z^p)^2} \frac{1}{\sqrt{\pi\sigma^2}} \exp\left[- \frac{(R_1 - R_0^p)^2}{\sigma^2} \right] \frac{\alpha_p^2}{1 - \lambda_p(\omega)} \\ & \times \mathcal{J}_{m_1^r}\left[\sqrt{\frac{2J_r^1}{\kappa_1}} k_r^p \right] \mathcal{J}_{m_1^r}\left[\sqrt{\frac{2J_r^2}{\kappa_1}} k_r^p \right] \mathcal{J}_{m_1^z}\left[\sqrt{\frac{2J_z^1}{\nu_1}} k_z^p \right] \mathcal{J}_{m_1^z}\left[\sqrt{\frac{2J_z^2}{\nu_1}} k_z^p \right] , \end{aligned} \tag{5.65}$$

where we introduced the shortening notations $1/\mathcal{D}_{\boldsymbol{m}_1,\boldsymbol{m}_1} = 1/\mathcal{D}_{\boldsymbol{m}_1,\boldsymbol{m}_1}(R_1, J_r^1, J_z^1, R_1, J_r^2, J_z^2, \omega)$, as well as $\kappa_1 = \kappa(R_1)$, $\nu_1 = \nu(R_1)$, and $k_z^p = k_z^{n_p}$. One should also note that the sum on k_ϕ^p was already executed thanks to the constraint from Eq. (5.59).

Following the same approach as in the collisionless case, let us replace the sums on k_r^p and R_0^p by continuous expressions. Equation (5.65) becomes

$$\frac{1}{\mathcal{D}_{\boldsymbol{m}_1,\boldsymbol{m}_1}} = \frac{G}{2\pi R_1 h} \sum_{n_p} \int \mathrm{d}k_r \, \frac{1}{k_r^2+(k_z^p)^2} \frac{\alpha_p^2}{1-\lambda_p(\omega)} \times \mathcal{J}_{m_1^r}\left[\sqrt{\frac{2J_r^1}{\kappa_1}}k_r\right] \mathcal{J}_{m_1^r}\left[\sqrt{\frac{2J_r^2}{\kappa_1}}k_r\right] \mathcal{J}_{m_1^z}\left[\sqrt{\frac{2J_z^1}{\nu_1}}k_z^p\right] \mathcal{J}_{m_1^z}\left[\sqrt{\frac{2J_z^2}{\nu_1}}k_z^p\right]. \tag{5.66}$$

In Eq. (5.66), there still remains a sum on the vertical index n_p. As already described in Eq. (5.49) for the collisionless case, one may follow two possible strategies to complete the evaluation of the susceptibility coefficients. If the disc is sufficiently thick, one can replace the sum on n_p by a continuous integral over k_z. In the limit of a thin disc, one should however keep the discrete sum in Eq. (5.66). In the next calculations, let us follow the first continuous approach. The second approach is presented in Appendix 5.C, and we show once again that these two approaches are fully consistent one with another, and that they also allow for the recovery of the razor-thin results from Sect. 3.6.

Using the vertical step distance $\Delta k_z \simeq \pi/h$ from Eq. (5.17) and assuming that the function in the r.h.s. of Eq. (5.66) vary on scales larger than Δk_z, one may use once again Riemann sum formula to rewrite Eq. (5.66) as

$$\frac{1}{\mathcal{D}_{\boldsymbol{m}_1,\boldsymbol{m}_1}} = \frac{G}{2\pi^2 R_1} \int \mathrm{d}k_r \mathrm{d}k_z \, \frac{1}{k_r^2+k_z^2} \frac{\alpha_{k_r,k_z}^2}{1-\lambda_{k_r,k_z}(\omega)} \times \mathcal{J}_{m_1^r}\left[\sqrt{\frac{2J_r^1}{\kappa_1}}k_r\right] \mathcal{J}_{m_1^r}\left[\sqrt{\frac{2J_r^2}{\kappa_1}}k_r\right] \mathcal{J}_{m_1^z}\left[\sqrt{\frac{2J_z^1}{\nu_1}}k_z\right] \mathcal{J}_{m_1^z}\left[\sqrt{\frac{2J_z^2}{\nu_1}}k_z\right]. \tag{5.67}$$

Such an explicit expression of the dressed susceptibility coefficients constitutes the main result of this section. Equation (5.67) relates the gravitational susceptibility of the disc to known analytic functions of its actions via simple regular quadratures. Following Eq. (5.57), one may further simplify Eq. (5.67) by relying on the approximation of the small denominators. It becomes

$$\frac{1}{\mathcal{D}_{\boldsymbol{m}_1,\boldsymbol{m}_1}} = \frac{G}{2\pi^2 R_1} \frac{|\mathcal{V}_{\mathrm{max}}|}{(k_r^{\mathrm{max}})^2+(k_z^{\mathrm{max}})^2} \frac{\alpha_{\mathrm{max}}^2}{1-\lambda_{\mathrm{max}}} \times \mathcal{J}_{m_1^r}\left[\sqrt{\frac{2J_r^1}{\kappa_1}}k_r^{\mathrm{max}}\right] \mathcal{J}_{m_1^r}\left[\sqrt{\frac{2J_r^2}{\kappa_1}}k_r^{\mathrm{max}}\right] \mathcal{J}_{m_1^z}\left[\sqrt{\frac{2J_z^1}{\nu_1}}k_z^{\mathrm{max}}\right] \mathcal{J}_{m_1^z}\left[\sqrt{\frac{2J_z^2}{\nu_1}}k_z^{\mathrm{max}}\right]. \tag{5.68}$$

This approximation can be improved by rather performing the integrations in Eq. (5.67) for $(k_r, k_z) \in \mathcal{V}_{\mathrm{max}}$. This approach allows for a more precise determination of the diffusion flux but is more numerically demanding. Such an improved approximation does not alter the principal conclusions drawn in the upcoming applications. Finally, for m_1^z odd, the antisymmetric analogs of the previous expressions of the dressed susceptibility coefficients are straightforward to obtain thanks to the substitution $\alpha \to \beta$.

As a final step, let us now compute the Balescu-Lenard drift and diffusion coefficients from Eqs. (2.69) and (2.70). Restricting oneself only to local resonances and using the shortened notation from Eq. (3.82), one can write the collisional drift coefficients as

$$A_{\boldsymbol{m}_1}(\boldsymbol{J}_1) = -\frac{8\pi^4 \mu}{(\boldsymbol{m}_1 \cdot \boldsymbol{\Omega}_1)'} \int \mathrm{d}J_r^2 \mathrm{d}J_z^2 \, \frac{\boldsymbol{m}_1 \cdot \partial F / \partial \boldsymbol{J}(J_\phi^1, J_r^2, J_z^2)}{|\mathcal{D}_{\boldsymbol{m}_1, \boldsymbol{m}_1}(\boldsymbol{J}_1, \boldsymbol{J}_2, \boldsymbol{m}_1 \cdot \boldsymbol{\Omega}_1)|^2}, \tag{5.69}$$

while the diffusion coefficients are given by

$$D_{\boldsymbol{m}_1}(\boldsymbol{J}_1) = \frac{8\pi^4 \mu}{(\boldsymbol{m}_1 \cdot \boldsymbol{\Omega}_1)'} \int \mathrm{d}J_r^2 \mathrm{d}J_z^2 \, \frac{F(J_\phi^1, J_r^2, J_z^2)}{|\mathcal{D}_{\boldsymbol{m}_1, \boldsymbol{m}_1}(\boldsymbol{J}_1, \boldsymbol{J}_2, \boldsymbol{m}_1 \cdot \boldsymbol{\Omega}_2)|^2}. \tag{5.70}$$

In Eqs. (5.69) and (5.70), the susceptibility coefficients are given by Eq. (5.67), or even Eq. (5.68) when using the approximation of the small denominators. In particular, because of the restriction to local resonances, they always have to be evaluated for $J_\phi^2 = J_\phi^1$. Note that in the case where the DF is a quasi-isothermal DF as in Eq. (5.5), and where the susceptibility coefficients are computed via the approximation of the small denominators from Eq. (5.68), the integrations w.r.t. J_r^2 and J_z^2 in Eqs. (5.69) and (5.70) may be computed explicitly (see Fouvry et al. (2017)). The simple and tractable expressions of the collisional drift and diffusion coefficients obtained in Eqs. (5.69) and (5.70) constitute an important result of this section. Let us finally insist on the fact that the application of the thick WKB approximation to the Balescu-Lenard equation is self-contained and that no ad hoc fittings were required. Except for the explicit calculation of the thickened amplification eigenvalues in Eq. (5.30), the previous calculations are not limited to the quasi-isothermal DF from Eq. (5.5). The collisional drift and diffusion coefficients from Eqs. (5.69) and (5.70) are valid for any tepid disc's DF, provided that one can rely on the epicyclic angle-action mapping from Eq. (5.4).

5.7 Application to Disc Thickening

Let us now implement the previous thick WKB collisionless and collisional diffusion equations in order to investigate the various resonant processes at play during the secular evolution of a thick stellar disc. Let us first emphasise that the self-consistent description of the secular dynamics of a self-gravitating thickened disc is a challenging task that raises many difficulties. There exist no generic angle-action coordinates in the thickened geometry, nor appropriate basis elements, nor methods to compute the properties of the disc's collective effects. The previous WKB framework allows for the resolution of all these difficulties, at the cost of additional assumptions, e.g., epicyclic approximation, tightly wound perturbations, etc. This framework appears therefore as a legitimate first step to investigate from first principles the intricate dynamics of thickened discs. In Sect. 5.7.1, we present the considered thick disc model. In Sect. 5.7.2, we show how our formalism allows us to recover qualitatively the secular formation of vertical resonant ridges observed in the numerical experiments of Solway et al. (2012). Sections 5.7.3 and 5.7.4 respectively consider the associated diffusion timescales as well as the secular in-plane diffusion. In Sect. 5.7.5, we consider the mechanism of disc thickening via the resonant diffusion induced by central decaying bars. Finally, in Sect. 5.7.6, we show how one can account for the joint evolution of giant molecular clouds (GMCs) and how they hasten the secular diffusion. Let us first describe the considered disc model.

5.7.1 A Thickened Disc Model

In order to setup a model of thickened stellar disc, we follow the model recently considered in Solway et al. (2012) (hereafter So12). We follow specifically the numerical parameters from their simulation UCB keeping only the most massive of its two components. This simulation is particularly relevant in the context of secular diffusion, as it dealt with an unperturbed isolated stable and stationary thick stellar disc. On secular timescales, this thick disc developed spontaneously sequences of transient spirals seeded by the disc's discreteness and, only on the very long-term, a central bar. This disc should be seen as a thickened version of the razor-thin Mestel disc presented in Sect. 3.7.1. Let us start from the razor-thin surface density of the Mestel disc Σ_{M} introduced in Eq. (3.85). Then, assuming a given vertical profile shape, one can thicken Σ_{M} to construct a density ρ_{M}. Indeed, let us define the $3D$ density $\rho_{\mathrm{M}}(R, z)$ as

$$\rho_{\mathrm{M}}(R, z) = \Sigma_{\mathrm{M}}(R) \, \frac{1}{4 z_0(R)} \, \mathrm{sech}^2\!\left[\frac{z}{2 z_0(R)}\right] . \tag{5.71}$$

Equation (5.71) corresponds to a Spitzer profile (Spitzer 1942), where we introduced the local thickness z_0 of the disc. It satisfies $\int \mathrm{d}z \, \rho_{\mathrm{M}}(R, z) = \Sigma_{\mathrm{M}}(R)$. This profile corresponds to an isothermal vertical distribution, i.e. a vertical statistical and thermodynamical equilibrium. Let us note that at this stage, one could have used alternative vertical profiles, e.g., exponential. The results presented thereafter can straightforwardly be applied to other vertical profiles, provided one adapts accordingly the relations between h, z_0, and σ_z/ν obtained in Eqs. (5.11) and (5.74). After having constructed the system's total density, one can numerically determine the associated thickened potential ψ_{M} given by $\psi_{\mathrm{M}}(\boldsymbol{x}) = -\int \mathrm{d}\boldsymbol{x}_1 \, G \rho_{\mathrm{M}}(\boldsymbol{x}_1)/|\boldsymbol{x} - \boldsymbol{x}_1|$. Thanks to the disc's axisymmetry, this can be rewritten as

$$\psi_M(R,z) = \int dR_1 dz_1 \, \frac{-4GR_1\rho_M(R_1,z_1)}{\sqrt{(R-R_1)^2+(z-z_1)^2}} \, \mathcal{F}_{\rm ell}\!\left[\frac{\pi}{2}, -\frac{4RR_1}{(R-R_1)^2+(z-z_1)^2}\right], \tag{5.72}$$

where $\mathcal{F}_{\rm ell}[\phi,m] = \int_0^\phi d\phi' [1-m\sin^2(\phi')]^{-1/2}$ is the elliptic integral of the first kind. Thanks to the numerical calculation of the thickened total potential ψ_M, one may then rely on the epicyclic approximation from Sect. 5.2 to construct the mapping $R_g \mapsto J_\phi$, as well as the three intrinsic frequencies Ω_ϕ, κ and ν. Once these elements are determined, the angle-action mapping from Eq. (5.4) is fully characterised. For a sufficiently thin disc, one expects these mappings to be close to the ones obtained in Eqs. (3.86) and (3.87) for the razor-thin case.

Let us emphasise that the equilibrium value of the vertical velocity dispersion σ_z is directly constrained by the thickened mean density profile ρ_M. Indeed, the one-dimensional vertical Jeans equation (see, e.g., Eq. (4.271) in Binney and Tremaine (2008)) imposes

$$\frac{\partial(\rho_M \sigma_z^2)}{\partial z} = -\rho_M \frac{\partial \psi_M}{\partial z}, \tag{5.73}$$

where we assume that σ_z only depends on R. Differentiating once this relation w.r.t. z and evaluating it at $z=0$, one immediately gets for a Spitzer profile the relation

$$\frac{\sigma_z(R)}{\nu(R)} = \sqrt{2}\, z_0(R). \tag{5.74}$$

As a consequence, once the scale height of the disc z_0 and the vertical frequency ν have been determined, the value of the velocity dispersion σ_z immediately follows from the constraint of vertical equilibrium. The previous determinations of the system's intrinsic frequencies required the use of the system's total potential ψ_M. However, here we are interested in the dynamics of the active component of the disc, the stars, whose surface density $\Sigma_{\rm star}$ is only one component of the total surface density Σ_M. As was done in Eqs. (3.90) and (3.91) in the razor-thin case, we introduce two taper functions $T_{\rm inner}$ and $T_{\rm outer}$ to deal with the inner singularity and the infinite extent of the system, as well as an active fraction ξ, so that $\Sigma_{\rm star}$ is given by Eq. (3.92), and is illustrated in Fig. 5.10. Using the same units than So12's

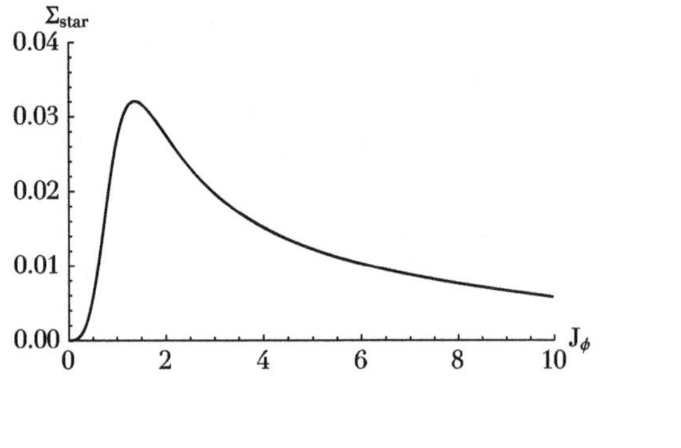

Fig. 5.10 Illustration of the active surface density $\Sigma_{\rm star}$ for the thickened disc presented in Sect. 5.7.1. As a result of the tapering functions, the disc's self-gravity is turned off in its inner and outer regions

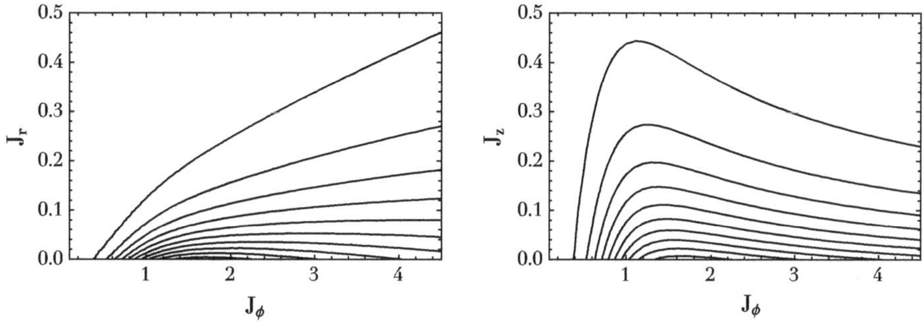

Fig. 5.11 *Left panel* Illustration of the contours of the initial quasi-isothermal DF from Eq. (5.5) in the plane $(J_\phi, J_r, J_z = 0)$. Contours are spaced linearly between 95% and 5% of the maximum of the DF. *Right panel* Contours of the quasi-isothermal DF in the plane $(J_\phi, J_r = 0, J_z)$ following the same conventions as in the *left panel*

UCB simulation, our numerical parameters are given by $V_0 = G = R_\mathrm{i} = 1$, $R_\mathrm{o} = 15$, $\nu_\mathrm{t} = 4$, $\mu_\mathrm{t} = 6$, $R_\mathrm{max} = 25$, $\sigma_r = 0.227$, and $\xi = 0.4$. Finally, to mimic So12's vertical profile, we use for the Spitzer profile from Eq. (5.71) a constant scale height given by $z_0 = 0.26$. One can also straightforwardly estimate the total active mass of the system as $M_\mathrm{tot} = 5.8$. Using these numerical parameters, the shape of the quasi-isothermal DF F_star from Eq. (5.5) is illustrated in Fig. 5.11.

It is important to note that So12's simulation was limited to the harmonic sector $0 \leq m_\phi \leq 8$, except $m_\phi = 1$ to avoid decentring effects. In our analysis, in order to clarify and simplify the dynamical mechanisms at play, we impose an even more drastic limitation to the potential perturbations, and we restrict ourselves only to $m_\phi = 2$. In addition to this azimuthal restriction, all our analyses are also limited to only 9 different resonance vectors $\boldsymbol{m} = (m_\phi, m_r, m_z)$. Indeed, we assume $m_\phi = 2$, $m_r \in \{-1, 0, 1\}$ and $m_z \in \{-1, 0, 1\}$. Among these resonances, one can identify the corotation resonance (COR) as $\boldsymbol{m} = (2, 0, 0)$, the radial (resp. vertical) inner Lindblad resonance (rILR) (resp. vILR) as $\boldsymbol{m} = (2, -1, 0)$ (resp. $\boldsymbol{m} = (2, 0, -1)$), and the radial (resp. vertical) outer Lindblad resonance (rOLR) (resp. vOLR) as $\boldsymbol{m} = (2, 1, 0)$ (resp. $\boldsymbol{m} = (2, 0, 1)$). Having computed the intrinsic frequencies $\boldsymbol{\Omega}$ and specified the considered resonance vectors $\boldsymbol{m}$, one can study the behaviour of the resonance frequencies $\omega = \boldsymbol{m} \cdot \boldsymbol{\Omega}$ as a function of the position within the disc, as illustrated in Fig. 5.12. These frequencies correspond to the frequencies for which the amplification eigenvalues and the perturbation autocorrelation from Eq. (5.56) have to be evaluated.

When simulating the previous quasi-stationary and stable thick disc on secular timescales, So12 (private communication) observed sequences of transient spirals within the disc, which on the long-term led to an irreversible diffusion in action space of the system's DF, and especially to a thickening of the stellar disc. In order to probe such diffusion features, one can consider the marginal distribution of vertical action J_z as a function of the guiding radius R_g within the disc. To do so, let us define the function $F_Z(R_\mathrm{g}, J_z, t)$ as

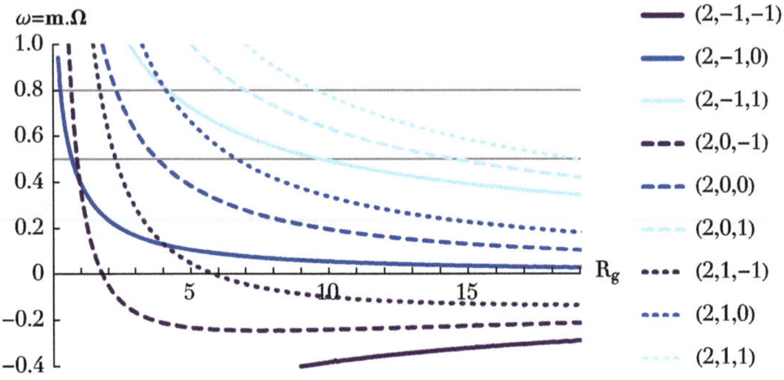

Fig. 5.12 Illustration of the behaviour of the intrinsic frequencies $\omega = \boldsymbol{m}\cdot\boldsymbol{\Omega}$ as a function of the position within the disc (given by R_g) and the resonance vector $\boldsymbol{m} = (m_\phi, m_r, m_z)$. The *grey* lines correspond to the pattern speed $m_p\Omega_p$ introduced in the bar perturbations from Eq. (5.87) and considered in Fig. 5.22

$$F_Z(R_g, J_z, t) = \int d\boldsymbol{\theta}' d\boldsymbol{J}'\, \delta_D(R_g - R_g')\, \delta_D(J_z - J_z')\, F(\boldsymbol{J}', t)$$
$$= (2\pi)^3 \frac{dJ_\phi}{dR_g} \int dJ_r'\, F(R_g, J_r', J_z, t)\,. \tag{5.75}$$

Following Eqs. (2.34) and (2.72), and rewriting both collisionless and collisional diffusion equations as $\partial F/\partial t = \text{div}(\boldsymbol{\mathcal{F}}_{\text{tot}})$, one can straightforwardly estimate the time variations of F_Z as

$$\frac{\partial F_Z}{\partial t} = (2\pi)^3 \frac{dJ_\phi}{dR_g} \int dJ_r'\, \text{div}(\boldsymbol{\mathcal{F}}_{\text{tot}})(R_g, J_r', J_z, t)\,. \tag{5.76}$$

Equation (5.76) can be rewritten as the divergence of a flux $\boldsymbol{\mathcal{F}}_Z = (\mathcal{F}_Z^\phi, \mathcal{F}_Z^z)$ defined in the (J_ϕ, J_z)−plane, so that

$$\frac{\partial F_Z(J_\phi, J_z)}{\partial t} = \left(\frac{\partial}{\partial J_\phi}, \frac{\partial}{\partial J_z}\right)\cdot \boldsymbol{\mathcal{F}}_Z = \frac{\partial \mathcal{F}_Z^\phi}{\partial J_\phi} + \frac{\partial \mathcal{F}_Z^z}{\partial J_z}\,, \tag{5.77}$$

where we introduced the flux components $(\mathcal{F}_Z^\phi, \mathcal{F}_Z^z)$ as

$$\mathcal{F}_Z^\phi = (2\pi)^3 \int dJ_r'\, \mathcal{F}_{\text{tot}}^\phi(J_\phi, J_r', J_z) \quad ; \quad \mathcal{F}_Z^z = (2\pi)^3 \int dJ_r'\, \mathcal{F}_{\text{tot}}^z(J_\phi, J_r', J_z)\,. \tag{5.78}$$

In Eq. (5.78), we introduced the components of the total diffusion flux $\boldsymbol{\mathcal{F}}_{\text{tot}}$ in the (J_ϕ, J_r, J_z) space as $\boldsymbol{\mathcal{F}}_{\text{tot}} = (\mathcal{F}_{\text{tot}}^\phi, \mathcal{F}_{\text{tot}}^r, \mathcal{F}_{\text{tot}}^z)$.

As observed in So12's simulations, the initial contours of F_Z are illustrated in the left panel of Fig. 5.13, while their long-term evolution is illustrated in the right panel of the same figure. Comparing the two panels of Fig. 5.13, one can indubitably note

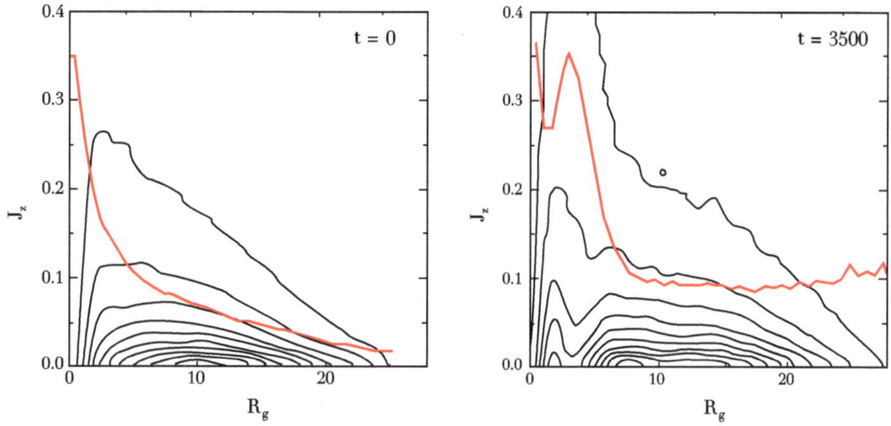

Fig. 5.13 Illustration of the evolution of the function F_Z from Eq. (5.75), as observed in the direct numerical simulation UCB1 of So12. *Left panel* Initial contours of $F_Z(R_g, J_z, t)$ for $t = 0$. Such a representation illustrates the distribution of vertical actions J_z as a function of the position within the disc given by the guiding radius R_g. Contours are spaced linearly between 95% and 5% of the function maximum. The *red* curve gives the mean value of J_z for a given R_g. *Right panel* Same as in the *left panel* but at a much later stage of the evolution $t = 3500$. In the inner regions of the disc, one clearly notes the formation on secular timescales of a narrow ridge of enhanced vertical actions J_z

the spontaneous formation on secular times of a narrow ridge of enhanced vertical actions in the inner regions of the disc, characterised by an increase of the mean value of the vertical actions in these regions. Such a feature is the direct vertical equivalent of what was observed in the radial direction in the razor-thin simulations presented in Fig. 3.5. Let us note that the disc considered in So12 was purposely designed to be linearly stable, quasi-stationary, isolated, and unperturbed. So12 also explicitly checked that this disc did not develop any spiral mode or bar instability for hundreds of dynamical time. The only fluctuations in the disc are then due to weak transient spiral arms seeded by the disc's finite number of particles. In such a context, the ridge from Fig. 5.13 is therefore a signature of the spontaneous thickening of the disc sourced by its intrinsic shot noise amplified by self-gravity.

Let us now investigate in Sect. 5.7.2 how the thickened WKB limits of the collisionless and collisional diffusion equations obtained in Sects. 5.5 and 5.6 allow us to explain such a feature. Secular evolution being a slow process, as in the previous chapters, we will restrict ourselves here to the computation of the initial diffusion fluxes, $\boldsymbol{\mathcal{F}}_{\mathrm{tot}}$, at the time $t = 0$.

5.7.2 Shot-Noise-Driven Resonant Disc Thickening

In order to compute the diffusion fluxes associated with the collisionless and collisional diffusion equations, the first step is to study the properties of the system's self-gravity. To do so, let us consider the amplification eigenvalues $\lambda(k_r, k_z)$ from Eq. (5.30), thanks to which one may perform the approximation of the small denominators. For a given position J_ϕ and a given resonance vector $\boldsymbol{m}$, we illustrate in Fig. 5.14 the behaviour of the function $(k_r, k_z) \mapsto \lambda(k_r, k_z)$. Such a behaviour allows us to identify a region $\mathcal{V}_{\max}(\boldsymbol{m}, J_\phi)$ of maximum amplification over which the integrations on k_r and k_z may be performed in Eqs. (5.56) and (5.67). Figure 5.15 illustrates the importance of the system's self-gravity by representing the behaviour of the function $J_\phi \mapsto 1/(1-\lambda_{\max}(\boldsymbol{m}, J_\phi))$ for different resonance vectors $\boldsymbol{m}$. Following our characterisation of the disc's amplification, let us now compute in turn the induced collisionless diffusion (Sect. 5.7.2.1) as well as the collisional one (Sect. 5.7.2.2), and investigate if such approaches are able to recover the secular formation of the vertical resonant ridges observed in Fig. 5.13 via direct $N-$body simulations.

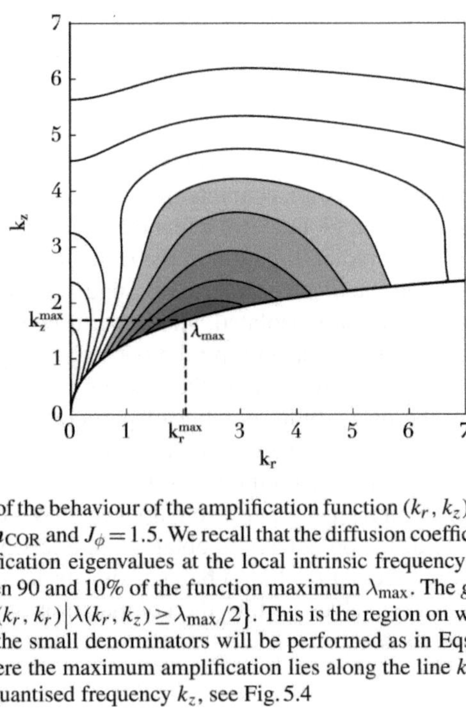

Fig. 5.14 Illustration of the behaviour of the amplification function $(k_r, k_z) \mapsto \lambda(k_r, k_z)$ as obtained in Eq. (5.30), for $\boldsymbol{m} = \boldsymbol{m}_{\mathrm{COR}}$ and $J_\phi = 1.5$. We recall that the diffusion coefficients generically require to compute the amplification eigenvalues at the local intrinsic frequency $\omega = \boldsymbol{m} \cdot \boldsymbol{\Omega}$. Contours are spaced linearly between 90 and 10% of the function maximum $\lambda_{\max}$. The *grey* domain corresponds to the region $\mathcal{V}_{\max} = \{(k_r, k_r) | \lambda(k_r, k_z) \geq \lambda_{\max}/2\}$. This is the region on which the integrations for the approximation of the small denominators will be performed as in Eqs. (5.57) and (5.68). One can finally note that here the maximum amplification lies along the line $k_z = k_z^1(k_r)$, i.e. along the line of the minimum quantised frequency k_z, see Fig. 5.4

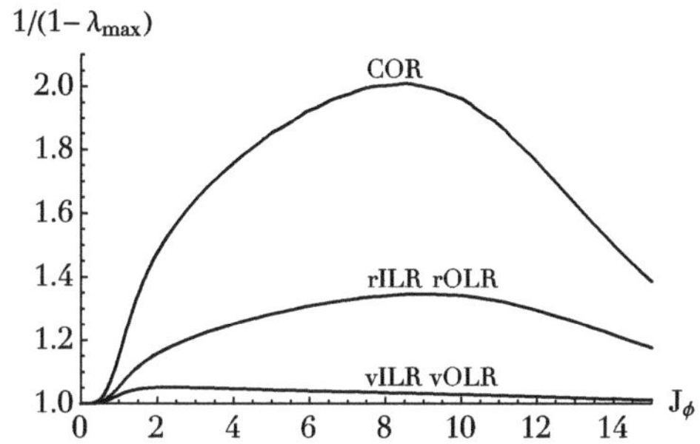

Fig. 5.15 Illustration of the dependence of the amplification factor $1/(1-\lambda_{\max}(\boldsymbol{m}, J_\phi))$, as given by Eq. (5.30), for various resonances $\boldsymbol{m}$ as a function of the position within the disc given by J_ϕ. One can note that the amplification associated with the corotation (COR) is always larger than the ones associated with the other resonances. As expected from the taper functions of Eq. (3.90), self-gravity is turned off in the inner and outer regions of the disc

5.7.2.1 Collisionless Forced Thickening

As a first to approach to understand the formation of the vertical ridge observed in Fig. 5.13, let us rely on the WKB limit of the collisionless diffusion equation obtained in Sect. 5.5. So12 considered an isolated disc, so that in order to use our collisionless formalism, one should assume some form for the perturbation power spectrum $\widehat{\mathcal{C}}[m_\phi, \omega, R_{\mathrm{g}}, k_r, k_z]$ that sources Eq. (5.56). Following the same approximation than the one considered in the razor-thin equation (3.93), let us assume that the source of perturbation comes from the system's internal Poisson shot noise due to the finite number of stars. In the galactic context, such perturbations could also mimic the perturbations by compact gas clouds within the disc (see Sect. 5.7.6). With such a Poisson shot noise, the intrinsic potential fluctuations vary like $\delta\psi^{\mathrm{e}} \propto \sqrt{\Sigma_{\mathrm{star}}}$. For simplicity, we only keep the dependence w.r.t. R_{g} and neglect any dependence w.r.t. ω, k_r and k_z in the autocorrelation $\widehat{\mathcal{C}}$ from Eq. (5.56). Up to a normalisation, let us therefore assume for our illustration purposes that the autocorrelation of the external perturbations takes the simple form

$$\widehat{\mathcal{C}}[m_\phi, \omega, R_{\mathrm{g}}, k_r, k_z] = \delta_{m_\phi}^2 \, \Sigma_{\mathrm{star}}(R_{\mathrm{g}}) \,. \tag{5.79}$$

As discussed in the end of Sect. 5.7.1, we restrict potential perturbations to the sole harmonic sector $m_\phi = 2$, and the same restriction applies to $\widehat{\mathcal{C}}$, hence the Kronecker symbol $\delta_{m_\phi}^2$. Of course, one should keep in mind that Poisson shot noise is not per se an external perturbation, as it is induced by the disc's constituents themselves. In order to account in a more rigorous and self-consistent way for these intrinsic finite$-N$ effects, one has to rely on the inhomogeneous Balescu-Lenard equation. This will be the focus of Sect. 5.7.2.2. In Eq. (5.79), having no dependence w.r.t. ω implies in particular that for a given location in the disc, all resonances undergo

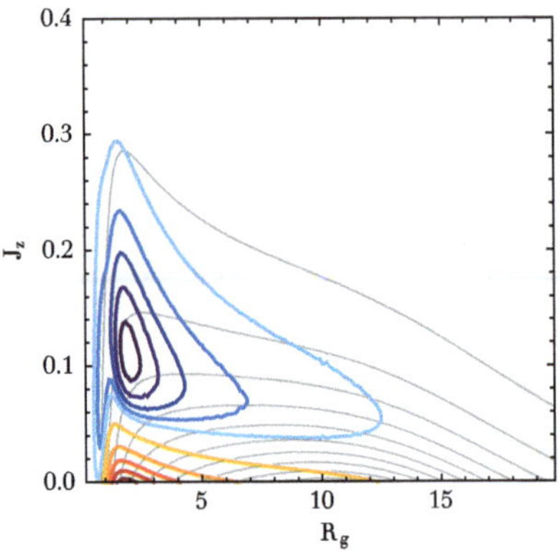

Fig. 5.16 Illustration of the initial contours of $\partial F_Z/\partial t\big|_{t=0}$ as predicted by the WKB collisionless diffusion equation from Sect. 5.5, when considering a secular forcing sourced by Poisson shot noise as in Eq. (5.79). *Red* contours, for which $\partial F_Z/\partial t\big|_{t=0} < 0$, are associated with regions from which the particles will be depleted and are spaced linearly between 90 and 10% of the function minimum. *Blue* contours, for which $\partial F_Z/\partial t\big|_{t=0} > 0$, are associated with regions where the number of orbits will increase during the diffusion and are spaced linearly between 90 and 10% of the function maximum. The background contours illustrate the initial contours of $F_Z(t=0)$ spaced linearly between 95 and 5% of the function maximum and computed for the quasi-isothermal DF from Eq. (5.5)

the same perturbations, even if they are not associated with the same local resonant frequencies $\omega = \boldsymbol{m}\cdot\boldsymbol{\Omega}$.

Relying on the previous estimation of the disc's amplification eigenvalues and on our assumption for the perturbation power spectrum, one can compute the collisionless diffusion coefficients from Eq. (5.57) and the associated collisionless diffusion flux $\boldsymbol{\mathcal{F}}_{\text{tot}}$ from Eq. (2.34). The initial time variations of F_Z from Eq. (5.76) can then be estimated. The initial contours of $\partial F_Z/\partial t\big|_{t=0}$ are illustrated in Fig. 5.16. In this figure, one qualitatively recovers the formation of a resonant ridge of increased vertical actions in the inner region of the disc, as observed in So12. This illustrates how the Poisson shot noise induced by the finite number of particles and approximated by Eq. (5.79), can indeed be the source of a secular disc thickening. This qualitative agreement between the numerical measurements from Fig. 5.13 and the collisionless WKB predictions from Fig. 5.16 is rewarding considering the various approximations required to obtain the thickened WKB limit of the collisionless diffusion equation.

Relying on the same collisionless approach, let us briefly investigate how the disc's gravitational susceptibility may impact its secular dynamics. To do so, let us consider the effect of varying the fraction of mass in the disc, by changing the value of the active fraction ξ (see Eq. (3.91)). The dependence of the system's collisionless response with ξ is illustrated in Fig. 5.17. As expected, as one increases the disc's self-gravity, the dressing of the perturbations gets stronger which subsequently hastens the orbital diffusion.

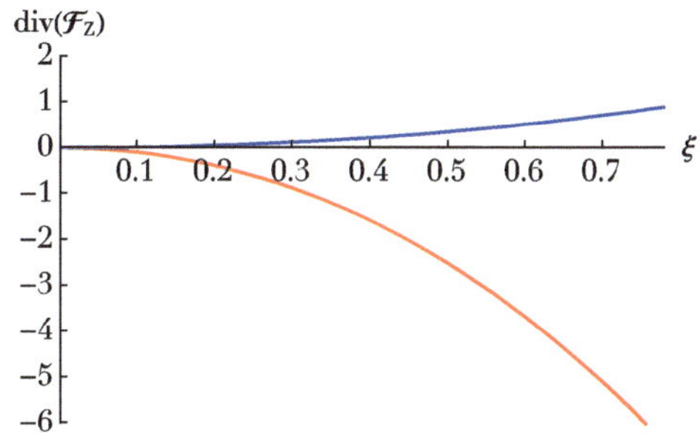

Fig. 5.17 Illustration of the dependence of the system's collisionless secular response to Poisson shot noise, as one varies the disc's active fraction ξ. The units for the vertical axis were rescaled to clarify the presentation. The *blue* line corresponds to the maximum value of div($\mathcal{F}_Z$), while the *red* line corresponds to the minimum value of div($\mathcal{F}_Z$). The larger the disc's active fraction ξ, the stronger the disc's gravitational susceptibility and therefore the faster the diffusion. For $\xi \gtrsim 0.8$, the disc becomes dynamically unstable. See Fig. 2 in Weinberg (1993) for a similar illustration of the crucial role of collective effects in accelerating orbital diffusion.

5.7.2.2 Collisional Thickening

In the previous section, we investigated how the WKB collisionless diffusion equation could explain the vertical ridge observed in Fig. 5.13. This essentially relied on treating the intrinsic Poisson shot noise as an external perturbation, via Eq. (5.79). In order to account in a self-consistent manner for these internal and self-induced perturbations, one should rely on the WKB Balescu-Lenard equation derived in Sect. 5.6. Thanks to the previous estimation of the disc's amplification eigenvalues, one can straightforwardly compute the disc's dressed susceptibility coefficients given by Eq. (5.68). One may then compute the collisional drift and diffusion coefficients from Eqs. (5.69) and (5.70), and the associated total collisional diffusion flux $\mathcal{F}_{\rm tot}$. As the particles' mass scales like $\mu = M_{\rm tot}/N$, let us rather consider the quantity $N\mathcal{F}_{\rm tot}$ which is independent of N. As defined in Eq. (5.77), one can then compute the collisional diffusion flux $|N\mathcal{F}_Z|$ in the (J_ϕ, J_z)−plane. We illustrate in Fig. 5.18 the initial contours of $|N\mathcal{F}_Z|(t = 0)$. In Fig. 5.18, one can note that the diffusion flux $N\mathcal{F}_Z$ is maximum in the disc's inner region. Let us note that both Figs. 5.16 and 5.18, which were obtained respectively in a collisionless or collisional approach, are in qualitative agreement and both predict an increase of the vertical actions in the inner regions as was observed in direct numerical simulations. The crude assumption for the Poisson shot noise in Eq. (5.79) used with the collisionless diffusion equation allowed us to mimic the results from the collisional Balescu-Lenard formalism, for which the spectral properties of the internal Poisson shot noise are self-consistently accounted for.

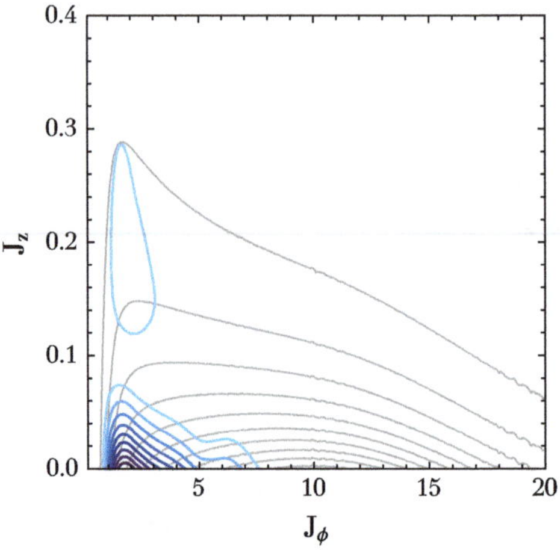

Fig. 5.18 Illustration of the norm of the collisional diffusion flux $|N\mathcal{F}_Z|(t\!=\!0)$ in the (J_ϕ, J_z)−plane, as predicted by the thickened WKB limit of the Balescu-Lenard equation. The *blue* contours are spaced linearly between 90 and 10% of the maximum norm. The background contours correspond to the initial contours of $F_Z(t\!=\!0)$ spaced linearly between 95 and 5% of the function maximum and computed for the quasi-isothermal DF from Eq. (5.5). One can clearly note the presence of an enhanced diffusion flux in the inner regions of the disc, compatible with the localised increase in vertical actions observed in Fig. 5.13

5.7.2.3 Vertical Kinetic Heating

In order to better assess the secular increase in vertical actions induced by finite−N effects, let us now consider the associated increase in the disc's vertical velocity dispersion. Indeed, in observations, disc thickening is best probed by considering the evolution of the vertical velocity dispersion $\varsigma_z^2(R_g, t)\!=\!\langle v_z^2\rangle(R_g, t)$, which can be computed as

$$\varsigma_z^2(R_g, t) = \frac{\displaystyle\int \mathrm{d}\boldsymbol{\theta}'\mathrm{d}\boldsymbol{J}'\,\delta_D(R_g - R_g')\,F(\boldsymbol{J}', t)\,(v_z')^2}{\displaystyle\int \mathrm{d}\boldsymbol{\theta}'\mathrm{d}\boldsymbol{J}'\,\delta_D(R_g - R_g')\,F(\boldsymbol{J}', t)} = \nu(R_g)\,\frac{\displaystyle\int \mathrm{d}J_r'\mathrm{d}J_z'\,F(R_g, J_r', J_z', t)\,J_z'}{\displaystyle\int \mathrm{d}J_r'\mathrm{d}J_z'\,F(R_g, J_r', J_z', t)}\,, \tag{5.80}$$

where to obtain the second equality, we relied on the the epicyclic approximation from Eq. (5.4) which gives $v_z^2\!=\!2J_z\nu\sin^2(\theta_z)$. For $t\!=\!0$, the system's DF is given by the quasi-isothermal DF from Eq. (5.5) and one recovers $\varsigma_z^2(R_g, t\!=\!0)\!=\!\sigma_z^2(R_g)$. The initial time derivative of ς_z^2 can also be computed. It reads

$$\left.\frac{\partial \varsigma_z^2}{\partial t}\right|_{t=0} = \nu\,\frac{\displaystyle\int \mathrm{d}J_r'\mathrm{d}J_z'\,J_z'\left.\frac{\partial F}{\partial t}\right|_{t=0} - \frac{\sigma_z^2}{\nu}\int \mathrm{d}J_r'\mathrm{d}J_z'\left.\frac{\partial F}{\partial t}\right|_{t=0}}{\displaystyle\int \mathrm{d}J_r'\mathrm{d}J_z'\,F(t\!=\!0)}\,, \tag{5.81}$$

where $\partial F/\partial t = \mathrm{div}(\boldsymbol{\mathcal{F}}_{\mathrm{tot}})$ is given by the diffusion equations (either collisionless or collisional). Finally, as $\partial \varsigma_z^2/\partial t = 2\varsigma_z \partial \varsigma_z/\partial t$, one can compute the expected increase in the vertical velocity dispersion ς_z resulting from the disc's intrinsic Poisson shot noise. This is illustrated in Fig. 5.19, where we represent $\varsigma_z(R_{\mathrm{g}}, t) \simeq \sigma_z(R_{\mathrm{g}}) + t\, \partial \varsigma_z/\partial t|_{t=0}$ as predicted by both collisionless and collisional formalisms. As was observed in Figs. 5.16 and 5.18, the vertical velocity dispersion also demonstrates that the most significant increase in vertical velocity dispersion occurs in the inner regions of the disc. This illustrates once again how the self-induced Poisson shot noise can indeed be the source of a disc thickening on secular timescales. Such a mechanism is qualitatively captured by both collisionless and collisional WKB diffusion equations. These qualitative agreements are all the more satisfactory in view of the various assumptions introduced throughout the derivations to obtain analytical and explicit expressions for both collisionless and collisional diffusion fluxes. Recall finally that a strength of the Balescu-Lenard formalism is that it is self-contained and does not involve any ad hoc fittings of the system's perturbations. Following the calculation of the induced collisional increase in ς_z presented in the right panel of Fig. 5.19, one may now compare the typical timescale of collisional diffusion predicted by the thick WKB Balescu-Lenard equation with the one observed in numerical simulations. This is the purpose of the next section.

5.7.3 *Diffusion Timescale*

Our previous estimations of the collisional diffusion flux $N\boldsymbol{\mathcal{F}}_{\mathrm{Z}}$ now allow us to compare the timescale of appearance of the ridge predicted by the thickened WKB Balescu-Lenard equation with the time during which So12's simulation was performed. Following Sect. 3.7.3, let us therefore compare the rescaled times of diffusion $\Delta\tau$, as defined in Eq. (3.95).

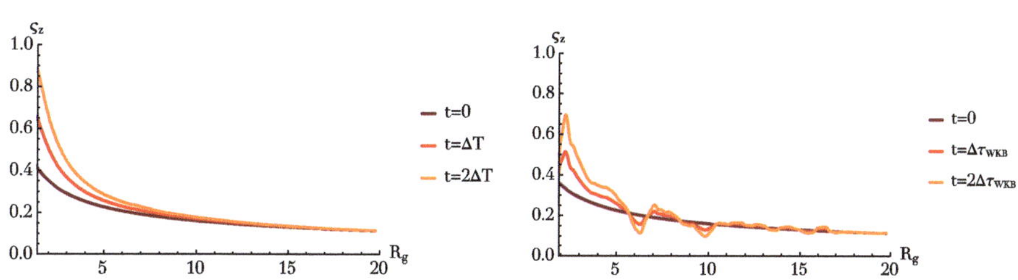

Fig. 5.19 Illustration of the increase in the vertical velocity dispersion ς_z induced by the intrinsic Poisson shot noise. *Left panel* Prediction for the collisionless WKB diffusion equation from Sect. 5.5, when approximating the Poisson shot noise with Eq. (5.79). For $t = 0$, one has $\varsigma_z(R_{\mathrm{g}}, t = 0) = \sigma_z(R_{\mathrm{g}})$, while for later times (here, ΔT is an arbitrary timestep), we relied on the approximation $\varsigma_z(R_{\mathrm{g}}, t) \simeq \sigma_z(R_{\mathrm{g}}) + t\, \partial \varsigma_z/\partial t|_{t=0}$ and on Eq. (5.81). *Right panel* Same as the left panel but for the thickened WKB limit of the collisional Balescu-Lenard equation derived in Sect. 5.6. Here, $\Delta\tau_{\mathrm{WKB}}$ is a timescale introduced in Sect. 5.7.3

The right panel of Fig. 5.13 was obtained in So12's simulation after a time $\Delta t_{\mathrm{So12}} = 3500$ in a simulation with $N = 2\times 10^5$ particles. As a consequence, the vertical ridge was observed in So12 after a rescaled time $\Delta\tau_{\mathrm{So12}} = \Delta t_{\mathrm{So12}}/N \simeq 2\times 10^{-2}$. When looking at the mean evolution of J_z in Fig. 5.13, one can note that during the rescaled time $\Delta\tau_{\mathrm{So12}}$, the mean vertical action in the inner region of the disc was approximately doubled. This time may then be compared with the typical time necessary to lead to a similar increase via the Balescu-Lenard equation. The epicyclic approximation from Sect. 5.2 immediately gives $v_z^2 = 2\nu J_z \sin^2(\theta_z)$, so that $\varsigma_z^2 = \nu\langle J_z\rangle$. Hence, doubling the mean vertical action $\langle J_z\rangle$ amounts to multiplying the vertical velocity dispersion ς_z by $\sqrt{2}$. The right panel of Fig. 5.19 gives us that such an increase of ς_z is reached after a rescaled time $\Delta\tau_{\mathrm{WKB}} \simeq 10^3$. Comparing the numerically measured rescaled time $\Delta\tau_{\mathrm{So12}}$ to the thick WKB Balescu-Lenard predictions, one therefore gets

$$\frac{\Delta\tau_{\mathrm{So12}}}{\Delta\tau_{\mathrm{WKB}}} \simeq 2\times 10^{-5}\,. \tag{5.82}$$

Note that the disagreement obtained here between the measured and the predicted timescales is even larger than what was obtained in Eq. (3.97) in the razor-thin case, when considering radial diffusion in razor-thin stellar discs. The initial timescale discrepancy from Eq. (3.97) was resolved in Eq. (4.31) by resorting to a global evaluation of the Balescu-Lenard diffusion flux. We subsequently showed in Sect. 4.3.3 that this discrepancy was caused by the incompleteness of the WKB basis, which cannot correctly capture swing amplification (illustrated in Fig. 3.14), the strong amplification of unwinding perturbations. The present thickened WKB formalism suffers from the same flaws, as illustrated in the timescale mismatch from Eq. (5.82). Even if the lack of any loosely wound contributions to the disc's susceptibility leads to such a significant mismatch, the diffusion features recovered in Figs. 5.18 and 5.19 illustrate however how the thickened WKB limit of the Balescu-Lenard equation still allows for an explicit qualitative description of the long-term evolution of discrete self-gravitating thick discs induced by their intrinsic Poisson shot noise.

5.7.4 Radial Migration

Let us now detail how the previous results are also in agreement with what was presented in Sect. 3.7.2 in the context of razor-thin discs. In order to study the diffusion in the (R_{g}, J_r)–plane, similarly to Eq. (5.75), let us define the function $F_{\mathrm{R}}(R_{\mathrm{g}}, J_r, t)$ as

$$\begin{aligned} F_{\mathrm{R}}(R_{\mathrm{g}}, J_z, t) &= \int \mathrm{d}\boldsymbol{\theta}'\mathrm{d}\boldsymbol{J}'\, \delta_{\mathrm{D}}(R_{\mathrm{g}} - R_{\mathrm{g}}')\, \delta_{\mathrm{D}}(J_r - J_r')\, F(\boldsymbol{J}', t) \\ &= (2\pi)^3 \frac{\mathrm{d}J_\phi}{\mathrm{d}R_{\mathrm{g}}} \int \mathrm{d}J_z'\, F(R_{\mathrm{g}}, J_r, J_z', t)\,. \end{aligned} \tag{5.83}$$

As in Eq. (5.76), the time derivative of F_{R} reads

$$\frac{\partial F_{\mathrm{R}}}{\partial t} = (2\pi)^3 \frac{\mathrm{d}J_\phi}{\mathrm{d}R_{\mathrm{g}}} \int \mathrm{d}J_z' \, \mathrm{div}(\boldsymbol{\mathcal{F}}_{\mathrm{tot}})(R_{\mathrm{g}}, J_r, J_z', t) \,. \tag{5.84}$$

Similarly to Eq. (5.77), the associated diffusion in the (J_ϕ, J_r)−plane is straightforwardly captured by the flux $\boldsymbol{\mathcal{F}}_{\mathrm{R}} = (\mathcal{F}_{\mathrm{R}}^\phi, \mathcal{F}_{\mathrm{R}}^r)$ with

$$\frac{\partial F_{\mathrm{R}}(J_\phi, J_r)}{\partial t} = \left(\frac{\partial}{\partial J_\phi}, \frac{\partial}{\partial J_r}\right) \cdot \boldsymbol{\mathcal{F}}_{\mathrm{R}} = \frac{\partial \mathcal{F}_{\mathrm{R}}^\phi}{\partial J_\phi} + \frac{\partial \mathcal{F}_{\mathrm{R}}^r}{\partial \mathcal{F}_{\mathrm{R}}^r} \,, \tag{5.85}$$

where the flux components $(\mathcal{F}_{\mathrm{R}}^\phi, \mathcal{F}_{\mathrm{R}}^r)$ read

$$\mathcal{F}_{\mathrm{R}}^\phi = (2\pi)^3 \int \mathrm{d}J_z' \, \mathcal{F}_{\mathrm{tot}}^\phi(J_\phi, J_r, J_z') \quad ; \quad \mathcal{F}_{\mathrm{R}}^r = (2\pi)^3 \int \mathrm{d}J_z' \, \mathcal{F}_{\mathrm{tot}}^r(J_\phi, J_r, J_z') \,. \tag{5.86}$$

As in Eq. (5.78), we introduced here the total diffusion flux $\boldsymbol{\mathcal{F}}_{\mathrm{tot}}$ in the (J_ϕ, J_r, J_z)−plane as $\boldsymbol{\mathcal{F}}_{\mathrm{tot}} = (\mathcal{F}_{\mathrm{tot}}^\phi, \mathcal{F}_{\mathrm{tot}}^r, \mathcal{F}_{\mathrm{tot}}^z)$. Because it is marginalised over J_z', the function F_{R} allows us to get rid of the vertical dependence of the diffusion. It mimics the razor-thin measurements presented in Sect. 3.7.2.

Relying on the shot noise perturbation from Eq. (5.79), Fig. 5.20 illustrates the initial contours of $\partial F_{\mathrm{R}}/\partial t|_{t=0}$ predicted by the thickened WKB limit of the collisionless diffusion equation. In Fig. 5.20, one predicts the formation in the (R_{g}, J_r)−plane of a narrow ridge of resonant orbits in the inner regions of the disc along the direction of the rILR. One therefore recovers the same feature as observed in the razor-thin Fig. 3.9.

Similarly, one can perform the same predictions by relying on the thickened WKB limit of the collisional Balescu-Lenard equation. This is illustrated in Fig. 5.21 where we represent the initial contours of $|N\boldsymbol{\mathcal{F}}_{\mathrm{R}}|(t=0)$. Even with the collisional approach, one also recovers the formation of an inner narrow ridge of radial diffusion aligned with the rILR resonance. This is in qualitative agreement with what was observed in the razor-thin Fig. 3.13. These results illustrate once again how the razor-thin and thickened WKB formalisms are indeed in agreement, as emphasised in Appendix 5.C.

5.7.5 *Thickening Induced by Bars*

In order to investigate other possible mechanisms of secular thickening, let us now consider a different source of perturbations driving the WKB collisionless diffusion coefficients from Eq. (5.57). Rather than focusing on the effect of Poisson shot noise, let us now study the secular effect of a stochastic series of central bars on the disc thickness. We therefore assume that the autocorrelation $\widehat{\mathcal{C}}$ of the external perturbations takes the form

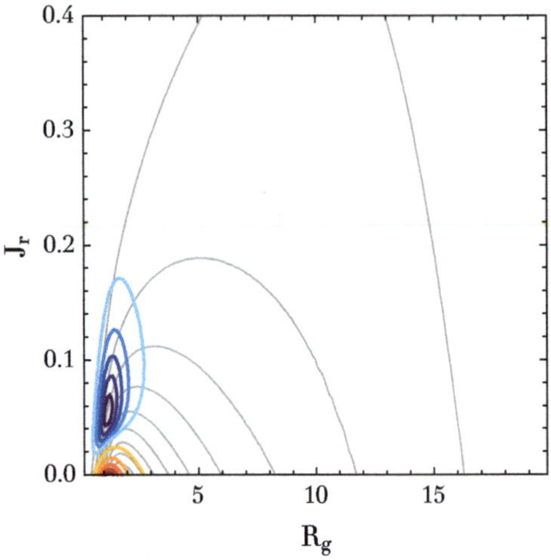

Fig. 5.20 Illustration of the initial contours of $\partial F_{\mathrm{R}}/\partial t|_{t=0}$ from Eq. (5.84), computed via the WKB collisionless diffusion equation from Sect. 5.5, when considering a secular forcing sourced by Poisson shot noise approximated with Eq. (5.79). We use the same conventions as in Fig. 5.16. The background contours illustrate the initial contours of $F_{\mathrm{R}}(t=0)$. They are spaced linearly between 95% and 5% of the function maximum and are computed for the quasi-isothermal DF from Eq. (5.5). This figure should be compared to Fig. 3.9 corresponding to the razor-thin case, for which we also recovered the formation of a narrow ridge of increased radial actions in the inner region of the disc along the direction of the rILR

$$\widehat{\mathcal{C}}[m_\phi, \omega, R_{\mathrm{g}}, k_r, k_z] = \delta_{m_\phi}^{m_{\mathrm{p}}} \, A_{\mathrm{b}}(R_{\mathrm{g}}) \, \exp\left[-\frac{(\omega - m_{\mathrm{p}}\Omega_{\mathrm{p}})^2}{2\sigma_{\mathrm{p}}^2} \right], \tag{5.87}$$

where $m_{\mathrm{p}}=2$ is the bar's pattern number, Ω_{p} its typical pattern speed, and $\sigma_{\mathrm{p}} \sim 1/T_{\mathrm{b}} \sim (1/\Omega_{\mathrm{p}})(\partial \Omega_{\mathrm{p}}/\partial t)$, with T_{b} the typical bar's lifetime, characterises the typical decay time of the bar frequency. The slower Ω_{p} evolves, the smaller σ_{p}, and therefore the narrower the frequency window in Eq. (5.87). In Eq. (5.87), we also introduced $A_{\mathrm{b}}(R_{\mathrm{g}})$ an amplitude factor varying with the position in the disc, which aims at describing the radial profile and extension of the bar. Let us underline that Eq. (5.87) is a rather crude description, as we neglected any dependence w.r.t. the frequencies k_r and k_z. We study perturbations imposed by various series of bars characterised by $\Omega_{\mathrm{p}} \in \{0.4, 0.25\}$ and $\sigma_{\mathrm{p}} \in \{0.03, 0.06\}$. Finally, in order to focus our interest on the intermediate regions of the disc, belonging neither to the bulge nor the bar, we consider $A_{\mathrm{b}}(R_{\mathrm{g}}) = H[R_{\mathrm{g}} - R_{\mathrm{cut}}]$, with $H[x]$ a Heaviside function, such that $H[x]=1$ for $x \geq 0$ and 0 otherwise, and $R_{\mathrm{cut}}=2.5$ is a truncation radius below which the bar is present. The initial contours of $\partial F_{\mathrm{Z}}/\partial t|_{t=0}$ for these various choices of bar perturbations are illustrated in Fig. 5.22.

The various panels presented in Fig. 5.22 first allow us to note how the additional dependence on ω present in Eq. (5.87) tends to localise the ridge of enhanced thickness. Let us also emphasise one important property of the collisionless diffusion

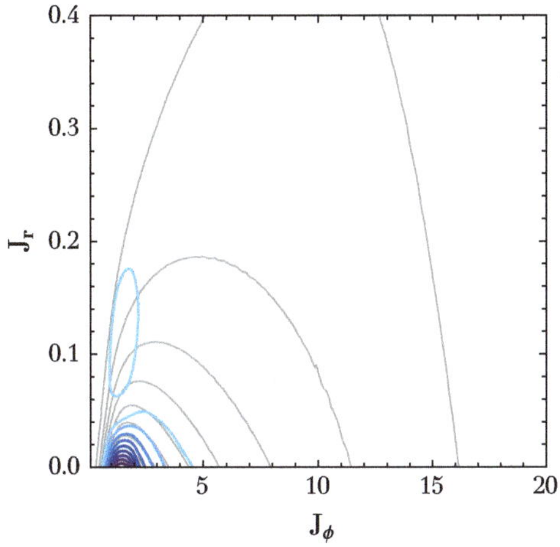

Fig. 5.21 Illustration of the norm of the collisional diffusion flux $|N\mathcal{F}_\mathrm{R}|(t=0)$ in the (J_ϕ, J_r)−plane, as predicted by the thickened WKB limit of the Balescu-Lenard equation derived in Sect. 5.6. The *blue* contours are spaced linearly between 90 and 10% of the maximum norm. The background contours correspond to the initial contours of $F_\mathrm{R}(t=0)$. They are spaced linearly between 95% and 5% of the function maximum and are computed for the quasi-isothermal DF from Eq. (5.5). One clearly notes the presence of an enhanced diffusion flux in the inner region of the disc towards larger radial actions. This figure should be compared to Fig. 3.13 corresponding to the razor-thin case, which also predicted the formation of a narrow ridge of enhanced radial action in the inner regions of the disc along the direction of the rILR

coefficients from Eq. (5.57), which is the fact that the orbital diffusion is strongly affected by the dynamical properties of the perturbing bars. Comparing the left-hand panels of Fig. 5.22 with the right-hand ones, one immediately recovers that the slower the bars, the further out the diffusion. As Ω_p decreases, the ridge move outwards, i.e. particles resonating with slower bars are located further out in the disc. Comparing the top panels of Fig. 5.22 with the bottom ones, one recovers that the more long-lived the bars, the narrower the diffusion features. As σ_p decreases, the different ridges get sharper and do not overlay anymore. If the pattern speeds of the bars decrease rapidly, the associated perturbations swipe a broader temporal frequency range, and therefore perturb a larger number of particles, hence the wider ridges. Finally, the position of the various ridges observed in Fig. 5.22 can be straightforwardly predicted thanks to Fig. 5.12, which illustrates the dependence of the various resonance frequencies $\omega = \boldsymbol{m}\cdot\boldsymbol{\Omega}$ as a function of the position in the disc. In order to allow for a resonant diffusion, one should match the frequency of the bars perturbation, $m_\mathrm{p}\Omega_\mathrm{p}$, with the local orbital frequency $\boldsymbol{m}\cdot\boldsymbol{\Omega}$. Different resonances, i.e. different resonance vectors $\boldsymbol{m}$, are then associated with different locations in the disc, as can be seen in Fig. 5.12. Because the previous shot noise perturbations from Eq. (5.79) and the bar ones from Eq. (5.87) do not have the same spectral structure, the diffusion features predicted in Figs. 5.16 and 5.22 are significantly different. This underlines the critical role played

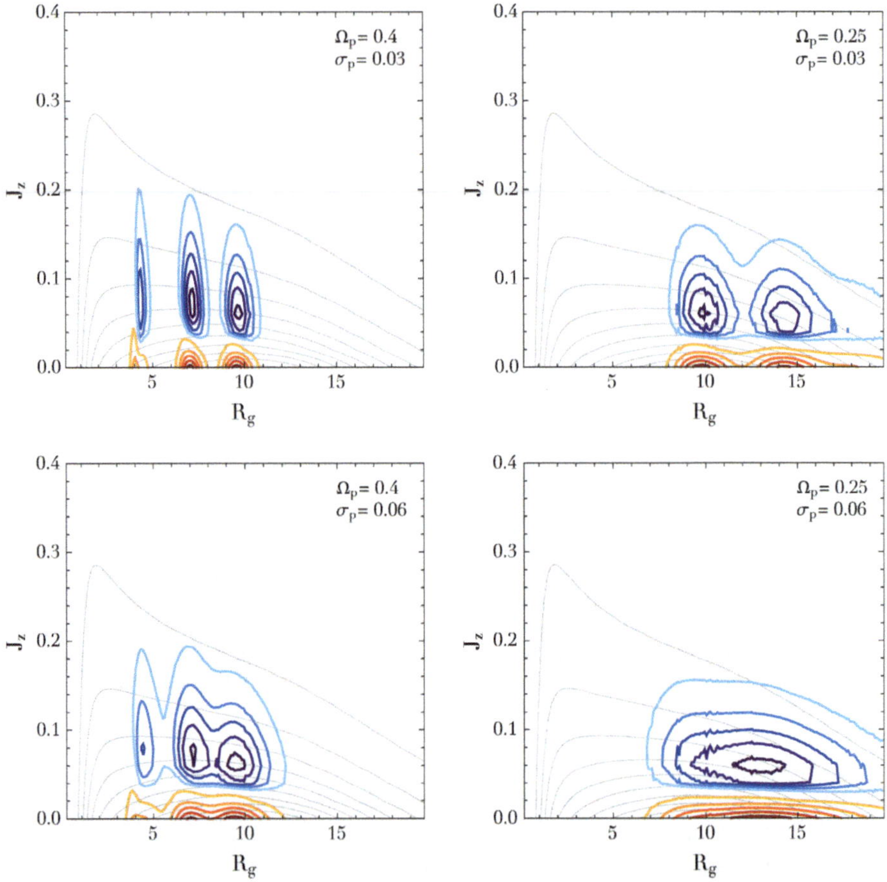

Fig. 5.22 Illustration of the initial contours of $\partial F_Z/\partial t|_{t=0}$ using the same conventions as in Fig. 5.16. Here, we consider a secular collisionless forcing by a series of bars, whose perturbations are approximated by Eq. (5.87), for various precession rates Ω_p and temporal decays σ_p. The diffusion in the disc's inner regions has been turned off by considering a perturbation amplitude $A_b(R_g) = H[R_g - R_{cut}]$ with $R_{cut} = 2.5$. One can predict the positions of the various resonance radii thanks to the behaviours of the various intrinsic frequencies $\omega = \boldsymbol{m} \cdot \boldsymbol{\Omega}$ illustrated in Fig. 5.12. *Top-left panel* $\Omega_p = 0.4$ and $\sigma_p = 0.03$, i.e. long-lived fast bars. *Top-right panel* $\Omega_p = 0.25$ and $\sigma_p = 0.03$, i.e. long-lived slow bars. *Bottom-left panel* $\Omega_p = 0.4$ and $\sigma_p = 0.06$, i.e. short-lived fast bars. *Bottom-right panel* $\Omega_p = 0.25$ and $\sigma_p = 0.06$, i.e. short-lived slow bars

by the perturbations' spectral characteristics in shaping the collisionless diffusion coefficients.

As seen in Fig. 5.22, the process of secular thickening induced by bar-like perturbations can have a very clear chemo-dynamical signature in the radial distribution of stars at a given age and velocity dispersion. The structure of a disc's stellar DF will be mainly shaped by two competing mechanisms: gas inflow will continuously regenerate a cold component of stars within a razor-thin disc, while potential fluctuations within the disc will trigger both radial and vertical migrations in regions which resonate with the perturbations. As a consequence, the distributions of stellar ages, metallicities, radial and vertical velocities will reflect the net effect of all these simultaneous processes. See the end of Chap. 3 for a brief discussion on how chemistry can be incorporated in these formalisms. These will be affected by the disc's underlying orbital structure, the spectral properties of the perturbations, the rate of star formation, the gas infall within the disc, etc.

5.7.6 GMCs Triggered Thickening

In a realistic thick galactic disc, one does not expect the self-induced diffusion of stars alone to drive the disc's thickening within a Hubble time, the number of stars being too large to lead to an efficient collisional heating. However, if one accounts for the joint evolution of the disc's giant molecular clouds (GMCs), one has to update the previous predictions for the collisional timescale of diffusion. Indeed, this second population of less numerous but more massive particles can significantly hasten the secular diffusion. So12 gives a possible scaling to physical units as

$$R_{\mathrm{i}} = 0.75\,\mathrm{kpc} \quad ; \quad \tau_0 = \frac{R_{\mathrm{i}}}{V_0} = 3.0\,\mathrm{Myr}\,. \tag{5.88}$$

For a typical Milky Way like galaxy, the number of stars scales like $N_{\mathrm{MW}} \simeq 10^{11}$. As a consequence, the rescaled time of collisional thickening $\Delta\tau_{\mathrm{So12}} \simeq 2\times 10^{-2}$ measured in So12's simulation becomes for a Milky Way like galaxy

$$\Delta t_{\mathrm{MW}} \simeq 6\times 10^{6}\,\mathrm{Gyr} \simeq 6\times 10^{5}\, t_{\mathrm{Hub.}}\,, \tag{5.89}$$

where we introduced the Hubble time $t_{\mathrm{Hub.}} \simeq 10\,\mathrm{Gyr}$. This estimate shows that the mechanism of self-induced collisional thickening discussed in Sect. 5.7.2.2 is not sufficiently efficient to be relevant per se for a Milky Way like galaxy. However, it has been suggested (e.g., Spitzer and Schwarzschild 1953; Wielen 1977; Lacey 1984; Binney and Lacey 1988; Jenkins 1992; Ida et al. 1993; Shiidsuka and Ida 1999; Hänninen and Flynn 2002; Aumer et al. 2016) that in stellar discs, the joint evolution of the stars and a population of forming and dissolving GMCs could be responsible for the disc's thickening as a result of local deflections. As already emphasised in Eq. (2.76), an important strength of the Balescu-Lenard formalism is that it also allows for the simultaneous description of the dynamics of multiple components. This multi-component equation accounts at the same time for transient spiral structures and non-local resonant encounters between these various components. Note that the resonant diffusion captured by the Balescu-Lenard equation is a different mechanism from the close encounters associated with the scattering mechanism from Spitzer and Schwarzschild (1953). We refer to Chavanis (2013) for a detailed discussion on the links between the self-consistent Balescu-Lenard equation and other kinetic theories based on two-body encounters. Let us now briefly discuss how the joint evolution of stars and GMCs could enable a thickening of stellar discs on a much shorter timescale.

Let us follow the notations from Sect. 2.3.6 where we presented the multi-component Balescu-Lenard equation. Let us assume that the disc contains a total mass $M_{\mathrm{tot}}^{\star}$ of $N_{\star}$ stars of individual mass $\mu_{\star}$ described by the DF $F^{\star}$. In addition, we suppose that the disc also contains a total mass $M_{\mathrm{tot}}^{\mathrm{G}}$ of N_{G} GMCs of individual mass μ_{G} described by the DF F^{G}. In order to simplify our presentation, we will assume that the stars and the GMCs are distributed according to a similar distribution (in reality, the GMCs are typically dynamically colder). Because of their respective normalisations, the DFs then satisfy

$$F^{\mathrm{G}} = \frac{M_{\mathrm{tot}}^{\mathrm{G}}}{M_{\mathrm{tot}}^{\star}} F^{\star} . \tag{5.90}$$

The total collisional drift and diffusion coefficients $A_{\boldsymbol{m}_1}^{\mathrm{tot}}$ and $D_{\boldsymbol{m}_1}^{\mathrm{tot}}$ from Eq. (2.82) may then be estimated as

$$A_{\boldsymbol{m}_1}^{\mathrm{tot}} = (1+\alpha_A)\, A_{\boldsymbol{m}_1}^{\star} \quad ; \quad D_{\boldsymbol{m}_1}^{\mathrm{tot}} = (1+\alpha_D)\, \mu_{\star}\, D_{\boldsymbol{m}_1}^{\star} , \tag{5.91}$$

where we introduced as $A_{\boldsymbol{m}_1}^{\star}$ and $D_{\boldsymbol{m}_1}^{\star}$ the drift and diffusion coefficients of the stars' population when considered alone. In Eq. (5.91), we also introduced the dimensionless quantities α_A and α_D as

$$\alpha_A = \frac{M_{\mathrm{tot}}^{\mathrm{G}}}{M_{\mathrm{tot}}^{\star}} = \frac{\mu_{\mathrm{G}}\, N_{\mathrm{G}}}{\mu_{\star}\, N_{\star}} \quad ; \quad \alpha_D = \frac{\mu_{\mathrm{G}} M_{\mathrm{tot}}^{\mathrm{G}}}{\mu_{\star} M_{\mathrm{tot}}^{\star}} = \left(\frac{\mu_{\mathrm{G}}}{\mu_{\star}}\right)^2 \frac{N_{\mathrm{G}}}{N_{\star}} . \tag{5.92}$$

When accounting simultaneously for the presence of stars and of GMCs, the multi-component Balescu-Lenard equation (2.81) gives us the evolution of the stars' DF $F^{\star}$ as

$$\frac{\partial F^{\star}}{\partial t} = \frac{\partial}{\partial \boldsymbol{J}_1} \cdot \left[\boldsymbol{m}_1 \mu_{\star} \left\{ (1+\alpha_A) A_{\boldsymbol{m}_1}^{\star} F^{\star} + (1+\alpha_D) D_{\boldsymbol{m}_1}^{\star} \boldsymbol{m}_1 \cdot \frac{\partial F^{\star}}{\partial \boldsymbol{J}_1} \right\} \right] , \tag{5.93}$$

where we did not write the dependence w.r.t. $\boldsymbol{J}_1$ to simplify the notations. In Eq. (5.93), the case without GMCs can be recovered by assuming $\alpha_A = \alpha_D = 0$. Murray (2011) gives the typical current properties of the Milky Way's GMCs as

$$\mu_{\mathrm{G}} \simeq 10^5\, M_{\odot} \quad ; \quad N_{\mathrm{G}} \simeq 10^4 \quad ; \quad M_{\mathrm{tot}}^{\mathrm{G}} \simeq 10^9\, M_{\odot} . \tag{5.94}$$

A more involved modelling of the GMC population should also account for the expected secular variability of this population, due to the exponential decay of the disc's star formation and the rapid disappearance of GMCs. For a Milky Way like galaxy with $N_{\star} \simeq 10^{11}$ and $\mu_{\star} \simeq 1 M_{\odot}$, Eq. (5.92) gives us

$$\alpha_A \simeq 10^{-2} \quad ; \quad \alpha_D \simeq 10^3 \,, \tag{5.95}$$

so that relying on the fact that $\alpha_A \ll 1$ and $\alpha_D \gg 1$, Eq. (5.93) becomes

$$\frac{\partial F^\star}{\partial t} = \frac{\partial}{\partial \boldsymbol{J}_1} \cdot \left[\boldsymbol{m}_1 \mu_\star \left\{ A^\star_{\boldsymbol{m}_1} F^\star + \alpha_D D^\star_{\boldsymbol{m}_1} \boldsymbol{m}_1 \cdot \frac{\partial F^\star}{\partial \boldsymbol{J}_1} \right\} \right] . \tag{5.96}$$

As a consequence, the joint evolution of the GMCs tends to boost the diffusion coefficients both in absolute terms as well as w.r.t. the drift ones. Because $\alpha_D \gg 1$, the GMCs act as a catalyst and can significantly hasten the diffusion of the stars and therefore the thickening of the stellar disc. Indeed, the multi-component Balescu-Lenard equation captures the effects of multiple resonant deflections of stars by GMCs, leading to a diffusion of the lighter stellar population towards larger J_z, while the GMCs sink in. Let us assume that this selective boost of the diffusion component w.r.t. the drift directly translates to the timescale of thickening. We therefore write

$$\Delta t_{\mathrm{G}+\star} = \frac{\Delta t_\star}{\alpha_D} \,, \tag{5.97}$$

where $\Delta t_\star$ corresponds to the timescale of spontaneous thickening when only stars are considered, while $\Delta t_{\mathrm{G}+\star}$ corresponds to the case where the joint evolution of the GMCs is also accounted for. As a consequence, the presence of the GMCs, which are less numerous but more massive than the stars, can therefore significantly impact how stars diffuse compared to the case where they diffuse alone. Note that these considerations are generic and independent from the thickened WKB approximation developed in this chapter. Applied to Eq. (5.89), the diffusion boost from Eq. (5.97) leads to

$$\Delta t_{\mathrm{MW+G}} \simeq 6 \times 10^2 \, t_{\mathrm{Hub.}} \,, \tag{5.98}$$

where $\Delta t_{\mathrm{MW+G}}$ corresponds to the timescale of self-induced thickening of a Milky Way like galaxy when the joint evolution of the GMCs is also accounted for. In Eq. (5.98), let us underline how the joint presence of the GMCs tends to significantly hasten the thickening of stellar discs induced by discrete resonant encounters. However, we note that despite this boost, such a spontaneous thickening remains too slow to be significant during the lifetime of a Milky Way like galaxy. This analysis would therefore tend to show that the self-induced mechanism of secular collisional thickening induced by finite$-N$ fluctuations, captured by the Balescu-Lenard equation and studied numerically in So12, even when boosted by the presence of the GMCs, is not sufficiently rapid to lead to a significant thickening of a Milky Way like galaxy on a Hubble time. Aumer et al. (2016) recently reached a similar conclusion by studying the quiescent growth of isolated discs in numerical simulations.

Inspired by this consideration on the role played by GMCs, let us now perform the same calculations in the case of razor-thin discs and update the timescale of collisional radial diffusion presented in Sect. 3.7.3. Following the results from Sellwood (2012), we showed that the ridge in the $(J_\phi, J_r)-$plane observed in Fig. 3.5 appeared after a time $\Delta t_{\mathrm{S12}}^{\mathrm{radial}} = 1500$ for $N = 5 \times 10^7$ particles. The associated rescaled time of diffusion is then given by $\Delta \tau_{\mathrm{S12}}^{\mathrm{radial}} = 3 \times 10^{-5}$. Thanks to the physical units from Eq. (5.88), for a Milky Way like galaxy and accounting only for the stellar component, the radial ridge would appear after a time $\Delta t_{\mathrm{MW}}^{\mathrm{radial}} = 10^3\, t_{\mathrm{Hub.}}$. Accounting for the GMCs diffusion acceleration obtained in Eq. (5.97) would hasten the radial diffusion so that in a Milky Way like galaxy, the radial ridge would appear on a timescale of the order of $\Delta t_{\mathrm{MW+G}}^{\mathrm{radial}} \simeq \Delta t_{\mathrm{MW}}^{\mathrm{radial}}/(10^3) \simeq t_{\mathrm{Hub.}}$ As a conclusion, while we had showed in Eq. (5.98), that the simultaneous presence of the GMCs was still not sufficient to allow for the appearance of a vertical ridge on the typical lifetime of a Milky Way like galaxy, such an accelerated self-induced mechanism appears as fast enough to induce a radial ridge in a Milky Way like galaxy's DF on a Hubble time. This could lead to a signature in the Milky Way's DF, soon probed by the GAIA spacecraft.

5.8 Conclusion

In this chapter, we presented applications of the two diffusion formalisms (collisionless and collisional) in the context of thickened stellar discs. Relying on the epicyclic approximation (Sect. 5.2) and the construction of a thickened WKB basis (Sects. 5.3 and 5.4), we derived the thick WKB limit of these two equations (Sects. 5.5 and 5.6), by assuming that only radially tightly wound transient spiral perturbations are sustained by the disc. We introduced in particular an ad hoc vertical cavity in order to solve Poisson's equation in a closed form. This yielded simple double quadratures for the collisionless diffusion coefficients in Eq. (5.56), as well for the collisional drift and diffusion coefficients in Eqs. (5.69) and (5.70). These simple expressions provided us with a straightforward tool to estimate the locations of maximum diffusion within a thick stellar disc. The use of an improved thick WKB approximation also allowed us to derive in Eq. (5.38) a new scale height dependent thickened Toomre's parameter.

We applied in Sect. 5.7 these two formalisms to a shot noise perturbed tepid stable thick disc. The estimated diffusion fluxes predict the formation in the inner region of the disc of a vertical ridge of resonant orbits towards larger actions, in qualitative agreement with the ridges observed in the direct $N-$body simulations from Solway et al. (2012). Let us note that these diffusion frameworks extend the findings of Binney and Lacey (1988) to the self-gravitating case, as here we treat in a coherent and self-consistent manner the collective dressing of the perturbations, the associated spiral response and the induced thickening. This is the appropriate approach to account self-consistently and simultaneously for churning, blurring (Schönrich and Binney 2009), and thickening. We noted a discrepancy in the diffusion timescale predicted by this formalism (Eq. (5.82)), which was interpreted as being due to the WKB approximation that does not account for loosely wound perturbations and their associated strong swing amplification.

These applications illustrated that potential fluctuations within the disc induce a vertical bending of a subset of resonant orbits, leading to an increase in the vertical velocity dispersion. This generically offers a mechanism allowing for stellar discs to thicken on secular timescales, driven by their own intrinsic Poisson shot noise, or by a set of dynamically dragged central bars, or catalysed by the joint evolution of GMCs. When considering the effects of GMCs (Sect. 5.7.6), we showed that such a self-induced thickening mechanism remains still too slow to lead to a significant secular thickening on cosmic times of a Milky Way like galaxy (see D'Onghia et al. (2013) and references therein for a discussion on the effects of GMCs on spiral activity). Determining which of these processes are the dominant ones in the secular thickening of stellar discs depends directly on the relative amplitudes of the various external and internal potential fluctuations which can source the diffusion. For example, the statistical properties of external perturbations can be quantified beforehand in numerical simulations. All these mechanisms should have clear signatures in the vertical metallicity gradients to be observed in detail by GAIA. This offers a promising way of weighing the relative importance of these processes.

Finally, we relied on various approximations, which we now recall. We enforced the epicyclic approximation as well as the plane parallel Schwarzschild approximation to build an integrable model of thickened stellar disc. To solve Poisson's equation vertically, the vertical edge of the disc was approximated with a sharp edge. The radial components were described within the WKB approximation, i.e. assumed to be radially tightly wound. When computing the disc's self-gravity, we neglected the vertical action gradients of the DF w.r.t. the radial ones, and also assumed that the orbits were closed on resonance. Finally, when implementing the dressed collisionless diffusion, we assumed some partially ad hoc external source of perturbations to describe the disc's internal shot noise or sequences of central decaying bars.

5.8.1 Future Works

Having exhibited in detail how one could compute the characteristics of the secular diffusion in thickened axisymmetric discs, one could now extend these approaches in various ways. One first side product of the thickened WKB approximation is the derivation in Eq. (5.38) of a new generalised thickened Q parameter. In order to assess the quality of this stability parameter, it would be of interest to investigate via numerical simulations how accurately such a parameter can predict the presence of local axisymmetric instabilities in thickened stellar discs. One difficulty with such a numerical investigation is the preparation of the disc's initial conditions, thanks to which one aims at setting up a disc initially as close as possible to an equilibrium.

A possible improvement of the present WKB approach would be to implement anharmonic corrections in the vertical oscillations to better account for the stiffness of the vertical potential. This would require to improve the thickened epicyclic approximation from Sect. 5.2. As was emphasised by the timescale comparison from Eq. (5.82), in order to correctly account for the system's self-gravity, one should eventually get rid of the WKB approximation, to capture the contributions associated with strongly amplified loosely wound perturbations. This was already a challenge in the case of razor-thin discs (see Chap. 4), and its implementation for thick discs would be all the more difficult, as one does not have explicit angle-action coordinates for thick discs beyond the epicyclic approximation. In order to construct such coordinates, one can rely on the torus machine to build perturbatively a mapping of action space from an integrable model to a non-integrable one via fits of generating functions (Kaasalainen and Binney 1994, b, 2016). Once these coordinates are constructed, one would then have to solve the exact field equations, construct an appropriate basis of potentials, and deal with the full response matrix. Should chaos become important in such systems, one could finally resort to the dual stochastic Langevin rewriting (see Appendix 6.C) to account for the associated chaotic diffusion. Integrating forward in time the Balescu-Lenard equation remains still a challenging numerical problem, in particular because of the self-consistency requirement. As the diffusion occurs, the system's orbital structure gets distorted, so that the system's drift and diffusion coefficients have to be recomputed to account for the system's new DF. One venue to perform such an integration in time is to resort to the associated Langevin rewriting, which describes the stochastic dynamics of one test star instead of the diffusion of the system's whole DF.

One could finally revisit the bar-like perturbation crudely approximated in Eq. (5.87) to account for more realistic perturbations. This would require to perform statistical measurements of bars formations and dissolutions in cosmological simulations. Similarly, one could investigate more generally the statistical properties of cosmic noise, following the steps of Aubert and Pichon (2007), and infer how efficiently such externally-driven perturbations can thicken stellar discs. Finally, inspired by Chap. 6, one could also investigate the secular thickening of debris protoplanetary of galactocentric discs in the quasi-Keplerian regime.

Appendix

A Antisymmetric Basis

In Sect. 5.3, we restricted ourselves to the construction of the symmetric thick WKB basis elements. One can proceed similarly for the antisymmetric ones. Assuming $\psi_z(-z) = -\psi_z(z)$, the ansatz from Eq. (5.12) immediately imposes $D = -A$ and $C = -B$, while the continuity conditions from Eq. (5.13) become

$$\begin{cases} A\,\mathrm{e}^{-k_r h} = 2\mathrm{i}B\sin(k_z h)\,, \\ k_r\,A\,\mathrm{e}^{-k_r h} = -2\mathrm{i}k_z B\cos(k_z h)\,. \end{cases} \tag{5.99}$$

Similarly to Eq. (5.14), we obtain the antisymmetric quantisation relation

$$\tan(k_z h) = -\frac{k_z}{k_r}\,, \tag{5.100}$$

which is illustrated in Fig. 5.4. One can also note that the antisymmetric elements also follow the typical step distance Δk_z obtained in Eq. (5.17). Following Eq. (5.18), the full expression of the antisymmetric elements is given by

$$\psi^{[k_\phi,k_r,R_0,n]}(R,\phi,z) = \mathcal{A}\,\psi_r^{[k_\phi,k_r,R_0]}(R,\phi) \begin{cases} \sin(k_z^n z) & \text{if } |z|\leq h\,, \\ \mathrm{e}^{krh}\sin(k_z^n h)\,\mathrm{e}^{-k_r|z|} & \text{if } z\geq h\,, \\ -\,\mathrm{e}^{k_r h}\sin(k_z^n h)\,\mathrm{e}^{-k_r|z|} & \text{if } |z|\leq h\,, \end{cases} \tag{5.101}$$

and

$$\rho^{[k_\phi,k_r,R_0,n]}(R,\phi,z) = -\frac{k_r^2+(k_z^n)^2}{4\pi G}\,\psi^{[k_\phi,k_r,R_0,n]}(R,\phi,z)\,\Theta\!\left[\frac{z}{h}\right]. \tag{5.102}$$

Similarly to Eq. (5.20), the amplitude of the antisymmetric elements is given by

$$\mathcal{A} = \sqrt{\frac{G}{R_0 h(k_r^2+(k_z^n)^2)}}\,\beta_n\,, \tag{5.103}$$

where, in analogy with Eq. (5.21), β_n is a numerical prefactor reading

$$\beta_n = \sqrt{\frac{2}{1-\sin(2k_z^n h)/(2k_z^n h)}}\,. \tag{5.104}$$

As illustrated in Fig. 5.4, let us note that the antisymmetric quantisation relation (5.100) imposes for the antisymmetric vertical frequency to satisfy $k_z^1 > \pi/(2h)$, and in this domain, one has $1.3 \leq \beta_n \leq 1.5$. Similarly to Eq. (5.22), the Fourier transformed antisymmetric basis elements are given by

$$\psi_{\boldsymbol{m}}^{[k_\phi,k_r,R_0,n]}(\boldsymbol{J}) = \delta_{m_\phi}^{k_\phi}\,\delta_{m_z}^{\mathrm{odd}}\,\mathcal{A}\,\mathrm{e}^{\mathrm{i}k_r R_\mathrm{g}}\,\mathrm{i}^{m_z-1-m_r}\,\mathcal{B}_{R_0}(R_\mathrm{g})\,\mathcal{J}_{m_r}\!\left[\sqrt{\tfrac{2J_r}{\kappa}}\,k_r\right]\mathcal{J}_{m_z}\!\left[\sqrt{\tfrac{2J_z}{\nu}}\,k_z^n\right]. \tag{5.105}$$

B A Diagonal Response Matrix

In this Appendix, let us detail why we may assume, as in Eq. (5.23), that the disc's response matrix is diagonal in the thickened WKB limit. Let us first note that because the symmetric (resp. antisymmetric) Fourier transformed basis elements from Eq. (5.22) (resp. (5.105)) involve a $\delta_{m_z}^{\mathrm{even}}$ (resp. $\delta_{m_z}^{\mathrm{odd}}$), the response matrix coefficients from Eq. (2.17) are equal to zero as soon as the two considered basis elements do not share the same vertical symmetry. We may therefore treat separately the symmetric and antisymmetric cases.

The thickened WKB basis elements introduced in Eq. (5.6) depend on four indices $[k_\phi, k_r, R_0, n]$. Following the same argument as in Sect. 3.4 for the razor-thin case, we may assume that the response matrix is diagonal w.r.t. the indices $[k_\phi, k_r, R_0]$. As a consequence, it then only remains to check whether or not for a given set $[k_\phi, k_r, R_0]$, the response matrix is diagonal w.r.t. the index k_z^n. The expression (5.24) of the symmetric diagonal basis elements is straightforward to generalise to the non-diagonal ones and gives

$$\widehat{\mathbf{M}}_{pq} = \frac{2\pi G \Sigma \alpha_p \alpha_q}{h\kappa^2 \sqrt{(1+(k_z^p/k_r)^2)(1+(k_z^q/k_r)^2)}} \sum_{\ell_z \text{even}} \exp\!\left[-\frac{(k_z^p)^2+(k_z^q)^2}{2\nu^2/\sigma_z^2}\right] \mathcal{I}_{\ell_z}\!\left[\frac{k_z^p k_z^q}{\nu^2/\sigma_z^2}\right] \times \frac{1}{(1-s_{\ell_z}^2)} \left\{ \mathcal{F}(s_{\ell_z}, \chi_r) - \ell_z \frac{\nu}{\sigma_z^2} \frac{\sigma_r^2}{\kappa} \mathcal{G}(s_{\ell_z}, \chi_r) \right\}. \tag{5.106}$$

As already underlined in Eq. (5.28), starting from Eq. (5.106), it is straightforward to obtain the expression of the associated antisymmetric non-diagonal coefficients thanks to the substitution $\alpha \to \beta$ and the restriction of the sum on ℓ_z to odd values. Because it is a symmetric matrix, showing that the response matrix is diagonal amounts to proving that for $p \neq q$, one has $\widehat{\mathbf{M}}_{pq} \ll \widehat{\mathbf{M}}_{pp}$. In order to perform such a comparison, let us focus in Eq. (5.106) on the quantities which depend on k_z^p and k_z^q. We introduce the quantity $K_{pq}^{(\ell_z)}$ as

$$K_{pq}^{(\ell_z)} = \frac{1}{\sqrt{(1+(k_z^p/k_r)^2)(1+(k_z^q/k_r)^2)}} \exp\!\left[-\frac{(k_z^p)^2+(k_z^q)^2}{2\nu^2/\sigma_z^2}\right] \mathcal{I}_{\ell_z}\!\left[\frac{k_z^p k_z^q}{\nu^2/\sigma_z^2}\right]. \tag{5.107}$$

One can note that the definition from Eq. (5.107) does not involve the prefactors α_p and α_q, as they are always of order unity. In addition, Eq. (5.107) does not involve the terms $\mathcal{F}(s_{\ell_z}, \chi_r)$, $\mathcal{G}(s_{\ell_z}, \chi_r)$, and $1/(1-s_{\ell_z})$ from Eq. (5.106), as they do not depend on the choices of k_z^p and k_z^q. Figure 5.23 illustrates the behaviours of the reduction functions $s_{\ell_z} \mapsto \mathcal{F}(s_{\ell_z}, \chi_r)$, $\mathcal{G}(s_{\ell_z}, \chi_r)$ defined in Eq. (5.27). We note in Fig. 5.23 that these functions are ill-defined when computed for integer values of s_{ℓ_z}. In order to regularise these diverging behaviours a small imaginary part is added to s_{ℓ_z}. While this procedure works for exactly integer values, this does not however prevent the divergences of $\mathcal{F}$ and $\mathcal{G}$ in the neighbourhood of integers. As illustrated in Fig. 5.23, in order to avoid these divergences, let us assume that the functions $\mathcal{F}$ and $\mathcal{G}$ can be approximated by the smooth functions

$$\mathcal{F}(s_{\ell_z}, \chi_r) \simeq f_r \quad ; \quad \mathcal{G}(s_{\ell_z}, \chi_r) \simeq -g_r s_{\ell_z}, \tag{5.108}$$

where f_r and g_r do not depend on s_{ℓ_z}. As already underlined in Eq. (5.29), when computing the collisionless diffusion coefficients from Eq. (2.32) or the dressed susceptibility coefficients from Eq. (2.50), the frequency ω should be considered at resonance so that $\omega = \boldsymbol{m} \cdot \boldsymbol{\Omega}$. Following Eq. (5.29), the value of s_{ℓ_z} is either an integer (for $\ell_z = m_z$) or far from one provided that ν/κ is of high rational order. This distance

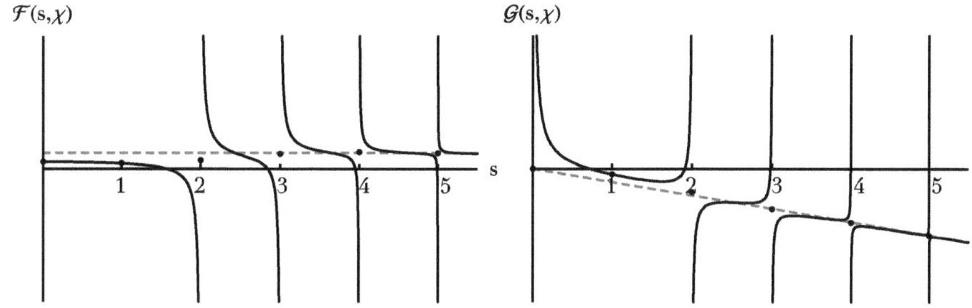

Fig. 5.23 Illustration for $\chi=1$ of the behaviour of the reduction functions $s\mapsto\mathcal{F}(s,\chi)$ (*left panel*) and $s\mapsto\mathcal{G}(s,\chi)$ (*right panel*) given by the *black* curves, along with their approximations from Eq. (5.108) given by the *grey* lines. One should note the divergences of these functions in the neighbourhood of integers. However, these functions are well defined when evaluated for integer values of s, provided one considers $\lim_{\eta\to 0}\mathrm{Re}[\mathcal{F}(n+\mathrm{i}\eta,\chi)]$ (similarly for $\mathcal{G}$), as illustrated with the *black* dots

from the exact resonance justifies the approximations from Eq. (5.108). Thanks to these approximations, the sum on ℓ_z in Eq. (5.106) may then be cut out according to the resulting powers of ℓ_z. In order to prove that for $p\neq q$, one has $\widehat{\mathbf{M}}_{pq}\ll\widehat{\mathbf{M}}_{pp}$, one should therefore prove that

$$S_\gamma(p,q)=\sum_{\ell_z}\frac{\ell_z^\gamma K_{pq}^{(\ell_z)}}{1-s_{\ell_z}^2}\ll S_\gamma(p,p)\,, \tag{5.109}$$

where the power index γ is such that $\gamma\in\{0,1,2\}$.

In order to further non-dimensionalise the problem, let us introduce the typical dynamical height of the disc, $d=\sigma_z/\nu$, as well as the dimensionless quantities

$$\ell_p=k_z^p d\quad;\quad \ell_q=k_z^q d\quad;\quad \ell_r=k_r d\,, \tag{5.110}$$

which allow us to rewrite Eq. (5.107) as

$$K_{pq}^{(\ell_z)}=\frac{\mathcal{I}_{\ell_z}\big[\ell_p\ell_q\big]\,\mathrm{e}^{-(\ell_p^2+\ell_q^2)/2}}{\sqrt{(1+(\ell_p/\ell_r)^2)(1+(\ell_q/\ell_r)^2)}}\,. \tag{5.111}$$

As was already illustrated in Fig. 5.4, let us recall that the fundamental symmetric frequency is significantly different from the other quantised frequencies (both symmetric and antisymmetric), as it is the only frequency inferior to $\pi/(2h)$. In order to emphasise this very specific property, in this Appendix only, let us renumber the vertical indices p, such that $p=0$ corresponds to the quantised fundamental symmetric mode, while $p\geq 1$ corresponds to the rest of the quantised frequencies, all superior to $\pi/(2h)$. With such a choice, the numbering of the antisymmetric elements only starts at $p=1$. Following Fig. 5.4, one has the inequalities

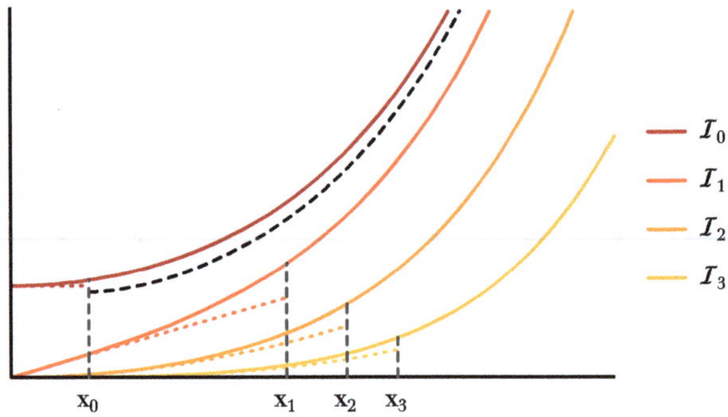

Fig. 5.24 Illustration of the asymptotic behaviours of the modified Bessel function $\mathcal{I}_n$ as given by Eq. (5.113). The full lines are the four first Bessel functions, along with their polynomial approximations in zero (*dashed* curves). The *black dashed* curve is their common exponential approximation. The transition between the two approximations is given by the quantity x_n

$$0 < \ell_0 < \frac{\pi}{2\sqrt{2}} \;\; ; \;\; \frac{(p-\frac{1}{2})\,\pi}{\sqrt{2}} < \ell_p < \frac{(p+\frac{1}{2})\,\pi}{\sqrt{2}} \;\; (\text{for}\; p \geq 1)\,, \tag{5.112}$$

where, following Eq. (5.74) for the Spitzer profile, we relied on the relation $h=2d$, with h the height of the WKB sharp cavity (see Fig. 5.3). Similarly, one has the relation $\ell_r = (k_r h)/\sqrt{2}$.

Let us note that the expression (5.111) of $K_{pq}^{(n)}$ involves a modified Bessel function $\mathcal{I}_n[\ell_p \ell_q]$ that needs as well to be approximated carefully. Equivalents in 0 and $+\infty$ of these Bessel functions are immediately given by

$$\mathcal{I}_n(x) \underset{0}{\sim} \frac{1}{n!}\left(\frac{x}{2}\right)^n \;\; ; \;\; \mathcal{I}_n(x) \underset{+\infty}{\sim} \frac{\mathrm{e}^x}{\sqrt{2\pi x}}\,. \tag{5.113}$$

As illustrated in Fig. 5.24, for a given value of n and x, one has to determine which approximation (polynomial or exponential) is relevant for $\mathcal{I}_n(x)$. Let us therefore define for each $n \geq 0$, the quantity x_n such that for $x \leq x_n$ (resp. $x \geq x_n$), one uses the asymptotic development from Eq. (5.113) in 0 (resp. $+\infty$). Because in the expression (5.111) the Bessel functions are only evaluated in $\ell_p \ell_q$, for p and q given, there exists an integer n_{pq} such that

$$\forall\, \ell_z < n_{pq}\,,\;\; \mathcal{I}_{\ell_z}\big[\ell_p \ell_q\big] \simeq \frac{\mathrm{e}^{\ell_p \ell_q}}{\sqrt{2\pi \ell_p \ell_q}} \;\; ; \;\; \forall\, \ell_z \geq n_{pq}\,,\;\; \mathcal{I}_{\ell_z}\big[\ell_p \ell_q\big] \simeq \frac{1}{\ell_z!}\left(\frac{\ell_p \ell_q}{2}\right)^{\ell_z}. \tag{5.114}$$

In Fig. 5.24, let us finally note that, except for $\ell_z = 0$, the exponential approximation of the Bessel function is significantly bigger than the actual value of $\mathcal{I}_{\ell_z}$. This does not impact the upcoming discussion, as, when proving $\widehat{\mathbf{M}}_{pq} \ll \widehat{\mathbf{M}}_{pp}$, the exponential approximation is applied for $\widehat{\mathbf{M}}_{pq}$ alone, or for $\widehat{\mathbf{M}}_{pq}$ and $\widehat{\mathbf{M}}_{pp}$ simultaneously with similar errors, so that the comparisons between the approximations also hold for the exact values. Following Eq. (5.109), a naive approach to compare the terms $S_\gamma(p,q)$ and $S_\gamma(p,p)$ would be to compare the sum on ℓ_z term by term, i.e. to prove that $K_{pq}^{(\ell_z)} \ll K_{pp}^{(\ell_z)}$ for all ℓ_z. However, this is not sufficient and one should therefore be more cautious. In Eq. (5.109), one cuts out the sum on ℓ_z appearing in $S_\gamma(p,q)$ between three different contributions, for which one can straightforwardly show:

- For the first terms, with $|\ell_z| < n_{pp}$ and $|\ell_z| < n_{pq}$:

$$K_{pq}^{(\ell_z)} \ll K_{pp}^{(1)} \,.$$

- For the intermediate terms, with $n_{pp} \leq |\ell_z| < n_{pq}$:

$$\sum_{n_{pp} \leq |\ell_z| < n_{pq}} \frac{\ell_z^\gamma K_{pq}^{(\ell_z)}}{1 - s_{\ell_z}^2} \ll K_{pp}^{(1)} \,.$$

- For the last terms, with $|\ell_z| \geq n_{pq}$:

$$\sum_{|\ell_z| \geq n_{pq}} \frac{\ell_z^\gamma K_{pq}^{(\ell_z)}}{1 - s_{\ell_z}^2} \ll K_{pp}^{(1)} \,.$$

This last relation holds whenever $k_r h \gtrsim 0.03$, but gets violated for $q=0$ in the limit of a razor-thin disc. The previous comparisons are straightforward to obtain thanks to the step distances between consecutive basis elements from Eq. (5.112) and the use of the approximations of the Bessel functions from Eq. (5.113). The combination of these relations shows that for $k_r h \gtrsim 0.03$, for all p and q, one has $\widehat{\mathbf{M}}_{pq} \ll \widehat{\mathbf{M}}_{pp}$. The same conclusion also holds for $k_r h \lesssim 0.03$, but only for $q \neq 0$. We therefore reached the following conclusions:

- The antisymmetric response matrix can always be assumed to be diagonal.
- For $k_r h \gtrsim 0.03$, the symmetric matrix response can be assumed to be diagonal
- For $k_r h \lesssim 0.03$, i.e. in the limit of a razor-thin disc, the symmetric response matrix takes the form of an arrowhead matrix.

As a last step of this Appendix, let us finally justify why for a sufficiently thin disc, for which the symmetric response matrix takes the form of an arrowhead matrix, the diagonal response matrix can still be assumed to be diagonal. In this limit, the symmetric response matrix takes the form

$$\widehat{\mathbf{M}} = \begin{pmatrix} \alpha & z_1 & \cdots & z_n \\ z_1 & d_1 & & \\ \vdots & & \ddots & \\ z_n & & & d_n \end{pmatrix} , \tag{5.115}$$

where thanks to the previous calculations, one has the comparison relations $\alpha \gg z_i$ and $z_i \gg d_i$. Let us assume that $\forall i$, $z_i \neq 0$ and that $\forall i \neq j$, $d_i \neq d_j$. Following O'Leary

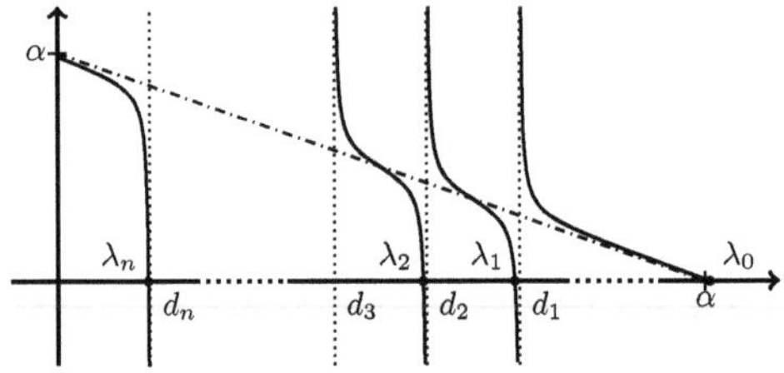

Fig. 5.25 Illustration of the behaviour of the function $\lambda \mapsto f_{\widehat{\mathbf{M}}}(\lambda)$, whose roots are the eigenvalues of the arrowhead response matrix from Eq. (5.115)

and Stewart (1990), it can be shown that the eigenvalues $(\lambda_i)_{0 \leq i \leq n}$ of the arrowhead matrix from Eq. (5.115) are the $(n+1)$ solutions of the equation

$$f_{\widehat{\mathbf{M}}}(\lambda) = \alpha - \lambda - \sum_{i=1}^{n} \frac{z_i^2}{d_i - \lambda} = 0\,. \tag{5.116}$$

In addition, provided that the d_i are in descending order, these eigenvalues are interlaced so that

$$\lambda_0 > d_1 > \lambda_1 > \ldots > d_n > \lambda_n\,. \tag{5.117}$$

Finally, the eigenvectors $\boldsymbol{x}_i$ associated with the eigenvalue λ_i are proportional to

$$\boldsymbol{x}_i = \left(1\,;\ \frac{z_1}{\lambda_i - d_1}\,;\ \ldots;\ \frac{z_j}{\lambda_i - d_j}\,;\ \ldots;\ \frac{z_n}{\lambda_i - d_n}\right). \tag{5.118}$$

Accounting for the comparison relations $\alpha \gg z_i$ and $z_i \gg d_i$, Fig. 5.25 illustrates the behaviour of the function $\lambda \mapsto f_{\widehat{\mathbf{M}}}(\lambda)$ from Eq. (5.116).

In order to justify why the arrowhead response matrix from Eq. (5.115) may be considered as diagonal, one has to justify that, despite its first line and column, the matrix eigenvalues remain close to the matrix coefficients, so that

$$\lambda_0 \simeq \alpha \quad \text{and} \quad \lambda_i \simeq d_i \ (\text{for}\, i \geq 1)\,. \tag{5.119}$$

In addition, one must also ensure that the associated eigenvectors $\boldsymbol{x}_i$ remain close to the natural basis elements so that

$$\boldsymbol{x}_i \simeq (0\,;\ \ldots;\ 1\,;\ 0\,;\ \ldots)\,, \tag{5.120}$$

where the non-zero index is at the ith position. As illustrated in Fig. 5.25, the determination of the eigenvalues λ_i requires to solve Eq. (5.116), which may be rewritten as

$$1 - \frac{\lambda_i}{\alpha} - \sum_{i=1}^{n} \frac{(z_i/\alpha)^2}{(d_i/\alpha) - (\lambda_i/\alpha)} = 0\,. \tag{5.121}$$

Because we have $(z_i/\alpha) \ll 1$, in order for Eq. (5.121) to be satisfied, one must necessarily either have $\lambda_i/\alpha \simeq 1$ or $((d_i/\alpha)-(\lambda_i/\alpha)) \ll 1$. It then follows immediately that $\lambda_0 \simeq \alpha$ and $\lambda_i \simeq d_i$. Equation (5.119) therefore holds and the matrix eigenvalues λ_i remain close to the matrix diagonal coefficients $(\alpha, d_1, ..., d_n)$. The eigenvectors $\boldsymbol{x}_i$ from Eq. (5.118) may then be rewritten as

$$\boldsymbol{x}_i = \left(1\,;\, \frac{(z_1/\alpha)^2}{(\lambda_i/\alpha)-(d_1/\alpha)} \frac{1}{(z_1/\alpha)}\,;\, ...\,;\, \frac{(z_j/\alpha)^2}{(\lambda_i/\alpha)-(d_j/\alpha)} \frac{1}{(z_j/\alpha)}\,;\, ...\right). \tag{5.122}$$

Let us consider the first eigenvector associated with $i=0$. Following Eq. (5.119), one has $\lambda_0 \simeq \alpha$, so that, because $d_j \ll \alpha$, the generic term from Eq. (5.122) becomes

$$\frac{(z_j/\alpha)^2}{(\lambda_0/\alpha)-(d_j/\alpha)} \frac{1}{(z_j/\alpha)} \simeq \frac{(z_j/\alpha)}{1} \ll 1\,, \tag{5.123}$$

where we relied on the fact that $z_j \ll \alpha$. As consequence, for $i=0$ in Eq. (5.122), all the terms except the first one are negligible in front of 1, and one gets $\boldsymbol{x}_0 \simeq (1;\, 0;\, ...;\, 0)$. Similarly, in Eq. (5.122), one can consider the case $i \neq 0$, for which the ith term of Eq. (5.122) takes the form

$$\frac{(z_i/\alpha)^2}{(\lambda_i/\alpha)-(d_i/\alpha)} \frac{1}{(z_i/\alpha)} \simeq \frac{1}{(z_i/\alpha)} \gg 1\,, \tag{5.124}$$

where we relied on the same argument as in Eq. (5.121). It states that for $i \neq 0$, there is only one dominant term in the sum from Eq. (5.121), given by $\frac{(z_i/\alpha)^2}{(d_i/\alpha)-(\lambda_i/\alpha)} \simeq 1$. As a consequence, for $i \neq 0$, the eigenvector $\boldsymbol{x}_i$ from Eq. (5.122) is dominated by its ith coefficient and the eigenvector may therefore be assumed to be proportional to $(0;\, ...;\, 1;\, 0;\, ...)$, where the non-zero index is at the ith position. We may therefore assume that the response matrix eigenvectors remain close to the natural basis elements. As a conclusion, even in the limit of a razor-thin disc, the symmetric arrowhead response matrix from Eq. (5.115) may still be assumed to be diagonal. We therefore justified why one may limit oneself to the diagonal coefficients of the response matrix, as in Eq. (5.23). The thickened WKB basis elements therefore allowed us to diagonalise the disc's response matrix. This is a crucial step in the explicit calculations of the collisionless and collisional diffusion fluxes as shown in Sects. 5.5 and 5.6.

C From Thick to Thin

In this Appendix, let us detail how one can recover all the razor-thin expressions obtained in Chap. 3, starting from the thickened WKB basis.

C.1 The Collisionless Case

Let us first consider the case of the collisionless diffusion presented in Sect. 5.5 and show one now may compute the collisionless diffusion coefficients when the disc is too thin to rely on the continuous expression from Eq. (5.50). We will show that this second approach is fully consistent with the one used in Eq. (5.50). We will also show how one can recover the razor-thin expressions previously obtained in Sect. 3.5.

We noted in Eq. (5.49) that in order to use Riemann sum formula w.r.t. the index k_z^p, one should ensure that the typical step distance $\Delta k_z \simeq \pi / h$ from Eq. (5.17) remains sufficiently small compared to the scale on which the function $k_z \mapsto g_s(k_z)$ varies. In the limit of a thinner disc, one has $h \to 0$, so that $\Delta k_z \to +\infty$. As a consequence, the continuous approximation cannot be used anymore, and one should keep the discrete sum on the quantised k_z^p in Eq. (5.49). Of course, it is also within this limit of a thinner disc, that one can recover the razor-thin results from Sect. 3.5.

Starting from Eq. (5.49), the expression (5.52) of the symmetric collisionless diffusion coefficients becomes

$$\begin{aligned} D_{\boldsymbol{m}}^{\mathrm{sym}}(\boldsymbol{J}) = \delta_{m_z}^{\mathrm{even}} \frac{1}{(2\pi)^2} \sum_{n_p, n_q} \int \mathrm{d}k_r^p \, \mathcal{J}_{m_r}\!\left[\sqrt{\tfrac{2J_r}{\kappa}}\, k_r^p\right] \mathcal{J}_{m_z}\!\left[\sqrt{\tfrac{2J_z}{\nu}}\, k_z^{n_p}(k_r^p)\right] \frac{\alpha_p^2}{1-\lambda_p} \\ \times \int \mathrm{d}k_r^q \, \mathcal{J}_{m_r}\!\left[\sqrt{\tfrac{2J_r}{\kappa}}\, k_r^q\right] \mathcal{J}_{m_z}\!\left[\sqrt{\tfrac{2J_z}{\nu}}\, k_z^{n_q}(k_r^q)\right] \frac{\alpha_q^2}{1-\lambda_q} \, \widehat{C}_{\delta\psi^{\mathrm{e}}}[m_\phi, \boldsymbol{m}\cdot\boldsymbol{\Omega}, R_{\mathrm{g}}, k_r^p, k_r^q, k_z^{n_p}(k_r^p), k_z^{n_q}(k_r^q)] \, , \end{aligned} \tag{5.125}$$

where the perturbation autocorrelation, $\widehat{C}_{\delta\psi^{\mathrm{e}}}$, was introduced in Eq. (5.51). Let us recall that the antisymmetric analog of Eq. (5.125) is straightforward to obtain thanks to the substitutions $\alpha_p \to \beta_p$ and $\delta_{m_z}^{\mathrm{even}} \to \delta_{m_z}^{\mathrm{odd}}$. For the antisymmetric case, as already noted in Eq. (3.73), one should pay attention to the fact that the perturbation autocorrelation involves the odd-restricted vertical Fourier transform of the potential perturbations. As in Eq. (5.55), the next step of the calculation is to diagonalise the perturbation autocorrelation, where one should pay attention to the fact that $k_z = k_z(k_r, n)$ is no longer a free variable but should be seen as a function of the considered k_r and n. Following Appendix D in Fouvry et al. (2017), Eq. (5.55) becomes here

$$\left\langle \widehat{\delta\psi^{\mathrm{e}}}_{m_\phi, k_r^1, k_z^{n_1}}[R_{\mathrm{g}}, \omega_1] \, \widehat{\delta\psi^{\mathrm{e}}}^{*}_{m_\phi, k_r^2, k_z^{n_2}}[R_{\mathrm{g}}, \omega_2] \right\rangle = 2\pi h \delta_{\mathrm{D}}(\omega_1 - \omega_2) \, \delta_{\mathrm{D}}(k_r^1 - k_r^2) \, \delta_{n_1}^{n_2} \, \widehat{C}[m_\phi, \omega_1, R_{\mathrm{g}}, k_r^1, k_z^{n_1}] \, , \tag{5.126}$$

where the diagonalisation w.r.t. the vertical dependence is captured by the Kronecker symbol $\delta_{n_1}^{n_2}$. This diagonalised autocorrelation allows us to rewrite the diffusion coefficients from Eq. (5.125) as

$$D_{\boldsymbol{m}}^{\mathrm{sym}}(\boldsymbol{J}) = \delta_{m_z}^{\mathrm{even}} \frac{1}{4h} \sum_{n_p} \int \mathrm{d}k_r^p \, \mathcal{J}_{m_r}^2\!\left[\sqrt{\tfrac{2J_r}{\kappa}} k_r^p\right] \mathcal{J}_{m_z}^2\!\left[\sqrt{\tfrac{2J_z}{\nu}} k_z^{n_p}(k_r^p)\right] \left[\frac{\alpha_p^2}{1-\lambda_p}\right]^2 \widehat{\mathcal{C}}[m_\phi, \boldsymbol{m}\cdot\boldsymbol{\Omega}, R_{\mathrm{g}}, k_r^p, k_z^{n_p}(k_r^p)] \,. \tag{5.127}$$

Equation (5.127) is the direct discrete equivalent of Eq. (5.56), and both expressions are in full agreement. Indeed, starting from Eq. (5.127), the continuous expression w.r.t. k_z^p can immediately be recovered by using Riemann sum formula with the step distance $\Delta k_z \simeq \pi/h$ from Eq. (5.17). Similarly to Eq. (5.57), one can also simplify Eq. (5.127) thanks to the approximation of the small denominators, which gives here

$$D_{\boldsymbol{m}}^{\mathrm{sym}}(\boldsymbol{J}) = \delta_{m_z}^{\mathrm{even}} \frac{1}{4h} \sum_{n_p} \Delta k_r^{n_p} \, \mathcal{J}_{m_r}^2\!\left[\sqrt{\tfrac{2J_r}{\kappa}} k_{r,n_p}^{\mathrm{max}}\right] \mathcal{J}_{m_z}^2\!\left[\sqrt{\tfrac{2J_z}{\nu}} k_{z,n_p}^{\mathrm{max}}\right] \left[\frac{(\alpha_{n_p}^{\mathrm{max}})^2}{1-\lambda_{n_p}^{\mathrm{max}}}\right]^2 \widehat{\mathcal{C}}[m_\phi, \boldsymbol{m}\cdot\boldsymbol{\Omega}, R_{\mathrm{g}}, k_{r,n_p}^{\mathrm{max}}, k_{z,n_p}^{\mathrm{max}}] \,. \tag{5.128}$$

In Eq. (5.128), for a given value of the index n_p, we consider the behaviour of the function $k_r^p \mapsto \lambda(k_r^p, k_z^{n_p}(k_r^p))$, and assume that it reaches a maximum value $\lambda_{n_p}^{\mathrm{max}}$ for $k_r = k_{r,n_p}^{\mathrm{max}}$ on a domain of typical extension $\Delta k_r^{n_p}$. In Eq. (5.128), we also used the shortening notation $k_{z,n_p}^{\mathrm{max}} = k_z^{n_p}(k_{r,n_p}^{\mathrm{max}})$. The antisymmetric analogs of Eqs. (5.127) and (5.128) are straightforward to obtain by considering the antisymmetric quantised frequencies k_z from Eq. (5.100) and performing the substitution $\alpha_p \to \beta_p$. As already emphasised in Eq. (5.56), one should pay attention to the fact that in these antisymmetric analogs, $\widehat{\mathcal{C}}$ involves an even-restricted vertical Fourier transform of the autocorrelation, despite the fact that one is interested in antisymmetric diffusion coefficients.

Starting from the discrete expression of the diffusion coefficients obtained in Eq. (5.127), let us now illustrate how one can recover the razor-thin WKB diffusion coefficients from Sect. 3.5 by considering the limit of a thinner disc. As already noted in Fig. 5.4, let us recall that except for the fundamental symmetric frequency $k_{z,\mathrm{s}}^1$, one always has $k_z^n > \pi/(2h)$. As a consequence, in the infinitely thin limit, for which $h \to 0$, one has $k_z^n \to +\infty$, except for $k_{z,\mathrm{s}}^1$. Let us also recall that in Eq. (5.127), the dependence of $\widehat{\mathcal{C}}[k_z^p]$ takes the form

$$\widehat{\mathcal{C}}[k_z^p] = \int_{-2h}^{2h} \mathrm{d}v \, \widehat{\mathcal{C}}[v] \, \cos[k_z^p v] \,. \tag{5.129}$$

One therefore gets the majoration $|\widehat{\mathcal{C}}[k_z^p]| \leq 4h\, \widehat{\mathcal{C}}_{\mathrm{max}}$, which, in the razor-thin limit, cancels out the prefactor $1/(4h)$ present in Eq. (5.127). Because $\forall n \geq 0$, $\lim_{x\to+\infty} \mathcal{J}_n(x) = 0$, it immediately follows from Eq. (5.127) that

$$\lim_{\mathrm{thin}} D_{\boldsymbol{m}}^{\mathrm{anti}}(\boldsymbol{J}) = 0 \,. \tag{5.130}$$

In addition, Eq. (5.127) also implies that for symmetric diffusion coefficients, the sum on n_p may be limited to the only fundamental term $n_p = 1$. Equation (5.16) gives us that in the razor-thin limit, one has $k_{z,\mathrm{s}}^1 \sim \sqrt{k_r/h}$. Equation (5.127) therefore also implies that for $m_z \neq 0$, one has $\lim_{\mathrm{thin}} D_{\boldsymbol{m}}^{\mathrm{sym}} = 0$. Therefore, in the infinitely thin limit, only the symmetric diffusion coefficients for $m_z=0$ does not vanish. In addition, from Eq. (5.127), it is also straightforward to obtain that in order to have a non-vanishing symmetric diffusion coefficient, one should also restrict oneself to $J_z=0$. In the razor-thin limit for $m_z=0$ and $J_z=0$, one can therefore write

$$\lim_{\text{thin}} D_{\boldsymbol{m}}^{\text{sym}}(\boldsymbol{J}) = \lim_{\text{thin}} \frac{1}{4h}\int \mathrm{d}k_r^p \, \mathcal{J}_{m_r}^2\!\left[\sqrt{\tfrac{2J_r}{\kappa}}k_r^p\right] \left[\frac{\alpha_1^2}{1-\lambda_p}\right]^2 \widehat{\mathcal{C}}[m_\phi, \boldsymbol{m}\cdot\boldsymbol{\Omega}, R_{\mathrm{g}}, k_r^p, k_{z,\mathrm{s}}^1]\,. \tag{5.131}$$

The definition of the prefactor α_p in Eq. (5.21) immediately gives us $\lim_{\text{thin}} \alpha_1=1$. In addition, we also obtained in Eq. (5.34) that $\lim_{\text{thin}} \lambda_p=\lambda_p^{\text{thin}}$. The last step of the present calculation is to study, in the razor-thin limit, the behaviour of the term $\widehat{\mathcal{C}}[k_{z,\mathrm{s}}^1]$ from Eq. (5.129). Equation (5.129) takes the form of an integral of length $4h$ of a function oscillating at the frequency $k_{z,\mathrm{s}}^1 \simeq \sqrt{k_r/h}$. In this interval, the number of oscillations of the fluctuating term is of order $k_{z,\mathrm{s}}^1 h \sim \sqrt{k_r h}$, so that in the razor-thin limit the number of oscillations of the function $v \mapsto \cos[k_{z,\mathrm{s}}^1 v]$ tends to 0. In the razor-thin limit, Eq. (5.129) then becomes $\lim_{\text{thin}} \widehat{\mathcal{C}}[k_{z,\mathrm{s}}^1]=4h\widehat{\mathcal{C}}[v=0]$. Using this relation in Eq. (5.131), one finally gets

$$\lim_{\text{thin}} D_{\boldsymbol{m}}^{\text{sym}}(\boldsymbol{J}) = \int \mathrm{d}k_r^p \, \mathcal{J}_{m_r}^2\!\left[\sqrt{\tfrac{2J_r}{\kappa}}k_r^p\right] \left[\frac{1}{1-\lambda_{\text{thin}}}\right]^2 \widehat{\mathcal{C}}_{\text{thin}}[m_\phi, \boldsymbol{m}\cdot\boldsymbol{\Omega}, R_{\mathrm{g}}, k_r^p]\,, \tag{5.132}$$

where $\widehat{\mathcal{C}}_{\text{thin}}[m_\phi, \boldsymbol{m}\cdot\boldsymbol{\Omega}, R_{\mathrm{g}}, k_r^p]$ stands for the local razor-thin power spectrum of the external perturbations in the equatorial plane as defined in Eq. (3.67) in the razor-thin case. In Eq. (5.132), we fully recovered the razor-thin result previously obtained in Eq. (3.68).

C.2 The Collisional Case

Let us now follow the same approach for the collisional diffusion. We will especially show how one should estimate the system's susceptibility coefficients in the case where the disc is too thin to rely on the continuous expression from Eq. (5.67), and how this approach allows for the recovery of the razor-thin results previously obtained in Sect. 3.6.

As already noted in the previous section, let us recall that in the razor-thin limit, for which $h \to 0$, the quantised vertical frequencies k_z^n are such that $k_z^n \to +\infty$, except for the fundamental symmetric frequency $k_{z,\mathrm{s}}^1$. In the expression (5.66) of the dressed susceptibility coefficients, let us also note the presence of a prefactor $1/h$, so that in the razor-thin limit, one has to study the behaviour of a term of the form

$$\frac{1}{h}\frac{1}{k_r^2+(k_z^{n_p})^2} \xrightarrow[\text{thin}]{} \begin{cases} \dfrac{1}{k_r} & \text{if } k_z^{n_p}=k_{z,\mathrm{s}}^1\,, \\ 0 & \text{if } k_z^{n_p}\neq k_{z,\mathrm{s}}^1\,. \end{cases} \tag{5.133}$$

In the razor-thin limit, because all the other terms appearing in Eq. (5.66) are bounded, one therefore gets

$$\lim_{\text{thin}} \frac{1}{\mathcal{D}^{\text{anti}}_{\boldsymbol{m}_1,\boldsymbol{m}_1}} = 0\,. \tag{5.134}$$

In addition, in the razor-thin limit, the sum on n_p appearing in Eq. (5.66) may also be limited to the only fundamental term $n_p = 1$. Moreover, in order to have non-vanishing susceptibility coefficients, as already justified in the collisionless case, only symmetric diffusion coefficients associated with $m_1^z = 0$ and $J_z^1 = 0$ will not vanish in the razor-thin limit. Finally, let us recall that in the razor-thin limit, one has $\lim_{\text{thin}} \lambda_p = \lambda_{\text{thin}}$ and $\lim_{\text{thin}} \alpha_1 = 1$. Thanks to these restrictions, in the razor-thin limit, the symmetric susceptibility coefficients from Eq. (5.66) become

$$\lim_{\text{thin}} \frac{1}{\mathcal{D}^{\text{sym}}_{\boldsymbol{m}_1,\boldsymbol{m}_1}} \sim \frac{1}{\mathcal{D}^{\text{thin}}_{\boldsymbol{m}_1,\boldsymbol{m}_1}} \, \mathcal{J}_0\!\left[\sqrt{\frac{2J_z^2}{\nu_1}}\, k^1_{z,\text{s}}\right], \tag{5.135}$$

where $1/\mathcal{D}^{\text{thin}}_{\boldsymbol{m}_1,\boldsymbol{m}_1}$ stands for the razor-thin WKB susceptibility coefficients obtained in Eq. (3.80).

In order to recover the razor-thin WKB Balescu-Lenard diffusion flux, let us now consider the expression (5.69) of the thickened WKB drift coefficients and study their behaviour in the razor-thin limit. Let us first rewrite the thick quasi-isothermal DF from Eq. (5.5) as

$$F_{\text{thick}}(J_\phi^1, J_r^1, J_z^1) = F_{\text{thin}}(J_\phi^1, J_r^1)\, \frac{\nu_1}{2\pi\sigma_z^2}\, \exp\!\left[-\frac{\nu_1 J_z^1}{\sigma_z^2}\right], \tag{5.136}$$

where we wrote F_{thin} for the razor-thin quasi-isothermal DF from Eq. (5.69). In order to illustrate the gist of this calculation, let us now focus only the remaining dependences w.r.t. J_z^2 in Eq. (5.69). This corresponds to an expression of the form

$$\frac{\nu_1}{2\pi\sigma_z^2}\!\int\! \mathrm{d}J_z^2\, \exp\!\left[-\frac{\nu_1 J_z^2}{\sigma_z^2}\right] \mathcal{J}_0^2\!\left[\sqrt{\frac{2J_z^2}{\nu_1}}\, k^1_{z,\text{s}}\right] = \frac{1}{2\pi}\, \mathcal{I}_0\!\left[\frac{(k^1_{z,\text{s}})^2}{\nu_1^2/\sigma_z^2}\right] \exp\!\left[-\frac{(k^1_{z,\text{s}})^2}{\nu_1^2/\sigma_z^2}\right] \underset{\text{thin}}{\longrightarrow} \frac{1}{2\pi}\,, \tag{5.137}$$

where we relied on the formula from Eq. (3.42), as well as on the fact that in the razor-thin limit $(k^1_{z,\text{s}})^2/(\nu_1^2/\sigma_z^2) \sim h \to 0$. Using Eq. (5.137) into the general expression (5.69) of the drift coefficients, one gets

$$\lim_{\text{thin}} A^{\text{sym}}_{\boldsymbol{m}_1}(\boldsymbol{J}_1) = -\frac{4\pi^3\mu}{(\boldsymbol{m}_1\cdot\boldsymbol{\Omega}_1)'}\!\int\! \mathrm{d}J_r^2\, \frac{\boldsymbol{m}_1\cdot\partial F_{\text{thin}}/\partial \boldsymbol{J}(J_\phi^1, J_r^2)}{|\mathcal{D}^{\text{thin}}_{\boldsymbol{m}_1,\boldsymbol{m}_1}(J_\phi^1, J_r^1, J_\phi^1, J_r^2, \boldsymbol{m}_1\cdot\boldsymbol{\Omega}_1)|^2}\,, \tag{5.138}$$

where one has to restrict oneself to $m_1^z = 0$ and $J_z^1 = 0$. Following the same approach, the razor-thin limit of the collisional diffusion coefficients from Eq. (5.70) is straightforward to compute and reads

$$\lim_{\text{thin}} D_{m_1}^{\text{sym}}(\boldsymbol{J}_1) = \frac{4\pi^3\mu}{(\boldsymbol{m}_1\cdot\boldsymbol{\Omega}_1)'} \int \mathrm{d}J_r^2 \, \frac{F_{\text{thin}}(J_\phi^1, J_r^2)}{|\mathcal{D}_{\boldsymbol{m}_1,\boldsymbol{m}_1}^{\text{thin}}(J_\phi^1, J_r^1, J_\phi^1, J_r^2, \boldsymbol{m}_1\cdot\boldsymbol{\Omega}_1)|^2} . \tag{5.139}$$

This concludes our calculations, as we note that the two razor-thin limits obtained in Eqs. (5.138) and (5.139) are in full agreement with the razor-thin results previously obtained in Eqs. (3.83) and (3.84).

References

Abadi, M. G., Navarro, J. F., Steinmetz, M., & Eke, V. R. (2003). *ApJ*, *597*, 21.
Aubert, D., & Pichon, C. (2007). *MNRAS*, *374*, 877.
Aumer, M., Binney, J., & Schönrich, R. (2016). *MNRAS*, *459*, 3326.
Barbanis, B., & Woltjer, L. (1967). *ApJ*, *150*, 461.
Binney, J., & Lacey, C. (1988). *MNRAS*, *230*, 597.
Binney, J., & Tremaine, S. (2008). *Galactic dynamics* (2nd ed.). Princeton University Press.
Bird, J. C., Kazantzidis, S., & Weinberg, D. H. (2012). *MNRAS*, *420*, 913.
Bournaud, F., Elmegreen, B. G., & Martig, M. (2009). *ApJ*, *707*, L1.
Bovy, J., Rix, H.-W., Liu, C., et al. (2012). *ApJ*, *753*, 148.
Brook, C. B., Kawata, D., Gibson, B. K., & Freeman, K. C. (2004). *ApJ*, *612*, 894.
Burstein, D. (1979). *ApJ*, *234*, 829.
Carlberg, R. G., & Sellwood, J. A. (1985). *ApJ*, *292*, 79.
Chavanis, P.-H. (2013). *A&A*, *556*, A93.
Comerón, S., Elmegreen, B. G., Knapen, J. H., et al. (2011). *ApJ*, *741*, 28.
Courant, R., & Hilbert, D. (1953). *Methods of mathematical physics* (Vol. 1). New York: Interscience.
Di Matteo, P., Lehnert, M. D., Qu, Y., & van Driel, W. (2011). *A&A*, *525*, L3.
D'Onghia, E., Vogelsberger, M., & Hernquist, L. (2013). *ApJ*, *766*, 34.
Eisenstein, D. J., Weinberg, D. H., Agol, E., et al. (2011). *AJ*, *142*, 72.
Fouvry, J.-B., Pichon, C., Chavanis, P.-H., & Monk, L. (2017). *MNRAS*, *471*, 2642.
Freeman, K. C. (1987). *ARA&A*, *25*, 603.
Gilmore, G., & Reid, N. (1983). *MNRAS*, *202*, 1025.
Grand, R. J. J., Springel, V., Gómez, F. A., et al. (2016). *MNRAS*, *459*, 199.
Griv, E., & Gedalin, M. (2012). *MNRAS*, *422*, 600.
Hänninen, J., & Flynn, C. (2002). *MNRAS*, *337*, 731.
Haywood, M. (2008). *MNRAS*, *388*, 1175.
Ida, S., Kokubo, E., & Makino, J. (1993). *MNRAS*, *263*, 875.
Ivezić, Ž., Sesar, B., Jurić, M., et al. (2008). *ApJ*, *684*, 287.
Jenkins, A. (1992). *MNRAS*, *257*, 620.
Jurić, M., Ivezić, Ž., Brooks, A., et al. (2008). *ApJ*, *673*, 864.
Kaasalainen, M., & Binney, J. (1994a). *Phys. Rev. Lett.*, *73*, 2377.
Kaasalainen, M., & Binney, J. (1994b). *MNRAS*, *268*, 1033.
Kaasalainen, M., & Binney, J. (2016). *MNRAS*, *456*, 1982.
Lacey, C. (1984). *MNRAS*, *208*, 687.
Loebman, S. R., Roškar, R., Debattista, V. P., et al. (2011). *ApJ*, *737*, 8.
Lynden-Bell, D., & Kalnajs, A. J. (1972). *MNRAS*, *157*, 1.
Meza, A., Navarro, J. F., Abadi, M. G., & Steinmetz, M. (2005). *MNRAS*, *359*, 93.
Minchev, I., Chiappini, C., & Martig, M. (2014). *A&A*, *572*, A92.
Minchev, I., & Famaey, B. (2010). *ApJ*, *722*, 112.
Minchev, I., Famaey, B., Quillen, A. C., et al. (2012). *A&A*, *548*, A127.

Minchev, I., Martig, M., Streich, D., et al. (2015). *ApJ*, *804*, L9.
Minchev, I., & Quillen, A. C. (2006). *MNRAS*, *368*, 623.
Monari, G., Famaey, B., & Siebert, A. (2016). *MNRAS*, *457*, 2569.
Mould, J. (2005). *AJ*, *129*, 698.
Murray, N. (2011). *ApJ*, *729*, 133.
Noguchi, M. (1998). *Nature*, *392*, 253.
O'Leary, D., & Stewart, G. (1990). *J. Comput. Phys.*, *90*, 497.
Purcell, C. W., Bullock, J. S., Tollerud, E. J., Rocha, M., & Chakrabarti, S. (2011). *Nature*, *477*, 301.
Quillen, A. C., Minchev, I., Bland-Hawthorn, J., & Haywood, M. (2009). *MNRAS*, *397*, 1599.
Quinn, P. J., Hernquist, L., & Fullagar, D. P. (1993). *ApJ*, *403*, 74.
Romeo, A. B. (1992). *MNRAS*, *256*, 307.
Schönrich, R., & Binney, J. (2009a). *MNRAS*, *396*, 203.
Schönrich, R., & Binney, J. (2009b). *MNRAS*, *399*, 1145.
Sellwood, J. A. (2012). *ApJ*, *751*, 44.
Sellwood, J. A., & Binney, J. (2002). *MNRAS*, *336*, 785.
Sellwood, J. A., & Carlberg, R. G. (1984). *ApJ*, *282*, 61.
Shiidsuka, K., & Ida, S. (1999). *MNRAS*, *307*, 737.
Solway, M., Sellwood, J. A., & Schönrich, R. (2012). *MNRAS*, *422*, 1363.
Spitzer, L. Jr. (1942). *ApJ*, *95*, 329.
Spitzer, L. Jr., & Schwarzschild, M. (1953). *ApJ*, *118*, 106.
Toth, G., & Ostriker, J. P. (1992). *ApJ*, *389*, 5.
Vandervoort, P. O. (1970). *ApJ*, *161*, 87.
Villalobos, Á., & Helmi, A. (2008). *MNRAS*, *391*, 1806.
Weinberg, M. D. (1993). *ApJ*, *410*, 543.
Wielen, R. (1977). *A&A*, *60*, 263.
Yoachim, P., & Dalcanton, J. J. (2006). *AJ*, *131*, 226.

Chapter 6
Quasi-Keplerian Systems

6.1 Introduction

The previous chapters focused on the dynamics of stellar discs, either razor-thin or thickened. For these systems, we explored two regimes of secular diffusion either collisionless or collisional, depending on whether fluctuations are external or internal. In this chapter, we focus on another family of self-gravitating systems, for which a large set of particles orbits a dominant massive object. This corresponds for example to stars bound to a central supermassive black hole in galactic nuclei, or to protoplanetary debris discs encircling a central star. As will be emphasised in the upcoming discussions, such systems, because they are dominated by one central object, have the peculiarity of being dynamically degenerate. This requires some adjustments to tailor the previous diffusion formalisms. Let us first discuss the main properties of such systems.

Stars in a stellar cluster surrounding a dominant supermassive black hole (BH) evolve in a quasi-Keplerian potential. As a consequence, their orbits take the form of ellipses, which conserve their spatial orientation for many orbital periods, as illustrated in Fig. 6.1. This is a signature of the dynamical degeneracy of the Keplerian potential. The stellar cluster may then be represented as a system of massive Keplerian wires, for which the mass of each star is smeared out along the elliptic path followed by its quasi-Keplerian orbit. Such ideas were first developed in Rauch and Tremaine (1996), which introduced the concept of "resonant relaxation" by noting that wire-wire interactions greatly enhance the relaxation of the stars' angular momenta w.r.t. conventional estimates which do not account for the coherence of stars' orbits over many dynamical times and consider only uncorrelated two-body encounters. See Alexander (2005, 2017) for a review of the various stellar processes occurring in the vicinity of supermassive black holes.

A detailed understanding of the relaxation processes occurring in galactic nuclei is important in order to predict the rates of tidal disruptions of stars by BHs (e.g.,

---

The work presented in this chapter is based on Fouvry et al. (2017a).

J.-B. Fouvry, *Secular Evolution of Self-Gravitating Systems Over Cosmic Age*, Springer Theses, DOI 10.1007/978-3-319-66760-7_6

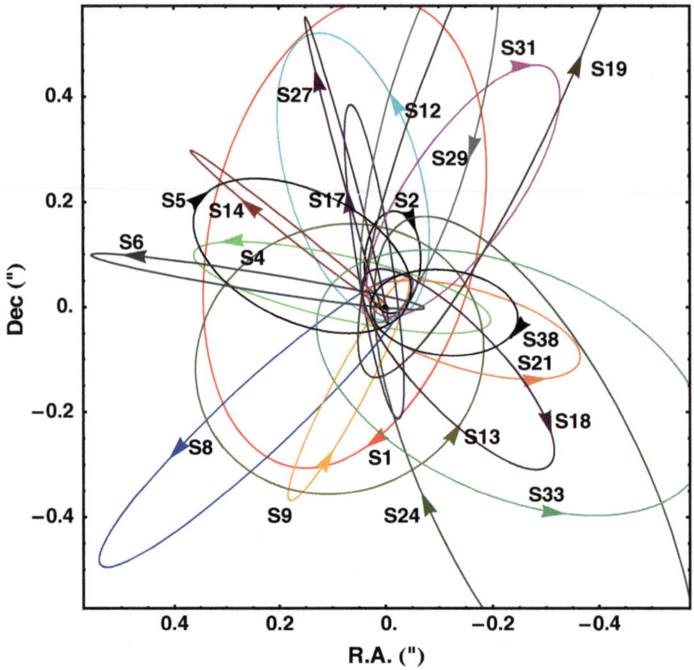

Fig. 6.1 Extracted from Fig. 16 of Gillessen et al. (2009). Observations of the individual trajectories of twenty stars orbiting in the vicinity of Sgr A*, the supermassive black hole at the centre of the Milky Way. Because of the dominant mass of the central BH, stars follow quasi-Keplerian orbits

Rauch and Tremaine 1996; Rauch and Ingalls 1998), the merging rates of binary supermassive BHs (e.g., Yu 2002), or the rate of gravitational wave emissions from star-BH interactions (e.g., Hopman and Alexander 2006; Merritt et al. 2011). Resonant relaxation also appears as the appropriate framework to understand some of the features of young stellar populations found in the centre of our own Galaxy (e.g., Kocsis and Tremaine 2011).

As for stellar discs, a first way to study the secular dynamics of quasi-Keplerian stellar clusters is to rely on direct $N-$body simulations. However, in this context, gaining physical insights from these simulations is challenging, as various complex dynamical processes are intimately entangled there. In addition, because of the significative breadth of timescales in these systems between the fast Keplerian motion and cosmic times, the computational costs of these simulations are such that one can typically only run a few realisations, limited to a relatively small number N of particles. Moreover, this cannot be scaled up easily to astrophysical systems, as different dynamical mechanisms scale differently with N (Heggie and Hut 2003). When focusing on resonant relaxation, one can improve these simulations by using $N-$wires code (e.g., Kocsis and Tremaine 2015), in which stars are replaced by orbit-averaged Keplerian wires.

A complementary approach to understand and describe the dynamics of such systems is to rely on tools from kinetic theory. Especially, in order to account for effects induced by the system's finite number of particles, the Balescu–Lenard formalism, presented in Sect. 2.3, appears as perfectly well suited. However, in the context of quasi-Keplerian systems, the application of the Balescu–Lenard formalism in its original form raises two additional difficulties, which ask for a particular attention. The first difficulty comes from the fact that one has to describe the dynamics of a system within a possibly non-inertial set of coordinates. This requires to pay a careful attention to canonical changes of coordinates as will be emphasised in Sect. 6.2. The second difficulty arises from the intrinsic dynamical degeneracies of the Keplerian problem, i.e. the fact that the Keplerian frequencies $\boldsymbol{\Omega}_{\mathrm{Kep}}$ satisfy commensurability conditions of the form $\boldsymbol{n}\cdot\boldsymbol{\Omega}_{\mathrm{Kep}}\simeq 0$, for some vectors of integers $\boldsymbol{n}=(n_1, n_2, n_3)$, as will be discussed in Sect. 6.3. Indeed, the Balescu–Lenard formalism in its original form assumes that resonances are localised in action space and are not degenerate. As a consequence, it must be re-examined before it can be applied to the degeneracies inherent to quasi-Keplerian systems.

In the upcoming sections, we will show how one can account for these degeneracies in the case of a cluster of N particles orbiting a massive, possibly relativistic, central body. This will require to first average the equations of motion over the fast Keplerian angle associated with the orbital motion of stars around the BH. Once such an averaging is carried out, we will emphasise how the generic Balescu–Lenard formalism applies straightforwardly and yields the associated degenerate secular collisional equation. As will be detailed in the upcoming sections, this equation captures the drift and diffusion of particles' actions induced by their mutual resonant interaction at the frequency shifts present in addition to the mean Keplerian dynamics, e.g., possibly induced by the cluster's self-gravity or relativistic effects. This new equation will be shown to be ideally suited to describe the secular evolution of a large set of particles orbiting a massive central object, by capturing the secular effects of sequences of polarised wire-wire interactions (associated with scalar or vector resonant relaxation) on the underlying cluster's orbital structure.

This chapter is organised as follows. Section 6.2 specifies the BBGKY hierarchy to systems with a finite number of particles orbiting a central massive body, by using canonical coordinates to account adequately for the motion due to the central body. Section 6.3 describes the angle-action coordinates appropriate for such quasi-Keplerian systems and discusses how the dynamical degeneracies should be dealt with. Section 6.4 averages the corresponding dynamical equations over the fast Keplerian angles and discusses the newly obtained set of coupled evolution equations. Section 6.5 presents in detail the degenerate one- and multi-component Keplerian Balescu–Lenard equations, whose important properties will be discussed. Finally, in Sect. 6.6, we present some applications of this new degenerate collisional formalism, respectively to razor-thin axisymmetric discs, $3D$ spherical clusters, and to understand the suppression of resonant relaxation as stars move closer to the central BH, a phenomenon coined the Schwarzschild barrier.

6.2 The Associated BBGKY Hierarchy

Let us consider a set of N stars of individual mass μ, orbiting a central BH of mass $M_\bullet$. We assume the system to be quasi-Keplerian so that defining the total stellar mass $M_\star = \mu N$, one has

$$\varepsilon = \frac{M_\star}{M_\bullet} \ll 1\,. \tag{6.1}$$

We place ourselves within an inertial frame and denote as $\boldsymbol{X}_\bullet$ the position of the BH and $\boldsymbol{X}_i$ the position of the ith star. The total Hamiltonian of the system reads

$$H = \frac{\boldsymbol{P}_\bullet^2}{2M_\bullet} + \sum_{i=1}^{N} \frac{\boldsymbol{P}_i^2}{2\mu} + \mu M_\bullet \sum_{i=1}^{N} U(|\boldsymbol{X}_i - \boldsymbol{X}_\bullet|) + \mu^2 \sum_{i<j}^{N} U(|\boldsymbol{X}_i - \boldsymbol{X}_j|) + \mu M_\star \sum_{i=1}^{N} \Phi_{\mathrm{rel}}(\boldsymbol{X}_i - \boldsymbol{X}_\bullet)\,, \tag{6.2}$$

in which we introduced the canonical momenta as $\boldsymbol{P}_\bullet = M_\bullet \dot{\boldsymbol{X}}_\bullet$ and $\boldsymbol{P}_i = \mu \dot{\boldsymbol{X}}_i$. We also introduced as $U(|\boldsymbol{X}|)$ the binary interaction potential, i.e. $U(|\boldsymbol{X}| = -G/|\boldsymbol{X}|)$ in the gravitational context. Let us now detail the various interaction terms appearing in Eq. (6.2). The first two terms correspond to the kinetic energy of the BH and the stars. The third term corresponds to the Keplerian potential of the BH, while the fourth term captures the pairwise interactions among stars. Finally, the last term of Eq. (6.2) accounts for the relativistic correction forces occurring in the vicinity of the BH, such as the Schwarzschild and Lense-Thirring precessions, as detailed in Appendix 6.A. Let us emphasise the normalisation prefactor $\mu M_\star$ of these relativistic corrections, which was introduced for later convenience. One can note that Eq. (6.2) does not contain any additional external potential contributions. As such contributions may offset the system and introduce non-trivial inertial effects, they were not accounted for to clarify the presentation (see item III of Sect. 6.4 for a discussion of how such external contributions may also drive the system's secular dynamics). The Hamiltonian from Eq. (6.2) is therefore the direct equivalent, in the context of quasi-Keplerian systems, of the Hamiltonian introduced in Eq. (2.36) when deriving the non-degenerate inhomogeneous Balescu–Lenard equation.

Following the method from Appendix 2.A, our aim is now to derive an appropriate BBGKY hierarchy for the Hamiltonian from Eq. (6.2), to get a better grasp of how finite$-N$ effects may source the long-term evolution of quasi-Keplerian systems. To do so, let us first rewrite the Hamiltonian from Eq. (6.2) as N decoupled Kepler Hamiltonians plus some perturbations. Such dynamical problems dominated by one central body are extensively studied in the context of planetary dynamics. We follow Duncan et al. (1998) to perform a canonical change of coordinates to a new set of coordinates, the democratic heliocentric coordinates. Let us define the new coordinates $(\boldsymbol{x}_\bullet, \boldsymbol{x}_1, \ldots, \boldsymbol{x}_N)$ as

$$\boldsymbol{x}_\bullet = \frac{1}{M_{\mathrm{tot}}} \bigg[M_\bullet \boldsymbol{X}_\bullet + \sum_{i=1}^{N} \mu \boldsymbol{X}_i \bigg] \quad ; \quad \boldsymbol{x}_i = \boldsymbol{X}_i - \boldsymbol{X}_\bullet\,. \tag{6.3}$$

In Eq. (6.3), we introduced the total mass of the system $M_{\rm tot} = M_\bullet + M_\star$, and one should pay attention to the fact that this differs from the definition of $M_{\rm tot}$ used in the previous sections. These new coordinates are such that $\boldsymbol{x}_\bullet$ corresponds to the position of the system's centre of mass, while $\boldsymbol{x}_i$ gives the location of the ith star w.r.t. the BH. These relations can easily be inverted as

$$\boldsymbol{X}_\bullet = \boldsymbol{x}_\bullet - \frac{1}{M_{\rm tot}} \sum_{i=1}^{N} \mu \boldsymbol{x}_i \quad ; \quad \boldsymbol{X}_i = \boldsymbol{x}_\bullet + \boldsymbol{x}_i - \frac{1}{M_{\rm tot}} \sum_{j=1}^{N} \mu \boldsymbol{x}_j \,. \tag{6.4}$$

Following Duncan et al. (1998), the associated canonical momenta $(\boldsymbol{p}_\bullet, \boldsymbol{p}_1, \ldots, \boldsymbol{p}_N)$ are given by

$$\boldsymbol{p}_\bullet = \boldsymbol{P}_\bullet + \sum_{i=1}^{N} \boldsymbol{P}_i \quad ; \quad \boldsymbol{p}_i = \boldsymbol{P}_i - \frac{\mu}{M_{\rm tot}} \bigg[\boldsymbol{P}_\bullet + \sum_{j=1}^{N} \boldsymbol{P}_j \bigg] . \tag{6.5}$$

These new canonical coordinates allow us to rewrite the Hamiltonian from Eq. (6.2) as

$$H = \sum_{i=1}^{N} \bigg[\frac{\boldsymbol{p}_i^2}{2\mu} + \mu M_\bullet U(|\boldsymbol{x}_i|) + \mu M_\star \Phi_{\rm rel}(\boldsymbol{x}_i) \bigg] + \mu^2 \sum_{i<j} U(|\boldsymbol{x}_i - \boldsymbol{x}_j|) + \frac{\boldsymbol{p}_\bullet^2}{2M_{\rm tot}} + \frac{1}{2M_\bullet} \bigg[\sum_{i=1}^{N} \boldsymbol{p}_i \bigg]^2 . \tag{6.6}$$

In Eq. (6.6), let us note that the two first terms correspond to N coupled Kepler problems (with relativistic corrections), completed with the presence of the two last additional kinetic terms. The coordinates being canonical, the evolution of the total momentum $\boldsymbol{p}_\bullet$ is given by Hamilton's equation $\dot{\boldsymbol{p}}_\bullet = -\partial H / \partial \boldsymbol{x}_\bullet = 0$. Without loss of generality, let us therefore assume that $\boldsymbol{p}_\bullet = 0$. The evolution of the system's barycentre is then given by $\dot{\boldsymbol{x}}_\bullet = \partial H / \partial \boldsymbol{p}_\bullet = \boldsymbol{p}_\bullet / M_{\rm tot} = 0$, so that we may set as well $\boldsymbol{x}_\bullet = 0$. Let us finally introduce the notation $\boldsymbol{v}_n = \boldsymbol{p}_n / \mu \neq \dot{\boldsymbol{x}}_n$, so that the Hamiltonian from Eq. (6.6) becomes

$$H = \sum_{i=1}^{N} \bigg[\frac{\mu}{2} \boldsymbol{v}_i^2 + \mu M_\bullet U(|\boldsymbol{x}_i|) + \mu M_\star \Phi_{\rm rel}(\boldsymbol{x}_i) \bigg] + \mu^2 \sum_{i<j}^{N} U(|\boldsymbol{x}_i - \boldsymbol{x}_j|) + \frac{\mu^2}{2M_\bullet} \bigg[\sum_{i=1}^{N} \boldsymbol{v}_i \bigg]^2 . \tag{6.7}$$

Let us emphasise how the Hamiltonian from Eq. (6.7) is similar to the one considered in Eq. (2.36) to describe isolated long-range systems. In Eq. (6.7), one should also note the presence of two additional potential contributions due to the central BH and the relativistic corrections. These only affect each particle individually, which makes them easy to deal with. A second difference comes from the additional kinetic terms in Eq. (6.7) associated with the change of coordinates from Eq. (6.3). As will be justified in Sect. 6.4, we will be in a position to neglect these contributions at the order considered in our kinetic developments.

Starting from the Hamiltonian from Eq. (6.7), let us now proceed as in Sect. 2.A to derive the associated BBGKY hierarchy. The upcoming calculations being very similar to the ones presented in Sect. 2.A, we mainly emphasise here the important changes in the quasi-Keplerian context. Following the normalisation convention from Eq. (2.94), one can obtain a statistical description of the system by considering its $N-$body probability distribution function $P_N(\Gamma_1, \ldots, \Gamma_N, t)$, where we introduced the phase space coordinates $\Gamma = (\boldsymbol{x}, \boldsymbol{v})$. The dynamics of P_N is fully given by Liouville's equation (2.95), which reads

$$\frac{\partial P_N}{\partial t} + \sum_{i=1}^{N} \left[\dot{\boldsymbol{x}}_i \cdot \frac{\partial P_N}{\partial \boldsymbol{x}_i} + \dot{\boldsymbol{v}}_i \cdot \frac{\partial P_N}{\partial \boldsymbol{v}_i} \right] = 0 \,. \tag{6.8}$$

Here, the dynamics of individual particles is given by Hamilton's equations $\mu \, \mathrm{d}\boldsymbol{x}_i/\mathrm{d}t = \partial H/\partial \boldsymbol{v}_i$ and $\mu \, \mathrm{d}\boldsymbol{v}_i/\mathrm{d}t = -\partial H/\partial \boldsymbol{x}_i$, for the total Hamiltonian H from Eq. (6.7). Following Eq. (2.97) and the associated conventions, let us define the system's reduced PDFs P_n, and subsequently the reduced DFs f_n following Eq. (2.99) and its normalisations. The generic BBGKY equation (2.100) for f_n becomes here in the quasi-Keplerian context

$$\begin{aligned} \frac{\partial f_n}{\partial t} &+ \sum_{i=1}^{n} \bigg\{ \bigg[\boldsymbol{v}_i + \frac{\varepsilon}{N} \sum_{j=1}^{n} \boldsymbol{v}_j \bigg] \cdot \frac{\partial f_n}{\partial \boldsymbol{x}_i} + \bigg[M_\bullet \boldsymbol{\mathcal{F}}_{i0} + \mu \sum_{j=1, j \neq i}^{n} \boldsymbol{\mathcal{F}}_{ij} + M_\star \boldsymbol{\mathcal{F}}_{i\mathrm{r}} \bigg] \cdot \frac{\partial f_n}{\partial \boldsymbol{v}_i} \bigg\} \\ &+ \sum_{i=1}^{n} \int \mathrm{d}\Gamma_{n+1} \left[\frac{1}{M_\bullet} \boldsymbol{v}_{n+1} \cdot \frac{\partial f_{n+1}}{\partial \boldsymbol{x}_i} + \boldsymbol{\mathcal{F}}_{i,n+1} \cdot \frac{\partial f_{n+1}}{\partial \boldsymbol{v}_i} \right] = 0 \,. \end{aligned} \tag{6.9}$$

In Eq. (6.9), we introduced the force exerted by particle j on particle i as $\mu \boldsymbol{\mathcal{F}}_{ij} = -\mu \partial U_{ij}/\partial \boldsymbol{x}_i$, with the shortening notation $U_{ij} = U(|\boldsymbol{x}_i - \boldsymbol{x}_j|)$. We also wrote the force exerted by the BH on particle i as $M_\bullet \boldsymbol{\mathcal{F}}_{i0} = -M_\bullet \partial U_{i0}/\partial \boldsymbol{x}_i$, with $U_{i0} = U(|\boldsymbol{x}_i|)$. Finally, the force associated with the relativistic corrections on particle i was written as $M_\star \boldsymbol{\mathcal{F}}_{i\mathrm{r}} = -M_\star \partial \Phi_{\mathrm{rel}}/\partial \boldsymbol{x}_i$. As expected from the presence of the additional kinetic terms in Eq. (6.7), one can note that Eq. (6.9) differs in particular from Eq. (2.100) via two additional kinetic contributions.

In the evolution equation (6.9), in order to isolate the contributions arising from correlations among stars, let us follow Eqs. (2.101) and (2.102), and introduce the cluster representation of the DFs. Here, we are especially interested in the system's $1-$body DF as well as in its $2-$ and $3-$body correlation functions g_2 and g_3. Relying on the normalisations obtained in Eq. (2.103), and because the individual mass of the stars scales like $\mu \sim 1/N$, one immediately has $|f_1| \sim 1$, $|g_2| \sim 1/N$, and $|g_3| \sim 1/N^2$. Thanks to this cluster decomposition and starting from Eq. (6.9), one can write the two first equations of the BBGKY hierarchy. The first equation (2.104) becomes here

$$\begin{aligned} \frac{\partial f_1}{\partial t} &+ \left[\boldsymbol{v}_1 + \frac{\varepsilon}{N} \boldsymbol{v}_1 \right] \cdot \frac{\partial f_1}{\partial \boldsymbol{x}_1} + \left[M_\bullet \boldsymbol{\mathcal{F}}_{10} + M_\star \boldsymbol{\mathcal{F}}_{1\mathrm{r}} \right] \cdot \frac{\partial f_1}{\partial \boldsymbol{v}_1} + \left[\int \mathrm{d}\Gamma_2 \, \boldsymbol{\mathcal{F}}_{12} f_1(\Gamma_2) \right] \cdot \frac{\partial f_1}{\partial \boldsymbol{v}_1} + \int \mathrm{d}\Gamma_2 \, \boldsymbol{\mathcal{F}}_{12} \cdot \frac{\partial g_2(\Gamma_1, \Gamma_2)}{\partial \boldsymbol{v}_1} \\ &+ \frac{1}{M_\bullet} \frac{\partial f_1}{\partial \boldsymbol{x}_1} \cdot \int \mathrm{d}\Gamma_2 \, \boldsymbol{v}_2 f_1(\Gamma_2) + \frac{1}{M_\bullet} \int \mathrm{d}\Gamma_2 \, \boldsymbol{v}_2 \cdot \frac{\partial g_2(\Gamma_1, \Gamma_2)}{\partial \boldsymbol{x}_1} = 0 \,. \end{aligned} \tag{6.10}$$

Similarly, the second equation (2.105) becomes

$$\begin{aligned}
&\frac{1}{2}\frac{\partial g_2}{\partial t} + \left[\boldsymbol{v}_1 + \frac{\varepsilon}{N}(\boldsymbol{v}_1 + \boldsymbol{v}_2)\right]\cdot\frac{\partial g_2}{\partial \boldsymbol{x}_1} + \frac{\varepsilon}{N}\boldsymbol{v}_2\cdot\frac{\partial f_1}{\partial \boldsymbol{x}_1} f_1(\Gamma_2) + \left[M_\bullet \boldsymbol{\mathcal{F}}_{10} + M_\star \boldsymbol{\mathcal{F}}_{1\mathrm{r}}\right]\cdot\frac{\partial g_2}{\partial \boldsymbol{v}_1} + \left[\int \mathrm{d}\Gamma_3\, \boldsymbol{\mathcal{F}}_{13} f_1(\Gamma_3)\right]\cdot\frac{\partial g_2}{\partial \boldsymbol{v}_1} \\
&+ \mu \boldsymbol{\mathcal{F}}_{12}\cdot\frac{\partial f_1}{\partial \boldsymbol{v}_1} f_1(\Gamma_2) + \left[\int \mathrm{d}\Gamma_3\, \boldsymbol{\mathcal{F}}_{13} g_2(\Gamma_2, \Gamma_3)\right]\cdot\frac{\partial f_1}{\partial \boldsymbol{v}_1} + \frac{1}{M_\bullet}\frac{\partial f_1}{\partial \boldsymbol{x}_1}\cdot\int \mathrm{d}\Gamma_3\, \boldsymbol{v}_3 g_2(\Gamma_2, \Gamma_3) + \frac{1}{M_\bullet}\frac{\partial g_2}{\partial \boldsymbol{x}_1}\cdot\int \mathrm{d}\Gamma_3\, \boldsymbol{v}_3 f_1(\Gamma_3) \\
&+ \mu \boldsymbol{\mathcal{F}}_{12}\cdot\frac{\partial g_2}{\partial \boldsymbol{v}_1} + \int \mathrm{d}\Gamma_3\, \boldsymbol{\mathcal{F}}_{13}\cdot\frac{\partial g_3(\Gamma_1, \Gamma_2, \Gamma_3)}{\partial \boldsymbol{v}_1} + \frac{1}{M_\bullet}\int \mathrm{d}\Gamma_3\, \boldsymbol{v}_3\cdot\frac{\partial g_3(\Gamma_1, \Gamma_2, \Gamma_3)}{\partial \boldsymbol{x}_1} + (1 \leftrightarrow 2) = 0 ,
\end{aligned} \tag{6.11}$$

where $(1 \leftrightarrow 2)$ stands for the permutation of indices 1 and 2 and applies to all preceding terms.

As we are interested in first-order collisional effects, let us proceed as in Eqs. (2.107) and (2.108), and truncate Eqs. (6.10) and (6.11) at order $1/N$. At this stage, let us emphasise that such quasi-Keplerian systems involve two small parameters, namely $1/N$ associated with the system's discreteness, and $\varepsilon = M_\star/M_\bullet$ capturing the dominance of the BH on the stars' individual dynamics. As the upcoming calculations will emphasise, we will perform kinetic developments where we keep only small terms of order ε and ε/N, while higher order corrections will be neglected. In Eq. (6.10), we note that all the terms are at least of order $1/N$ and should therefore all be kept. In Eq. (6.11), the first two lines are of order $1/N$ (except for the correction $(\varepsilon/N)(\boldsymbol{v}_1 + \boldsymbol{v}_2)\cdot\partial g_2/\partial \boldsymbol{x}_1$ which may be neglected) and should be kept, while all terms from the third line are of order $1/N^2$ and may therefore be neglected. As already noted in Eq. (2.105), we note that the first term of the third line of Eq. (6.11), while being of order $1/N^2$ can still get arbitrary large as particles 1 and 2 get closer. This term captures strong collisions and is not accounted for in the present formalism. In addition to these truncations, and in order to consider terms of order 1, let us finally define the system's 1–body DF F and its 2–body autocorrelation $\mathcal{C}$ as

$$F = \frac{f_1}{M_\star} \quad ; \quad \mathcal{C} = \frac{g_2}{\mu M_\star} . \tag{6.12}$$

One should pay attention to the fact that these normalisations differ from the generic ones introduced in Eq. (2.106). Finally, in order to highlight the different orders of magnitude of the various components present in the problem, let us rescale as well some of the quantities appearing in Eqs. (6.10) and (6.11). We first rescale the binary interaction potential, U, by using the mass of the central BH, so that

$$\boldsymbol{\mathcal{F}}_{ij} = -\frac{\partial U_{ij}}{\partial \boldsymbol{x}_i} \quad ; \quad U_{ij} = -\frac{G M_\bullet}{|\boldsymbol{x}_i - \boldsymbol{x}_j|} . \tag{6.13}$$

In addition, the potential $\Phi_{\mathrm{r}} = \Phi_{\mathrm{rel}}$ associated with the relativistic corrections is rescaled so that

$$\boldsymbol{\mathcal{F}}_{i\mathrm{r}} = -\frac{\partial \Phi_{\mathrm{r}}}{\partial \boldsymbol{x}_i} \quad ; \quad \Phi_{\mathrm{r}} \to \frac{\Phi_{\mathrm{r}}}{M_\bullet} \quad ; \quad \boldsymbol{\mathcal{F}}_{i\mathrm{r}} \to \frac{\boldsymbol{\mathcal{F}}_{i\mathrm{r}}}{M_\bullet} . \tag{6.14}$$

As a result of these various truncations and renormalisations, the first BBGKY equation (6.10) becomes

$$\frac{\partial F}{\partial t}+\left[\boldsymbol{v}_1+\frac{\varepsilon}{N}\boldsymbol{v}_1\right]\cdot\frac{\partial F}{\partial \boldsymbol{x}_1}+\boldsymbol{\mathcal{F}}_{10}\cdot\frac{\partial F}{\partial \boldsymbol{v}_1}+\varepsilon\left[\int \mathrm{d}\Gamma_2\,\boldsymbol{\mathcal{F}}_{12}F(\Gamma_2)\right]\cdot\frac{\partial F}{\partial \boldsymbol{v}_1}+\varepsilon\boldsymbol{\mathcal{F}}_{1\mathrm{r}}\cdot\frac{\partial F}{\partial \boldsymbol{v}_1}+\frac{\varepsilon}{N}\int \mathrm{d}\Gamma_2\,\boldsymbol{\mathcal{F}}_{12}\cdot\frac{\partial \mathcal{C}(\Gamma_1,\Gamma_2)}{\partial \boldsymbol{v}_1}$$
$$+\varepsilon\frac{\partial F}{\partial \boldsymbol{x}_1}\cdot\int \mathrm{d}\Gamma_2\,\boldsymbol{v}_2F(\Gamma_2)+\frac{\varepsilon}{N}\int \mathrm{d}\Gamma_2\,\boldsymbol{v}_2\cdot\frac{\partial \mathcal{C}(\Gamma_1,\Gamma_2)}{\partial \boldsymbol{x}_1}=0\,. \tag{6.15}$$

Similarly, the second BBGKY equation (6.11) becomes

$$\frac{1}{2}\frac{\partial \mathcal{C}}{\partial t}+\boldsymbol{v}_1\cdot\frac{\partial \mathcal{C}}{\partial \boldsymbol{x}_1}+\boldsymbol{\mathcal{F}}_{10}\cdot\frac{\partial \mathcal{C}}{\partial \boldsymbol{x}_1}+\varepsilon\left[\int \mathrm{d}\Gamma_3\,\boldsymbol{\mathcal{F}}_{13}F(\Gamma_3)\right]\cdot\frac{\partial \mathcal{C}}{\partial \boldsymbol{v}_1}+\varepsilon\boldsymbol{\mathcal{F}}_{1\mathrm{r}}\cdot\frac{\partial \mathcal{C}}{\partial \boldsymbol{v}_1}$$
$$+\varepsilon\boldsymbol{\mathcal{F}}_{12}\cdot\frac{\partial F}{\partial \boldsymbol{v}_1}F(\Gamma_2)+\varepsilon\left[\int \mathrm{d}\Gamma_3\,\boldsymbol{\mathcal{F}}_{13}\mathcal{C}(\Gamma_2,\Gamma_3)\right]\cdot\frac{\partial F}{\partial \boldsymbol{v}_1}$$
$$+\varepsilon\boldsymbol{v}_2\cdot\frac{\partial F}{\partial \boldsymbol{x}_1}F(\Gamma_2)+\varepsilon\frac{\partial F}{\partial \boldsymbol{x}_1}\cdot\int \mathrm{d}\Gamma_3\,\boldsymbol{v}_3\mathcal{C}(\Gamma_2,\Gamma_3)+\varepsilon\frac{\partial \mathcal{C}}{\partial \boldsymbol{x}_1}\cdot\int \mathrm{d}\Gamma_3\,\boldsymbol{v}_3F(\Gamma_3)+(1\leftrightarrow 2)=0\,. \tag{6.16}$$

One should of course note how Eqs. (6.15) and (6.16) are similar to the associated generic ones obtained in Eqs. (2.107) and (2.108). Differences arise from the contributions from the central BH and the relativistic corrections as well as from the additional kinetic terms present in the Hamiltonian from Eq. (6.7). As will be shown in Sect. 6.4, once averaged over the BH-induced Keplerian motion, these kinetic corrections will not come into play at the order considered in the kinetic developments.

The next step of the calculations is now to rewrite Eqs. (6.15) and (6.16) within appropriate angle-action coordinates to capture in a simple manner the dominant mean Keplerian motion induced by the central BH. One subtlety with such Keplerian dynamics comes from the dynamical degeneracies of the associated Keplerian orbital frequencies. These degeneracies have to be handled with care as we will now detail.

6.3 Degenerate Angle-Action Coordinates

In Eqs. (6.15) and (6.16), one notes the presence of a dominant advection term $\boldsymbol{v}_1\cdot\partial/\partial\boldsymbol{x}_1+\boldsymbol{\mathcal{F}}_{10}\cdot\partial/\partial\boldsymbol{v}_1$ associated with the Keplerian motion induced by the central BH. The next step of our derivation is now to introduce the appropriate angle-action coordinates to capture this integrable Keplerian motion. Following Sect. 1.3, let us remap the physical coordinates $(\boldsymbol{x},\boldsymbol{v})$ to the Keplerian angle-action ones $(\boldsymbol{\theta},\boldsymbol{J})$. The Keplerian orbital frequencies associated with these coordinates are then given by

$$\dot{\boldsymbol{\theta}}=\boldsymbol{\Omega}_{\mathrm{Kep}}(\boldsymbol{J})=\frac{\partial H_{\mathrm{Kep}}}{\partial \boldsymbol{J}}\,, \tag{6.17}$$

where H_{Kep} stands for the Hamiltonian associated with the Keplerian motion due to the BH. Of course, various choices of angle-action coordinates are possible. For $3D$

spherical potential, the usual angle-action coordinates (Binney and Tremaine 2008) are given by

$$(\boldsymbol{J}, \boldsymbol{\theta}) = (J_1, J_2, J_3, \theta_1, \theta_2, \theta_3) = (J_r, L, L_z, \theta_1, \theta_2, \theta_3)\,, \tag{6.18}$$

where J_r and L are respectively the radial action and the magnitude of the angular momentum, while L_z is its projection along the $z-$axis (see Appendix 4.D). The Keplerian Hamiltonian then becomes $H_{\rm Kep} = H_{\rm Kep}(J_r + L)$. Another choice of $3D$ angle-action coordinates is given by the Delaunay variables (Sridhar and Touma 1999; Binney and Tremaine 2008) reading

$$(\boldsymbol{J}, \boldsymbol{\theta}) = (I, L, L_z, w, g, h)\,. \tag{6.19}$$

In Eq. (6.19), we introduced as $(I = J_r + L, L, L_z)$ the three actions of the system, while (w, g, h) are the associated angles. Here, the angles have straightforward interpretations in terms of the orbital elements of the Keplerian ellipses: w stands for the orbital phase or mean anomaly, g is the angle from the ascending node to the periapsis, while h is the longitude of the ascending node. Within these variables, the Keplerian Hamiltonian becomes $H_{\rm Kep} = H_{\rm Kep}(I)$, so that the angles g and h become integrals of motion, while the angle w advances at the frequency $\dot{w} = \Omega_{\rm Kep} = \partial H_{\rm Kep}/\partial I$. Because of the existence of these additional conserved quantities, the Keplerian potential is considered to be dynamically degenerate. This can have some crucial consequences on the long-term behaviour of the system, as we will now detail.

To clarify the upcoming discussions, let us note as d the dimension of the considered physical space, e.g., $d = 2$ for a razor-thin disc. Within this space, we consider an integrable potential ψ_0 and one associated angle-action mapping $(\boldsymbol{x}, \boldsymbol{v}) \mapsto (\boldsymbol{\theta}, \boldsymbol{J})$. A potential is said to be degenerate if there exists $\boldsymbol{n} \in \mathbb{Z}^d$ such that

$$\forall \boldsymbol{J}\,,\; \boldsymbol{n} \cdot \boldsymbol{\Omega}(\boldsymbol{J}) = 0\,, \tag{6.20}$$

where it is important for the vector $\boldsymbol{n}$ to be independent of $\boldsymbol{J}$, for the degeneracy to be global. See Fig. 1.5 for an illustration of resonant orbits. Of course, a given potential may have more than one such degeneracy, and we denote as k the degree of degeneracy of the potential, i.e. the number of linearly independent vectors $\boldsymbol{n}$ satisfying Eq. (6.20). Let us consider for example the $3D$ angle-action coordinates from Eq. (6.18). The associated frequencies and degeneracy vectors are given by

$$\boldsymbol{\Omega}_{\rm 3D} = (\Omega_{\rm Kep}, \Omega_{\rm Kep}, 0) \;\;\Rightarrow\;\; \boldsymbol{n}_1 = (1, -1, 0) \text{ and } \boldsymbol{n}_2 = (0, 0, 1)\,, \tag{6.21}$$

so that $k = 2$. Similarly, for the $3D$ Delaunay angle-action variables from Eq. (6.19) one can write

$$\boldsymbol{\Omega}_{\rm Del} = (\Omega_{\rm Kep}, 0, 0) \;\;\Rightarrow\;\; \boldsymbol{n}_1 = (0, 1, 0) \text{ and } \boldsymbol{n}_2 = (0, 0, 1)\,, \tag{6.22}$$

so that one gets as well $k=2$, i.e. the degree of degeneracy of the potential is independent of the chosen angle-action coordinates. Because of their simpler degeneracy vectors $\boldsymbol{n}_1$ and $\boldsymbol{n}_2$, the Delaunay variables from Eq. (6.19) appear as a more appropriate choice than the usual coordinates from Eq. (6.18).

As a final remark, let us emphasise that for a given degenerate potential, one can always remap the system's angle-action coordinates to get simpler dynamical degeneracies. As an illustration, let us assume that in our initial choice of angle-action coordinates $(\boldsymbol{\theta}, \boldsymbol{J})$, the system's degeneracies are captured by the k degeneracy vectors $\boldsymbol{n}_1, \ldots, \boldsymbol{n}_k$. Thanks to a linear change of coordinates $(\boldsymbol{\theta}, \boldsymbol{J}) \mapsto (\boldsymbol{\theta}', \boldsymbol{J}')$, one can always construct new angle-action coordinates within which the k degeneracy vectors take the simple form $\boldsymbol{n}_i' = \boldsymbol{e}_i$, where $\boldsymbol{e}_i$ are the natural basis elements of $\mathbb{Z}^d$. Indeed, following Morbidelli (2002), because the vectors $\boldsymbol{n}_i$ are by definition linearly independent, we may complete this family with $d-k$ vectors $\boldsymbol{n}_{k+1}, \ldots, \boldsymbol{n}_d \in \mathbb{Z}^d$ to construct a basis over $\mathbb{Q}^d$. Defining the transformation matrix $\boldsymbol{\mathcal{A}}$ of determinant 1

$$\boldsymbol{\mathcal{A}} = \big(\boldsymbol{n}_1, \ldots, \boldsymbol{n}_d\big)^{\mathrm{t}} / |(\boldsymbol{n}_1, \ldots, \boldsymbol{n}_d)| \, , \tag{6.23}$$

we introduce new angle-action coordinates $(\boldsymbol{\theta}', \boldsymbol{J}')$ as

$$\boldsymbol{\theta}' = \boldsymbol{\mathcal{A}} \cdot \boldsymbol{\theta} \quad ; \quad \boldsymbol{J}' = (\boldsymbol{\mathcal{A}}^{\mathrm{t}})^{-1} \cdot \boldsymbol{J} \, . \tag{6.24}$$

It is straightforward to check that $(\boldsymbol{\theta}', \boldsymbol{J}')$ are indeed new angle-action coordinates, for which $\boldsymbol{J}'$ are conserved and $\boldsymbol{\theta}' \in [0, 2\pi]$. In addition, within these new coordinates, the system's degeneracies are immediately characterised by the k vectors $\boldsymbol{n}_i' = \boldsymbol{e}_i$. The new intrinsic frequencies then satisfy $\Omega_i' = 0$ for $1 \leq i \leq k$: the degeneracies of the potential became simpler.

In all the upcoming calculations, we will always consider such simpler angle-action coordinates, for which the additional conserved quantities are straightforward to obtain. Let us finally introduce the notations

$$\boldsymbol{\theta}^{\mathrm{s}} = (\theta_1, \ldots, \theta_k) \; ; \; \boldsymbol{\theta}^{\mathrm{f}} = (\theta_{k+1}, \ldots, \theta_d) \; ; \; \boldsymbol{J}^{\mathrm{s}} = (J_1, \ldots, J_k) \; ; \; \boldsymbol{J}^{\mathrm{f}} = (J_{k+1}, \ldots, J_d) \; ; \; \boldsymbol{\mathcal{E}} = (\boldsymbol{J}, \boldsymbol{\theta}^{\mathrm{s}}) \, . \tag{6.25}$$

In Eq. (6.25), $\boldsymbol{\theta}^{\mathrm{s}}$ and $\boldsymbol{J}^{\mathrm{s}}$ correspond to the slow angles and actions, while $\boldsymbol{\theta}^{\mathrm{f}}$ and $\boldsymbol{J}^{\mathrm{f}}$ correspond to the fast angles and actions. Finally, we introduced as $\boldsymbol{\mathcal{E}}$ the vector of all the conserved quantities for the underlying dynamics. In the case of a Keplerian potential, this corresponds to a Keplerian elliptical wire. For a degenerate potential, the slow angles are the ones for which the intrinsic frequencies are equal to 0, while these frequencies are non-zero for the fast angles. Let us finally define the degenerate angle-average w.r.t. the fast angles as

$$\overline{F}(\boldsymbol{J}, \boldsymbol{\theta}^{\mathrm{s}}) = \!\int\! \frac{\mathrm{d}\boldsymbol{\theta}^{\mathrm{f}}}{(2\pi)^{d-k}} \, F(\boldsymbol{J}, \boldsymbol{\theta}^{\mathrm{s}}, \boldsymbol{\theta}^{\mathrm{f}}) \, . \tag{6.26}$$

Let us now use these angle-action coordinates to rewrite the two evolution equations (6.15) and (6.16). Because they were tailored for the Keplerian dynamics, these coordinates allow us to rewrite the Keplerian advection term as

$$\mathbf{v}_1 \cdot \frac{\partial}{\partial \mathbf{x}_1} + \boldsymbol{\mathcal{F}}_{10} \cdot \frac{\partial}{\partial \mathbf{v}_1} = \boldsymbol{\Omega}_{\mathrm{Kep}} \cdot \frac{\partial}{\partial \boldsymbol{\theta}} . \tag{6.27}$$

In addition, let us emphasise that the angle average from Eq. (6.26) is such that the advection term from Eq. (6.27) immediately vanishes when averaged, so that

$$\overline{\boldsymbol{\Omega}_{\mathrm{Kep}} \cdot \frac{\partial F}{\partial \boldsymbol{\theta}}} = \int \frac{\mathrm{d}\theta_{k+1}}{2\pi} \ldots \frac{\mathrm{d}\theta_d}{2\pi} \sum_{i=k+1}^{d} \Omega^i_{\mathrm{Kep}}(\mathbf{J}) \frac{\partial F}{\partial \theta_i} = 0 . \tag{6.28}$$

Finally, the coordinates mapping $(\mathbf{x}, \mathbf{v}) \mapsto (\boldsymbol{\theta}, \mathbf{J})$ being canonical, infinitesimal volumes are conserved so that $\mathrm{d}\Gamma = \mathrm{d}\mathbf{x}\mathrm{d}\mathbf{v} = \mathrm{d}\boldsymbol{\theta}\mathrm{d}\mathbf{J}$. Poisson brackets are also preserved, so that for any functions $G_1(\mathbf{x}, \mathbf{v})$ and $G_2(\mathbf{x}, \mathbf{v})$ one has

$$\big[G_1, G_2\big] = \frac{\partial G_1}{\partial \mathbf{x}} \cdot \frac{\partial G_2}{\partial \mathbf{v}} - \frac{\partial G_1}{\partial \mathbf{v}} \cdot \frac{\partial G_2}{\partial \mathbf{x}} = \frac{\partial G_1}{\partial \boldsymbol{\theta}} \cdot \frac{\partial G_2}{\partial \mathbf{J}} - \frac{\partial G_1}{\partial \mathbf{J}} \cdot \frac{\partial G_2}{\partial \boldsymbol{\theta}} . \tag{6.29}$$

In order to shorten the notations, let us finally introduce the rescaled self-consistent potential of the stars Φ as

$$\Phi(\mathbf{x}_1) = \int \mathrm{d}\Gamma_2 \, U_{12} \, F(\Gamma_2) \quad ; \quad -\frac{\partial \Phi}{\partial \mathbf{x}_1} = \int \mathrm{d}\Gamma_2 \, \boldsymbol{\mathcal{F}}_{12} \, F(\Gamma_2) . \tag{6.30}$$

One may now rewrite Eq. (6.15) within these new coordinates as

$$\begin{aligned} & \frac{\partial F}{\partial t} + \boldsymbol{\Omega}^1_{\mathrm{Kep}} \cdot \frac{\partial F}{\partial \boldsymbol{\theta}_1} + \varepsilon\big[F, \Phi + \Phi_{\mathrm{r}}\big] + \frac{\varepsilon}{N} \int \mathrm{d}\Gamma_2 \, \big[\mathcal{C}(\Gamma_1, \Gamma_2), U_{12}\big]_{(1)} \\ & + \frac{\varepsilon}{N}\bigg[F, \frac{\mathbf{v}_1^2}{2}\bigg] + \varepsilon\bigg[F, \mathbf{v}_1 \cdot \int \mathrm{d}\Gamma_2 \, \mathbf{v}_2 F(\Gamma_2)\bigg] + \frac{\varepsilon}{N} \int \mathrm{d}\Gamma_2 \, \big[\mathcal{C}(\Gamma_1, \Gamma_2), \mathbf{v}_1 \cdot \mathbf{v}_2\big]_{(1)} = 0 , \end{aligned} \tag{6.31}$$

where we wrote $\boldsymbol{\Omega}^1_{\mathrm{Kep}} = \boldsymbol{\Omega}_{\mathrm{Kep}}(\mathbf{J}_1)$, and introduced the notation

$$\big[G_1(\Gamma_1, \Gamma_2), G_2(\Gamma_1, \Gamma_2)\big]_{(1)} = \frac{\partial G_1}{\partial \boldsymbol{\theta}_1} \cdot \frac{\partial G_2}{\partial \mathbf{J}_1} - \frac{\partial G_1}{\partial \mathbf{J}_1} \cdot \frac{\partial G_2}{\partial \boldsymbol{\theta}_1} , \tag{6.32}$$

corresponding to the Poisson bracket w.r.t. the variables 1. In Eq. (6.31), we gathered on the second line all the terms associated with the additional kinetic terms present in the Hamiltonian from Eq. (6.7). As shown in Sect. 6.4, these terms once averaged over the fast Keplerian motion will be negligible at the order considered here. Equation (6.16) can also straightforwardly be rewritten as

$$
\begin{aligned}
&\frac{1}{2}\frac{\partial \mathcal{C}}{\partial t} + \boldsymbol{\Omega}_{\mathrm{Kep}}^{1}\cdot\frac{\partial \mathcal{C}}{\partial \boldsymbol{\theta}_1} + \varepsilon\big[\mathcal{C}(\Gamma_1,\Gamma_2),\Phi+\Phi_{\mathrm{r}}\big]_{(1)} + \varepsilon\big[F(\Gamma_1)F(\Gamma_2),U_{12}\big]_{(1)} + \varepsilon\!\int\!\mathrm{d}\Gamma_3\,\mathcal{C}(\Gamma_2,\Gamma_3)\,\big[F(\Gamma_1),U_{13}\big]_{(1)} \\
&+\varepsilon\big[F(\Gamma_1),\boldsymbol{v}_1\cdot\boldsymbol{v}_2F(\Gamma_2)\big]_{(1)} + \varepsilon\Big[F(\Gamma_1),\boldsymbol{v}_1\cdot\!\int\!\mathrm{d}\Gamma_3\,\boldsymbol{v}_3\mathcal{C}(\Gamma_2,\Gamma_3)\Big]_{(1)} + \varepsilon\Big[\mathcal{C}(\Gamma_1,\Gamma_2),\boldsymbol{v}_1\cdot\!\int\!\mathrm{d}\Gamma_3\,\boldsymbol{v}_3F(\Gamma_3)\Big]_{(1)} \\
&+(1\leftrightarrow 2)=0\,,
\end{aligned}
\tag{6.33}
$$

where the terms present in the second line are the ones associated with the additional kinetic terms from Eq. (6.7).

The rewriting from Eq. (6.31) allows us to easily identify the various timescales of the problem. These are: (i) the Keplerian dynamical timescale $T_{\mathrm{Kep}}=1/\Omega_{\mathrm{Kep}}$ associated with the dominant BH-induced Keplerian dynamics and captured by the advection term $\boldsymbol{\Omega}_{\mathrm{Kep}}^{1}\cdot\partial F/\partial\boldsymbol{\theta}_1$, (ii) the secular collisionless timescale of evolution $T_{\mathrm{sec}}=\varepsilon^{-1}T_{\mathrm{Kep}}$ associated with the potential term $\varepsilon\big[\Phi+\Phi_{\mathrm{r}}\big]$ due to the stars' self-consistent potential as well as the relativistic corrections, and finally (iii) the collisional timescale of relaxation $T_{\mathrm{relax}}=N\,T_{\mathrm{sec}}$ associated with the last term of the first line of Eq. (6.31). Having obtained Eqs. (6.31) and (6.33), which describe the joint evolution of the system's $1-$body DF and its $2-$body autocorrelation, we will show in the next section how one may get rid of the BH-induced Keplerian dynamics via an appropriate degenerate angle-average.

6.4 Averaging the Evolution Equations

As the Keplerian dynamics due to the BH is much faster than the one associated with all the other potential contributions, rather than considering the stars as point particles, let us describe them as massive elliptical wires, for which the mass of the star is smeared out along the elliptic path of its Keplerian orbit. This is the exact purpose of the degenerate angle-average from Eq. (6.26). As noted in Eq. (6.28), such an average naturally cancels out any contributions associated with the BH Keplerian advection term. Let us start from Eq. (6.31) and multiply it by $\int\mathrm{d}\boldsymbol{\theta}^{\mathrm{f}}/(2\pi)^{d-k}$. In order to estimate the average of the various terms that occur in Eq. (6.31), let us finally assume that the system's DF, F, can be decomposed as

$$
F=\overline{F}+\epsilon f \quad \text{with} \quad f\sim O(1) \;\; \text{and} \;\; \overline{f}=0\,, \tag{6.34}
$$

where $\epsilon\ll 1$ is an additional small parameter of order $1/N$. The ansatz from Eq. (6.34) is the crucial assumption of the present derivation. Indeed, contrary to what was generically discussed in Figs. 1.7 and 1.8 w.r.t. the mechanisms of phase mixing or violent relaxation, the BH's domination on the dynamics strongly limits the efficiency of such mechanisms to allow for a rapid dissolution of any $\boldsymbol{\theta}^{\mathrm{f}}-$dependence. Here, we therefore assume that the ansatz from Eq. (6.34) was satisfied because in its initial state the system was already phase mixed.

Relying on this ansatz, let us now discuss in turn how the various terms appearing in Eq. (6.31) can be dealt with once averaged over the fast Keplerian angle. In the first Poisson bracket of Eq. (6.31), let us recall that the self-consistent potential Φ introduced in Eq. (6.30) should be seen as a functional of the system's DF F. As a consequence, this term becomes

$$\begin{aligned}\varepsilon\,\overline{\left[F,\,\Phi(F)+\Phi_{\mathrm{r}}\right]} &= \varepsilon\,\overline{\left[\overline{F}+\epsilon f,\,\Phi(\overline{F}+\epsilon f)+\Phi_{\mathrm{r}}\right]}\\ &= \varepsilon\overline{\left[\overline{F},\,\Phi(\overline{F})+\Phi_{\mathrm{r}}\right]}+O(\varepsilon\epsilon)\\ &= (2\pi)^{d-k}\varepsilon\left[\overline{F},\,\overline{\Phi}(\overline{F})+\overline{\Phi}_{\mathrm{r}}\right]+O(\varepsilon\epsilon)\,. \end{aligned} \tag{6.35}$$

In Eq. (6.35), we introduced the system's averaged self-consistent potential $\overline{\Phi}$ as

$$\overline{\Phi}(\boldsymbol{\mathcal{E}}_1)=\int\mathrm{d}\boldsymbol{\mathcal{E}}_2\,\overline{F}(\boldsymbol{\mathcal{E}}_2)\,\overline{U}_{12}(\boldsymbol{\mathcal{E}}_1,\boldsymbol{\mathcal{E}}_2)\,, \tag{6.36}$$

where for clarity, the notation for the self-consistent potential was shortened as $\overline{\Phi}=\overline{\Phi}(\overline{F})$. In Eq. (6.36), we also introduced the (doubly) averaged interaction potential $\overline{U}_{12}$ as

$$\overline{U}_{12}(\boldsymbol{\mathcal{E}}_1,\boldsymbol{\mathcal{E}}_2)=\int\frac{\mathrm{d}\boldsymbol{\theta}_1^{\mathrm{f}}}{(2\pi)^{d-k}}\frac{\mathrm{d}\boldsymbol{\theta}_2^{\mathrm{f}}}{(2\pi)^{d-k}}\,U_{12}(\Gamma_1,\Gamma_2)\,. \tag{6.37}$$

This potential describes the pairwise interaction potential between two Keplerian wires of coordinates $\boldsymbol{\mathcal{E}}_1$ and $\boldsymbol{\mathcal{E}}_2$. Finally, we also defined the averaged potential $\overline{\Phi}_{\mathrm{r}}$ as

$$\overline{\Phi}_{\mathrm{r}}(\boldsymbol{\mathcal{E}})=\frac{1}{(2\pi)^{d-k}}\int\frac{\mathrm{d}\boldsymbol{\theta}^{\mathrm{f}}}{(2\pi)^{d-k}}\,\Phi_{\mathrm{r}}(\Gamma)\,, \tag{6.38}$$

where we introduced the prefactor $1/(2\pi)^{d-k}$ for convenience. See Appendix 6.A for the expression of the relativistic precession frequencies. In Eq. (6.35), let us note that at first-order in ε and zeroth order in ϵ, the self-consistent potential of the system has to be computed while only considering the averaged system's DF $\overline{F}$. In order to deal with the second Poisson bracket in Eq. (6.31), we perform on the 2−body autocorrelation $\mathcal{C}$ the same double average than the one introduced in Eq. (6.37). Similarly to Eq. (6.34), let us assume that the 2−body autocorrelation can be developed as

$$\mathcal{C}=\overline{\mathcal{C}}+\epsilon c\quad\text{with}\;\; c\sim O(1)\;\;\text{and}\;\;\overline{c}=0\,. \tag{6.39}$$

At first-order in ε and zeroth order in ϵ, the third term from Eq. (6.31) can then be rewritten as

$$\frac{\varepsilon}{N}\int\mathrm{d}\Gamma_2\,\overline{\left[\mathcal{C}(\Gamma_1,\Gamma_2),\,U_{12}\right]_{(1)}}=\frac{\varepsilon\,(2\pi)^{d-k}}{N}\int\mathrm{d}\boldsymbol{\mathcal{E}}_2\left[\overline{\mathcal{C}}(\boldsymbol{\mathcal{E}}_1,\boldsymbol{\mathcal{E}}_2),\,\overline{U}_{12}\right]_{(1)}\,. \tag{6.40}$$

Finally, let us deal with all the additional kinetic terms present in the second line of Eq. (6.31). Once averaged over the fast Keplerian angle, and considering only terms at first-order in ε and zeroth order in ϵ, these various terms involve the quantities

$$\int \mathrm{d}\boldsymbol{\theta}_1^{\mathrm{f}} \, \boldsymbol{v}_1 = 0 \;\; ; \;\; \int \mathrm{d}\boldsymbol{\theta}_1^{\mathrm{f}} \, \frac{\boldsymbol{v}_1^2}{2} \propto H_{\mathrm{Kep}}(\boldsymbol{J}_1^{\mathrm{f}}) \, . \tag{6.41}$$

In Eq. (6.41), the first identity comes the fact that Keplerian orbits are closed, so that the mean displacement over one orbit is zero, while the second identity is a direct consequence of the virial theorem. Because these terms either vanish or do not depend on the slow coordinates $\boldsymbol{\theta}^{\mathrm{s}}$ and $\boldsymbol{J}^{\mathrm{s}}$, at the order considered here, they will not contribute to the dynamics once averaged over the fast Keplerian angle. As a conclusion, keeping only terms of order ε and ε/N, Eq. (6.31) becomes

$$\frac{\partial \overline{F}}{\partial t} + \varepsilon (2\pi)^{d-k} \big[\overline{F}, \overline{\Phi} + \overline{\Phi}_{\mathrm{r}} \big] + \frac{\varepsilon (2\pi)^{d-k}}{N} \int \mathrm{d}\boldsymbol{\mathcal{E}}_2 \, \big[\overline{\mathcal{C}}(\boldsymbol{\mathcal{E}}_1, \boldsymbol{\mathcal{E}}_2), \overline{U}_{12} \big]_{(1)} = 0 \, . \tag{6.42}$$

Because Eq. (6.42) was obtained via an average over the fast angles, one can note in this equation that all the functions appearing in the Poisson brackets only depend on $\boldsymbol{\mathcal{E}}_1 = (\boldsymbol{J}_1, \boldsymbol{\theta}_1^{\mathrm{s}})$. As a consequence, the Poisson brackets from Eq. (6.29) take here the shortened form

$$\big[G_1(\boldsymbol{\mathcal{E}}), G_2(\boldsymbol{\mathcal{E}}) \big] = \frac{\partial G_1}{\partial \boldsymbol{\theta}^{\mathrm{s}}} \cdot \frac{\partial G_2}{\partial \boldsymbol{J}^{\mathrm{s}}} - \frac{\partial G_1}{\partial \boldsymbol{J}^{\mathrm{s}}} \cdot \frac{\partial G_2}{\partial \boldsymbol{\theta}^{\mathrm{s}}} \, , \tag{6.43}$$

so that only derivatives w.r.t. the slow coordinates come into play. Let us now define the rescaled time τ as

$$\tau = (2\pi)^{d-k} \, \varepsilon t \, , \tag{6.44}$$

so that Eq. (6.42) finally becomes

$$\frac{\partial \overline{F}}{\partial \tau} + \big[\overline{F}, \overline{\Phi} + \overline{\Phi}_{\mathrm{r}} \big] + \frac{1}{N} \int \mathrm{d}\boldsymbol{\mathcal{E}}_2 \, \big[\overline{\mathcal{C}}(\boldsymbol{\mathcal{E}}_1, \boldsymbol{\mathcal{E}}_2), \overline{U}_{12} \big]_{(1)} = 0 \, . \tag{6.45}$$

Let us follow a similar angle-averaging procedure to deal with the second equation (6.33) of the BBGKY hierarchy. Let us therefore multiply Eq. (6.33) by $1/(2\pi)^{2(d-k)} \int \mathrm{d}\boldsymbol{\theta}_1^{\mathrm{f}} \mathrm{d}\boldsymbol{\theta}_2^{\mathrm{f}}$ and rely on the assumptions from Eqs. (6.34) and (6.39). Keeping only terms of order ε, Eq. (6.33) finally becomes

$$\begin{aligned} & \frac{1}{2} \frac{\partial \overline{\mathcal{C}}}{\partial \tau} + \big[\overline{\mathcal{C}}(\boldsymbol{\mathcal{E}}_1, \boldsymbol{\mathcal{E}}_2), \overline{\Phi}(\boldsymbol{\mathcal{E}}_1) + \overline{\Phi}_{\mathrm{r}}(\boldsymbol{\mathcal{E}}_1) \big]_{(1)} + \frac{\big[\overline{F}(\boldsymbol{\mathcal{E}}_1) \overline{F}(\boldsymbol{\mathcal{E}}_2), \overline{U}_{12} \big]_{(1)}}{(2\pi)^{d-k}} \\ & + \int \mathrm{d}\boldsymbol{\mathcal{E}}_3 \, \overline{\mathcal{C}}(\boldsymbol{\mathcal{E}}_2, \boldsymbol{\mathcal{E}}_3) \, \big[\overline{F}(\boldsymbol{\mathcal{E}}_1), \overline{U}_{13} \big]_{(1)} + (1 \leftrightarrow 2) = 0 \, , \end{aligned} \tag{6.46}$$

where once again, one can note that all the additional kinetic terms present in the second line of Eq. (6.33) vanish at the order considered here. Equations (6.45) and (6.46) are the main results of this section. They describe the coupled evolutions of the system's averaged DF $\overline{F}$ and 2−body correlation $\overline{\mathcal{C}}$ driven by finite−N effects. Let us already underline the strong analogies between these two equations and the non-degenerate equations (2.107) and (2.108). A rewriting of the degenerate equations (6.45) and (6.46) was also recently obtained in Sridhar and Touma (2016a, b), following Gilbert's method (Gilbert 1968). Starting from Eq. (6.45), one can now investigate at least four different dynamical regimes of evolution, as we now detail:

I Considering Eq. (6.45), the system of Keplerian wires could be initially far from a quasi-stationary equilibrium, so that $[\overline{F}, \overline{\Phi} + \overline{\Phi}_{\rm r}] \neq 0$. It is then expected that the system will undergo a first collisionless phase of violent relaxation (Lynden-Bell 1967), allowing it to rapidly reach a quasi-stationary equilibrium. See Fig. 1.8 and the associated discussion, for an illustration of the classical violent relaxation in self-gravitating systems. We do not investigate this process further here. However, we assume that this collisionless relaxation of Keplerian wires is efficient enough, so that the system rapidly reaches a quasi-stationary stable state. Following this initial violent phase, the system's dynamics is then driven by a much slower secular evolution, either collisionless (item III) or collisional (item IV).

II For a given stationary DF of Keplerian wires, Eq. (6.45) also captures the system's gravitational susceptibility, so that one could also investigate the possible existence of collisionless dynamical instabilities through the equation $\partial \overline{F}/\partial \tau + [\overline{F}, \overline{\Phi} + \overline{\Phi}_{\rm r}] = 0$. See Appendix 4.C for an illustration of dynamical instabilities in non-degenerate stellar discs. We do not investigate such instabilities further here. However, similarly to what was assumed for non-degenerate systems, we suppose that throughout its evolution, the system, while still being able to amplify and dress perturbations, always remains dynamically stable w.r.t. the collisionless dynamics. See, e.g., Tremaine (2005); Polyachenko et al. (2007); Jalali and Tremaine (2012) for examples of stability investigations in the quasi-Keplerian context.

III After the system has reached a quasi-stationary stable state, one may now study the system's secular evolution along quasi-stationary stable equilibria. As was presented in detail in Sect. 2.2, a first way to induce a long-term evolution is via the presence of external stochastic fluctuations. In order to describe such externally-induced secular collisionless evolution, one would start from Eq. (6.45), neglect the contributions associated with the collisional term in $1/N$ in Eq. (6.45), and look for the long-term effects of external perturbations. This would correspond to the specification to degenerate quasi-Keplerian systems of the secular collisionless stochastic forcing considered in Sect. 2.2. In the case of quasi-Keplerian systems, one additional difficulty comes from the canonical change of coordinates we had to perform in Eq. (6.4) to emphasise the properties of the dominant BH-induced Keplerian dynamics. Adding external perturbations may offset the system and introduce non-trivial inertial forces, which should be dealt with carefully. We do not present thereafter the specification of such externally forced secular dynamics to the case of quasi-Keplerian systems.

IV During its secular evolution along quasi-stationary equilibria, the dynamics of an isolated system of Keplerian wires may also be driven by finite$-N$ fluctuations. Such a self-induced collisional dynamics was presented in detail in Sect. 2.3, in the context of non-degenerate inhomogeneous systems. In the present quasi-Keplerian context, this amounts to neglecting any effects associated with external stochastic perturbations and consider the contributions coming from the $1/N$ collisional term in Eq. (6.45). Characterising this collisional term requires to consider as well Eq. (6.46), which describes the dynamics of the system's fluctuations. In Sect. 6.5, we present in detail how this approach may be pursued for quasi-Keplerian systems. We then derive the analog of the Balescu–Lenard equation (2.67) in the context of degenerate dynamical systems, such as galactic nuclei. As discussed in the next section, such a diffusion equation sourced by finite$-N$ fluctuations captures the known mechanism of resonant relaxation (Rauch and Tremaine 1996) of particular relevance for galactic nuclei. See Bar-Or and Alexander (2014) for a similar study of the effect of finite$-N$ stochastic internal forcings via the so-called $\eta-$formalism.

Let us finally note that one could also investigate the secular dynamics of a quasi-stationary non-axisymmetric set of eccentric orbits orbiting a central black hole. This would correspond for example to the expected lopsided configuration of M31's galactic centre (Tremaine 1995). In order to derive the Balescu–Lenard equation associated with such a configuration, one would first have to find new angle-action variables for which item II would be satisfied, and then proceed further with the formalism. We also emphasised in item II, the role played by the system's self-gravity, whose importance varies with the context. Indeed, depending on the mass of the stellar cluster, one expects that there exists a regime for which the system's self-induced orbital precession is significant, but the wires' self-gravity remains too weak to induce a collisionless instability. In such a regime, accounting for the self-induced polarisation of the wires becomes important in item III and IV. This motivates the upcoming derivation of the degenerate Balescu–Lenard equation.

6.5 The Degenerate Balescu–Lenard Equation

Assuming that the evolution of the system is driven by finite$-N$ effects, let us now illustrate how one may derive the inhomogeneous degenerate Balescu–Lenard equation. This equation captures the long-term effects of the $1/N$ collisional contribution in Eq. (6.45). It is assumed that the system is isolated and undergoes no external perturbations. In order to ease the derivation of the closed kinetic equation satisfied by $\overline{F}$, we rely on the strong analogies between the present quasi-Keplerian case and the non-degenerate equations considered in Appendix 2.B. As already underlined in Sect. 2.3.1 when deriving the Balescu–Lenard equation, we rely on the adiabatic approximation that the system secularly relaxes through a series of collisionless equilibria. In the present quasi-Keplerian context, these collisionless equilibria correspond to stationary (and stable) steady states of the collisionless advection term $\left[\overline{F}, \overline{\Phi} + \overline{\Phi}_{\rm r}\right]$ from Eq. (6.45). Let us therefore assume that throughout its evolution, the system's DF satisfies

$$\forall \tau\,, \; \left[\overline{F}(\tau), \overline{\Phi}(\tau) + \overline{\Phi}_{\rm r}(\tau)\right] = 0\,. \tag{6.47}$$

As already underlined in item I of the previous section, it is expected that such collisionless equilibria are rapidly reached by the system (on a few $T_{\rm sec}$) through an efficient out-of-equilibrium violent relaxation. In addition, we also assume that the symmetry of the system is such that collisionless equilibria are of the form

$$\overline{F}(\boldsymbol{J}, \boldsymbol{\theta}^{\rm s}, \tau) = \overline{F}(\boldsymbol{J}, \tau)\,, \tag{6.48}$$

so that throughout its evolution, the system's averaged DF $\overline{F}$ does not depend on the slow angles $\boldsymbol{\theta}^{\rm s}$. At this stage, let us note that in the present quasi-Keplerian context, for which additional conserved quantities other than the actions $\boldsymbol{J}$ are available (namely the slow angles $\boldsymbol{\theta}^{\rm s}$), the assumption from Eq. (6.48) limits the breadth of collisionless equilibria which can considered. For example, lopsided collisionless equilibria, such as the one expected in M31, cannot be considered. Despite the assumption from Eq. (6.48), let us emphasise however that the system's averaged autocorrelation $\overline{\mathcal{C}}$, which evolves according to Eq. (6.46), still depends on the two slow angles $\boldsymbol{\theta}_1^{\rm s}$ and $\boldsymbol{\theta}_2^{\rm s}$. Finally, let us assume that the system's symmetry also guarantees that

$$\overline{F} = \overline{F}(\boldsymbol{J}) \quad \Rightarrow \quad \overline{\Phi} = \overline{\Phi}(\boldsymbol{J}) \;\; \text{and} \;\; \overline{\Phi}_{\rm r} = \overline{\Phi}_{\rm r}(\boldsymbol{J})\,. \tag{6.49}$$

One should note that the previous assumptions, while being restrictive, are still satisfied, among others, for two important cases namely razor-thin axisymmetric discs (see Sect. 6.6.1) and $3D$ spherical systems (see Sect. 6.6.2). When assuming Eqs. (6.48) and (6.49), the equilibrium condition from Eq. (6.47) is then immediately satisfied. Finally, let us introduce the total precession frequencies $\boldsymbol{\Omega}^{\rm s}$ as

$$\boldsymbol{\Omega}^{\rm s}(\boldsymbol{J}) = \frac{\partial[\overline{\Phi} + \overline{\Phi}_{\rm r}]}{\partial \boldsymbol{J}^{\rm s}}\,. \tag{6.50}$$

These frequencies capture the precession of the slow angles $\boldsymbol{\theta}^{\rm s}$, i.e. the precession of the Keplerian wires induced by the joint contributions from the system's self-consistent potential $\overline{\Phi}$ and the relativistic corrections $\overline{\Phi}_{\rm r}$. Let us note that these frequencies do not involve the Keplerian frequencies from Eq. (6.17) and are therefore a priori non-degenerate. The two evolution equations (6.45) and (6.46) can then be rewritten as

$$\frac{\partial \overline{F}}{\partial \tau} + \frac{1}{N}\!\int\! \mathrm{d}\boldsymbol{\mathcal{E}}_2 \left[\mathcal{C}(\boldsymbol{\mathcal{E}}_1, \boldsymbol{\mathcal{E}}_2), \overline{U}_{12}\right]_{(1)} = 0\,, \tag{6.51}$$

and

$$\frac{1}{2}\frac{\partial \overline{\mathcal{C}}}{\partial t} + \boldsymbol{\Omega}_1^{\mathrm{s}} \cdot \frac{\partial \overline{\mathcal{C}}(\boldsymbol{\mathcal{E}}_1, \boldsymbol{\mathcal{E}}_2)}{\partial \boldsymbol{\theta}_1^{\mathrm{s}}} - \frac{1}{(2\pi)^{d-k}} \frac{\partial \overline{F}}{\partial \boldsymbol{J}_1^{\mathrm{s}}} \cdot \frac{\partial \overline{U}_{12}}{\partial \boldsymbol{\theta}_1^{\mathrm{s}}}$$
$$- \int \mathrm{d}\boldsymbol{\mathcal{E}}_3 \, \overline{\mathcal{C}}(\boldsymbol{\mathcal{E}}_2, \boldsymbol{\mathcal{E}}_3) \frac{\partial \overline{F}}{\partial \boldsymbol{J}_1^{\mathrm{s}}} \cdot \frac{\partial \overline{U}_{13}}{\partial \boldsymbol{\theta}_1^{\mathrm{s}}} + (1 \leftrightarrow 2) = 0 \, . \tag{6.52}$$

At this stage, let us emphasise how the two coupled evolution equations (6.51) and (6.52) are similar to the non-degenerate equations (2.107) and (2.108). The only differences correspond to changes in the prefactors, as well as to the fact that only derivatives w.r.t. the slow angles and actions $\boldsymbol{\theta}^{\mathrm{s}}$ and $\boldsymbol{J}^{\mathrm{s}}$ are present in the quasi-Keplerian context. Relying on these strong similarities, we may follow the same method as in Appendix 2.B to derive the kinetic equation satisfied by the averaged DF $\overline{F}$. Because of these analogies, we do not repeat here this derivation, but refer to Appendix B in Fouvry et al. (2017a) for a detailed presentation of this derivation. As a brief summary, let us recall the main steps of this calculation. The first step is to solve Eq. (6.52) to obtain the system autocorrelation $\overline{\mathcal{C}}$ as a functional of the system's $1-$body DF $\overline{F}$. To do so, one relies on Bogoliubov's ansatz, which assumes that $\overline{F}$ evolves on a timescale much larger than the one associated with $\overline{\mathcal{C}}$. Injecting this inverted expression of $\overline{\mathcal{C}}$ in Eq. (6.51), one finally obtains the closed kinetic equation satisfied by $\overline{F}$ only. This is the degenerate inhomogeneous Balescu–Lenard equation, that we will now present in detail.

6.5.1 The One-Component Balescu–Lenard Equation

Once the two coupled equations (6.51) and (6.52) are solved, one gets the degenerate inhomogeneous Balescu–Lenard equation reading

$$\frac{\partial \overline{F}}{\partial \tau} = \frac{\pi (2\pi)^{2k-d}}{N} \frac{\partial}{\partial \boldsymbol{J}_1^{\mathrm{s}}} \cdot \Bigg[\sum_{\boldsymbol{m}_1^{\mathrm{s}}, \boldsymbol{m}_2^{\mathrm{s}}} \boldsymbol{m}_1^{\mathrm{s}} \int \mathrm{d}\boldsymbol{J}_2 \, \frac{\delta_{\mathrm{D}}(\boldsymbol{m}_1^{\mathrm{s}} \cdot \boldsymbol{\Omega}_1^{\mathrm{s}} - \boldsymbol{m}_2^{\mathrm{s}} \cdot \boldsymbol{\Omega}_2^{\mathrm{s}})}{|\mathcal{D}_{\boldsymbol{m}_1^{\mathrm{s}}, \boldsymbol{m}_2^{\mathrm{s}}}(\boldsymbol{J}_1, \boldsymbol{J}_2, \boldsymbol{m}_1^{\mathrm{s}} \cdot \boldsymbol{\Omega}_1^{\mathrm{s}})|^2}$$
$$\times \bigg(\boldsymbol{m}_1^{\mathrm{s}} \cdot \frac{\partial}{\partial \boldsymbol{J}_1^{\mathrm{s}}} - \boldsymbol{m}_2^{\mathrm{s}} \cdot \frac{\partial}{\partial \boldsymbol{J}_2^{\mathrm{s}}} \bigg) \overline{F}(\boldsymbol{J}_1, \tau) \, \overline{F}(\boldsymbol{J}_2, \tau) \Bigg] . \tag{6.53}$$

Let us first strongly emphasise how this degenerate Balescu–Lenard equation resembles the non-degenerate one from Eq. (2.67). Let us recall now some important properties of this diffusion equation. In Eq. (6.53), we noted as d the dimension of the physical space, and k as the number of dynamical degeneracies of the underlying zeroth-order potential (here, the Keplerian potential induced by the BH). The r.h.s. of Eq. (6.53) is the degenerate inhomogeneous Balescu–Lenard collision operator. It captures the secular diffusion of Keplerian wires induced by finite$-N$ fluctuations, i.e. it describes the distortion of these wires as their actions diffuse through their self-interaction. Of course, because it was obtained thanks to a kinetic development at order $1/N$, the r.h.s. of Eq. (6.53) vanishes in the limit $N \to +\infty$. Equation (6.53) also encompasses a resonance condition via the Dirac delta $\delta_{\mathrm{D}}(\boldsymbol{m}_1^{\mathrm{s}} \cdot \boldsymbol{\Omega}_1^{\mathrm{s}} - \boldsymbol{m}_2^{\mathrm{s}} \cdot \boldsymbol{\Omega}_2^{\mathrm{s}})$ (with the shortened notation $\boldsymbol{\Omega}_i^{\mathrm{s}} = \boldsymbol{\Omega}^{\mathrm{s}}(\boldsymbol{J}_i)$), where $\boldsymbol{m}_1^{\mathrm{s}}, \boldsymbol{m}_2^{\mathrm{s}} \in \mathbb{Z}^k$ are integer resonance vectors. It is important to note here that this resonance condition only involves the precession frequencies of the Keplerian wires. As already noted previously, for a given resonance vector $\boldsymbol{m}_1^{\mathrm{s}}$, the diffusion in action space will occur along the discrete direction given by this vector. One should finally interpret the integration over the dummy variable $\boldsymbol{J}_2$ as a scan of action space looking for regions where the resonance condition is satisfied. These resonant distant encounters between precessing Keplerian wires are the drivers of the collisional evolution. In analogy with Fig. 2.5, we illustrate in Fig. 6.2 this resonance condition on precession frequencies. We also note that Eq. (6.53) involves the antisymmetric operator, $\boldsymbol{m}_1^{\mathrm{s}} \cdot \partial/\partial \boldsymbol{J}_1^{\mathrm{s}} - \boldsymbol{m}_2^{\mathrm{s}} \cdot \partial/\partial \boldsymbol{J}_2^{\mathrm{s}}$, which when applied to the quadratic term $\overline{F}(\boldsymbol{J}_1)\overline{F}(\boldsymbol{J}_2)$ "weighs" the relative number of pairwise resonant orbits caught in the resonant configuration. Because it accounts for collective effects, i.e. the dressing of fluctuations by the system's susceptibility, Eq. (6.53) involves the dressed susceptibility coefficients $1/\mathcal{D}_{\boldsymbol{m}_1^{\mathrm{s}},\boldsymbol{m}_2^{\mathrm{s}}}(\boldsymbol{J}_1, \boldsymbol{J}_2, \omega)$. In the quasi-Keplerian context, the dressed susceptibility coefficients from Eq. (2.50) become

$$\frac{1}{\mathcal{D}_{\boldsymbol{m}_1^{\mathrm{s}},\boldsymbol{m}_2^{\mathrm{s}}}(\boldsymbol{J}_1, \boldsymbol{J}_2, \omega)} = \sum_{p,q} \overline{\psi}^{(p)}_{\boldsymbol{m}_1^{\mathrm{s}}}(\boldsymbol{J}_1) \big[\mathbf{I} - \widehat{\mathbf{M}}(\omega)\big]^{-1}_{pq} \overline{\psi}^{(q)*}_{\boldsymbol{m}_2^{\mathrm{s}}}(\boldsymbol{J}_2) \, , \tag{6.54}$$

where $\mathbf{I}$ stands for the identity matrix and $\widehat{\mathbf{M}}$ is the system's averaged response matrix. In the quasi-Keplerian context, the response matrix from Eq. (2.17) becomes

$$\widehat{\mathbf{M}}_{pq}(\omega) = (2\pi)^k \sum_{\boldsymbol{m}^{\mathrm{s}}} \int \mathrm{d}\boldsymbol{J} \, \frac{\boldsymbol{m}^{\mathrm{s}} \cdot \partial \overline{F}/\partial \boldsymbol{J}^{\mathrm{s}}}{\omega - \boldsymbol{m}^{\mathrm{s}} \cdot \boldsymbol{\Omega}^{\mathrm{s}}} \, \overline{\psi}^{(p)*}_{\boldsymbol{m}^{\mathrm{s}}}(\boldsymbol{J}) \, \overline{\psi}^{(q)}_{\boldsymbol{m}^{\mathrm{s}}}(\boldsymbol{J}) \, . \tag{6.55}$$

One can note that in Eqs. (6.54) and (6.55), we had to rely on the matrix method (see Sect. 2.2.2) to relate the DF's perturbations to the induced potential perturbations. In the non-degenerate case, this requires the introduction of a biorthogonal basis of potentials and densities $(\psi^{(p)}, \rho^{(p)})$ satisfying Eq. (2.12). In the degenerate quasi-Keplerian context, we rely on the same method and introduce a basis of potentials and densities satisfying similarly

$$\psi^{(p)}(\boldsymbol{x}) = \int \mathrm{d}\boldsymbol{x}' \, \rho^{(p)}(\boldsymbol{x}') \, U(|\boldsymbol{x} - \boldsymbol{x}'|) \quad ; \quad \int \mathrm{d}\boldsymbol{x} \, \psi^{(p)}(\boldsymbol{x}) \, \rho^{(q)*}(\boldsymbol{x}) = -\delta_p^q \, . \tag{6.56}$$

Equation (6.56) is identical to Eq. (2.12), except for the fact that Eq. (6.56) involves the rescaled interaction potential U from Eq. (6.13). Once the basis elements $\psi^{(p)}$ specified, one can define their average $\overline{\psi}^{(p)}$ as in Eq. (6.26). Finally, following the convention from Eq. (2.6), we define their Fourier transform w.r.t. the slow angles $\boldsymbol{\theta}^{\mathrm{s}}$ as

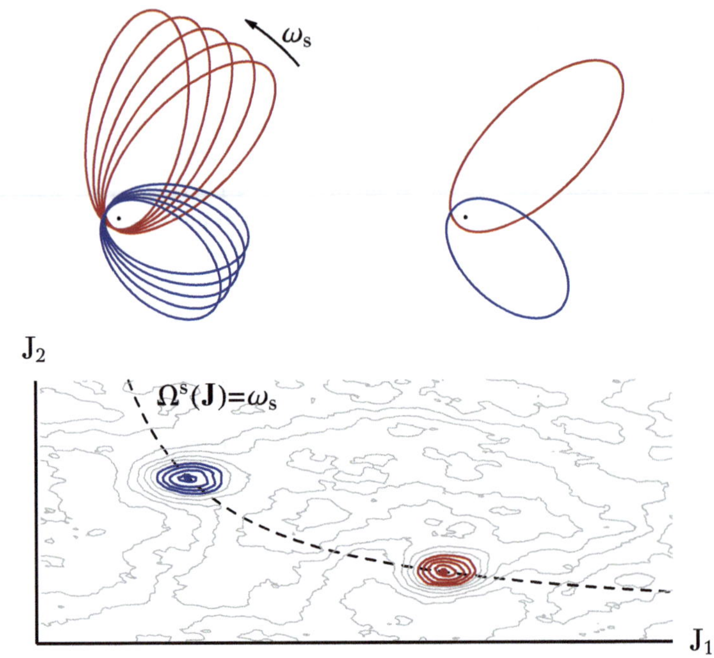

Fig. 6.2 Illustration of the resonance condition $\delta_D(\boldsymbol{m}_1^s\cdot\boldsymbol{\Omega}_1^s-\boldsymbol{m}_2^s\cdot\boldsymbol{\Omega}_2^s)$ appearing in the degenerate inhomogeneous Balescu–Lenard equation (6.53) in the case of a razor-thin axisymmetric quasi-Keplerian disc. *Top-left panel* A set of two resonant Keplerian wires precessing at the same frequency ω_s. *Top-right panel* The same two wires in the rotating frame at frequency ω_s in which the two orbits are in resonance. *Bottom panel* Fluctuations of the system's DF in action space caused by finite$-N$ effects and showing overdensities for the *blue* and *red* wires. The *dashed* line corresponds to the critical resonant line in action space along which the resonance condition $\Omega^s(\boldsymbol{J})=\omega_s$ is satisfied. The two wires satisfy a resonance condition for their precession frequencies. Uncorrelated sequences of such resonant interactions will lead to a secular diffusion of the system's orbital structure following Eq. (6.53). These resonances are non-local in the sense that the two resonant orbits need not be close in position nor in action space. As emphasised in Sect. 6.6.1, in razor-thin axisymmetric discs, the system's symmetry enforces $m_1^s=m_2^s$, i.e. the two orbits are caught in the same resonance

$$\overline{\psi}^{(p)}(\boldsymbol{\mathcal{E}})=\sum_{\boldsymbol{m}^s}\overline{\psi}^{(p)}_{\boldsymbol{m}^s}(\boldsymbol{J})\,\mathrm{e}^{\mathrm{i}\boldsymbol{m}^s\cdot\boldsymbol{\theta}^s}\quad;\quad\overline{\psi}^{(p)}_{\boldsymbol{m}^s}(\boldsymbol{J})=\int\frac{\mathrm{d}\boldsymbol{\theta}^s}{(2\pi)^k}\,\overline{\psi}^{(p)}(\boldsymbol{\mathcal{E}})\,\mathrm{e}^{-\mathrm{i}\boldsymbol{m}^s\cdot\boldsymbol{\theta}^s}\,. \tag{6.57}$$

Inspired by the various rephrasings presented in Sect. 2.3.4, it is straightforward to rewrite Eq. (6.53) as an anisotropic self-consistent non-linear diffusion equation, by introducing the associated drift and diffusion coefficients. The non-degenerate equation (2.68) becomes for quasi-Keplerian systems

$$\frac{\partial\overline{F}}{\partial\tau}=\frac{\partial}{\partial\boldsymbol{J}_1^s}\cdot\bigg[\sum_{\boldsymbol{m}_1^s}\boldsymbol{m}_1^s\bigg(A_{\boldsymbol{m}_1^s}(\boldsymbol{J}_1)\,\overline{F}(\boldsymbol{J}_1)+D_{\boldsymbol{m}_1^s}(\boldsymbol{J}_1)\,\boldsymbol{m}_1^s\cdot\frac{\partial\overline{F}}{\partial\boldsymbol{J}_1^s}\bigg)\bigg], \tag{6.58}$$

where $A_{\boldsymbol{m}_1^{\mathrm{s}}}(\boldsymbol{J}_1)$ and $D_{\boldsymbol{m}_1^{\mathrm{s}}}(\boldsymbol{J}_1)$ stand respectively for the degenerate drift and diffusion coefficients associated with a given resonance $\boldsymbol{m}_1^{\mathrm{s}}$. The Balescu–Lenard equation being a self-consistent integro-differential equation, these drift and diffusion coefficients secularly depend on the system's averaged DF, but this was not written out explicitly to simplify the notations. Similarly to Eqs. (2.69) and (2.70), the degenerate drift and diffusion coefficients appearing in Eq. (6.58) are given by

$$\begin{aligned} A_{\boldsymbol{m}_1^{\mathrm{s}}}(\boldsymbol{J}_1) &= -\frac{\pi(2\pi)^{2k-d}}{N} \sum_{\boldsymbol{m}_2^{\mathrm{s}}} \int \mathrm{d}\boldsymbol{J}_2 \, \frac{\delta_{\mathrm{D}}(\boldsymbol{m}_1^{\mathrm{s}}\cdot\boldsymbol{\Omega}_1^{\mathrm{s}} - \boldsymbol{m}_2^{\mathrm{s}}\cdot\boldsymbol{\Omega}_2^{\mathrm{s}})}{|\mathcal{D}_{\boldsymbol{m}_1^{\mathrm{s}},\boldsymbol{m}_2^{\mathrm{s}}}(\boldsymbol{J}_1, \boldsymbol{J}_2, \boldsymbol{m}_1^{\mathrm{s}}\cdot\boldsymbol{\Omega}_1^{\mathrm{s}})|^2} \, \boldsymbol{m}_2^{\mathrm{s}}\cdot\frac{\partial \overline{F}}{\partial \boldsymbol{J}_2^{\mathrm{s}}} \, , \\ D_{\boldsymbol{m}_1^{\mathrm{s}}}(\boldsymbol{J}_1) &= \frac{\pi(2\pi)^{2k-d}}{N} \sum_{\boldsymbol{m}_2^{\mathrm{s}}} \int \mathrm{d}\boldsymbol{J}_2 \, \frac{\delta_{\mathrm{D}}(\boldsymbol{m}_1^{\mathrm{s}}\cdot\boldsymbol{\Omega}_1^{\mathrm{s}} - \boldsymbol{m}_2^{\mathrm{s}}\cdot\boldsymbol{\Omega}_2^{\mathrm{s}})}{|\mathcal{D}_{\boldsymbol{m}_1^{\mathrm{s}},\boldsymbol{m}_2^{\mathrm{s}}}(\boldsymbol{J}_1, \boldsymbol{J}_2, \boldsymbol{m}_1^{\mathrm{s}}\cdot\boldsymbol{\Omega}_1^{\mathrm{s}})|^2} \, \overline{F}(\boldsymbol{J}_2) \, . \end{aligned} \tag{6.59}$$

As presented in Sect. 2.3.5, when neglecting collective effects (i.e. when neglecting the last term of Eq. (6.52)), the degenerate inhomogeneous Balescu–Lenard equation (6.53) becomes the degenerate inhomogeneous Landau equation reading

$$\begin{aligned} \frac{\partial \overline{F}}{\partial \tau} = \frac{\pi(2\pi)^{2k-d}}{N} \frac{\partial}{\partial \boldsymbol{J}_1^{\mathrm{s}}} \cdot \bigg[\sum_{\boldsymbol{m}_1^{\mathrm{s}},\boldsymbol{m}_2^{\mathrm{s}}} \boldsymbol{m}_1^{\mathrm{s}} \int \mathrm{d}\boldsymbol{J}_2 \, \delta_{\mathrm{D}}(\boldsymbol{m}_1^{\mathrm{s}}\cdot\boldsymbol{\Omega}_1^{\mathrm{s}} - \boldsymbol{m}_2^{\mathrm{s}}\cdot\boldsymbol{\Omega}_2^{\mathrm{s}}) \big| A_{\boldsymbol{m}_1^{\mathrm{s}},\boldsymbol{m}_2^{\mathrm{s}}}(\boldsymbol{J}_1, \boldsymbol{J}_2) \big|^2 \\ \times \bigg(\boldsymbol{m}_1^{\mathrm{s}}\cdot\frac{\partial}{\partial \boldsymbol{J}_1^{\mathrm{s}}} - \boldsymbol{m}_2^{\mathrm{s}}\cdot\frac{\partial}{\partial \boldsymbol{J}_2^{\mathrm{s}}} \bigg) \overline{F}(\boldsymbol{J}_1, \tau)\, \overline{F}(\boldsymbol{J}_2, \tau) \bigg] , \end{aligned} \tag{6.60}$$

where, as in Eq. (2.74), we introduced the averaged bare susceptibility coefficients $A_{\boldsymbol{m}_1^{\mathrm{s}},\boldsymbol{m}_2^{\mathrm{s}}}(\boldsymbol{J}_1, \boldsymbol{J}_2)$ as

$$A_{\boldsymbol{m}_1^{\mathrm{s}},\boldsymbol{m}_2^{\mathrm{s}}}(\boldsymbol{J}_1, \boldsymbol{J}_2) = \int \frac{\mathrm{d}\boldsymbol{\theta}_1^{\mathrm{s}}}{(2\pi)^k} \frac{\mathrm{d}\boldsymbol{\theta}_2^{\mathrm{s}}}{(2\pi)^k} \, \overline{U}_{12}(\boldsymbol{\mathcal{E}}_1, \boldsymbol{\mathcal{E}}_2) \, \mathrm{e}^{-\mathrm{i}(\boldsymbol{m}_1^{\mathrm{s}}\cdot\boldsymbol{\theta}_1^{\mathrm{s}} - \boldsymbol{m}_2^{\mathrm{s}}\cdot\boldsymbol{\theta}_2^{\mathrm{s}})} \, , \tag{6.61}$$

where the averaged rescaled wire-wire interaction potential $\overline{U}_{12}$ was introduced in Eq. (6.37).

As a final remark, let us note that the degenerate Balescu–Lenard equation (6.53) (similarly for the associated Landau equation (6.60)), while defined on the full action space $\boldsymbol{J} = (\boldsymbol{J}^{\mathrm{s}}, \boldsymbol{J}^{\mathrm{f}})$, does not allow for changes in the fast actions $\boldsymbol{J}^{\mathrm{f}}$. Indeed, let us define the marginal DF, $P_{\overline{F}}$, as $P_{\overline{F}} = \int \mathrm{d}\boldsymbol{J}^{\mathrm{s}} \overline{F}(\boldsymbol{J})$. Equation (6.53) then gives

$$\frac{\partial P_{\overline{F}}}{\partial \tau} = 0 \, . \tag{6.62}$$

As a consequence, in the degenerate context, the collisional secular diffusion only occurs in the directions $\boldsymbol{J}^{\mathrm{f}} = \mathrm{cst}$. Such a conservation of the individual fast actions of the particles is a direct consequence of the adiabatic invariance of these actions, whose associated intrinsic frequencies are much faster than the precession frequencies involved in the collisional diffusion.

6.5.2 The Multi-Component Balescu–Lenard Equation

Similarly to what was presented in Sect. 2.3.6, one may also generalise the degenerate Balescu–Lenard equation (6.53) to a system of multiple components. This is of particular relevance for quasi-Keplerian systems such as galactic nuclei, for which one expects that the joint presence of multiple type of stars or black holes orbiting a central supermassive black hole could be of importance for the system's fate, by inducing for example relative segregation.

As in Eq. (2.76), let us assume that the considered system is made of various components indexed by the letters "a" and "b". The particles of the component "a" have an individual mass μ_{a} and follow the DF F^{a}. As detailed in Appendix 6.B (which also details all the normalisations used), the evolution of each DF is then given by the multi-component degenerate inhomogeneous Balescu–Lenard equation reading

$$\frac{\partial \overline{F^{\mathrm{a}}}}{\partial \tau} = \pi (2\pi)^{2k-d} \frac{\partial}{\partial \boldsymbol{J}_1^{\mathrm{s}}} \cdot \Bigg[\sum_{\boldsymbol{m}_1^{\mathrm{s}}, \boldsymbol{m}_2^{\mathrm{s}}} \boldsymbol{m}_1^{\mathrm{s}} \int \mathrm{d}\boldsymbol{J}_2 \, \frac{\delta_{\mathrm{D}}(\boldsymbol{m}_1^{\mathrm{s}} \cdot \boldsymbol{\Omega}_1^{\mathrm{s}} - \boldsymbol{m}_2^{\mathrm{s}} \cdot \boldsymbol{\Omega}_2^{\mathrm{s}})}{|\mathcal{D}_{\boldsymbol{m}_1^{\mathrm{s}}, \boldsymbol{m}_2^{\mathrm{s}}}(\boldsymbol{J}_1, \boldsymbol{J}_2, \boldsymbol{m}_1^{\mathrm{s}} \cdot \boldsymbol{\Omega}_1^{\mathrm{s}})|^2} \times \sum_{\mathrm{b}} \bigg\{ \eta_{\mathrm{b}} \boldsymbol{m}_1^{\mathrm{s}} \cdot \frac{\partial \overline{F^{\mathrm{a}}}}{\partial \boldsymbol{J}_1^{\mathrm{s}}} \overline{F^{\mathrm{b}}}(\boldsymbol{J}_2) - \eta_{\mathrm{a}} \overline{F^{\mathrm{a}}}(\boldsymbol{J}_1) \, \boldsymbol{m}_2^{\mathrm{s}} \cdot \frac{\partial \overline{F^{\mathrm{b}}}}{\partial \boldsymbol{J}_2^{\mathrm{s}}} \bigg\} \Bigg] . \tag{6.63}$$

In Eq. (6.63), we introduced the dimensionless relative mass $\eta_{\mathrm{a}} = \mu_{\mathrm{a}}/M_\star$, where $M_\star = \sum_{\mathrm{a}} M_\star^{\mathrm{a}}$ is the total active mass of the system. Here, η_{a}, which scales like $1/N_{\mathrm{a}}$ with N_{a} the number of particles of the component "a", plays the role of a small parameter and replaces the prefactor $1/N$ present in the one-component equation (6.53). In the multi-component case, the dressed susceptibility coefficients are still given by Eq. (6.54). However, as already noted in Eq. (2.77), the system's response matrix now encompasses all the active components of the system, so that Eq. (6.55) becomes

$$\widehat{\mathbf{M}}_{pq}(\omega) = (2\pi)^k \sum_{\boldsymbol{m}^{\mathrm{s}}} \int \mathrm{d}\boldsymbol{J} \, \frac{\boldsymbol{m}^{\mathrm{s}} \cdot \partial (\sum_{\mathrm{b}} \overline{F^{\mathrm{b}}}) / \partial \boldsymbol{J}^{\mathrm{s}}}{\omega - \boldsymbol{m}^{\mathrm{s}} \cdot \boldsymbol{\Omega}^{\mathrm{s}}} \, \overline{\psi}_{\boldsymbol{m}^{\mathrm{s}}}^{(p)*}(\boldsymbol{J}) \, \overline{\psi}_{\boldsymbol{m}^{\mathrm{s}}}^{(q)}(\boldsymbol{J}) . \tag{6.64}$$

In the limit where only one component is considered, one has $\eta_{\mathrm{a}} = 1/N_{\mathrm{a}}$, and Eq. (6.63) returns to the single mass Balescu–Lenard equation (6.53). The multi-component Balescu–Lenard equation (6.63) describes the evolution of the component "a". It differs from the one-component Balescu–Lenard equation (6.53) via the weight η_{a}, as well as via the sum over all the other components "b" weighted by the factor η_{b}.

Similarly to the non-degenerate equation (2.78), one can rewrite Eq. (6.63) as an anisotropic non-linear diffusion equation reading

$$\frac{\partial \overline{F^{\mathrm{a}}}}{\partial \tau} = \frac{\partial}{\partial \boldsymbol{J}_1^{\mathrm{s}}} \cdot \bigg[\sum_{\boldsymbol{m}_1^{\mathrm{s}}} \boldsymbol{m}_1^{\mathrm{s}} \sum_{\mathrm{b}} \bigg\{ \eta_{\mathrm{a}} A^{\mathrm{b}}_{\boldsymbol{m}_1^{\mathrm{s}}}(\boldsymbol{J}_1)\, \overline{F^{\mathrm{a}}}(\boldsymbol{J}_1) + \eta_{\mathrm{b}} D^{\mathrm{b}}_{\boldsymbol{m}_1^{\mathrm{s}}}(\boldsymbol{J}_1)\, \boldsymbol{m}_1^{\mathrm{s}} \cdot \frac{\partial \overline{F^{\mathrm{a}}}}{\partial \boldsymbol{J}_1^{\mathrm{s}}} \bigg\} \bigg], \tag{6.65}$$

where we introduced the multi-component degenerate drift and diffusion coefficients $A^{\mathrm{b}}_{\boldsymbol{m}_1^{\mathrm{s}}}(\boldsymbol{J}_1)$ and $D^{\mathrm{b}}_{\boldsymbol{m}_1^{\mathrm{s}}}(\boldsymbol{J}_1)$, which depend in particular on the component "b" used as the underlying DF to estimate them. Similarly to the non-degenerate drift and diffusion coefficients from Eqs. (2.79) and (2.80), they read

$$\begin{aligned} A^{\mathrm{b}}_{\boldsymbol{m}_1^{\mathrm{s}}}(\boldsymbol{J}_1) &= -\pi (2\pi)^{2k-d} \sum_{\boldsymbol{m}_2^{\mathrm{s}}} \int \mathrm{d}\boldsymbol{J}_2 \, \frac{\delta_{\mathrm{D}}(\boldsymbol{m}_1^{\mathrm{s}} \cdot \boldsymbol{\Omega}_1^{\mathrm{s}} - \boldsymbol{m}_2^{\mathrm{s}} \cdot \boldsymbol{\Omega}_2^{\mathrm{s}})}{|\mathcal{D}_{\boldsymbol{m}_1^{\mathrm{s}}, \boldsymbol{m}_2^{\mathrm{s}}}(\boldsymbol{J}_1, \boldsymbol{J}_2, \boldsymbol{m}_1^{\mathrm{s}} \cdot \boldsymbol{\Omega}_1^{\mathrm{s}})|^2} \, \boldsymbol{m}_2^{\mathrm{s}} \cdot \frac{\partial \overline{F^{\mathrm{b}}}}{\partial \boldsymbol{J}_2^{\mathrm{s}}} , \\ D^{\mathrm{b}}_{\boldsymbol{m}_1^{\mathrm{s}}}(\boldsymbol{J}_1) &= \pi (2\pi)^{2k-d} \sum_{\boldsymbol{m}_2^{\mathrm{s}}} \int \mathrm{d}\boldsymbol{J}_2 \, \frac{\delta_{\mathrm{D}}(\boldsymbol{m}_1^{\mathrm{s}} \cdot \boldsymbol{\Omega}_1^{\mathrm{s}} - \boldsymbol{m}_2^{\mathrm{s}} \cdot \boldsymbol{\Omega}_2^{\mathrm{s}})}{|\mathcal{D}_{\boldsymbol{m}_1^{\mathrm{s}}, \boldsymbol{m}_2^{\mathrm{s}}}(\boldsymbol{J}_1, \boldsymbol{J}_2, \boldsymbol{m}_1^{\mathrm{s}} \cdot \boldsymbol{\Omega}_1^{\mathrm{s}})|^2} \, \overline{F^{\mathrm{b}}}(\boldsymbol{J}_2) . \end{aligned} \tag{6.66}$$

Finally, following Eq. (2.81), a final enlightening rewriting of Eq. (6.65) is possible as

$$\frac{\partial \overline{F^{\mathrm{a}}}}{\partial \tau} = \frac{\partial}{\partial \boldsymbol{J}_1^{\mathrm{s}}} \cdot \bigg[\sum_{\boldsymbol{m}_1^{\mathrm{s}}} \bigg\{ \eta_{\mathrm{a}} \, A^{\mathrm{tot}}_{\boldsymbol{m}_1^{\mathrm{s}}}(\boldsymbol{J}_1)\, \overline{F^{\mathrm{a}}}(\boldsymbol{J}_1) + D^{\mathrm{tot}}_{\boldsymbol{m}_1^{\mathrm{s}}}(\boldsymbol{J}_1)\, \boldsymbol{m}_1^{\mathrm{s}} \cdot \frac{\partial \overline{F^{\mathrm{a}}}}{\partial \boldsymbol{J}_1^{\mathrm{s}}} \bigg\} \bigg], \tag{6.67}$$

where, similarly to Eq. (2.82), we introduced the total drift and diffusion coefficients $A^{\mathrm{tot}}_{\boldsymbol{m}_1^{\mathrm{s}}}(\boldsymbol{J}_1)$ and $D^{\mathrm{tot}}_{\boldsymbol{m}_1^{\mathrm{s}}}(\boldsymbol{J}_1)$ as

$$A^{\mathrm{tot}}_{\boldsymbol{m}_1^{\mathrm{s}}}(\boldsymbol{J}_1) = \sum_{\mathrm{b}} A^{\mathrm{b}}_{\boldsymbol{m}_1^{\mathrm{s}}}(\boldsymbol{J}_1) \quad ; \quad D^{\mathrm{tot}}_{\boldsymbol{m}_1^{\mathrm{s}}}(\boldsymbol{J}_1) = \sum_{\mathrm{b}} \eta_{\mathrm{b}} \, D^{\mathrm{b}}_{\boldsymbol{m}_1^{\mathrm{s}}}(\boldsymbol{J}_1) . \tag{6.68}$$

In Eq. (6.67), let us emphasise that the total drift coefficients are multiplied by the dimensionless mass η_{a} of the considered component. This essentially captures the process of segregation, when multiple masses are involved, as components with larger individual masses will secularly tend towards narrower steady states. This can be seen for example in Eq. (6.67) by seeking asymptotic stationary states to Eq. (6.67) by nulling the curly brace in its r.h.s.

As a final remark, let us note that the demonstration in Sect. 2.3.7 that the non-degenerate Balescu–Lenard equation (2.67) satisfies a H-theorem for Boltzmann's entropy naturally extends to the degenerate case. Therefore, the degenerate Balescu–Lenard equation (2.67) and Landau equation (6.60) satisfy a H-theorem for the system's entropy defined similarly to Eq. (2.83) as $S(\tau) = -\int \mathrm{d}\boldsymbol{J}_1 s(\overline{F}(\boldsymbol{J}_1, \tau))$. Finally, following Eq. (2.91), the multi-component degenerate Balescu–Lenard equation (6.63) satisfies similarly a H-theorem for the system's total entropy defined as $S_{\mathrm{tot}}(\tau) = -\int \mathrm{d}\boldsymbol{J}_1 \sum_{\mathrm{a}} (1/\eta_{\mathrm{a}}) s(\overline{F^{\mathrm{a}}}(\boldsymbol{J}_1, \tau))$.

6.6 Applications

In the previous sections, we detailed how one could describe the secular dynamics of a system composed of a finite number of particles orbiting a central massive object. In the derivation of the degenerate Balescu–Lenard equation (6.53), we especially assumed in Eq. (6.49), that the symmetry of the considered system was such that DFs $\overline{F}$ depending only on the actions would lead to self-consistent potentials $\overline{\Phi}$ depending as well only on the actions. Let us now examine in turn some more specific configurations for which the assumption from Eq. (6.49) is indeed satisfied, and discuss how the previous results can be further extended for these specific geometries. This will allow us to underline the wealth of possible physical implications one can draw from this general framework. Sections 6.6.1 and 6.6.2 will respectively consider the cases of razor-thin axisymmetric discs and $3D$ spherical systems, while in Sect. 6.6.3, we will detail how the present formalism allows us to recover the phenomenon of the relativistic Schwarzschild barrier (Merritt et al. 2011) recently discovered in $N-$body simulations.

6.6.1 Razor-Thin Axisymmetric Discs

As a first case of interest, let us specialise the Balescu–Lenard equation (6.53) to razor-thin axisymmetric discs. The dimension of the physical space is $d=2$, while the number of degeneracies of the Keplerian dynamics is $k=1$. For such systems, the resonance condition from Eq. (6.53) becomes a simpler $1D$ resonance condition naively reading $m_1^{\mathrm{s}}\Omega_1^{\mathrm{s}} - m_2^{\mathrm{s}}\Omega_2^{\mathrm{s}} = 0$. Let us now detail how the Balescu–Lenard equation (6.53) can be further simplified in the case of razor-thin discs, as a consequence of additional symmetries of the pairwise interaction potential. For razor-thin discs, the Delaunay angle-action coordinates from Eq. (6.19) become

$$(\boldsymbol{J}, \boldsymbol{\theta}) = (J_1, J_2, \theta_1, \theta_2) = (J^{\mathrm{s}}, J^{\mathrm{f}}, \theta_1, \theta_2) = (L, I, g, w)\,. \tag{6.69}$$

Introducing the polar coordinates (R, ϕ), the rescaled interaction potential U_{12} from Eq. (6.13) can be written as

$$U_{12} = -\frac{G M_\bullet}{|\boldsymbol{x}_1 - \boldsymbol{x}_2|} = -\frac{G M_\bullet}{\sqrt{R_1^2 + R_1^2 - 2 R_1 R_2 \cos(\phi_1 - \phi_2)}}\,. \tag{6.70}$$

Following Eqs. (3.28a) and (3.28b) of Binney and Tremaine (2008), the mapping from the physical polar coordinates to the $2D$ Delaunay angle-action coordinates reads

$$R = a(1 - e\cos(\eta)) \quad ; \quad \phi = g + f\,, \tag{6.71}$$

where we introduced the semi-major axis a, the eccentricity e, the true anomaly f, and the eccentric anomaly η as

$$a = \frac{I^2}{GM_\bullet} \;\; ; \;\; e = \sqrt{1-(L/I)^2} \;\; ; \;\; f = \tan^{-1}\left[\frac{\sqrt{1-e^2}\sin(\eta)}{\cos(\eta)-e}\right] \;\; ; \;\; w = \eta - e\sin(\eta) \, . \tag{6.72}$$

These mappings allow us to rewrite the interaction potential from Eq. (6.70) as

$$U_{12} = U(g_1-g_2, w_1, w_2, \boldsymbol{J}_1, \boldsymbol{J}_2) \implies \overline{U}_{12} = \overline{U}(g_1-g_2, \boldsymbol{J}_1, \boldsymbol{J}_2) \, . \tag{6.73}$$

Because of this dependence, the bare susceptibility coefficients from Eq. (6.61) satisfy

$$A_{m_1^s, m_2^s}(\boldsymbol{J}_1, \boldsymbol{J}_2) = \delta_{m_1^s}^{m_2^s} \, A_{m_1^s, m_1^s}(\boldsymbol{J}_1, \boldsymbol{J}_2) \, . \tag{6.74}$$

Let us now show that a similar relation also holds for the dressed susceptibility coefficients from Eq. (6.54). When computing the response matrix for razor-thin stellar discs in Chap. 4, we already emphasised in Eq. (4.5) that the basis elements for a razor-thin disc may generically be written as

$$\psi^{(p)}(R, \phi) = \mathrm{e}^{\mathrm{i}\ell^p} \, \mathcal{U}_{n^p}^{\ell^p}(R) \, , \tag{6.75}$$

where ℓ^p and n^p are two integer indices, and $\mathcal{U}_n^\ell$ are radial functions (see Fig. 4.6 for an illustration of possible radial functions). With the decomposition from Eq. (6.75), one can note that the azimuthal and radial dependences have been disentangled. In the mapping from Eq. (6.71), only the azimuthal angle ϕ depends on the slow angle g, so that the Fourier transformed averaged basis elements from Eq. (6.57) satisfy

$$\overline{\psi}_{m^s}^{(p)}(\boldsymbol{J}) = \delta_{\ell^p}^{m^s} \, \overline{\psi}_{m^s}^{(p)}(\boldsymbol{J}) \, . \tag{6.76}$$

As a consequence, the system's response matrix from Eq. (6.55) satisfies

$$\widehat{\mathbf{M}}_{pq}(\omega) = \delta_{\ell^p}^{\ell^q} \, \widehat{\mathbf{M}}_{pq}(\omega) \, . \tag{6.77}$$

The two properties from Eqs. (6.76) and (6.77) finally allow us to rewrite the dressed susceptibility coefficients from Eq. (6.54) as

$$\frac{1}{\mathcal{D}_{m_1^s, m_2^s}(\boldsymbol{J}_1, \boldsymbol{J}_2, \omega)} = \delta_{m_1^s}^{m_2^s} \, \frac{1}{\mathcal{D}_{m_1^s, m_1^s}(\boldsymbol{J}_1, \boldsymbol{J}_2, \omega)} \, , \tag{6.78}$$

so that both the bare and dressed susceptibility coefficients satisfy similar relations.

The two additional symmetry properties from Eqs. (6.74) and (6.78) allow us to simplify the resonance condition of the Balescu–Lenard equation (6.53). For razor-thin discs, one can write

$$\frac{\partial \overline{F}}{\partial \tau} = \frac{\pi}{N} \frac{\partial}{\partial J_1^{\mathrm{s}}} \left[\int \mathrm{d}\boldsymbol{J}_2 \, \frac{\delta_{\mathrm{D}}(\Omega^{\mathrm{s}}(\boldsymbol{J}_1) - \Omega^{\mathrm{s}}(\boldsymbol{J}_2))}{|\mathcal{D}_{\mathrm{tot}}(\boldsymbol{J}_1, \boldsymbol{J}_2)|^2} \left(\frac{\partial}{\partial J_1^{\mathrm{s}}} - \frac{\partial}{\partial J_2^{\mathrm{s}}} \right) \overline{F}(\boldsymbol{J}_1) \, \overline{F}(\boldsymbol{J}_2) \right], \tag{6.79}$$

where we introduced the system's unique total dressed susceptibility coefficient as

$$\frac{1}{|\mathcal{D}_{\mathrm{tot}}(\boldsymbol{J}_1, \boldsymbol{J}_2)|^2} = \sum_{m_1^{\mathrm{s}}} \frac{|m_1^{\mathrm{s}}|}{|\mathcal{D}_{m_1^{\mathrm{s}}, m_1^{\mathrm{s}}}(\boldsymbol{J}_1, \boldsymbol{J}_2, m_1^{\mathrm{s}} \Omega^{\mathrm{s}}(\boldsymbol{J}_1))|^2} \, . \tag{6.80}$$

If one neglects collective effects, Eq. (6.79) becomes the associated Landau equation, for which the total dressed susceptibility coefficient $1/|\mathcal{D}_{\mathrm{tot}}(\boldsymbol{J}_1, \boldsymbol{J}_2)|^2$ should be replaced by the bare one $|A_{\mathrm{tot}}(\boldsymbol{J}_1, \boldsymbol{J}_2)|^2$ reading

$$|A_{\mathrm{tot}}(\boldsymbol{J}_1, \boldsymbol{J}_2)|^2 = \sum_{m_1^{\mathrm{s}}} |m_1^{\mathrm{s}}| \, |A_{m_1^{\mathrm{s}}, m_1^{\mathrm{s}}}(\boldsymbol{J}_1, \boldsymbol{J}_2)|^2 \, . \tag{6.81}$$

The Landau analog of Eq. (6.79) for razor-thin axisymmetric discs with the bare susceptibility coefficients from Eq. (6.81) was also recently derived in Sridhar and Touma (2017) via Gilbert's equation.

Thanks to these additional symmetries, the degenerate Balescu–Lenard equation (6.79) for razor-thin discs involves a simpler resonance condition, which constrains resonant encounters to occur only between Keplerian wires caught in the same resonance, as illustrated in Fig. 6.2. Finally, in order to compute effectively the diffusion flux from Eq. (6.79), one can follow the exact same approach as detailed in Sect. 4.2.5 to deal with the resonance condition. We do not repeat here this method. More details can be found in Sect. 6.1 of Fouvry et al. (2017a). Thanks to Eqs. (6.36) and (6.103), one can compute the two quasi-stationary potentials $\overline{\Phi}$ and $\overline{\Phi}_{\mathrm{r}}$, which respectively capture the contributions from the self-induced potential as well as the relativistic corrections. One can then estimate the associated precession frequencies Ω^{s}, thanks to which the critical resonant lines $\gamma(\omega) = \{\boldsymbol{J} \,|\, \Omega^{\mathrm{s}}(\boldsymbol{J}) = \omega\}$ can be determined. These curves characterise the set of all orbits which precess at the same frequency ω. As already emphasised in Eq. (4.29), the calculation of the diffusion flux then only involves a simple one-dimensional integral of a regular integrand along these resonant lines. In the context of quasi-Keplerian systems, one expects two additional difficulties. The first one is associated with the calculation of the wire-wire interaction potential $\overline{U}_{12}$ from Eq. (6.37), which exhibits a diverging behaviour as one considers the interaction of nearby wires (see, e.g., Touma et al. 2009). The second difficulty arises from the calculation of the system's response matrix given by Eq. (6.55), which, as already emphasised in Sect. 4.2.3, can be a cumbersome and delicate task.

6.6.2 Spherical Clusters

Let us now specify the degenerate Balescu–Lenard equation (6.53) to $3D$ spherical systems. The dimension of the physical space is $d=3$, while the number of degeneracies of the Keplerian dynamics is $k=2$. As a consequence, the resonance condition from Eq. (6.53) becomes two-dimensional. In the $3D$ context, the Delaunay variables from Eq. (6.19) become

$$(\boldsymbol{J}, \boldsymbol{\theta}) = (J_1^{\mathrm{s}}, J_2^{\mathrm{s}}, J_3^{\mathrm{f}}, \theta_1^{\mathrm{s}}, \theta_2^{\mathrm{s}}, \theta_3^{\mathrm{f}}) = (L, L_z, I, g, h, w)\,, \tag{6.82}$$

where, as previously, g stands for the angle from the ascending node to the periapsis, h for the longitude of the ascending node, and w for the mean anomaly, i.e. the Keplerian orbital phase. As in the previous section, let us now detail how the $3D$ spherical geometry allows us to simplify the degenerate Balescu–Lenard equation. Within the spherical coordinates (R, θ, ϕ), the rescaled interaction potential U from Eq. (6.13) can be written as

$$U_{12} = -\frac{GM_\bullet}{|\boldsymbol{x}_1-\boldsymbol{x}_2|} = -GM_\bullet\bigg[R_1^2+R_2^2-2R_1R_2\bigg\{\sin(\theta_1)\sin(\theta_2)\cos(\phi_1-\phi_2)+\cos(\theta_1)\cos(\theta_2)\bigg\}\bigg]^{-1/2}. \tag{6.83}$$

Following Eq. (5.20) of Merritt (2015), the mapping from the physical spherical coordinates to the Delaunay angle-action ones takes the form

$$R = a(1-e\cos(\eta)) \;\;;\;\; \phi = h + \tan^{-1}\big[\cos(i)\tan(g+f)\big] \;\;;\;\; \theta = \cos^{-1}\big[\sin(i)\sin(g+f)\big], \tag{6.84}$$

where a, e, f and η were previously introduced in Eq. (6.72). We also introduced the orbit's inclination i as $\cos(i)=L_z/L$. When used in Eq. (6.83), these mappings immediately give the dependences

$$U_{12} = U(g_1, g_2, h_1-h_2, w_1, w_2, \boldsymbol{J}_1, \boldsymbol{J}_2) \implies \overline{U}_{12} = \overline{U}(g_1, g_2, h_1-h_2, \boldsymbol{J}_1, \boldsymbol{J}_2)\,. \tag{6.85}$$

Computing the averaged bare susceptibility coefficients from Eq. (6.61), one immediately gets

$$A_{\boldsymbol{m}_1^{\mathrm{s}},\boldsymbol{m}_2^{\mathrm{s}}}(\boldsymbol{J}_1, \boldsymbol{J}_2) = \delta_{m_{1,h}^{\mathrm{s}}}^{m_{2,h}^{\mathrm{s}}}\, A_{\boldsymbol{m}_1^{\mathrm{s}},\boldsymbol{m}_2^{\mathrm{s}}}(\boldsymbol{J}_1, \boldsymbol{J}_2)\,, \tag{6.86}$$

where we wrote the resonance vectors as $\boldsymbol{m}_1^{\mathrm{s}}=(m_{1,g}^{\mathrm{s}}, m_{1,h}^{\mathrm{s}})$, so that the coefficient $m_{1,h}^{\mathrm{s}}$ is the one associated with the slow angle h.

As in Sect. 6.6.1, let us briefly emphasise how such a property also holds for the dressed susceptibility coefficients from Eq. (6.54). For $3D$ systems, as already emphasised in Eq. (4.67), the basis elements may generically be cast under the form

$$\psi^{(p)}(R, \theta, \phi) = Y_{\ell^p}^{m^p}(\theta, \phi)\,\mathcal{U}_{n^p}^{\ell^p}(R)\,, \tag{6.87}$$

where ℓ^p, m^p, and n^p are three integer indices, Y_ℓ^m are the usual spherical harmonics, and $\mathcal{U}_n^\ell$ are radial functions. We note in the mappings from Eq. (6.84) that only the azimuthal angle ϕ depends on the slow angle h. Because the spherical harmonics are of the form $Y_\ell^m(\theta, \phi) \propto P_\ell^m(\cos(\theta))\, \mathrm{e}^{\mathrm{i}m\phi}$, where P_ℓ^m are associated Legendre polynomials, one immediately gets from Eq. (6.57) that the averaged Fourier transformed basis elements satisfy

$$\overline{\psi}^{(p)}_{\boldsymbol{m}^{\mathrm{s}}}(\boldsymbol{J}) = \delta_{m^p}^{m_h^{\mathrm{s}}}\, \overline{\psi}^{(p)}_{\boldsymbol{m}^{\mathrm{s}}}(\boldsymbol{J})\,. \tag{6.88}$$

The expression (6.55) of the system's response matrix then straightforwardly satisfies

$$\widehat{\mathbf{M}}_{pq}(\omega) = \delta_{m^p}^{m^q}\, \widehat{\mathbf{M}}_{pq}(\omega)\,. \tag{6.89}$$

The combination of the two properties from Eqs. (6.88) and (6.89) allow us to finally rewrite the dressed susceptibility coefficients from Eq. (6.54) as

$$\frac{1}{\mathcal{D}_{\boldsymbol{m}_1^{\mathrm{s}},\boldsymbol{m}_2^{\mathrm{s}}}(\boldsymbol{J}_1, \boldsymbol{J}_2, \omega)} = \delta_{m_{1,h}^{\mathrm{s}}}^{m_{2,h}^{\mathrm{s}}}\, \frac{1}{\mathcal{D}_{\boldsymbol{m}_1^{\mathrm{s}},\boldsymbol{m}_2^{\mathrm{s}}}(\boldsymbol{J}_1, \boldsymbol{J}_2, \omega)}\,, \tag{6.90}$$

so that they satisfy the same symmetry relation than the bare susceptibility coefficients.

Thanks to the two additional symmetries properties from Eqs. (6.86) and (6.90), one may now simplify the resonance condition from the Balescu–Lenard equation (6.53), so that in the context of $3D$ spherical systems it becomes

$$\begin{aligned} \frac{\partial \overline{F}}{\partial \tau} = \frac{2\pi^2}{N} \frac{\partial}{\partial \boldsymbol{J}_1^{\mathrm{s}}} \cdot \bigg[\sum_{\boldsymbol{m}_1^{\mathrm{s}}, m_{2,g}^{\mathrm{s}}} \boldsymbol{m}_1^{\mathrm{s}} \int \mathrm{d}\boldsymbol{J}_2\, & \frac{\delta_{\mathrm{D}}(\boldsymbol{m}_1^{\mathrm{s}} \cdot \boldsymbol{\Omega}_1^{\mathrm{s}} - (m_{2,g}^{\mathrm{s}}, m_{1,h}^{\mathrm{s}}) \cdot \boldsymbol{\Omega}_2^{\mathrm{s}})}{|\mathcal{D}_{\boldsymbol{m}_1^{\mathrm{s}}, (m_{2,g}^{\mathrm{s}}, m_{1,h}^{\mathrm{s}})}(\boldsymbol{J}_1, \boldsymbol{J}_2, \boldsymbol{m}_1^{\mathrm{s}} \cdot \boldsymbol{\Omega}_1^{\mathrm{s}})|^2} \\ & \times \bigg(\boldsymbol{m}_1^{\mathrm{s}} \cdot \frac{\partial}{\partial \boldsymbol{J}_1^{\mathrm{s}}} - (m_{2,g}^{\mathrm{s}}, m_{1,h}^{\mathrm{s}}) \cdot \frac{\partial}{\partial \boldsymbol{J}_2^{\mathrm{s}}} \bigg) \overline{F}(\boldsymbol{J}_1)\, \overline{F}(\boldsymbol{J}_2) \bigg]\,, \end{aligned} \tag{6.91}$$

where the resonance vectors were written as $\boldsymbol{m}_1^{\mathrm{s}} = (m_{1,g}^{\mathrm{s}}, m_{1,h}^{\mathrm{s}})$. It is straightforward to obtain the Landau equation associated with the Balescu–Lenard equation (6.91). This only amounts to neglecting collective effects, and therefore perform the substitution $1/|\mathcal{D}|^2 \to |A|^2$. Let us note that in the $3D$ case, the 1.5PN relativistic precession frequencies obtained in Eq. (6.104) depend on the action L_z, so that, at this stage, further simplifications of Eq. (6.91) are not possible anymore. To effectively evaluate the diffusion flux in Eq. (6.91), one may follow the method presented in Sect. 4.2.5, by identifying the system's critical surfaces of resonance. We do not detail these calculations here.

6.6.3 Relativistic Barrier Crossing

As a final discussion of the physical content of the Balescu–Lenard equation (6.53), let us now illustrate how this degenerate diffusion equation allows for a qualitative description of the Schwarzschild barrier encountered by stars as they diffuse towards the central BH. This Schwarzschild barrier was discovered in Merritt et al. (2011) via simulations of spherically symmetric clusters. Here, in order to clarify the upcoming discussion, we will consider the case of a razor-thin axisymmetric disc of stars, whose secular evolution is described by Eq. (6.79), but the same idea applies to the $3D$ case. In the resonance condition from Eq. (6.79), let us recall that the precession frequency Ω^{s}, as defined in Eq. (6.50), is composed of two components. The first one comes from the system's self-consistent potential (Eq. (6.36)) and reads

$$\Omega^{\mathrm{s}}_{\mathrm{self}}(L_1, I_1) = \frac{\partial}{\partial L_1}\bigg[\overline{\Phi}(L_1, I_1)\bigg] = \frac{\partial}{\partial L_1}\bigg[\int \mathrm{d}\mathcal{E}_2\, \overline{F}(\mathcal{E}_2)\, \overline{U}_{12}\bigg]. \tag{6.92}$$

The second contribution comes from the relativistic effects occurring in the vicinity of the BH. These precession frequencies are briefly recovered in Appendix 6.A. For a razor-thin disc, they read

$$\Omega^{\mathrm{s}}_{\mathrm{rel}}(L, I) = \frac{1}{2\pi}\frac{M_\bullet}{M_\star}\frac{(GM_\bullet)^4}{c^2}\bigg[-\frac{3}{I^3L^2} + \frac{GM_\bullet}{c}\frac{6s}{I^3L^3}\bigg]. \tag{6.93}$$

Let us now study how these precession frequencies depend on the distance to the central BH. Following the timescale comparisons of Kocsis and Tremaine (2011), one expects the relativistic precession frequency $\Omega^{\mathrm{s}}_{\mathrm{rel}}$ to dominate close to the BH (and diverge as stars get closer to capture), while the self-consistent one, $\Omega^{\mathrm{s}}_{\mathrm{self}}$, will be the largest in the vicinity of the considered disc. Figure 6.3 illustrates the typical behaviour of these precession frequencies. In order to induce a diffusion, the Balescu–Lenard equation (6.79) requires the resonance condition $\Omega^{\mathrm{s}}_{\mathrm{tot}}(\boldsymbol{J}_1)-\Omega^{\mathrm{s}}_{\mathrm{tot}}(\boldsymbol{J}_2)=0$ to be satisfied. In Fig. 6.3, we illustrate that for a given value of the precession frequency ω_{s}, one can identify the associated locations in the disc where the resonance condition is satisfied. Let us also recall that Eq. (6.79) involves the quadratic factor $\overline{F}(\boldsymbol{J}_1)\,\overline{F}(\boldsymbol{J}_2)$, i.e. the product of the system's DF in the two locations which are in resonance. As a consequence, because the disc is only located in the outer regions of the BH, the resonant coupling between two locations within the disc will be much stronger, than the resonant coupling involving one resonant location in the very inner regions of the system close to the BH. In Fig. 6.3, this corresponds to the fact that the resonant coupling between the two outer dots will be much larger than the couplings involving the inner dot in the vicinity of the BH. As stars migrate even closer to the BH, the situation gets even worse, because the required precession frequency to allow for a resonant coupling then becomes too large to resonate with any part of the disc. For such a situation, no efficient resonant couplings are possible and the diffusion is drastically suppressed. As a conclusion, the divergence of the relativis-

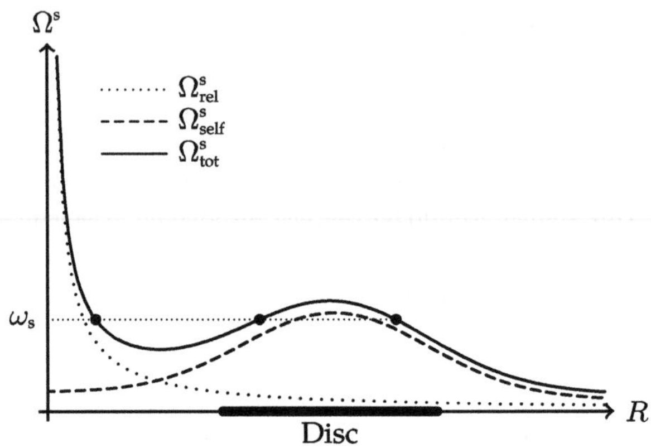

Fig. 6.3 Illustration of the typical dependence of the precession frequencies Ω^{s}_{self} and Ω^{s}_{rel} (Eqs. (6.92) and (6.93)) as a function of the distance to the central BH. The relativistic precession frequencies Ω^{s}_{rel} diverge as stars get closer to the BH, while the self-consistent precession frequencies Ω^{s}_{self} are typically the largest for stars in the neighbourhood of the considered disc. The *black* dots give all the locations in the disc whose precession frequency is equal to ω_{s}, as illustrated by the *dotted* horizontal line. Because these disc's locations are in resonance they will contribute to the Balescu–Lenard flux in Eq. (6.53). Equation (6.53) involves the product of the system's DF in the two resonating locations. As a consequence, here the resonant coupling between the two outer points, which both belong to the region where the disc dominates, will be much stronger, than the couplings involving the inner point, which does not belong to the core of the disc. As stars move inward, because of the relativistic corrections, their precession frequencies increase up to a point where it prevents any resonant coupling with the disc's region. This drastically suppresses the diffusion and induces a diffusion barrier

tic precession frequencies in the neighbourhood of the BH implies that stars whose orbits diffuse inwards closer to the BH will experience a steep rise in their own precession frequency. This prevents them from being able to resonate with the disc, leading to a strong suppression of any further inward diffusion. This is the so-called Schwarzschild barrier.

Such an explanation of the Schwarzschild barrier via the notion of resonant coupling is directly related to the explanation proposed in Bar-Or and Alexander (2014) relying on the concept of adiabatic invariance. In this picture, a test star may undergo resonant relaxation if the timescale of its relativistic precession is longer than the coherence time of the perturbations induced by the field stars and felt by this test star. As the typical coherence time of the perturbations scales like the inverse of the typical frequency of the field stars (which lie within the cluster), the requirement for an efficient resonant diffusion from the adiabatic invariance point of view is equivalent to the requirement from the point of view of the resonance condition of the Balescu–Lenard equation.

Let us illustrate this diffusion barrier in the neighbourhood of the BH by considering the orbit-averaged motion of individual Keplerian wires. The degenerate Balescu–Lenard equation (6.79) for razor-thin discs is a diffusion equation in action space, which describes self-consistently the evolution of the whole system's DF.

Instead of describing the dynamics of the system's DF, one could also be interested in describing the associated stochastic evolution of individual Keplerian wires. From the ensemble average of these individual dynamics, one should recover the self-consistent DF's diffusion equation (6.79). Following Appendix 6.C, let us rewrite Eq. (6.79) as

$$\frac{\partial \overline{F}}{\partial \tau} = \frac{\partial}{\partial L}\left[A(\boldsymbol{J},\tau)\,\overline{F}(\boldsymbol{J},\tau) + D(\boldsymbol{J},\tau)\frac{\partial \overline{F}}{\partial L}\right]. \tag{6.94}$$

Equation (6.94) is the self-consistent anisotropic Fokker–Planck equation which describes the evolution of the disc's DF as a whole. In Appendix 6.C, we show how one can obtain from Eq. (6.94) the corresponding stochastic Langevin equation, which captures the secular dynamics of individual test Keplerian wires. Let us therefore denote as $\mathcal{J}(\tau)=(\mathcal{L}(\tau),\mathcal{I}(\tau))$ the position at time τ of a test wire in the $2D$ action space $\boldsymbol{J}=(L,I)$. Following Eq. (6.128), the dynamics of this test wire takes the form

$$\frac{\mathrm{d}\mathcal{L}}{\mathrm{d}\tau} = h(\mathcal{J},\tau) + g(\mathcal{J},\tau)\,\Gamma(\tau) \quad ; \quad \frac{\mathrm{d}\mathcal{I}}{\mathrm{d}\tau} = 0\,. \tag{6.95}$$

In Eq. (6.95), the $1D$ Langevin coefficients h and g describe the diffusion of the wire in the $L-$direction. They follow from Eq. (6.130) and read

$$h = -A + \frac{\partial D}{\partial L} - \sqrt{D}\frac{\partial \sqrt{D}}{\partial L} \quad ; \quad g = \sqrt{D}\,, \tag{6.96}$$

while the stochastic Langevin force $\Gamma(\tau)$ follows the statistics from Eq. (6.129). In Eq. (6.95), as already underlined in Eq. (6.62), we recover the fact that the individual fast action $J^{\mathrm{f}}=I$ is conserved during the wire's resonant relaxation.

Let us insist on the fact that Eq. (6.95) is a rewriting of the Balescu–Lenard equation (6.79) to capture individual dynamics. The Langevin equation (6.95) therefore describes the diffusion of an individual test Keplerian wire embedded in the self-induced noisy environment described by the drift and diffusion coefficients from the Balescu–Lenard equation (6.79). Let us note that because the rewriting from Eq. (6.95) is a self-consistent rewriting of the system's dynamics, it could be used iteratively to integrate forward in time the Balescu–Lenard equation (6.79). Rather than having to integrate forward in time the system's DF as a whole, Eq. (6.95) only requires to integrate forward in time first-order stochastic differential equations. To do so, one would discretise Eq. (6.95) in time as $\mathcal{L}_{i+1}=\mathcal{L}_i + (\mathrm{d}\mathcal{L}/\mathrm{d}\tau)_i\Delta\tau$, while sampling initial $\mathcal{L}_0$ to match the system's initial DF. Let us emphasise that the individual stochastic equations (6.95) share some striking similarities with the individual Hamilton's equations associated with the total Hamiltonian from Eq. (6.7). However, the gain of the present Langevin rewriting is to allow for individual timesteps, $\Delta\tau$, orders of magnitude larger than the original ones required to solve for the trajectories of individual stars. Indeed, the Langevin equation focuses directly on the dynamics of Keplerian wires instead of stars themselves. The integration of the fast Keplerian orbital motion does not need to be performed. In addition, it also deals seamlessly with the relativistic corrections, which are already integrated upon the fast angles.

Let us finally illustrate qualitatively in Fig. 6.4 the dynamics of individual wires as given by Eq. (6.95). Figure 6.4 follows the representations from Bar-Or and Alexander (2016), by representing the diffusion of wires in the $(j, a) = (L/I, I^2/(GM_\bullet))$ space. As emphasised in Eq. (6.95), during the diffusion, the fast action I of the stars is conserved, so that they diffuse only in the j-direction, along $a = \mathrm{cst.}$ lines. Individual wires may resonate with other wires precessing at the same frequency, such as the blue and red wires in Fig. 6.4. However, as already illustrated in Fig. 6.3, because of the relativistic corrections, the precession frequencies diverge as stars get closer to the BH. This increase in the precession frequencies forbids then any resonant coupling between a star in the inner fast precessing region and stars belonging to the disc itself, where the precession frequencies are much smaller. Resonances becoming impossible, the diffusion is suppressed and wires cannot keep diffusing closer to the central BH. This strong suppression of the diffusion is the Schwarzschild barrier. The present explication of the Schwarzschild barrier is essentially the same

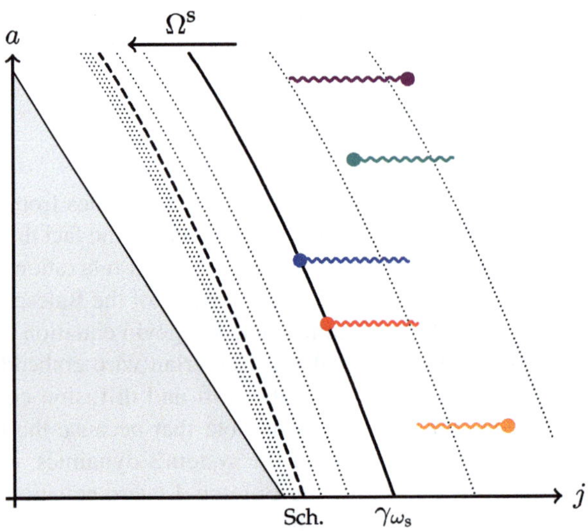

Fig. 6.4 Qualitative illustration of the individual diffusion of Keplerian wires in the $(j, a) = (L/I, I^2/(GM_\bullet))$ space, as given by the Langevin equation (6.95). The *grey* region corresponds to the capture region, within which stars inevitably sink into the BH. Because the fast action I is conserved during the diffusion (see Eq. (6.95)), wires' diffusion is one-dimensional, conserves a, and occurs only in the j-direction. The background contours illustrate the contour lines of the precession frequency, i.e. the function $(j, a) \mapsto \Omega^{\mathrm{s}}(j, a)$. As illustrated in Fig. 6.3, the precession frequencies drastically increase as wires approach the central BH, because of the contributions from the relativistic corrections. The *blue* and *red* wires precess at the same frequency ω_{s}, as they belong to the same critical line $\gamma_{\omega_{\mathrm{s}}}$. This allows them to resonate one with another. Because the precession frequencies diverge in the vicinity of the BH, such resonant couplings are much less likely as wires get closer to the BH. This effectively creates a diffusion barrier in action space, the so-called Schwarzschild barrier

than the one proposed in Bar-Or and Alexander (2014), which relied on the adiabatic invariance of the angular momentum induced by the fast relativistic precessions in the vicinity of the BH.

Our previous calculations explained the existence of a Schwarzschild barrier, which strongly suppresses the supply of tightly bound matter to the BH. As a final remark, let us note that the numerical analysis of Merritt et al. (2011) suggested that, in practice, this suppression is most probably tempered by simple two-body relaxation (not accounted for in the orbit-averaged approach followed in this section). Two-body relaxation then provides an additional mechanism to transport stars even closer to the BH, once resonant relaxation becomes inefficient. This was recently demonstrated in detail in Bar-Or and Alexander (2016), which showed that adiabatic invariance (i.e. the damping of resonant relaxation) limits the effects of resonant relaxation to a region well away of the loss lines. The dynamics of accretion of stars by the BH is then only very moderately affected by such resonant diffusions.

Finally, in Fouvry et al. (2017b), we recently presented a quantitative calculation of the drift and diffusion coefficients of the inhomogeneous multi-mass degenerate Landau equation (6.79), to describe the self-induced resonant relaxation of a discrete quasi-Keplerian axisymmetric razor-thin disc orbiting a supermassive BH. In this context, the total bare susceptibility coefficients, $|A_{\mathrm{tot}}(\boldsymbol{J}_1, \boldsymbol{J}_2)|^2$, from Eq. (6.81) can be computed following Gauss' method (Touma et al. 2009). This allows for an efficient and robust evaluation of the double angle-average from Eq. (6.37). An illustration of the stochastic dynamics of a test wire embedded in such a disc is illustrated in Fig. 6.5. When considering a discrete disc composed of multiple masses similarly distributed, the degenerate Landau equation predicts a mass segregation of the two components. Provided that the disc's DF increases with angular momentum, the population of lighter particles will gain eccentricity during this segregation. Finally,

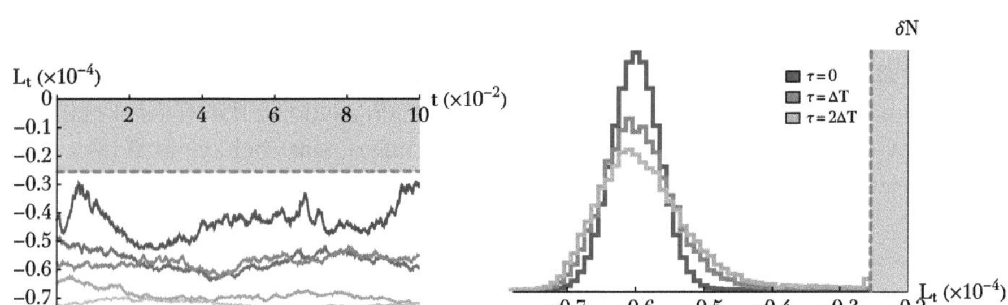

Fig. 6.5 Illustration extracted from Fouvry et al. (2017b), which represents the diffusion of the angular momentum L_{t} of a test star embedded in a discrete quasi-Keplerian axisymmetric razor-thin disc in the vicinity of a supermassive BH. *Left panel* Illustration of the stochastic relaxation of the angular momentum L_{t} for various initial conditions. Such a random walk is described by the Langevin equation (6.95). The Schwarzschild barrier, which prevents the diffusion towards too small angular momenta, is represented by the *grey* region. *Right panel* Illustration of the same diffusion for a population of test stars represented by an histogram. Solving for the particles' dynamics via the Langevin equation (6.95) allows for the integration forward in time of the Fokker–Planck equation (6.94), without resorting to direct N−body simulations

within this framework, one recovers quantitatively the quenching of the diffusion of a test star in the vicinity of the BH, due to the divergence of the relativistic precessions, i.e. the Schwarzschild barrier.

6.7 Conclusion

Supermassive BHs absorb stars and debris whose orbits reach the loss-cone (Frank and Rees 1976; Vasiliev and Merritt 2013), the region of phase space associated with unstable orbits, which take them directly into the BH or close enough to strongly interact with it. Such accretions affect the secular evolution of the BH's mass and spin, which is of particular interest to understand BHs' demographics and AGN feedback (Volonteri et al. 2016). These accretion events also provide information on stars, debris and gas blobs in the vicinity of the BH: for example, the continuous loss of stars can effectively reshape the central stellar distribution of the cluster (e.g., Genzel et al. 2000). All these processes have specific observable signatures, such as the binary capture and ejection of hyper-velocity stars (Hills 1988), tidal heating and disruption (Frank and Rees 1976), and eventually gravitational waves emission produced by inspiraling compact remnants (Abbott et al. 2016). These various mechanisms also offer the possibility for indirect detections of BHs, and for tests of general relativity in the strong field limit (Blanchet 2014). A new generation of interferometers, such as Gravity (Jocou et al. 2014), now have for primary goal to understand the dynamics of stars in the vicinity of supermassive BHs.

In this chapter, we presented how the generic Balescu–Lenard formalism can be tailored to describe the secular evolution of quasi-Keplerian systems, such as galactic nuclei, by appropriately dealing with the dynamical degeneracy of the mean motion. We therefore derived the collisional degenerate kinetic equation (6.53) describing the secular evolution of such systems at order $1/N$. Because purely Keplerian orbits do not precess, the dynamical evolution of these degenerate systems may significantly differ from that of fully self-gravitating systems, such as the stellar discs considered in the previous chapters. In the quasi-Keplerian context, stars behave as if they were smeared out onto their orbit-averaged Keplerian wires, and the secular collisional evolution of the system is then described by accounting for the dressed interactions between such wires. These wires undergo a resonant relaxation, sourced by the system's intrinsic Poisson shot noise, leading to the appearance of sequences of uncorrelated polarised density waves, whose net effect is to diffuse the system's orbital structure on secular timescales. The degenerate Balescu–Lenard equation (6.53) satisfies some essential properties. It is quadratic in the angle-averaged system's DF, accounts for the system's self-gravity as well as possible post-Newtonian corrections. This equation is sourced by the discreteness of the cluster and describes the resonant coupling between the system's wires. It can also account for a spectrum of masses via Eq. (6.63). The degenerate Balescu–Lenard equation (6.53) is therefore the quasilinear self-consistent master equation quantifying the effects of resonant relaxation. It provides a rich framework in which to describe the evolution of quasi-

Keplerian systems on cosmic times, such as galactic centres, or debris discs, which are an interesting venue in the context of planet formation (Tremaine 1998).

The principal ingredient used in the proposed derivation was the phase averaging of the two first equations of the BBGKY hierarchy over the fast angle associated with the BH-induced dominant Keplerian motion. Direct consequences of this phase average include that the associated fast actions are adiabatically conserved. As such, this description of the dynamics of Keplerian wires does not allow for the capture of the effects associated with mean motion resonances and direct 2−body relaxation. However, this is usually appropriate because the derivation ignored terms of order $O(1/N^2)$, so that it is valid for timescales of order $N T_{\mathrm{sec}}$. This timescale is expected to be much shorter than the 2−body relaxation time. In Sects. 6.6.1 and 6.6.2, we specified the degenerate Balescu–Lenard equation to razor-thin axisymmetric discs and spherical systems. Finally, in Sect. 6.6.3, we investigated in particular the properties of resonant relaxation in the vicinity of supermassive BHs. We showed that the degenerate Balescu–Lenard equation naturally captures the presence of the Schwarzschild barrier (see Fig. 6.4), where the efficiency of the resonant collisional diffusion is significantly suppressed.

Various recent papers have tackled as well the question of describing the long-term dynamics of quasi-Keplerian systems. The closest to the present derivation is the recent sequence of papers Sridhar and Touma (2016a, b), which obtained evolution equations equivalent to Eqs. (6.45) and (6.46), by following a different route based on the approach of Gilbert (1968), which itself extended the works of Balescu (1960); Lenard (1960) from plasma physics. In Sridhar and Touma (2017), they relied on the "passive response approximation" when considering razor-thin axisymmetric discs, which only allowed for the recovery of the $2D$ razor-thin bare susceptibility coefficients from Eq. (6.81) and the Landau version of Eq. (6.79).

Another very efficient way of modelling such quasi-Keplerian systems is by relying on Monte Carlo methods, for which the internal Poisson shot noise due to the finite number of wires is treated as an externally imposed perturbation (e.g., Madigan et al. 2011; Bar-Or and Alexander 2014). This is a very flexible method, especially if one wants to account for additional perturbations external to the cluster, such as those coming from the near neighbourhood of the BH. The η−formalism recently introduced in Bar-Or and Alexander (2014) and implemented in detail in Bar-Or and Alexander (2016) is one such scheme. After imposing plausible constraints on the power spectrum of the self-induced discreteness noise, they recovered the location of the Schwarzschild barrier (interpreted in terms of adiabatic invariance), and investigated the role of 2−body relaxation in the loss-cone problem. They showed in particular that on longer timescales, 2−body non-resonant relaxation erases the Schwarzschild barrier, and argued that resonant relaxation is effective only in a restricted region of action space away from the loss lines, so that its overall effect on plunge rates remains small.

These approaches suffer from two shortcomings, which are the need for ad hoc assumptions on the statistical characteristics of the cluster's shot noise, and the difficulty to account for the cluster's self-gravity. These two elements are self-consistently accounted for in the Balescu–Lenard equation. Finally, at the heart of the

$\eta-$formalism lies an important distinction between field and test stars. Indeed, the dynamics of the test stars are followed as they undergo the stochastic perturbations generated by the field stars. Such a split was also used in the recent restricted $N-$body simulations presented in Hamers et al. (2014). In these simulations, the motion of each field star is followed along its precessing Keplerian orbit (with a precession induced by both relativistic effects and the cluster's self-consistent potential), but interactions among field stars are ignored. The test stars are then followed by direct integration of their motion in the time-varying potential due to the field stars. Such a method is especially useful in order to characterise the typical properties of the stochastic perturbations generated by the field stars. Similarly to the $\eta-$formalism, this approach ignores interactions among field stars (and among test stars), and there is no back-influence of the test stars on the field ones. Let us finally note that in the course of this chapter, we also presented in Appendix 6.C a Langevin rewriting of the Balescu–Lenard equation. This approach combines the flexibility of Monte Carlo realisations with the self-consistent treatment offered by the Balescu–Lenard approach. A subsequent improvement of this stochastic rewriting lies in the possibility of adding to the dominant resonant relaxation, the secondary effects of two-body relaxation and gravitational waves losses.

6.7.1 Future Works

The previous specialisation of the Balescu–Lenard formalism to quasi-Keplerian systems offers possibilities for numerous follow-up works. Let us first note that here we mainly focused on the dynamics of galactic centres, but such methods could also be applied to the secular dynamics of protoplanetary systems and debris discs, which form another vast class of quasi-Keplerian systems.

In the presence of external perturbations, one should be in a position to generalise the collisionless formalism from Sect. 2.2 to account for the long-term effects of stochastic perturbations. The main difficulty here is that because the dynamics was described w.r.t. the central BH (see Sect. 6.2), such external perturbations may decentre the system and introduce non-trivial fictive forces, whose effects on secular timescales have to be carefully studied. Another generalisation of this formalism would be to consider the secular dynamics of quasi-stationary lopsided configurations. Because such configurations precess as a whole, this would first involve identifying new angle-action variables within which the system could be considered as quasi-stationary and then extend the formalism to such configurations.

In order to illustrate qualitatively the mechanisms described in this chapter, one would benefit from implementing the inhomogeneous degenerate Landau equation (i.e. without collective effects) for razor-thin axisymmetric discs. Most of the methods required for such a computation were already presented in Chap. 4. The most challenging part of such a computation is the wire-wire interaction potential (Touma et al. 2009), thanks to which the self-consistent precession frequencies as well as the bare susceptibility coefficients can be estimated. See Fouvry et al. (2017b) for recent developments. Similar investigations could also be pursued for $3D$ spherical systems.

We emphasised in Sect. 6.6.3 that as stars diffuse closer to the central BH, their precession frequencies increase up to a point where resonant relaxation becomes inefficient: this is the Schwarzschild barrier. The dynamics of stars in such a configuration is then driven by $2-$body relaxation effects, which allow stars to diffuse even further in. Such effects are induced by direct star-star interactions and cannot be accounted for in the present orbit-averaged approach which focuses only on wire-wire interactions. In order to get a better estimate of the infalling rate onto the BH, the next step would therefore be to improve the present formalism to also account for direct $2-$body relaxation, which has proven essential for the late stages of the diffusion.

In Appendix 6.C, we described how the self-consistent Balescu–Lenard could be rewritten as a stochastic Langevin equation to describe the collisional evolution of individual Keplerian wires. Such a rewriting offers a promising way to integrate forward in time the diffusion equation. As this directly involves Keplerian wires, one does not need to integrate the fast Keplerian motion induced by the central BH. This offers a significant timescale speed up for such $N-$wires implementation. This integration in time would also allow us to self-consistently account for the growth rate and spin up of the central BH as stars get absorbed. In this context, one could also investigate multi-component situations, where mass segregation is expected to play an important role.

Appendix

A Relativistic Precessions

In this Appendix, let us briefly detail the content of the averaged relativistic corrections encompassed by the potential $\overline{\Phi}_{\rm r}$ present in Eqs. (6.45) and (6.46). As we aim for explicit expressions of these corrections, let us use the $3D$ Delaunay variables from Eq. (6.19). In addition, we assume for simplicity that the spin of the BH is aligned with the $z-$direction and introduce its spin parameter $0 \leq s \leq 1$. We follow Merritt (2015) in order to recover explicit expressions for these precession frequencies.

The first relativistic correction is associated with a 1PN effect (i.e. a correction of order $1/c^2$), called the Schwarzschild precession. Equation (5.103) in Merritt (2015) gives us that during one Keplerian orbit of duration $T_{\rm Kep} = 2\pi/\Omega_{\rm Kep} = 2\pi I^3/(GM_\bullet)^2$, the slow angle g is modified by an amount

$$\Delta g_{\rm rel}^{\rm 1PN} = g(T_{\rm Kep}) - g(0) = \frac{6\pi G M_\bullet}{c^2 a(1-e^2)} \,. \tag{6.97}$$

This precession corresponds to a precession of the orbit's pericentre, while the orbit remains in its orbital plane. To the change from Eq. (6.97), one can straightforwardly associate an averaged precession frequency $\dot{g}_{\mathrm{rel}}^{\mathrm{1PN}} = \Delta g_{\mathrm{rel}}^{\mathrm{1PN}} / T_{\mathrm{Kep}}$ reading

$$\dot{g}_{\mathrm{rel}}^{\mathrm{1PN}} = \frac{3(GM_\bullet)^4}{c^2 I^3 L^2} = \frac{\partial H_{\mathrm{rel}}^{\mathrm{1PN}}}{\partial L}, \tag{6.98}$$

where the semi-major axis a and the eccentricity e respectively satisfy $a = I^2/(GM_\bullet)$ and $1-e^2 = (L/I)^2$. We also introduced the Hamiltonian $H_{\mathrm{rel}}^{\mathrm{1PN}}$ as

$$H_{\mathrm{rel}}^{\mathrm{1PN}}(I, L) = -\frac{3(GM_\bullet)^4}{c^2} \frac{1}{I^3 L}. \tag{6.99}$$

The next order relativistic corrections are associated with a 1.5PN effect (i.e. a correction of order $1/c^3$) called the Lense-Thirring precession. Following Eq. (5.118) of Merritt (2015), during one Keplerian orbit this effect leads to a precession of the slow angle g given by

$$\Delta g_{\mathrm{rel}}^{\mathrm{1.5PN}} = g(T_{\mathrm{Kep}}) - g(0) = -\frac{12\pi s}{c^3} \left[\frac{GM_\bullet}{(1-e^2)a} \right]^{3/2} \cos(i), \tag{6.100}$$

where we recall that we assume that the BH's spin is aligned with the z−direction. We also introduced the orbit's inclination i such that $L_z = L \cos(i)$. One can straightforwardly associate a precession frequency $\dot{g}_{\mathrm{rel}}^{\mathrm{1.5PN}} = \Delta g_{\mathrm{rel}}^{\mathrm{1.5PN}} / T_{\mathrm{Kep}}$ to this change, so that

$$\dot{g}_{\mathrm{rel}}^{\mathrm{1.5PN}} = -\frac{6s}{c^3} \frac{(GM_\bullet)^5 L_z}{I^3 L^4} = \frac{\partial H_{\mathrm{rel}}^{\mathrm{1.5PN}}}{\partial L}. \tag{6.101}$$

In Eq. (6.101), we introduced the Hamiltonian $H_{\mathrm{rel}}^{\mathrm{1.5PN}}$, which captures the corrections associated with the BH's spin as

$$H_{\mathrm{rel}}^{\mathrm{1.5PN}}(I, L, L_z) = \frac{2s(GM_\bullet)^5}{c^3} \frac{L_z}{I^3 L^3}. \tag{6.102}$$

Such a Hamiltonian also induces relativistic precessions w.r.t. the second slow angle h associated with the slow action L_z. We do not detail here how these precessions are indeed correctly described by the Hamiltonian $H_{\mathrm{rel}}^{\mathrm{1.5PN}}$.

Let us finally write the explicit expression of the averaged potential corrections $\overline{\Phi}_{\mathrm{r}}$ appearing in Eqs. (6.45) and (6.46). One has to pay a careful attention to the normalisation conventions introduced in Eqs. (6.2), (6.14), and (6.38). One gets

$$\overline{\Phi}_{\mathrm{r}}(I, L, L_z) = \frac{1}{(2\pi)^{d-k}} \frac{M_\bullet}{M_\star} \left[H_{\mathrm{rel}}^{\mathrm{1PN}}(I, L) + H_{\mathrm{rel}}^{\mathrm{1.5PN}}(I, L, L_z) \right]. \tag{6.103}$$

Following Eq. (6.50), this relativistic potential correction immediately leads to the associated precession frequencies $\boldsymbol{\Omega}^{\mathrm{s}}_{\mathrm{rel}}$ w.r.t. the slow angles $\boldsymbol{\theta}^{\mathrm{s}}$, which read

$$\boldsymbol{\Omega}^{\mathrm{s}}_{\mathrm{rel}} = \frac{\partial \overline{\Phi}_{\mathrm{r}}}{\partial \boldsymbol{J}^{\mathrm{s}}} = \frac{(GM_\bullet)^4}{(2\pi)^{d-k} c^2} \frac{M_\bullet}{M_\star} \frac{\partial}{\partial \boldsymbol{J}^{\mathrm{s}}} \bigg[-\frac{3}{I^3 L} + \frac{2GM_\bullet}{c} \frac{s L_z}{I^3 L^3} \bigg] . \tag{6.104}$$

Let us finally note that gravitational waves and the associated dissipations (Hopman and Alexander 2006) are not accounted for in Eq. (6.104), hence the possibility to obtain a Hamiltonian formulation for these precessions.

B Multi-component BBGKY Hierarchy

In this Appendix, let us detail how one can adapt the formalism presented in Sect. 6.2 to the case where the system is composed of multiple components. The different components are indexed by the letters "a", "b", etc. We assume that the component "a" is made of N_{a} particles of individual mass μ_{a}. The total mass of the component "a" is written as $M_\star^{\mathrm{a}}$. When accounting for multiple components and placing ourselves within the democratic heliocentric coordinates from Eq. (6.3), the system's total Hamiltonian from Eq. (6.7) becomes

$$\begin{aligned} H = & \sum_{\mathrm{a}} \sum_{i=1}^{N_{\mathrm{a}}} \frac{\mu_{\mathrm{a}}}{2} (\boldsymbol{v}_i^{\mathrm{a}})^2 + \sum_{\mathrm{a}} \mu_{\mathrm{a}} M_\bullet \sum_{i=1}^{N_{\mathrm{a}}} U(|\boldsymbol{x}_i^{\mathrm{a}}|) + \sum_{\mathrm{a}} \mu_{\mathrm{a}} M_\star \sum_{i=1}^{N_{\mathrm{a}}} \Phi_{\mathrm{r}}(\boldsymbol{x}_i^{\mathrm{a}}) \\ & + \sum_{\mathrm{a}} \mu_{\mathrm{a}}^2 \sum_{i<j}^{N_{\mathrm{a}}} U(|\boldsymbol{x}_i^{\mathrm{a}} - \boldsymbol{x}_j^{\mathrm{a}}|) + \sum_{\mathrm{a}<\mathrm{b}} \sum_{i=1}^{N_{\mathrm{a}}} \sum_{j=1}^{N_{\mathrm{b}}} \mu_{\mathrm{a}} \mu_{\mathrm{b}} U(|\boldsymbol{x}_i^{\mathrm{a}} - \boldsymbol{x}_j^{\mathrm{b}}|) + \frac{1}{2M_\bullet} \bigg[\sum_{\mathrm{a}} \mu_{\mathrm{a}} \sum_{i=1}^{N_{\mathrm{a}}} \boldsymbol{v}_i^{\mathrm{a}} \bigg]^2 , \end{aligned} \tag{6.105}$$

where we noted as $\Gamma_i^{\mathrm{a}} = (\boldsymbol{x}_i^{\mathrm{a}}, \boldsymbol{v}_i^{\mathrm{a}})$ the position and velocity of the ith particle of the component "a". In Eq. (6.105), the various terms are respectively the kinetic energy of the particles, the Keplerian potential due to the central BH, the relativistic potential corrections Φ_{r}, the self-gravity among a given component, the interactions between particles of different components, and finally the additional kinetic terms introduced by the change of coordinates from Eq. (6.3). One should also pay attention to the normalisation of the relativistic component Φ_{r}, as we wrote this potential as $\mu_{\mathrm{a}} M_\star \Phi_{\mathrm{r}}$, where we introduced the system's total active mass as $M_\star = \sum_{\mathrm{a}} M_\star^{\mathrm{a}}$ to have a writing similar to Eq. (6.7). Let us now introduce the system total PDF $P_{\mathrm{tot}}(\Gamma_1^{\mathrm{a}}, \ldots, \Gamma_{N_{\mathrm{a}}}^{\mathrm{a}}, \Gamma_1^{\mathrm{b}}, \ldots, \Gamma_{N_{\mathrm{b}}}^{\mathrm{b}}, \ldots)$ which gives the probability of finding at time t the particle 1 of the component "a" at position $\boldsymbol{x}_1^{\mathrm{a}}$ and velocity $\boldsymbol{v}_1^{\mathrm{a}}$, etc. We normalise P_{tot} following the convention from Eq. (2.94). Similarly to Eq. (6.8), P_{tot} evolves according to Liouville's equation which reads

$$\frac{\partial P_{\rm tot}}{\partial t} + \sum_{\rm a} \sum_{i=1}^{N_{\rm a}} \left[\dot{\boldsymbol{x}}_i^{\rm a} \cdot \frac{\partial P_{\rm tot}}{\partial \boldsymbol{x}_i^{\rm a}} + \dot{\boldsymbol{v}}_i^{\rm a} \cdot \frac{\partial P_{\rm tot}}{\partial \boldsymbol{v}_i^{\rm a}} \right] = 0 \,. \tag{6.106}$$

Following Eq. (2.97), we define the system's reduced PDFs $P_n^{{\rm a}_1,\ldots,{\rm a}_n}$ by integrating $P_{\rm tot}$ over all particles except n particles belonging respectively to the components ${\rm a}_1, \ldots, {\rm a}_n$. Our aim is now to write the two first equations of the associated BBGKY hierarchy. In order to clarify the upcoming calculations, let us from now on neglect any contributions associated with the last kinetic terms from Eq. (6.105). Indeed, in the single-component case, we justified in Eq. (6.41) that these terms, once averaged over the fast Keplerian motion, do not contribute to the system's dynamics at the order considered in our kinetic developments. To get the evolution equation for $P_1^{\rm a}$, one integrates Eq. (6.106) over all phase space coordinates except $\Gamma_1^{\rm a}$ and relies on the symmetry of $P_{\rm tot}$ w.r.t. interchanges of particles of the same component. One gets

$$\begin{aligned} & \frac{\partial P_1^{\rm a}}{\partial t} + \boldsymbol{v}_1^{\rm a} \cdot \frac{\partial P_1^{\rm a}}{\partial \boldsymbol{x}_1^{\rm a}} + \left[M_\bullet \boldsymbol{\mathcal{F}}_{1^{\rm a}0} + M_\star \boldsymbol{\mathcal{F}}_{1^{\rm a}{\rm r}} \right] \cdot \frac{\partial P_1^{\rm a}}{\partial \boldsymbol{v}_1^{\rm a}} \\ & + (N-1)\, \mu_{\rm a} \!\int\! {\rm d}\Gamma_2^{\rm a}\, \boldsymbol{\mathcal{F}}_{1^{\rm a}2^{\rm a}} \cdot \frac{\partial P_2^{\rm aa}}{\partial \boldsymbol{v}_1^{\rm a}} + \sum_{{\rm b}\neq{\rm a}} N_{\rm b}\, \mu_{\rm b} \!\int\! {\rm d}\Gamma_2^{\rm b}\, \boldsymbol{\mathcal{F}}_{1^{\rm a}2^{\rm b}} \cdot \frac{\partial P_2^{\rm ab}}{\partial \boldsymbol{v}_1^{\rm a}} = 0 \,. \end{aligned} \tag{6.107}$$

In Eq. (6.107), we used the same notations as in Eq. (6.9), and introduced as $\boldsymbol{\mathcal{F}}_{1^{\rm a}0}$ the force exerted by the BH on particle $1^{\rm a}$, $\boldsymbol{\mathcal{F}}_{1^{\rm a}{\rm r}}$ as the force acting on particle $1^{\rm a}$ due to the relativistic corrections, and finally $\boldsymbol{\mathcal{F}}_{ij}$ as the force between two particles. In order to get the second equation of the BBGKY hierarchy, one should proceed similarly and integrate Eq. (6.106) w.r.t. all particles except two. At this stage, two different cases should be investigated, depending on whether one considers $P_2^{\rm aa}$ or $P_2^{\rm ab}$ (with ${\rm a}\neq{\rm b}$). Let us first consider the diffusion equation satisfied by $P_2^{\rm aa}$, which ensues from Eq. (6.106) by integrating it w.r.t. all phase space coordinates except $\Gamma_1^{\rm a}$ and $\Gamma_2^{\rm a}$. It reads

$$\begin{aligned} & \frac{\partial P_2^{\rm aa}}{\partial t} + \boldsymbol{v}_1^{\rm a} \cdot \frac{\partial P_2^{\rm aa}}{\partial \boldsymbol{x}_1^{\rm a}} + \boldsymbol{v}_2^{\rm a} \cdot \frac{\partial P_2^{\rm aa}}{\partial \boldsymbol{x}_2^{\rm a}} + \mu_{\rm a} \boldsymbol{\mathcal{F}}_{1^{\rm a}2^{\rm a}} \cdot \frac{\partial P_2^{\rm aa}}{\partial \boldsymbol{v}_1^{\rm a}} + \mu_{\rm a} \boldsymbol{\mathcal{F}}_{2^{\rm a}1^{\rm a}} \cdot \frac{\partial P_2^{\rm aa}}{\partial \boldsymbol{v}_2^{\rm a}} \\ & + \left[M_\bullet \boldsymbol{\mathcal{F}}_{1^{\rm a}0} + M_\star \boldsymbol{\mathcal{F}}_{1^{\rm a}{\rm r}} \right] \cdot \frac{\partial P_2^{\rm aa}}{\partial \boldsymbol{v}_1^{\rm a}} + \left[M_\bullet \boldsymbol{\mathcal{F}}_{2^{\rm a}0} + M_\star \boldsymbol{\mathcal{F}}_{2^{\rm a}{\rm r}} \right] \cdot \frac{\partial P_2^{\rm aa}}{\partial \boldsymbol{v}_2^{\rm a}} \\ & + (N_{\rm a}-2)\, \mu_{\rm a} \!\int\! {\rm d}\Gamma_3^{\rm a} \left[\boldsymbol{\mathcal{F}}_{1^{\rm a}3^{\rm a}} \cdot \frac{\partial P_3^{\rm aaa}}{\partial \boldsymbol{v}_1^{\rm a}} + \boldsymbol{\mathcal{F}}_{2^{\rm a}3^{\rm a}} \cdot \frac{\partial P_3^{\rm aaa}}{\partial \boldsymbol{v}_2^{\rm a}} \right] \\ & + \sum_{{\rm b}\neq{\rm a}} N_{\rm b}\, \mu_{\rm b} \!\int\! {\rm d}\Gamma_3^{\rm b} \left[\boldsymbol{\mathcal{F}}_{1^{\rm a}3^{\rm b}} \cdot \frac{\partial P_3^{\rm aab}}{\partial \boldsymbol{v}_1^{\rm a}} + \boldsymbol{\mathcal{F}}_{2^{\rm a}3^{\rm b}} \cdot \frac{\partial P_3^{\rm aab}}{\partial \boldsymbol{v}_2^{\rm a}} \right] = 0 \,. \end{aligned} \tag{6.108}$$

When the two particles do not belong to the same component, the second equation of the hierarchy becomes

$$\frac{\partial P_2^{\mathrm{ab}}}{\partial t} + \boldsymbol{v}_1^{\mathrm{a}} \cdot \frac{\partial P_2^{\mathrm{ab}}}{\partial \boldsymbol{x}_1^{\mathrm{a}}} + \boldsymbol{v}_2^{\mathrm{b}} \cdot \frac{\partial P_2^{\mathrm{ab}}}{\partial \boldsymbol{x}_2^{\mathrm{b}}} + \mu_{\mathrm{b}} \boldsymbol{\mathcal{F}}_{1^{\mathrm{a}}2^{\mathrm{b}}} \cdot \frac{\partial P_2^{\mathrm{ab}}}{\partial \boldsymbol{v}_1^{\mathrm{a}}} + \mu_{\mathrm{a}} \boldsymbol{\mathcal{F}}_{2^{\mathrm{b}}1^{\mathrm{a}}} \cdot \frac{\partial P_2^{\mathrm{ab}}}{\partial \boldsymbol{v}_2^{\mathrm{b}}}$$
$$+ \left[M_{\bullet} \boldsymbol{\mathcal{F}}_{1^{\mathrm{a}}0} + M_{\star} \boldsymbol{\mathcal{F}}_{1^{\mathrm{a}}\mathrm{r}} \right] \cdot \frac{\partial P_2^{\mathrm{ab}}}{\partial \boldsymbol{v}_1^{\mathrm{a}}} + \left[M_{\bullet} \boldsymbol{\mathcal{F}}_{2^{\mathrm{b}}0} + M_{\star} \boldsymbol{\mathcal{F}}_{2^{\mathrm{b}}\mathrm{r}} \right] \cdot \frac{\partial P_2^{\mathrm{ab}}}{\partial \boldsymbol{v}_2^{\mathrm{b}}}$$
$$+ (N_{\mathrm{a}} - 1)\, \mu_{\mathrm{a}} \!\int\! \mathrm{d}\Gamma_3^{\mathrm{a}} \left[\boldsymbol{\mathcal{F}}_{1^{\mathrm{a}}3^{\mathrm{a}}} \cdot \frac{\partial P_3^{\mathrm{aba}}}{\partial \boldsymbol{v}_1^{\mathrm{a}}} + \boldsymbol{\mathcal{F}}_{2^{\mathrm{b}}3^{\mathrm{a}}} \cdot \frac{\partial P_3^{\mathrm{aba}}}{\partial \boldsymbol{v}_2^{\mathrm{b}}} \right] + (N_{\mathrm{b}} - 1)\, \mu_{\mathrm{b}} \!\int\! \mathrm{d}\Gamma_3^{\mathrm{b}} \left[\boldsymbol{\mathcal{F}}_{1^{\mathrm{a}}3^{\mathrm{b}}} \cdot \frac{\partial P_3^{\mathrm{abb}}}{\partial \boldsymbol{v}_1^{\mathrm{a}}} + \boldsymbol{\mathcal{F}}_{2^{\mathrm{b}}3^{\mathrm{b}}} \cdot \frac{\partial P_3^{\mathrm{abb}}}{\partial \boldsymbol{v}_2^{\mathrm{b}}} \right]$$
$$+ \sum_{\mathrm{c} \neq \mathrm{a,b}} N_{\mathrm{c}}\, \mu_{\mathrm{c}} \!\int\! \mathrm{d}\Gamma_3^{\mathrm{c}} \left[\boldsymbol{\mathcal{F}}_{1^{\mathrm{a}}3^{\mathrm{c}}} \cdot \frac{\partial P_3^{\mathrm{abc}}}{\partial \boldsymbol{v}_1^{\mathrm{a}}} + \boldsymbol{\mathcal{F}}_{2^{\mathrm{b}}3^{\mathrm{c}}} \cdot \frac{\partial P_3^{\mathrm{abc}}}{\partial \boldsymbol{v}_2^{\mathrm{b}}} \right] = 0 \,. \qquad (6.109)$$

Let us now adapt the definition of the reduced DFs from Eq. (2.99) to the multi-component case. We therefore introduce the system's renormalised DFs f_1^{a}, f_2^{ab}, and f_3^{abc} as

$$\begin{aligned}
&f_1^{\mathrm{a}} = \mu_{\mathrm{a}} N_{\mathrm{a}} P_1^{\mathrm{a}} \,; \\
&f_2^{\mathrm{aa}} = \mu_{\mathrm{a}}^2 N_{\mathrm{a}} (N_{\mathrm{a}} - 1) P_2^{\mathrm{aa}} \,;\; f_2^{\mathrm{ab}} = \mu_{\mathrm{a}} \mu_{\mathrm{b}} N_{\mathrm{a}} N_{\mathrm{b}} P_2^{\mathrm{ab}} \,; \qquad (6.110) \\
&f_3^{\mathrm{aaa}} = \mu_{\mathrm{a}}^3 N_{\mathrm{a}} (N_{\mathrm{a}} - 1)(N_{\mathrm{a}} - 2) P_3^{\mathrm{aaa}} \,;\; f_3^{\mathrm{aab}} = \mu_{\mathrm{a}}^2 \mu_{\mathrm{b}} N_{\mathrm{a}} (N_{\mathrm{a}} - 1) N_{\mathrm{b}} P_3^{\mathrm{aab}} \,;\; f_3^{\mathrm{abc}} = \mu_{\mathrm{a}} \mu_{\mathrm{b}} \mu_{\mathrm{c}} N_{\mathrm{a}} N_{\mathrm{b}} N_{\mathrm{c}} P_3^{\mathrm{abc}} \,,
\end{aligned}$$

where "a", "b", and "c" are associated with different components. These detailed normalisations allow us to rewrite Eq. (6.107) under the general form

$$\frac{\partial f_1^{\mathrm{a}}}{\partial t} + \boldsymbol{v}_1^{\mathrm{a}} \cdot \frac{\partial f_1^{\mathrm{a}}}{\partial \boldsymbol{x}_1^{\mathrm{a}}} + \left[M_{\bullet} \boldsymbol{\mathcal{F}}_{1^{\mathrm{a}}0} + M_{\star} \boldsymbol{\mathcal{F}}_{1^{\mathrm{a}}\mathrm{r}} \right] \cdot \frac{\partial f_1^{\mathrm{a}}}{\partial \boldsymbol{v}_1^{\mathrm{a}}} + \sum_{\mathrm{b}} \int \mathrm{d}\Gamma_2^{\mathrm{b}}\, \boldsymbol{\mathcal{F}}_{1^{\mathrm{a}}2^{\mathrm{b}}} \cdot \frac{\partial f_2^{\mathrm{ab}}}{\partial \boldsymbol{v}_1^{\mathrm{a}}} = 0 \,, \qquad (6.111)$$

where one should note that the sum over "b" runs for all components. Similarly, Eqs. (6.108) and (6.109) can both be cast under the same generic form reading

$$\frac{\partial f_2^{\mathrm{ab}}}{\partial t} + \boldsymbol{v}_1^{\mathrm{a}} \cdot \frac{\partial f_2^{\mathrm{ab}}}{\partial \boldsymbol{x}_1^{\mathrm{a}}} + \boldsymbol{v}_2^{\mathrm{b}} \cdot \frac{\partial f_2^{\mathrm{ab}}}{\partial \boldsymbol{x}_2^{\mathrm{b}}} + \mu_{\mathrm{b}} \boldsymbol{\mathcal{F}}_{1^{\mathrm{a}}2^{\mathrm{b}}} \cdot \frac{\partial f_2^{\mathrm{ab}}}{\partial \boldsymbol{v}_1^{\mathrm{a}}} + \mu_{\mathrm{a}} \boldsymbol{\mathcal{F}}_{2^{\mathrm{b}}1^{\mathrm{a}}} \cdot \frac{\partial f_2^{\mathrm{ab}}}{\partial \boldsymbol{v}_2^{\mathrm{b}}}$$
$$+ \left[M_{\bullet} \boldsymbol{\mathcal{F}}_{1^{\mathrm{a}}0} + M_{\star} \boldsymbol{\mathcal{F}}_{1^{\mathrm{a}}\mathrm{r}} \right] \cdot \frac{\partial f_2^{\mathrm{ab}}}{\partial \boldsymbol{v}_1^{\mathrm{a}}} + \left[M_{\bullet} \boldsymbol{\mathcal{F}}_{2^{\mathrm{b}}0} + M_{\star} \boldsymbol{\mathcal{F}}_{2^{\mathrm{b}}\mathrm{r}} \right] \cdot \frac{\partial f_2^{\mathrm{ab}}}{\partial \boldsymbol{v}_2^{\mathrm{b}}}$$
$$+ \sum_{\mathrm{c}} \int \mathrm{d}\Gamma_3^{\mathrm{c}} \left[\boldsymbol{\mathcal{F}}_{1^{\mathrm{a}}3^{\mathrm{c}}} \cdot \frac{\partial f_3^{\mathrm{abc}}}{\partial \boldsymbol{v}_1^{\mathrm{a}}} + \boldsymbol{\mathcal{F}}_{2^{\mathrm{b}}3^{\mathrm{c}}} \cdot \frac{\partial f_3^{\mathrm{abc}}}{\partial \boldsymbol{v}_2^{\mathrm{b}}} \right] = 0 \,, \qquad (6.112)$$

where we insist on the fact that Eq. (6.112) holds for both cases where "a" and "b" are equal or different, and that the sum over "c" runs for all components. Equations (6.111) and (6.112) are the direct multi-component analogs of the single-component BBGKY equation (6.9).

Following Eqs. (2.101) and (2.102), one can now introduce the cluster representation of the DFs, which in the multi-component context takes the form

$$f_2^{\mathrm{ab}}(\Gamma_1^{\mathrm{a}}, \Gamma_2^{\mathrm{b}}) = f_1^{\mathrm{a}}(\Gamma_1^{\mathrm{a}})\, f_1^{\mathrm{b}}(\Gamma_2^{\mathrm{b}}) + g_2^{\mathrm{ab}}(\Gamma_1^{\mathrm{a}}, \Gamma_2^{\mathrm{b}}) \,, \qquad (6.113)$$

and

$$
\begin{aligned}
f_3^{\mathrm{abc}}(\Gamma_1^{\mathrm{a}}, \Gamma_2^{\mathrm{b}}, \Gamma_3^{\mathrm{c}}) = & \, f_1^{\mathrm{a}}(\Gamma_1^{\mathrm{a}})\, f_1^{\mathrm{b}}(\Gamma_2^{\mathrm{b}})\, f_1^{\mathrm{c}}(\Gamma_3^{\mathrm{c}}) \\
& + f_1^{\mathrm{a}}(\Gamma_1^{\mathrm{a}})\, g_2^{\mathrm{bc}}(\Gamma_2^{\mathrm{b}}, \Gamma_3^{\mathrm{c}}) + f_1^{\mathrm{b}}(\Gamma_2^{\mathrm{b}})\, g_2^{\mathrm{ac}}(\Gamma_1^{\mathrm{a}}, \Gamma_3^{\mathrm{c}}) + f_1^{\mathrm{c}}(\Gamma_3^{\mathrm{c}})\, g_2^{\mathrm{ab}}(\Gamma_1^{\mathrm{a}}, \Gamma_2^{\mathrm{b}}) \\
& + g_3^{\mathrm{abc}}(\Gamma_1^{\mathrm{a}}, \Gamma_2^{\mathrm{b}}, \Gamma_3^{\mathrm{c}}) \, .
\end{aligned} \tag{6.114}
$$

As obtained in Eq. (2.103), let us assume that g_2^{ab} scales like the inverse of the number of particles, while g_3^{abc} scales like the square of the inverse of the number of particles. Relying on the decompositions from Eqs. (6.113) and (6.114), and keeping only terms of order $1/N_{\mathrm{a}}$ (where "a" runs over all components), one can rewrite the first equation (6.111) of the multi-component BBGKY hierarchy as

$$
\frac{\partial f_1^{\mathrm{a}}}{\partial t} + \boldsymbol{v}_1^{\mathrm{a}} \cdot \frac{\partial f_1^{\mathrm{a}}}{\partial \boldsymbol{x}_1^{\mathrm{a}}} + \bigg[M_{\bullet} \boldsymbol{\mathcal{F}}_{1^{\mathrm{a}}0} + M_{\star} \boldsymbol{\mathcal{F}}_{1^{\mathrm{a}}\mathrm{r}} + \sum_{\mathrm{b}} \int \mathrm{d}\Gamma_2^{\mathrm{b}}\, \boldsymbol{\mathcal{F}}_{1^{\mathrm{a}}2^{\mathrm{b}}}\, f_1^{\mathrm{b}}(\Gamma_2^{\mathrm{b}}) \bigg] \cdot \frac{\partial f_1^{\mathrm{a}}}{\partial \boldsymbol{v}_1^{\mathrm{a}}} + \sum_{\mathrm{b}} \int \mathrm{d}\Gamma_2^{\mathrm{b}}\, \boldsymbol{\mathcal{F}}_{1^{\mathrm{a}}2^{\mathrm{b}}} \cdot \frac{\partial g_2^{\mathrm{ab}}}{\partial \boldsymbol{v}_1^{\mathrm{a}}} = 0 \, , \tag{6.115}
$$

while the second equation (6.112) becomes

$$
\begin{aligned}
& \frac{\partial g_2^{\mathrm{ab}}}{\partial t} + \boldsymbol{v}_1^{\mathrm{a}} \cdot \frac{\partial g_2^{\mathrm{ab}}}{\partial \boldsymbol{x}_1^{\mathrm{a}}} + \boldsymbol{v}_2^{\mathrm{b}} \cdot \frac{\partial g_2^{\mathrm{ab}}}{\partial \boldsymbol{x}_2^{\mathrm{b}}} + \mu_{\mathrm{b}} \boldsymbol{\mathcal{F}}_{1^{\mathrm{a}}2^{\mathrm{b}}} \cdot \frac{\partial f_1^{\mathrm{a}}}{\partial \boldsymbol{v}_1^{\mathrm{a}}} f_1^{\mathrm{b}}(\Gamma_2^{\mathrm{b}}) + \mu_{\mathrm{a}} \boldsymbol{\mathcal{F}}_{2^{\mathrm{b}}1^{\mathrm{a}}} \cdot \frac{\partial f_1^{\mathrm{b}}}{\partial \boldsymbol{v}_2^{\mathrm{b}}} f_1^{\mathrm{a}}(\Gamma_1^{\mathrm{a}}) \\
& + \Big[M_{\bullet} \boldsymbol{\mathcal{F}}_{1^{\mathrm{a}}0} + M_{\star} \boldsymbol{\mathcal{F}}_{1^{\mathrm{a}}\mathrm{r}} \Big] \cdot \frac{\partial g_2^{\mathrm{ab}}}{\partial \boldsymbol{v}_1^{\mathrm{a}}} + \Big[M_{\bullet} \boldsymbol{\mathcal{F}}_{2^{\mathrm{b}}0} + M_{\star} \boldsymbol{\mathcal{F}}_{2^{\mathrm{b}}\mathrm{r}} \Big] \cdot \frac{\partial g_2^{\mathrm{ab}}}{\partial \boldsymbol{v}_2^{\mathrm{b}}} \\
& + \bigg[\sum_{\mathrm{c}} \int \mathrm{d}\Gamma_3^{\mathrm{c}}\, \boldsymbol{\mathcal{F}}_{1^{\mathrm{a}}3^{\mathrm{c}}}\, f_1^{\mathrm{c}}(\Gamma_3^{\mathrm{c}}) \bigg] \cdot \frac{\partial g_2^{\mathrm{ab}}}{\partial \boldsymbol{v}_1^{\mathrm{a}}} + \bigg[\sum_{\mathrm{c}} \int \mathrm{d}\Gamma_3^{\mathrm{c}}\, \boldsymbol{\mathcal{F}}_{2^{\mathrm{b}}3^{\mathrm{c}}}\, f_1^{\mathrm{c}}(\Gamma_3^{\mathrm{c}}) \bigg] \cdot \frac{\partial g_2^{\mathrm{ab}}}{\partial \boldsymbol{v}_2^{\mathrm{b}}} \\
& + \bigg[\sum_{\mathrm{c}} \int \mathrm{d}\Gamma_3^{\mathrm{c}}\, \boldsymbol{\mathcal{F}}_{1^{\mathrm{a}}3^{\mathrm{c}}}\, g_2^{\mathrm{bc}}(\Gamma_2^{\mathrm{b}}, \Gamma_3^{\mathrm{c}}) \bigg] \cdot \frac{\partial f_1^{\mathrm{a}}}{\partial \boldsymbol{v}_1^{\mathrm{a}}} + \bigg[\sum_{\mathrm{c}} \int \mathrm{d}\Gamma_3^{\mathrm{c}}\, \boldsymbol{\mathcal{F}}_{2^{\mathrm{b}}3^{\mathrm{c}}}\, g_2^{\mathrm{ac}}(\Gamma_1^{\mathrm{a}}, \Gamma_3^{\mathrm{c}}) \bigg] \cdot \frac{\partial f_1^{\mathrm{b}}}{\partial \boldsymbol{v}_2^{\mathrm{b}}} = 0 \, .
\end{aligned} \tag{6.116}
$$

Similarly to Eq. (6.12), let us now introduce the system's 1−body DF F^{a} and 2−body autocorrelation $\mathcal{C}^{\mathrm{ab}}$ as

$$
F^{\mathrm{a}} = \frac{f_1^{\mathrm{a}}}{M_{\star}} \quad ; \quad \mathcal{C}^{\mathrm{ab}} = \frac{g_2^{\mathrm{ab}}}{M_{\star}^2} \, . \tag{6.117}
$$

In Eq. (6.117), one should pay attention to the slight change in the normalisation of $\mathcal{C}^{\mathrm{ab}}$. This ensures a symmetric rescaling w.r.t. "a" and "b". Let us now follow Eqs. (6.13) and (6.14) to rescale the pairwise interaction potential as well as the relativistic corrections. Following these various renormalisations, Eq. (6.115) becomes

$$
\begin{aligned}
& \frac{\partial F^{\mathrm{a}}}{\partial t} + \boldsymbol{v}_1^{\mathrm{a}} \cdot \frac{\partial F^{\mathrm{a}}}{\partial \boldsymbol{x}_1^{\mathrm{a}}} + \boldsymbol{\mathcal{F}}_{1^{\mathrm{a}}0} \cdot \frac{\partial F^{\mathrm{a}}}{\partial \boldsymbol{v}_1^{\mathrm{a}}} + \varepsilon \bigg[\sum_{\mathrm{b}} \int \mathrm{d}\Gamma_2^{\mathrm{b}}\, \boldsymbol{\mathcal{F}}_{1^{\mathrm{a}}2^{\mathrm{b}}}\, F^{\mathrm{b}}(\Gamma_2^{\mathrm{b}}) \bigg] \cdot \frac{\partial F^{\mathrm{a}}}{\partial \boldsymbol{v}_1^{\mathrm{a}}} \\
& + \varepsilon \boldsymbol{\mathcal{F}}_{1^{\mathrm{a}}\mathrm{r}} \cdot \frac{\partial F^{\mathrm{a}}}{\partial \boldsymbol{v}_1^{\mathrm{a}}} + \varepsilon \sum_{\mathrm{b}} \int \mathrm{d}\Gamma_2^{\mathrm{b}}\, \boldsymbol{\mathcal{F}}_{1^{\mathrm{a}}2^{\mathrm{b}}} \cdot \frac{\partial \mathcal{C}^{\mathrm{ab}}}{\partial \boldsymbol{v}_1^{\mathrm{a}}} = 0 \, ,
\end{aligned} \tag{6.118}
$$

where we introduced the small parameter $\varepsilon = M_\star / M_\bullet = (\sum_{\mathrm{a}} M_\star^{\mathrm{a}})/M_\bullet$. Similarly, Eq. (6.116) becomes

$$
\begin{aligned}
&\frac{\partial \mathcal{C}^{\mathrm{ab}}}{\partial t} + \boldsymbol{v}_1^{\mathrm{a}} \cdot \frac{\partial \mathcal{C}^{\mathrm{ab}}}{\partial \boldsymbol{x}_1^{\mathrm{a}}} + \boldsymbol{v}_2^{\mathrm{b}} \cdot \frac{\partial \mathcal{C}^{\mathrm{ab}}}{\partial \boldsymbol{x}_2^{\mathrm{b}}} + \boldsymbol{\mathcal{F}}_{1^{\mathrm{a}}0} \cdot \frac{\partial \mathcal{C}^{\mathrm{ab}}}{\partial \boldsymbol{v}_1^{\mathrm{a}}} + \boldsymbol{\mathcal{F}}_{2^{\mathrm{b}}0} \cdot \frac{\partial \mathcal{C}^{\mathrm{ab}}}{\partial \boldsymbol{v}_2^{\mathrm{b}}} + \varepsilon \boldsymbol{\mathcal{F}}_{1^{\mathrm{a}}\mathrm{r}} \cdot \frac{\partial \mathcal{C}^{\mathrm{ab}}}{\partial \boldsymbol{v}_1^{\mathrm{a}}} + \varepsilon \boldsymbol{\mathcal{F}}_{2^{\mathrm{b}}\mathrm{r}} \cdot \frac{\partial \mathcal{C}^{\mathrm{ab}}}{\partial \boldsymbol{v}_2^{\mathrm{b}}} \\
&+ \varepsilon \eta_{\mathrm{b}} \boldsymbol{\mathcal{F}}_{1^{\mathrm{a}}2^{\mathrm{b}}} \cdot \frac{\partial F^{\mathrm{a}}}{\partial \boldsymbol{v}_1^{\mathrm{a}}} F^{\mathrm{b}}(\Gamma_2^{\mathrm{b}}) + \varepsilon \eta_{\mathrm{a}} \boldsymbol{\mathcal{F}}_{2^{\mathrm{b}}1^{\mathrm{a}}} \cdot \frac{\partial F^{\mathrm{b}}}{\partial \boldsymbol{v}_2^{\mathrm{b}}} F^{\mathrm{a}}(\Gamma_1^{\mathrm{a}}) \\
&+ \varepsilon \bigg[\sum_{\mathrm{c}} \int \mathrm{d}\Gamma_3^{\mathrm{c}} \boldsymbol{\mathcal{F}}_{1^{\mathrm{a}}3^{\mathrm{c}}} F^{\mathrm{c}}(\Gamma_3^{\mathrm{c}}) \bigg] \cdot \frac{\partial \mathcal{C}^{\mathrm{ab}}}{\partial \boldsymbol{v}_1^{\mathrm{a}}} + \varepsilon \bigg[\sum_{\mathrm{c}} \int \mathrm{d}\Gamma_3^{\mathrm{c}} \boldsymbol{\mathcal{F}}_{2^{\mathrm{b}}3^{\mathrm{c}}} F^{\mathrm{c}}(\Gamma_3^{\mathrm{c}}) \bigg] \cdot \frac{\partial \mathcal{C}^{\mathrm{ab}}}{\partial \boldsymbol{v}_2^{\mathrm{b}}} \\
&+ \varepsilon \bigg[\sum_{\mathrm{c}} \int \mathrm{d}\Gamma_3^{\mathrm{c}} \boldsymbol{\mathcal{F}}_{1^{\mathrm{a}}3^{\mathrm{c}}} \mathcal{C}^{\mathrm{bc}}(\Gamma_2, \Gamma_3^{\mathrm{c}}) \bigg] \cdot \frac{\partial F^{\mathrm{a}}}{\partial \boldsymbol{v}_1^{\mathrm{a}}} + \varepsilon \bigg[\sum_{\mathrm{c}} \int \mathrm{d}\Gamma_3^{\mathrm{c}} \boldsymbol{\mathcal{F}}_{2^{\mathrm{b}}3^{\mathrm{c}}} \mathcal{C}^{\mathrm{ac}}(\Gamma_1^{\mathrm{a}}, \Gamma_3^{\mathrm{c}}) \bigg] \cdot \frac{\partial F^{\mathrm{b}}}{\partial \boldsymbol{v}_2^{\mathrm{b}}} = 0 ,
\end{aligned}
\tag{6.119}
$$

where we introduced in the second line the small parameter $\eta_{\mathrm{a}} = \mu_{\mathrm{a}} / M_\star$ of order $1/N_{\mathrm{a}}$. Equations (6.118) and (6.119) are the direct multi-component equivalents of Eqs. (6.15) and (6.16).

As presented in Sect. 6.3, let us now rewrite the two previous BBGKY equations within the angle-action coordinates appropriate for the Keplerian motion induced by the central BH. Let us consequently perform the degenerate angle-average from Eq. (6.26) and assume that F^{a} and $\mathcal{C}^{\mathrm{ab}}$ satisfy the crucial assumptions from Eqs. (6.34) and (6.39). One can then rewrite Eq. (6.118) as

$$
\frac{\partial \overline{F^{\mathrm{a}}}}{\partial \tau} + \big[\overline{F^{\mathrm{a}}}, \overline{\Phi} + \overline{\Phi}_{\mathrm{r}} \big] + \sum_{\mathrm{b}} \int \mathrm{d}\boldsymbol{\mathcal{E}}_2 \, \big[\overline{\mathcal{C}^{\mathrm{ab}}}(\boldsymbol{\mathcal{E}}_1, \boldsymbol{\mathcal{E}}_2), \overline{U}_{12} \big]_{(1)} = 0 ,
\tag{6.120}
$$

where we introduced the rescaled time $\tau = (2\pi)^{d-k} \varepsilon t$ from Eq. (6.44) with $\varepsilon = M_\star / M_\bullet$. Following Eq. (6.36), we also introduced the total averaged self-consistent potential $\overline{\Phi}$ as

$$
\overline{\Phi} = \sum_{\mathrm{a}} \overline{\Phi^{\mathrm{a}}} ,
\tag{6.121}
$$

where the averaged potential $\overline{\Phi^{\mathrm{a}}}$ due to the component "a" follows from Eq. (6.36) and reads

$$
\overline{\Phi^{\mathrm{a}}}(\boldsymbol{\mathcal{E}}_1) = \int \mathrm{d}\boldsymbol{\mathcal{E}}_2 \, \overline{F^{\mathrm{a}}}(\boldsymbol{\mathcal{E}}_2) \, \overline{U}_{12}(\boldsymbol{\mathcal{E}}_1, \boldsymbol{\mathcal{E}}_2) .
\tag{6.122}
$$

In Eq. (6.122), we relied on the averaged wire-wire interaction potential $\overline{U}_{12}$ from Eq. (6.37). Following the same approach, Eq. (6.119) can be rewritten as

$$\frac{\partial \overline{C^{\mathrm{ab}}}}{\partial \tau} + \big[\overline{C^{\mathrm{ab}}}(\boldsymbol{\mathcal{E}}_1,\boldsymbol{\mathcal{E}}_2), \overline{\Phi}(\boldsymbol{\mathcal{E}}_1) + \overline{\Phi}_{\mathrm{r}}(\boldsymbol{\mathcal{E}}_1)\big]_{(1)} + \big[\overline{C^{\mathrm{ab}}}(\boldsymbol{\mathcal{E}}_1,\boldsymbol{\mathcal{E}}_2), \overline{\Phi}(\boldsymbol{\mathcal{E}}_2) + \overline{\Phi}_{\mathrm{r}}(\boldsymbol{\mathcal{E}}_2)\big]_{(2)}$$
$$+ \sum_{\mathrm{c}} \int \mathrm{d}\boldsymbol{\mathcal{E}}_3\, \overline{C^{\mathrm{bc}}}(\boldsymbol{\mathcal{E}}_2,\boldsymbol{\mathcal{E}}_3)\big[\overline{F^{\mathrm{a}}}(\boldsymbol{\mathcal{E}}_1), \overline{U}_{13}\big]_{(1)} + \sum_{\mathrm{c}} \int \mathrm{d}\boldsymbol{\mathcal{E}}_3\, \overline{C^{\mathrm{ac}}}(\boldsymbol{\mathcal{E}}_1,\boldsymbol{\mathcal{E}}_3)\big[\overline{F^{\mathrm{b}}}(\boldsymbol{\mathcal{E}}_2), \overline{U}_{23}\big]_{(2)}$$
$$+ \frac{1}{(2\pi)^{d-k}} \bigg\{ \eta_{\mathrm{b}} \big[\overline{F^{\mathrm{a}}}(\boldsymbol{\mathcal{E}}_1)\overline{F^{\mathrm{b}}}(\boldsymbol{\mathcal{E}}_2), \overline{U}_{12}\big]_{(1)} + \eta_{\mathrm{a}} \big[\overline{F^{\mathrm{a}}}(\boldsymbol{\mathcal{E}}_1)\overline{F^{\mathrm{b}}}(\boldsymbol{\mathcal{E}}_2), \overline{U}_{21}\big]_{(2)} \bigg\} = 0\,. \tag{6.123}$$

The two coupled evolution equations (6.120) and (6.123) are the direct multi-component equivalents of Eqs. (6.45) and (6.46). The main differences here are the changes in the mass prefactors in the last term (the source term) of Eq. (6.123). Indeed, it mixes the two small parameters $\eta_{\mathrm{a}} = \mu_{\mathrm{a}}/M_\star$ and $\eta_{\mathrm{b}} = \mu_{\mathrm{b}}/M_\star$. This change is the one which allows for mass segregation in multi-component systems, as briefly discussed in Sect. 6.5.2. Starting from Eqs. (6.120) and (6.123), one can then follow the method presented in Sect. 6.5 to derive the associated kinetic equation for $\overline{F^{\mathrm{a}}}$. This is the multi-component inhomogeneous degenerate Balescu–Lenard equation (6.63).

C From Fokker–Planck to Langevin

The degenerate inhomogeneous Balescu–Lenard equation (6.53) is a self-consistent integro-differential equation describing the evolution of the system's DF as a whole under the effect of its own graininess. Instead of describing the statistical dynamics of the full system's DF, one could be interested in characterising the individual dynamics of one test particle in this system. Following Risken (1996), let us recall how one may obtain the stochastic Langevin equation describing such an individual dynamics. Let us start from the generic writing of the degenerate Balescu–Lenard equation (6.58) written as an anisotropic Fokker–Planck equation. It reads

$$\frac{\partial \overline{F}}{\partial \tau} = \frac{\partial}{\partial \boldsymbol{J}^{\mathrm{s}}} \cdot \bigg[\boldsymbol{A}(\boldsymbol{J},\tau)\, \overline{F}(\boldsymbol{J},\tau) + \boldsymbol{D}(\boldsymbol{J},\tau) \cdot \frac{\partial \overline{F}}{\partial \boldsymbol{J}^{\mathrm{s}}} \bigg], \tag{6.124}$$

where, following the notations from Eq. (6.59), we introduced the system's total drift vector $\boldsymbol{A}(\boldsymbol{J},\tau)$ and diffusion tensor $\boldsymbol{D}(\boldsymbol{J},\tau)$ as

$$\boldsymbol{A}(\boldsymbol{J},\tau) = \sum_{\boldsymbol{m}^{\mathrm{s}}} \boldsymbol{m}^{\mathrm{s}}\, A_{\boldsymbol{m}^{\mathrm{s}}}(\boldsymbol{J},\tau) \quad ; \quad \boldsymbol{D}(\boldsymbol{J},\tau) = \sum_{\boldsymbol{m}^{\mathrm{s}}} \boldsymbol{m}^{\mathrm{s}} \otimes \boldsymbol{m}^{\mathrm{s}}\, D_{\boldsymbol{m}^{\mathrm{s}}}(\boldsymbol{J},\tau)\,. \tag{6.125}$$

Let us recall here that the Balescu–Lenard equation being self-consistent, the drift and diffusion coefficients depend secularly on the system's DF, $\overline{F}$, but this was not written out explicitly to shorten the notations. Following the notations from Eq. (4.94a) in Risken (1996), let us rewrite Eq. (6.124) as

$$\frac{\partial \overline{F}}{\partial \tau} = \frac{\partial}{\partial \boldsymbol{J}^{\mathrm{s}}} \cdot \bigg[- \boldsymbol{D}^{(1)}(\boldsymbol{J}, \tau)\, \overline{F}(\boldsymbol{J}, \tau) + \frac{\partial}{\partial \boldsymbol{J}^{\mathrm{s}}} \cdot \Big[\boldsymbol{D}^{(2)}(\boldsymbol{J}, \tau)\, \overline{F}(\boldsymbol{J}, \tau) \Big] \bigg] , \tag{6.126}$$

where we introduced the first- and second-order diffusion coefficients as

$$\boldsymbol{D}^{(1)}(\boldsymbol{J}, \tau) = -\boldsymbol{A}(\boldsymbol{J}, \tau) + \frac{\partial}{\partial \boldsymbol{J}^{\mathrm{s}}} \cdot \boldsymbol{D}(\boldsymbol{J}, \tau) \quad ; \quad \boldsymbol{D}^{(2)}(\boldsymbol{J}, \tau) = \boldsymbol{D}(\boldsymbol{J}, \tau) . \tag{6.127}$$

Here, let us emphasise that the diffusion of the Keplerian wires takes place in the full action domain $\boldsymbol{J}$, while Eq. (6.126) only involves gradients w.r.t. the slow actions $\boldsymbol{J}^{\mathrm{s}}$. This leads, amongst others, to the conservation of the fast actions $\boldsymbol{J}^{\mathrm{f}}$ during the resonant diffusion, as noted in Eq. (6.62). Of course, by enlarging the diffusion coefficients $\boldsymbol{D}^{(1)}$ and $\boldsymbol{D}^{(2)}$ with zero coefficients for all the adiabatically conserved fast actions $\boldsymbol{J}^{\mathrm{f}}$, it is straightforward to rewrite Eq. (6.126) as a diffusion equation in the full action space involving derivatives w.r.t. all action coordinates $\boldsymbol{J}$.

Let us now focus on the dynamics of a given test Keplerian wire. We denote as $\mathcal{J}(\tau)$ its position in action space a time τ. On secular timescales, this test particle undergoes an individual stochastic diffusion consistent with the system's averaged diffusion captured by the diffusion equation (6.126). This diffusion follows a stochastic Langevin equation reading

$$\frac{\mathrm{d}\mathcal{J}}{\mathrm{d}\tau} = \boldsymbol{h}(\mathcal{J}, \tau) + \boldsymbol{g}(\mathcal{J}, \tau) \cdot \boldsymbol{\Gamma}(\tau) , \tag{6.128}$$

where we introduced the Langevin vector and tensor $\boldsymbol{h}$ and $\boldsymbol{g}$, as well as the stochastic Langevin forces $\boldsymbol{\Gamma}(\tau)$, whose statistics are given by

$$\big\langle \boldsymbol{\Gamma}(\tau) \big\rangle = 0 \quad ; \quad \big\langle \boldsymbol{\Gamma}(\tau) \otimes \boldsymbol{\Gamma}(\tau') \big\rangle = 2 \boldsymbol{I} \, \delta_{\mathrm{D}}(\tau - \tau') , \tag{6.129}$$

with $\boldsymbol{I}$ the identity matrix. Following Eq. (3.124) of Risken (1996), let us finally express the Langevin coefficients from Eq. (6.128) as a function of the drift and diffusion coefficients appearing in Eq. (6.126). The second-order diffusion tensor $\boldsymbol{D}^{(2)}$ is definite positive, so that we may introduce as $\sqrt{\boldsymbol{D}}^{(2)}$ one of its square root. One then has the components relations

$$h_i = D_i^{(1)} - \sum_{j,k} \big(\sqrt{\boldsymbol{D}}^{(2)}\big)_{kj} \frac{\partial \big(\sqrt{\boldsymbol{D}}^{(2)}\big)_{ij}}{\partial x_k} \quad ; \quad g_{ij} = \big(\sqrt{\boldsymbol{D}}^{(2)}\big)_{ij} . \tag{6.130}$$

Thanks to Eq. (6.130), one can fully specify the detailed characteristics of the diffusion of an individual orbit as described by the Langevin equation (6.128). The self-consistency of the diffusion imposes to the diffusion coefficients $\boldsymbol{D}^{(1)}$ and $\boldsymbol{D}^{(2)}$, and therefore to the Langevin coefficients $\boldsymbol{h}$ and $\boldsymbol{g}$, to be updated as the system's DF $\overline{F}$ secularly changes. Let us finally emphasise that the previous presentation of the associated Langevin equation was made for quasi-Keplerian systems governed by the degenerate Balescu–Lenard equation (6.53). It is straightforward to follow the same approach to write the Langevin equation associated with the non-degenerate Balescu–Lenard equation (2.67), which can indeed also be cast as an anisotropic Fokker–Planck equation, as in Eq. (2.68).

References

Abbott, B. P., Abbott, R., Abbott, T. D., et al. (2016). *Phys. Rev. Lett.*, *116*, 061102.
Alexander, T. (2005). *Phys. Rep.*, *419*, 65.
Alexander, T. (2017). *ARA&A*, *55*, 1.
Balescu, R. (1960). *Phys. Fluids*, *3*, 52.
Bar-Or, B., & Alexander, T. (2014). *Class. Quantum Grav.*, *31*, 244003.
Bar-Or, B., & Alexander, T. (2016). *ApJ*, *820*, 129.
Binney, J., & Tremaine, S. (2008). *Galactic dynamics* (2nd ed.). Princeton University Press.
Blanchet, L. (2014). *Living Rev. Relativ.*, *17*.
Duncan, M. J., Levison, H. F., & Lee, M. H. (1998). *AJ*, *116*, 2067.
Fouvry, J.-B., Pichon, C., & Magorrian, J. (2017a). *A&A*, *598*, A71.
Fouvry, J.-B., Pichon, C., & Chavanis, P.-H. (2017b). *A&A*. (in press).
Frank, J., & Rees, M. J. (1976). *MNRAS*, *176*, 633.
Genzel, R., Pichon, C., Eckart, A., Gerhard, O. E., & Ott, T. (2000). *MNRAS*, *317*, 348.
Gilbert, I. H. (1968). *ApJ*, *152*, 1043.
Gillessen, S., Eisenhauer, F., Mikkola, S., & Trippe, S. (2009). *ApJ*, *692*, 1075.
Hamers, A. S., Portegies Zwart, S. F., & Merritt, D. (2014). *MNRAS*, *443*, 355.
Heggie, D., & Hut, P. (2003). *The gravitational million-body problem*. Cambridge University Press.
Hills, J. G. (1988). *Nature*, *331*, 687.
Hopman, C., & Alexander, T. (2006). *ApJ*, *645*, 1152.
Jalali, M. A., & Tremaine, S. (2012). *MNRAS*, *421*, 2368.
Jocou, L., Perraut, K., Moulin, T., et al. (2014). *Optical and infrared interferometry IV*, *9146*, 91461J.
Kocsis, B., & Tremaine, S. (2011). *MNRAS*, *412*, 187.
Kocsis, B., & Tremaine, S. (2015). *MNRAS*, *448*, 3265.
Lenard, A. (1960). *Annals of Physics*, *10*, 390.
Lynden-Bell, D. (1967). *MNRAS*, *136*, 101.
Madigan, A.-M., Hopman, C., & Levin, Y. (2011). *ApJ*, *738*, 99.
Merritt, D. (2015). *Astrophysical black holes*. Heidelberg: Springer.
Merritt, D., Alexander, T., Mikkola, S., & Will, C. M. (2011). *Phys. Rev. D*, *84*, 044024.
Morbidelli, A. (2002). *Modern celestial mechanics*. Taylor & Francis.
Polyachenko, E. V., Polyachenko, V. L., & Shukhman, I. G. (2007). *MNRAS*, *379*, 573.
Rauch, K. P., & Ingalls, B. (1998). *MNRAS*, *299*, 1231.
Rauch, K. P., & Tremaine, S. (1996). *New Astron.*, *1*, 149.
Risken, H. (1996). *The Fokker-Planck equation*. Berlin Heidelberg: Springer.
Sridhar, S., & Touma, J. R. (1999). *MNRAS*, *303*, 483.
Sridhar, S., & Touma, J. R. (2016a). *MNRAS*, *458*, 4129.

Sridhar, S., & Touma, J. R. (2016b). *MNRAS*, *458*, 4143.
Sridhar, S., & Touma, J. R. (2017). *MNRAS*, *465*, 1856.
Touma, J. R., Tremaine, S., & Kazandjian, M. V. (2009). *MNRAS*, *394*, 1085.
Tremaine, S. (1995). *AJ*, *110*, 628.
Tremaine, S. (1998). *AJ*, *116*, 2015.
Tremaine, S. (2005). *ApJ*, *625*, 143.
Vasiliev, E., & Merritt, D. (2013). *ApJ*, *774*, 87.
Volonteri, M., Dubois, Y., Pichon, C., & Devriendt, J. (2016). *MNRAS*, *460*, 2979.
Yu, Q. (2002). *MNRAS*, *331*, 935.

Chapter 7
Conclusion

7.1 Overview

Since the seminal works of Einstein, Langevin and Perrin, physicists understand how a drop of ink slowly diffuses in a glass of water. The fluctuations of the stochastic forces acting on water molecules drive the diffusion of the ink in the fluid. This is the archetype of the so-called fluctuation-dissipation theorem, which relates the rate of diffusion to the autocorrelation of the fluctuating forces. For galaxies, a similar process occurs but with two main differences related to the long-range nature of the gravitational force: (i) for the diffusion to be effective, stars need to resonate, i.e. present commensurable frequencies, otherwise they follow the mean path imposed by the mean field, (ii) the amplitude of the induced fluctuating forces are significantly boosted by collective effects, i.e. the fact that, because of self-gravity, each star polarises its neighbours. This thesis was concerned with studying the secular implications of this fluctuation-dissipation theorem by considering either externally-driven or self-induced fluctuations.

Self-gravitating systems are highly complex objects which undergo a wide variety of dynamical processes, depending on their internal "temperature", i.e. depending on whether they are pressure or centrifugally supported. Astrophysics is now in a position to investigate the secular dynamics of these systems. Cold systems are of particular interest, as they have the opportunity to reshuffle their orbital structure towards more likely configurations. First, the success of the ΛCDM model now offers a consistent paradigm in which to statistically characterise the cosmic environment. Self-gravitating systems should be seen as embedded in a lively environment, with which they interact throughout their lifetime. In addition, recent developments in kinetic theories now offer various self-consistent frameworks allowing for the description of these systems' secular dynamics. Whether they are external or internal, the long-term resonant effects of perturbations can be accounted for in detail. Let us also emphasise that the steady increase in computing power now allows for detailed simulations of ever greater resolution and complexity. It not only allows for simulations of isolated and idealised setups, but also for cosmological simulations

J.-B. Fouvry, *Secular Evolution of Self-Gravitating Systems Over Cosmic Age*,
Springer Theses, DOI 10.1007/978-3-319-66760-7_7

where environmental effects can play a role. Finally, upcoming observations, such as GAIA for the Milky Way or Gravity for the Galactic centre, will soon offer unprecedented detailed surveys of the phase space structure of these systems.

These joint progresses in the characterisation of the cosmic environment, the diffusion theory, the simulation power and the details of the observations offer the ideal context in which to study the secular dynamics of self-gravitating systems. While the seminal works of Goldreich, Lynden-Bell, Kalnajs, Toomre had opened the way to deriving a self-consistent linear response theory for self-gravitating systems, it should be emphasised that thanks to the recent works of Binney and Lacey (1988); Weinberg (2001a); Heyvearts (2010), etc., galactic dynamics has now entered a phase of quantitative statistical predictability on secular timescales. This thesis has contributed to illustrating how this line of work could be applied to varied challenges such as radial migration, disc thickening, or black hole feeding. Let us now detail the main conclusions drawn from each chapter.

In Chap. 2, we presented two complementary formalisms to describe the secular evolution of self-gravitating systems. Because they are assumed to be quasi-stationary and stable, such systems can only evolve driven by fluctuations. These may first originate from an external perturber, leading to a collisionless diffusion. Another source of fluctuations is associated with the system's own discreteness. This induces finite$-N$ effects, whose contributions on secular timescales are described by the inhomogeneous Balescu–Lenard equation. To derive these equations, we relied on angle-action coordinates to deal with the complexity of individual orbits. For both formalisms, we emphasised the importance of accounting for the system's self-gravity. It dresses perturbations and can very significantly hasten the system's secular evolution. These polarisation effects are especially important in cold dynamical systems such as stellar discs. These two frameworks allow for quantitative comparisons of the respective effects of nurture versus nature, i.e. externally-induced versus self-induced effects, on the long-term evolution of self-gravitating systems.

In Chap. 3, we considered the case of tepid razor-thin stellar discs. In order to overcome the two principal hurdles associated with the diffusion formalisms, namely the need to consider angle-action coordinates and the difficulty to estimate the system's non-local gravitational susceptibility, we relied on two additional assumptions. These were the epicyclic approximation, i.e. the restriction to cold quasi-circular orbits, as well as a tailored WKB approximation, i.e. the restriction to radially tightly wound perturbations. We illustrated how the WKB formalism offers simple quadratures for the diffusion flux in both collisionless and collisional frameworks. This provided us with a straightforward tool to estimate the loci of maximum diffusion within the disc. When applied to a discrete stable self-gravitating razor-thin stellar disc, we recovered qualitatively the formation of ridges of resonant orbits observed in numerical simulations. One additional strength of the Balescu–Lenard formalism is to offer explicit estimations of the collisional timescale of diffusion in the disc. We noted a discrepancy between the prediction from the WKB kinetic theory and the much shorter timescale inferred from numerical simulations. This was interpreted as due to the incompleteness of the WKB basis, which cannot account for the strong dressing of loosely wound perturbations, a well known linear mechanism coined swing amplification.

In Chap. 4, we returned to the case of discrete razor-thin stellar discs. In order to fully account for the disc's self-gravity, we implemented the matrix method. When combined with the collisional Balescu–Lenard equation, our prediction for the initial diffusion flux recovered the formation of the resonant ridge observed in numerical simulations. Because we had correctly taken into account self-gravity, we also matched the timescales of diffusion. To fully emphasise the relevance of the Balescu–Lenard formalism, we resorted to our own $N-$body simulations. We recovered the expected scalings of the system's response with the number of particles and the fraction of mass within the disc, as predicted by the collisional theory. When considered on even longer timescales, we recovered the mechanism of dynamical phase transition identified in Sellwood (2012). Stable and quasi-stationary systems can become dynamically unstable on the long-term, as the result of the slow, progressive, and irreversible build-up of collisional effects. This is a striking outcome of the large dynamical freedom given by the interactions captured by the second-order equation of the BBGKY hierarchy, which allows for spontaneous reshuffles of orbital structures towards states of higher entropy.

In Chap. 5, we investigated the secular dynamics of thickened stellar discs. Similarly to the razor-thin case, we devised a new thickened WKB approximation offering explicit expressions for both collisionless and collisional diffusion fluxes. As a side product, this formalism also offered a new thickened Q stability parameter. Following these derivations, we considered various mechanisms of secular thickening, such as internal Poisson shot noise, series of central bars, or even the diffusion acceleration induced by giant molecular clouds. We emphasised how each of these mechanisms have different signatures in the diffusion features appearing in the disc. As already noted in the razor-thin case, while qualitatively correct, one limitation of the thickened WKB approximation is the underestimation of the diffusion timescale, which can be significantly hastened by swing amplification in cold stellar discs.

In Chap. 6, we focused on quasi-Keplerian systems, and in particular galactic centres. The particularity of such systems is that their dynamics is mostly dominated by one central massive object. This makes the system dynamically degenerate. Individual particles follow Keplerian wires, which get slowly distorted on secular timescales. By paying careful attention to this degeneracy, we detailed how one could tailor the collisional formalism to describe the long-term evolution of such systems. We especially emphasised how this new kinetic equation is the master equation to describe the resonant relaxation of Keplerian wires. In the context of galactic centres, we illustrated how the divergence of the relativistic precession frequencies as stars move closer to the central black hole leads to a drastic reduction of the diffusion efficiency. This is the phenomenon of "Schwarzschild barrier" first observed in numerical simulations.

7.2 Outlook and Future Works

Throughout this thesis, we detailed at the end of each chapter some possible future works w.r.t. the themes investigated there. As a closing section for this thesis, let us replace these various prospects in a broader context.

The aim of this thesis was to offer galactic dynamics an analytical framework in which to describe evolutions on cosmic times. It appears now as a powerful approach because it allows for a tractable capture of numerous complex non-linear processes. In the continuation of the initial seminal works describing the linear response of self-gravitating systems, one now has at its disposal a self-consistent formulation to understand analytically the non-linear and secular response of these systems. The recent developments of kinetic theory offered us quite an unique opportunity: implementing for the first time in astrophysics a new diffusion equation, the inhomogeneous Balescu–Lenard equation. In the course of this manuscript, we emphasised how these approaches now provide us with the master equations to describe simultaneously and self-consistently a vast class of astrophysical processes. These include the mechanism of stellar migration (both churning and blurring) and disc thickening for stellar discs, but also resonant relaxation and BH feeding in galactic centres. Analytic galactic dynamics has entered the cosmic framework.

Such a formalism is rewarding both for its conceptual contributions but also for its practical usefulness. From the conceptual point of view, these approaches encompass all the wealth and complexity of self-gravitating systems' dynamics. The example of galactic centres illustrates it very clearly. This framework captures the non-trivial effects that a system's discreteness can have on the long-term, and provides a fascinating illustration of the fluctuation-dissipation theorem. It may also be used to study and understand entropy production. Secular dynamical phase transitions are as well an important prediction of this formalism. It describes how the slow and irreversible build up of collisional effects inevitably leads, on the long-term, to a destabilisation of secularly metastable states.

From the practical point of view, this framework can account for polarisation which accelerates considerably the diffusion in cold systems. It also naturally offers a new dressed multi-component Langevin rewriting, which allows for much larger timesteps. As an example, between two timesteps, a series of swing amplifications can take place: there is no need anymore to resolve them individually. Finally, it can be used to propagate statistics. From detailed measures of environmental perturbations, one can infer the typical evolutions that will be undergone by the systems. One can now treat statistically a galaxy on multiple orbital times, while accounting for the dynamical wealth associated with self-gravity.

When considering the long-term evolution of self-gravitating systems, an important dichotomy has to be drawn between self-induced and externally-induced secular dynamics. One should also pay attention to the system's initial reservoir of free energy, which differs greatly between, e.g., spirals and ellipticals. These distinctions allow us to disentangle the respective roles of nature vs. nurture in sourcing secular evolution, as one can quantify the diffusion signatures associated with different sources of fluctuations. These may then be compared to observations. For example, for stellar discs, one could investigate the expected diffusion associated with various sources of fluctuations: discreteness noise, clumps within the halo, central bars, or tidally induced spirals. Once all these mechanisms are statistically characterised, their predictions (e.g., for the metallicity-dispersion relation) could be compared to detailed observations of the structure of the Milky Way's DF, e.g., soon provided by GAIA. This would allow us to disentangle a posteriori the importance of these various mechanisms throughout the evolution of the Milky Way. This framework is expected to be very powerful to provide explicit predictions in the context of Galactic archeology (Binney 2013).

All the applications presented in this thesis were restricted to computing the initial diffusion flux at $t = 0^{+}$. In order to probe later stages of the secular evolution, one would have to integrate forward in time these diffusion equations. There are at least two anticipated difficulties. The first arises from the self-consistency of the diffusion equations. The system's drift and diffusion coefficients depend on the current value of the system's DF and have to be updated as the system evolves (Weinberg 2001b). The integration in time has to be made step by step, with successive updates of the system's DF, potential, and diffusion flux. Another difficulty is that these equations describe the diffusion of the system's DF as a whole. Integrating such a partial differential equation is a cumbersome numerical calculation, which most likely requires to rely on finite elements methods. An alternative approach, inspired from Monte Carlo simulations, follows from the Langevin rewriting of the diffusion equation presented in Sect. 6.C. One samples the system's DF with individual particles, and integrates the first-order stochastic ordinary differential equations describing the dynamics of test particles. With these equations, the involved timesteps are commensurable with a Hubble time. Such integrations of the diffusion could be used as valuable probes to validate the accuracy and robustness of $N-$body codes on secular timescales. Indeed, one of the only theoretical predictions to which $N-$body implementations are compared are derived from linear theory: one aims at recovering unstable modes in integrable systems (see Appendix 4.C). These tests can then only check the relevance of numerical codes on a few dynamical times. The formalisms developed in this thesis provide new test cases to quantify the validity of $N-$body implementations on secular timescales.

These various diffusion equations would also benefit from being generalised to describe wider classes of dynamical processes. As already detailed, one can naturally extend these approaches to account for multiple components and describe the corresponding expected mass segregation. See, e.g., Sect. 5.7.6 for an illustration of the role played by giant molecular clouds. Similarly, the system's DF can also be extended with additional parameters, such as metallicity. This drives the interplay between dynamics and chemistry. See Sect. 3.8.1 for an illustration of how to construct such extended DFs. In this regime, one accounts for the birth (and possibly death) of stars, i.e. for the possibility of sources and sinks of particles. Focusing on open systems is another promising regime for which these investigations should be pursued. Similarly, this formalism could be generalised to systems with a small or fluctuating number of "effective" particles, for example as a result of the progressive dissolution of overdensities.

Note that we assumed here integrability, i.e. the existence of angle-action coordinates. It was either guaranteed by the system's symmetry or by additional assumptions, such as the epicyclic approximation. When this is not the case, the system's dynamics may become chaotic and its secular evolution can possibly be driven by chaotic diffusion. For example, in the context of stellar discs, these chaotic effects are prone to play a role in the presence of a central bar. Indeed, the bar's potential makes the system's dynamics chaotic in some regions. Such secular dynamics associated with the formation of strong non-axisymmetric structures were not investigated in the present thesis. They definitely deserve a thorough investigation on their own.

Finally, in some regimes, the resonant orbital diffusion described by the Balescu–Lenard equation may vanish. This can for example occur in galactic centres, as illustrated by the Schwarzschild barrier driven by the divergence of the relativistic precession frequencies as stars move closer to the BH. This suppression can also be imposed by symmetry, e.g., the Balescu–Lenard collision operator vanishes for $1D$ homogeneous systems. Finally, it may occur on secular timescales if steady states for the Balescu–Lenard equation exist and can be reached by the system (though this is not always possible for self-gravitating systems). In such regimes, the dynamics is not described anymore by the Balescu–Lenard equation, and additional effects have to be accounted for. For galactic centres, this can be direct 2–body effects, i.e. star-star interactions, which allow stars to diffuse even closer to the BH. In addition to strong collisions, resonant effects associated with $1/N^2$ correlations can also drive the dynamics. This requires to consider the third equation of the BBGKY hierarchy and focus on slower effects associated with 3–body correlations. Generalising the Balescu–Lenard formalism to account for these two contributions is another interesting direction of improvement.

References

Binney, J. (2013). *New Astron. Rev.*, *57*, 29.
Binney, J., & Lacey, C. (1988). *MNRAS*, *230*, 597.
Heyvaerts, J. (2010). *MNRAS*, *407*, 355.
Sellwood, J. A. (2012). *ApJ*, *751*, 44.
Weinberg, M. D. (2001a). *MNRAS*, *328*, 311.
Weinberg, M. D. (2001b). *MNRAS*, *328*, 321.

MIX
Papier aus verantwortungsvollen Quellen
Paper from responsible sources
FSC® C105338

If you have any concerns about our products,
you can contact us on
ProductSafety@springernature.com

In case Publisher is established outside the EU,
the EU authorized representative is:
Springer Nature Customer Service Center GmbH
Europaplatz 3, 69115 Heidelberg, Germany

Printed by Libri Plureos GmbH
in Hamburg, Germany